The USE of NUTRIENTS in CROP PLANTS

The USE of NUTRIENTS in CROP PLANTS

N.K. Fageria

CRC Press
Taylor & Francis Group
Boca Raton London New York

CRC Press is an imprint of the
Taylor & Francis Group, an **informa** business

CRC Press
Taylor & Francis Group
6000 Broken Sound Parkway NW, Suite 300
Boca Raton, FL 33487-2742

First issued in paperback 2019

ISBN-13: 978-1-4200-7510-6 (hbk)
ISBN-13: 978-0-367-38688-7 (pbk)

Library of Congress Cataloging-in-Publication Data

Fageria, N. K., 1942-
  The use of nutrients in crop plants / author, N.K. Fageria.
    p. cm.
  Includes bibliographical references and index.
  ISBN 978-1-4200-7510-6 (hardback : alk. paper)
  1. Field crops--Nutrition. 2. Crops--Effect of minerals on. 3. Fertilizers. I. Title.

SB185.5.F345 2009
631.8'11--dc22   2008025445

Visit the Taylor & Francis Web site at
http://www.taylorandfrancis.com

and the CRC Press Web site at
http://www.crcpress.com

Contents

Preface

This book is the outgrowth of my more than 40 years' experience in research on mineral nutrition of crop plants. Its objective is to help bridge the gap between theoretical aspects of mineral nutrition and practical applicability of basic principles of fertilization and use efficiency of essential plant nutrients. Mineral nutrition played a significant role in improving crop yields in the 20th century, and its role in increasing crop yields will become even larger in the 21st century. This is due to the scarcity of natural resources (soil and water), higher cost of inorganic fertilizers, higher food demand by an increasing world population, environmental pollution concern regarding the use of inadequate rate, form, and methods of chemical fertilizers, and higher demand for quality food by consumers worldwide. Nutrient sufficiency is the basis of good human and animal health. Nutrient availability to the world population is primarily determined by the output of food produced from agricultural systems. If agricultural systems fail to provide adequate food in quantity and quality, there will be disorder in food security and chaos in the social systems, which will threaten peace and security. Under these conditions, improving food supply worldwide with adequate quantity and quality is fundamental. Supply of adequate mineral nutrients in adequate amount and proportion to higher plants will certainly determine such accomplishments. Hence this book provides information and discussion on maximizing essential nutrients uptake and use efficiency by food crops and improving their productivity without degrading the environment.

Justification for publishing this book lies in its format covering both theoretical and practical aspects of mineral nutrition of plants. The presentation of updated experimental data in the form of tables and figures makes the book more practical as well as more informative and attractive. It will serve as a reference book for those involved in research, teaching, and extension services and a textbook for senior- and graduate-level courses. Agricultural science is dynamic in nature, and fertilizer practices change with time due to release of new cultivators and other crop production practices in sustainable crop production systems. Inclusion of a large number of references of international dimension make this book a valuable tool for crop and soil professionals to maximize nutrient use efficiency in different agroecological regions. The majority of research data included in each chapter are the author's own work, providing evidence of plant responses to applied nutrients under field or greenhouse conditions. Hence the information is practical in nature. Comprehensive coverage of all essential plant nutrients with experimental results make this book unique and practical. The focus is on presenting in-depth and updated scientific information in the area of mineral nutrition. The information provided in this book will have a huge impact on management of inorganic and organic fertilizers, enhance the stability of agricultural systems, help agricultural scientists to maximize nutrient use efficiency, improve crop yields at lower cost, and help maintain a clean environment (air, water, and soil), all of which will contribute to the maintenance of sound human and animal health.

Preparing a book of this nature involves the assistance and cooperation of many people, to whom I am grateful. I also thank the National Rice and Bean Research Center of EMBRAPA, Brazil, for providing necessary facilities in writing the book. I express my appreciation to the publisher and share in their pride of a job well done. I dedicate this book with great respect to my late father, Goru Ram Fageria, and my mother, Dhaki Fageria; their hard work and dedication on a small farm in the Thar Desert of Rajasthan, India inspired my interest in higher education. Finally, I express sincere appreciation to my wife, Shanti; children, Rajesh, Satya Pal, and Savita; daughter-in-law, Neera; son-in-law, Ajay; and grandchildren, Anjit, Maia, and Sofia, for their understanding, patience, and strong encouragement, without which this book could not have been written.

N. K. Fageria
National Rice and Bean Research Center of EMBRAPA
Santo Antônio de Goiás
Brazil

Author

N. K. Fageria, doctor of science in agronomy, has been the senior research soil scientist at the National Rice and Bean Research Center, Empresa Brasileira de Pesquisa Agropecuária (EMBRAPA), since 1975. Dr. Fageria is a nationally and internationally recognized expert in the area of mineral nutrition of crop plants and has been a research fellow and ad hoc consultant of the Brazilian Scientific and Technological Research Council (CNPq) since 1989. Dr. Fageria is the author/coauthor of eight books and more than 250 scientific journal articles, book chapters, review articles, and technical bulletins. Dr. Fageria has written several review articles on nutrient management, enhancing nutrient use efficiency in crop plants, and ameliorating soil acidity by liming on tropical acid soils for sustainable crop production in *Advances in Agronomy*. He has been an invited speaker to several national and international congresses, symposia, and workshops. He is a member of the editorial board of the *Journal of Plant Nutrition* and the *Brazilian Journal of Plant Physiology* and has been a member of the international steering committee of symposia on plant-soil interactions at low pH since 1990.

1 Mineral Nutrition versus Yield of Field Crops

1.1 INTRODUCTION

Mineral nutrition—along with availability of water and cultivar; control of diseases, insects, and weeds; and socioeconomic conditions of the farmer—plays an important role in increasing crop productivity. Nutrient concentrations in soil solution have been of interest for many decades as indicators of soil fertility in agriculture (Hoagland et al., 1920). Mineral nutrition refers to the supply, availability, absorption, translocation, and utilization of inorganically formed elements for growth and development of crop plants. During the 20th century (1950 to 1990), grain yields of cereals (wheat, corn, and rice) tripled worldwide. Wheat yields in India, for example, increased by nearly 400% from 1960 to 1985, and yields of rice in Indonesia and China more than doubled. The vastly increased production resulted from high-yielding varieties, improved irrigation facilities, and use of chemical fertilizers, especially nitrogen. The results were significant in Asia and Latin America, where the term *green revolution* was used to describe the process (Brady and Weil, 2002). The increase in productivity of annual crops with the application of fertilizers and lime in the Brazilian *cerrado* (savanna) region during the 1970s and 1980s is another example of 20th-century expansion of the agricultural frontier in acid soils (Borlaug and Dowswell, 1997).

Stewart et al. (2005) reported that the average percentage of yield attributable to fertilizer generally ranged from about 40 to 60% in the United States and England and tended to be much higher in the tropics in the 20th century. Furthermore, the results of the Stewart et al. (2005) investigation indicate that the commonly cited generalization that at least 30 to 50% of crop yield is attributable to commercial fertilizer nutrient inputs is a reasonable, if not conservative, estimate. In addition, Stewart et al. (2005) reported that omission of N in corn declined yield of this crop by 41% and elimination of N in cotton production resulted in an estimated yield reduction of 37% in the United States. These authors also reported that if the effects of other nutrient inputs such as P and K had been measured, the estimated yield reductions would probably have been greater. Baligar et al. (2001) reported that as much as half of the rise in crop yields during the 20th century derived from increased use of fertilizers. The contribution of chemical fertilizers has reached 50 to 60% of the total increase in grain yields in China (Lu and Shi, 1998). Figure 1.1 and Figure 1.2 show a significant increase in grain yield of lowland rice with the application of nitrogen and phosphorus fertilizers in Brazilian Inceptisol. Nitrogen was responsible for 85% variation in grain yield and phosphorus was responsible for 90% variation in grain yield of rice. This indicates the importance of nitrogen and phosphorus in lowland

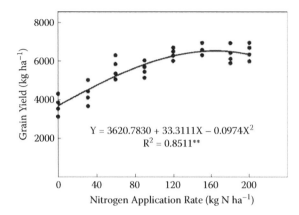

FIGURE 1.1   Relationship between nitrogen rate and grain yield of lowland rice grown on Brazilian Inceptisol (Fageria et al., 2008).

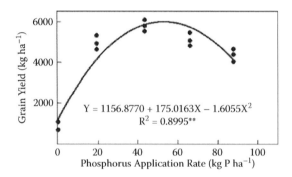

FIGURE 1.2   Relationship between phosphorus application rate and grain yield of lowland rice grown on Brazilian Inceptisol (Fageria et al., 2008).

rice production in Brazilian Inceptisols. Fageria and Baligar (2001) and Fageria et al. (1997) reported significant increases in grain yield of lowland rice with the application of nitrogen and phosphorus in Brazilian Inceptisols. Similarly, Fageria and Baligar (1997) reported that N, P, and Zn were the most yield-limiting nutrients for annual crop production in Brazilian Oxisols.

Raun and Johnson (1999) reported low N recovery efficiency in cereals worldwide, and deficiency of this nutrient for grain production of rice, wheat, sorghum, millet, barley, corn, and oats is very common in various parts of the world. Similarly, Fageria et al. (2003) reported deficiency of macro- and micronutrients in lowland rice around the world. Sumner and Noble (2003) reported that soil acidity is a problem in vast areas of the world and that liming is an effective practice to avoid deficiency of Ca and Mg and toxicity of Al^{3+} and H^+ ions. Fageria et al. (2002) reported that micronutrient deficiencies in crop plants are widespread because of (1) increased micronutrient demands from intensive cropping practices and adaptation of high-yielding cultivars, which may have higher micronutrient demand; (2) enhanced production of crops on marginal soils that contain low levels of essential nutrients; (3) increased use of high-analysis fertilizers with low amounts of micronutrient contamination; (4) decreased use of animal manures, composts, and crop residues; (5) use of soils that are inherently low in micronutrient reserves; and (6) involvement of natural and anthropogenic factors that limit adequate plant availability and create element imbalances.

Fageria and Baligar (2005) reported that soil infertility (due to natural element deficiencies or unavailability) is probably the single most important factor limiting crop yields worldwide. Application of macro- and micronutrient fertilizers has contributed substantially to the huge increase in world food production experienced during the 20th century. Loneragan (1997) reported that as much as 50% of the increase in crop yields worldwide during 20th century was due to use of chemical fertilizers. The role of mineral nutrition in increasing crop yields in the 21st century will be higher still, because world population is increasing rapidly and it is projected that there will be more than 8 billion people by the year 2025. Limited natural resources like land and water and stagnation in crop yields globally make food security a major challenge and opportunity for agricultural scientists in the 21st century. It is projected that food supply on the presently cultivated land must be doubled in the next two decades to meet the food demand of the growing world population (Cakmak, 2001).

To achieve food production at a desired level, use of chemical fertilizers and improvements in soil fertility are indispensable strategies. It is estimated that 60% of cultivated soils have nutrient deficiency/elemental toxicity problems and that about 50% of the world population suffers from micronutrient deficiencies (Cakmak, 2001). Furthermore, it is estimated that to meet future food needs, the total use of fertilizers will increase from 133 million tons per year in 1993 to about 200 million tons per year by 2030 (FAO, 2000). This scenario makes plant nutrition research a top priority in agriculture science to meet quality food demand in this millennium. Public concern about environmental quality and the long-term productivity of agroecosystems has emphasized the need to develop and implement management strategies that maintain soil fertility at an adequate level without degrading soil and water resources (Fageria et al., 1997). Most of the essential plant nutrients are also essential for human health and livestock production. The objective of this introductory chapter is to provide information on the history and importance of mineral nutrition in increasing crop yields, nutrient availability and requirements, and crop classification systems and to discuss yield and yield components for improving crop yields. This information may help in better planning mineral nutrition research and consequently improving crop yields.

1.2 HISTORY OF MINERAL NUTRITION RESEARCH

As a science, plant nutrition is a part of plant physiology. No one knows with certainty when humans first incorporated inorganic substances, manures, or wood ashes as fertilizer in soil to stimulate plant growth. However, it is documented in writings as early as 2500 BC that people recognized the richness and fertility of alluvial soils in valleys of the Tigris and Euphrates rivers (Tisdale et al., 1985). Forty-two centuries later, scientists were still trying to determine whether plant nutrients were derived from water, air, or soil ingested by plant roots. Early progress in the development of understanding of soil fertility and plant nutrition concepts was slow, although the Greeks and Romans made significant contributions in the years 800 to 200 BC (Westerman and Tucker, 1987; Fageria et al., 1997).

The theory of mineral nutrition of plants, which states that plants require mineral elements to develop, was postulated by German agronomist and chemist Carl Sprengel (1787–1859), who also formulated the law of the minimum (Van der Ploeg et al., 1999). Carl Sprengel in 1826 published an article in which the humus theory was refuted, and in 1828 he published another, extended journal article on soil chemistry and mineral nutrition of plants that contained in essence the law of the minimum (Van der Ploeg et al., 1999). However, in most of the publications on mineral nutrition of plants, the credit for developing the theory of mineral nutrition of plants and the law of the minimum goes to German chemist Justus von Liebig. Van der Ploeg et al. (1999) reported that to avoid a dispute on this subject, the Association of German Agricultural Experimental and Research Stations has given credit to both these scientists on this matter and created the Sprengel-Liebig Medal. Jean-Baptiste Boussingault (1802–87) from France and J. B. Lawes and J. H. Gilbert from Rothamsted Experiment Station, England, were other prominent pioneer agronomists of that time who contributed significantly to the development of the theory of mineral nutrition of plants and use of fertilizers in improving crop yields.

A significant contribution of Boussingault (1838) was the fixation of atmospheric nitrogen by leguminous plants. However, at that time he was not sure of legume contribution in nitrogen fixation. In 1886, German scientists Hellriegel and Wilfarth reported that legumes fix atmospheric nitrogen; however, the presence of symbiotic bacteria is essential for this process. These authors also concluded that nonleguminous plants do not fix atmospheric nitrogen and totally depend on nitrogen supplied by soil. This work provided final confirmation of the conclusion first reached by Boussingault in 1938 (Epstein and Bloom, 2005). The development of nutrient solution techniques for growing plants contributed significantly to the science of mineral nutrition. Credit for developing these techniques goes to German botanist Julius von Sachas (1860) and W. Knop in early 1860s. Hoagland (1884–1949) was the leading pioneer of the modern period of plant nutrition (Epstein and Bloom, 2005). Hoagland and Broyer (1936) developed nutrient solution, which is still in use with some modification for the study of mineral nutrition of plants.

During the 20th century, several scientists developed concepts that furthered understanding of nutrient availability to plants. Among these, Hoagland's (1922) study of oats plants in a pot yielded the concept of buffer power of soil nutrient availability (Okajima, 2001). Johnston and Hoagland (1929) also developed the intensity and capacity factors of nutrient availability. Hoagland and Broyer (1936) considered that selection or accumulation of nutrients depends on the aerobic metabolic process of roots; that is, to absorb nutrients, plants require the expenditure of energy against concentration and activity gradients. Such selective accumulation has been called "active or metabolic absorption" (Okajima, 2001). Bray (1954) proposed the nutrient mobility concept. According to this concept, the mobility of nutrients in soils is one of the most important factors in soil fertility relationships. The term *mobility* as used here means the overall process whereby nutrient ions reach the sorbing root surface, thereby making possible their absorption by plants. In the early part of the 20th century, Robertson (1907) and Spillman and Lang (1924) recognized that plant growth is affected by several factors and followed a sigmoid-type curve. The work of these researchers led to the development of the concept of the law of diminishing return. This law states that with each additional increment of a fertilizer, the increase in yield becomes smaller and smaller (Tucker, 1987). Readers who desire detailed knowledge of the history of mineral nutrition may refer to publications by Reed (1942), Browne (1944), Bodenheimer (1958), Fageria et al. (1997), Epstein (2000), Okajima (2001), Fageria (2005), and Epstein and Bloom (2005).

By 1873, von Liebig had identified the nutritional status of plants as one of the key factors regulating their susceptibility to diseases (Haneklaus et al., 2007). Though the role of individual nutrients in maintaining or promoting plant health received some attention in the 1960s and 1970s, research in the field of nutrient-induced resistance mechanisms has been limited by its complexity and a lack of recognition of its practical significance at a time when effective pesticides were available (Haneklaus et al., 2007).

1.3   NUTRIENT REQUIREMENTS FOR CROP PLANTS

Plants require 17 elements or essential nutrients for optimal growth and development. These nutrients are carbon (C), hydrogen (H), oxygen (O), nitrogen (N), phosphorus (P), potassium (K), calcium (Ca), magnesium (Mg), sulfur (S), zinc (Zn), copper (Cu), iron (Fe), manganese (Mn), boron (B), molybdenum (Mo), chlorine (Cl), and nickel (Ni). In addition, cobalt (Co) is cited as an essential micronutrient in many publications. Even though Co stimulates growth in certain plants, it is not considered essential according to the Arnon and Stout (1939) definition of essentiality. Essential nutrients may be defined as those without which plants cannot complete their life cycle, are irreplaceable by other elements, and are directly involved in plant metabolism (Fageria et al., 2002; Rice, 2007). Epstein and Bloom (2005) cited two criteria of essentiality of a nutrient. These criteria are (1) the nutrient is part of a molecule that is an intrinsic component of the structure or metabolism of a plant and (2) the plant shows abnormality in its growth and development when the nutrient in question is omitted from the growth medium compared with a plant not deprived of the nutrient from the growth medium. The C, H, and O are absorbed by plants from the air and from water, and the remaining essential nutrients from soil solution. Each of these essential chemical elements performs a specific biochemical or biophysical function within plant cells. Hence deficiency of even one of these elements can impair metabolism and interrupt normal development (Glass, 1989).

Based on the quantity required, nutrients are divided into macro- and micronutrients. Macronutrients are required in large quantities by plants compared to micronutrients. Micronutrients have also been called minor or trace elements, indicating that their concentrations in plant tissues are minor or in trace amounts relative to the concentrations of macronutrients. The higher quantity requirement of macronutrients for plants is associated with their role in making up the bulk of the carbohydrates, proteins, and lipids of plant cells, whereas micronutrients mostly participate in the enzyme activation process of the plant. Data related to the quantity of

TABLE 1.1

Accumulation of Macro- and Micronutrients in Upland Rice and Dry Bean Grown on Brazilian Oxisol

Yield/Nutrient Uptake	Upland Rice			Dry Bean		
	Shoot	Grain	Total	Shoot	Grain	Total
Dry wt./yield (kg ha^{-1})	6189	4434	10,623	2200	3409	5609
N (kg ha^{-1})	56	70	126	17	124	267
P (kg ha^{-1})	3	10	13	2	15	17
K (kg ha^{-1})	150	56	206	41	64	105
Ca (kg ha^{-1})	24	4	28	22	9	31
Mg (kg ha^{-1})	15	5	20	9	6	15
Zn (g ha^{-1})	161	138	299	62	123	185
Cu (g ha^{-1})	35	57	92	9	35	44
Mn (g ha^{-1})	1319	284	1603	31	49	80
Fe (g ha^{-1})	654	117	771	1010	275	1285
B (g ha^{-1})	53	30	83	—	—	—

Source: Adapted from Fageria et al. (2004a); Fageria et al. (2004b).

macro- and micronutrients accumulated in shoots and grains of upland rice (*Oryza sativa* L.) and dry beans (*Phaseolus vulgaris* L.) are presented in Table 1.1. Data in Table 1.1 show that macronutrient accumulation was much higher compared to micronutrient accumulation in cereal as well as legume crops. The order of nutrient accumulation in upland rice was K > N > Ca > Mg > P > Mn > Fe > Zn > Cu > B. Similarly, uptake of macro- and micronutrients in dry bean was in the order of N > K > Ca > P > Mg > Fe > Zn > Mn > Cu.

Macro- and micronutrient exportation to grain and requirements to produce 1 metric ton of grain of cereal, and legume species are presented in Table 1.2. Translocation of macro- and micronutrients was higher in the grain of dry bean compared to upland rice. One striking feature of these results is that K translocation to grain of upland rice was only 11% of the total uptake by plants. This means that most of the K (about 89%) in cereals may remain in the straw. Hence incorporation of straw of cereals into the soil after the harvest of cereal crops can be an important source of K supply to the succeeding crops. Requirements of N, P, and Ca were higher to produce 1 metric ton of grain of dry bean compared to upland rice. However, micronutrient requirements were lower for dry bean compared to upland rice to produce 1 metric ton of grain.

1.4 DIAGNOSTIC TECHNIQUES FOR NUTRITIONAL REQUIREMENTS

Nutrient requirements of crops depend on yield level, crop species, cultivar or genotypes within species, soil type, climatic conditions, and soil biology. Hence soil,

TABLE 1.2

Translocation of Macro- and Micronutrients to Grain and Requirement of These Elements to Produce 1 Metric Ton of Grain of Upland Rice and Dry Bean Grown on Brazilian Oxisols

	Upland Rice		Dry Bean	
Nutrient	Translocation to Grain (% of total uptake)	Requirement to Produce 1 Mg Grain in kg or g[a]	Translocation to Grain (% of total uptake)	Requirement to Produce 1 Mg Grain in kg or g[a]
N	55	28	88	37
P	77	3	90	4
K	11	40	61	27
Ca	16	6	28	8
Mg	27	4	41	4
Zn	46	65	67	48
Cu	62	20	79	11
Mn	18	351	61	21
Fe	15	169	21	333
B	36	18	—	—

[a] Macronutrients are in kilograms and micronutrients in grams.
Source: Adapted from Fageria et al. (2004a); Fageria et al. (2004b).

plant, and climatic factors and their interactions are involved in determining plant nutrient requirements. In addition to this, the economic value of a crop and the socio-economic conditions of the farmer also are important factors in determining the nutrient requirements of a crop. Diagnostic techniques for nutritional disorders are the methods for identifying nutrient deficiencies, toxicities, or imbalances in the soil-plant system (Fageria et al., 1997). Nutrient deficiencies in crop plants may occur due to soil erosion; leaching to lower profile; intensive cropping system; denitrification; soil acidity; immobilization; heavy liming of acid soils; infestation of diseases, insects, and weeds; water deficiency; and low application rates. Similarly, nutrient or elemental toxicity may occur due to excess, imbalance, and unfavorable environmental conditions.

Nutritional disorders are common in almost all field crops worldwide. The magnitude varies from crop to crop and region to region. Even some cultivars are more susceptible to nutritional deficiencies than others within a crop species (Fageria and Baligar, 2005b). The four methods of identifying nutrient disorders in crop plants are visual deficiency symptoms, soil test, plant tissue test, and crop responses to chemical fertilizers or organic manures. Among these methods, soil test is most common in most agroecosystems. These four approaches are becoming widely used separately or collectively as nutrient availability, deficiency, or sufficiency diagnostic aids. They are extremely helpful, but are not without limitations (Fageria and Baligar, 2005b). These methods are discussed in chapters dealing with individual nutrients.

1.5 ASSOCIATION BETWEEN NUTRIENT UPTAKE AND CROP YIELDS

From the viewpoint of sustainable agriculture, nutrient management ideally should provide a balance between nutrient inputs and outputs over the long term (Bacon et al., 1990; Heckman et al., 2003). In the establishment of a sustainable system, soil nutrient levels that are deficient are built up to levels that will support economic crop yields. To sustain soil fertility levels, nutrients that are removed by crop harvest or other losses from the system must be replaced annually or at least within the longer crop rotation cycle (Heckman et al., 2003). When nutrient buildup in soils exceeds plant removal, nutrient leaching and their removal in runoff become an environmental concern (Daniel et al., 1998; Sims et al., 1998; Heckman et al., 2003). Accurate values for crop nutrient removal are an important component of nutrient management planning and crop production (Heckman et al., 2003).

Agricultural production and productivity are directly linked with nutrient availability and uptake. To sustain high crop yields, the application of nutrients is required. Association between uptake of N in the grain of lowland rice grown on Brazilian Inceptisol and grain yield was highly significant ($P < 0.01$) and quadratic ($Y = 826.0022 + 113.6321X - 0.7141X^2$, $R^2 = 0.5099**$). Similarly, relationships between uptake of macro- and micronutrients in the grain of soybean and grain yield were highly significant ($P < 0.01$) and quadratic, except N (Table 1.3). In the case of N, association was linear and 95% variability in grain yield was due to accumulation of N in the grain. Based on R^2 values, it can be concluded that variation in soybean yield was higher due to uptake of N, P, K, Ca, and Mg compared to Zn, Cu, and Fe. Osaki et al. (1992) and Shinano et al. (1994) reported that amount of N accumulated in cereal and legume species showed a highly positive correlation with the total dry matter production at harvest. These authors further reported that N accumulation is one of the most important factors in improving yield of field crops.

TABLE 1.3

Relationship between Nutrient Uptake in Grain and Grain Yield of Soybean Grown on Brazilian Oxisol

Variable	Regression Equation	R^2
N vs. grain yield	$Y = 420.0452 + 12.5582X$	0.9525**
P vs. grain yield	$Y = -2872.9740 + 374.7647X - 5.054X^2$	0.9475**
K vs. grain yield	$Y = -1390.3840 + 104.8791X - 0.4625X^2$	0.9461**
Ca vs. grain yield	$Y = -972.8279 + 647.6862X - 20.3505X^2$	0.6672**
Mg vs. grain yield	$Y = -1341.5260 + 808.0470X - 28.4576X^2$	0.9504**
Zn vs. grain yield	$Y = -7575.7630 + 122.3149X - 0.3159X^2$	0.8996**
Cu vs. grain yield	$Y = -139.1487 + 109.5119X - 0.6986X^2$	0.8339**
Fe vs. grain yield	$Y = -925.7502 + 19.6478X - 0.0189X^2$	0.6715**

** Significant at the 1% probability level.

1.6 FACTORS AFFECTING NUTRIENT AVAILABILITY

Nutrient availability to plants is composed of several processes in the soil-plant system before a nutrient is absorbed or utilized by a plant. These processes include application of nutrient to soil or nutrient existing in the soil, transport from soil to plant roots, absorption by plant roots, transport to plant tops, and finally, utilization by plant in producing economic parts or organs. All these processes are affected by climatic, soil, and plant factors and their interactions. These factors vary from region to region and even within the same region. Hence availability of nutrients to plants is a very dynamic and complex process. Discussion about all the factors affecting nutrient availability to crop plants is beyond the scope of this chapter. However, synthesis of physical, chemical, and biological changes that occur in the rhizosphere, which significantly affect nutrient availability, is given in Figure 1.3. For a detailed discussion on the subject, readers may refer to the work of Marschner (1995), Fageria et al. (1997), Fageria (1992), Mengel et al. (2001), Brady and Weil (2002), Fageria and Baligar (2005b), Epstein and Bloom (2005), and Fageria and Stone (2006).

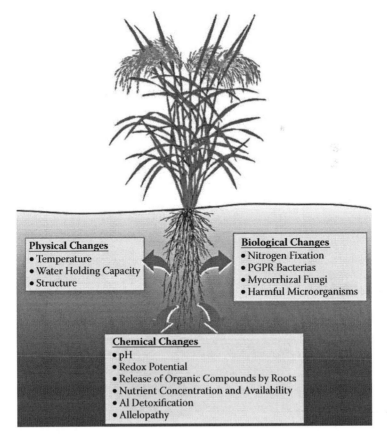

FIGURE 1.3 Physical, chemical, and biological changes in the rhizosphere (Fageria and Stone, 2006).

1.7 FIELD CROPS

Field crops, often referred to as agronomic crops, are crops grown on a large scale
for human consumption, livestock feeding, or raw materials for industrial products
(Fageria et al., 1997). Man's progress has been closely associated with growing field
crops. The food of mankind comes directly or indirectly from crop plants. Producing sufficient food or feed is considered a national priority for all nations in the
21st century. If possible, the best way to feed people or animals properly is to produce the food where it will be consumed rather than importing it. Choice of crops
and methods of cultivation have a profound effect on quantity as well as quality of
crop products. Malthus (1766–1834) was one of many who expressed concern over
man's food supply. He held that population tends to increase faster than its means of
subsistence can be made to increase. In 1998, Sir William Crookes, an English econ-
omist, predicted that by 1930 all available wheat lands in the United States would be
in use and that the United States would be driven to import and would be struggling
with Great Britain for the lion's share of the world's wheat supply. However, none of
these predictions came to pass during the 20th century. Food production increases
faster than global population. It can be concluded that the malnutrition or hunger
experienced in some parts of the world was not caused by a food shortage at the
global level. However, social, economic, or political factors might be responsible for
such occurrences in some regions, especially in the African continent.

1.7.1 CLASSIFICATION OF FIELD CROPS

Classification of field crop plants is essential for their use and for adopting appropriate
management practices to improve yields. Several classification systems exist for crop
classification. These classifications include (1) agronomic use, (2) botanical, (3) growth
habit, (4) special purpose, (5) forage crops, and (6) photorespiration. Among these
crop classification systems, agronomical use and botanical are the most dominant and
widely used by farmers as well as the agricultural scientific community.

1.7.1.1 Agronomic Use

Mankind bases the agronomic use classification of crop plants on their use. Crop
plants may be used for food, forage, fiber, oil, sugar, special uses, or medicinal
purposes. Cereals and legumes are prominent food crops. Table 1.4 presents a list
of crop plants according to their agronomic use. Some plants are listed under the
headings "Cereals" and "Legumes;" however, their grains have very high oil content.
Some of these plants are soybean, peanut, and corn. In this list, forage crops of special use and medicinal plants are not listed. Some examples of these plants are given
under the classifications of these plants.

1.7.1.2 Botanical

Botanical classification was one of the first effective systems for classifying plants. In
the 18th century, an eminent Swedish botanist, Carl von Linne (1708–1778), devised
a system known as the Linnaeus concept of classifying plants. Linne introduced
the natural system of classification, which seeks to arrange groupings on the basis

TABLE 1.4
Agronomic Classification of Important Crop Plants

Common Name	Scientific Name	Growth Habit	Photorespiration
	Cereals		
Rice	*Oryza sativa* L.	A	C_3
Wheat (common)	*Triticum aestivum* L.	A	C_3
Corn (maize)	*Zea mays* L.	A	C_4
Barley	*Hordeum vulgare* L.	A	C_3
Sorghum	*Sorghum bicolor* L. Moench	A	C_4
Rye	*Secale cereale* L.	A	C_3
Oats	*Avena sativa* L.	A	C_3
Triticale	*Triticale hexaploide* Lart.	A	C_3
Pearl millet	*Pennisetum glaucum* L R. Br.	A	C_4
	Legumes		
Dry bean	*Phaseolus vulgaris* L.	A	C_3
Soybean	*Glycine max* L. Merrill	A	C_3
Peanut (groundnut)	*Arachis hypogaea* L.	A	C_3
Bean (mung)	*Vigna radiata* L.	A	C_3
Bean (faba)	*Vicia faba* L.	A	C_3
Cowpea	*Vigna unguiculata* L. Walp.	A	C_3
Pea (common)	*Pisum sativum* L.	A	C_3
Chickpea	*Cicer arietinum* L.	A	C_3
Lentil	*Lens culinaris* Medikus	A	C_3
Pigeon pea	*Cajanus cajan* L. Millsp.	A	C_3
	Sugar Crops		
Sugarcane	*Saccharum officinarum* L.	P	C_4
Sugar beets	*Beta vulgaris* L.	B	C_3
	Fiber Crops		
Cotton	*Gossypium hirsutum* L.	A	C_3
Jute	*Corchorus olitorius* L.	A	C_3
Sisal	*Agave sisalana* Perr.	P	C_3
Kenaf	*Hibiscus cannabinua* L.	P	C_3
Ramie	*Boehmeria nivea* Gaud	P	C_3
	Oil Crops		
Sesame	*Sesamum idicum* L.	A	C_3
Sunflower	*Helianthus annuus* L.	A	C_3
Rapeseed (canola)	*Brassica napus* L.	A	C_3
Safflower	*Carthamus tinctorius* L.	A	C_3
	Roots and Tuber Crops		
Potato	*Solanum tuberosum* L	A, P	C_3
Taro	*Colocasia esculenta* L. Schott	P	C_3
Yams	*Dioscorea rotundata* Poir.	A	C_3

(continued on next page)

TABLE 1.4 (continued)
Agronomic Classification of Important Crop Plants

Common Name	Scientific Name	Growth Habit	Photorespiration
Sweet potato	*Ipomoea batatas* L.	A	C_3
Cassava	*Manihot esculenta* Crantz	P	C_3

Note: A = annual, P = perennial, C_3 = higher photorespiratory activity, and C_4 = lower photorespiratory activity.

of natural affinities and evolutionary tendencies rather than on mere similarities in appearance. Botanists have found that in the evolutionary process, reproductive organs like fruits, flowers, and seeds have changed far less under the influence of diverse environmental conditions than have the roots, stems, and leaves. Hence the botanical classification of the plants is based on structural differences and similarities in the morphology of reproductive parts, the organs least likely to be changed by the environment (Burger, 1984). Latin has been more widely studied than any other language; because of this, it has been used to provide the universal plant nomenclature. In this binomial system, the scientific name of each plant is composed of two words. The first word is the name of the genus to which the particular plant belongs; the second word is the name of the species. The *L.* or *Linn.*, which often follows the scientific name, denotes that this species was first named and described by the botanist Linnaeus.

In the botanical classification, crop plants are divided into two distinct groups based on characteristics of inflorescence or fruits. One is the grass family (Gramineae) and the other is the legume family (Leguminosae). Gramineae is a very large and specialized family of about 10,000 species (Cobley, 1976). The grass family includes important grain crops like wheat, rice, corn, barley, oats, and rye. It also includes many lawn, pasture, hay, marsh, and range grasses. Its members range in size from the low-growing, fine-leafed bent grasses to the giant bamboo, which has a woody stem and stretches 40 m or more. Grasses have a fibrous root system, cylindrical stems, tillers, leaves, and a head or panicle that develops at the end of a stem. Each fertile flower bears a single seed. The wheat head is an unbranched spike, whereas in rice it is a branched panicle.

Over two-thirds of edible dry matter continues to be provided by the cereals. Rice, corn, and wheat account for 54% of the edible dry matter, rather less than the legumes and oilseeds (about 12%) but rather more than sugar (5%; Evans, 1993). The cereals are also major suppliers of protein, providing about 54% of the total, followed by the legumes and oilseeds (21%), animal products (18%), and fruits and vegetables (4%; Evans, 1993). Rice, wheat, and corn seem likely to remain the staple foods of mankind through direct consumption and indirectly through their use as food grains (Evans, 1993).

After the grass family, the legume family is the most important group of crop plants for mankind. There are about 18,000 species in Leguminosae, and they are characterized by their fruits, which are pods or legumes, and by their (usually) alternate, compound, pinnate, or trifoliate leaves (Cobley, 1976). Legumes not only provide a large part of the protein for the world population but also contribute significantly to fixing atmospheric nitrogen that can be utilized by the plants. In addition, both the seeds and the leaves of legumes are characteristically high in nitrogen; this family of plants plays an important role in livestock feeding. The legume fruit is most commonly a pod, and the pod itself is called a legume. The term is derived from the Latin *legere*, which means "to gather." It originated from the fact that these fruits might be gathered without cutting the plants.

1.7.1.3 Growth Habit

Plants are classified on the basis of growth habit—that is, as annuals, biennials, or perennials. Annuals are the plants that complete their life cycle in one season. Rice, wheat, corn, and barley are typical examples of annual crop plants. Biennial plants require two seasons to complete their life cycle. White clover (*Melilotus officinalis* Lam.) and sugar beet are typical examples of biennial plants. Some plants complete their life cycle in more than two seasons. Such plants are known as perennial plants. Perennial plants may produce seeds each year, but they do not die with seed production. Typical examples of perennial plants are alfalfa (*Medicago sativa* L.) and white clover (*Trifolium repens* L.).

1.7.1.4 Forage Crops

Forage crops are important for feeding livestock, which is an important component of modern agricultural systems. A grassland ecosystem is an excellent example of a renewable resource, and, if properly managed, the system may be productive over a very long time. A first principle of modern pasture establishment is the association of grass with legume in any pasture to improve the quality of pasture forages (Fageria et al., 1997). Animals are grazed on the pastures as well as on dried forages known as hays. Green forages are also utilized for feeding livestock, and silage is forage preserved in succulent condition by a process of natural fermentation or by acidification. A soiling crop is one cut while still green and fed at once to livestock. Important forage crops are listed in Table 1.5.

1.7.1.5 Special Purpose

The special-purpose classification of crop plants is based on their specific use in the cropping systems. Some examples of special-purpose use of crops are catch crops, cover crops, and green manure crops. The importance of these crops in soil-ameliorating practice is increasing in recent years because of the high cost of chemical fertilizers, the increased risk of environmental pollution, and the need for sustainable cropping systems. Overall, these crops are used in cropping systems to conserve water and nutrients, protect the soil from erosion, control weeds, and improve the soil's physical, chemical, and biological properties and consequently crop yields. Table 1.6 shows a list of important catch crops, green manure crops, and cover crops.

TABLE 1.5

Important Forage Grasses and Legumes

Common Name	Scientific Name	Growth Habit	Photorespiration
	Grasses		
Bent grass (creeping)	*Agrostis palustris* Huds	P	C_4
Bermuda grass	*Cynodon dactylon* L. Pers.	P	C_4
Bluegrass (Kentucky)	*Poa pratensis* L.	P	C_3
Bromegrass (smooth)	*Bromus intermis* Leyss.	P	C_3
Orchard grass	*Dactylis glomerata* L.	P	C_3
Oats	*Avena sativa* L.	A	C_3
Bahia grass	*Paspalum notatum* Flugge	P	C_4
Buffel grass	*Cenchrus ciliaris* L. Link	P	C_4
Dallis grass	*Paspalum dilatatum* Poir.	P	C_4
Gamba grass	*Andropogon gayanus* Kunth	P	C_4
Guinea grass	*Panicum maximum* Jacq.	P	C_4
Jaragua	*Hyparrhenia rufa* Nees. Stapf.	P	C_4
Italian ryegrass	*Lolium multiflorum* Lam.	A	C_3
Limpograss	*Hemartharia altissima* Poir. Stapf. & ubbard	P	C_4
Napier grass	*Pennisetum purpureum* Schumach	P	C_4
Para grass	*Brachiaria mutica* Forsk Stafp.	P	C_4
Sudan grass	*Sorghum bicolor* drummondi	A	C_4
Surinam grass	*Brachiaria decumbens* Stapf.	P	C_4
Perennial ryegrass	*Lolium pernne* L.	P	C_3
Tall fescue	*Festuca arundinacea* schreb.	P	C_3
	Legumes		
Alfalfa	*Medicago sativa* L.	P	C_3
Birdfoot trefoil	*Lotus corniculatus* L.	P	C_3
Blue lupin	*Lupinus angustifolius* L.	A	C_3
Centro	*Centrosema pubescens* Benth	P	C_3
Ladino clover	*Trifolium repens* L.	P	C_3
Red clover	*Trifolium paratense* L.	P	C_3
Silverleaf desmodium	*Desmodium uncinatum* Jacq.	P	C_3
Sub clover	*Trifolium subterraneum* L.	A	C_3
Townsville stylo	*Stylosanthes humilis* H. B. K.	A	C_3
White lupin	*Lupinus albus* L.	A	C_3

Note: A = annual, P = perennial, C_3 = higher photorespiratory activity, and C_4 = lower photorespiratory activity.

1.7.1.6   Photorespiration

Crop plants are classified as C_3 and C_4 based on the pathways of carbon metabolism and their behavior in CO_2 uptake. Plants whose first carbon compound in photosynthesis consists of a three-carbon-atom chain are called C_3 plants, whereas C_4

TABLE 1.6

Major Catch Crops, Green Manure Crops, and Cover Crops for Tropical and Temperate Regions

Tropical Region		Temperate Region	
Common Name	Scientific Name	Common Name	Scientific Name
Sunn hemp	*Crotalaria juncea* L.	Hairy vetch	*Vicia villosa* Roth
Sesbania	*Sesbania aculeata* Retz Poir	Barrel medic	*Medicago truncatula* Gaertn
Sesbania	*Sesbania rostrata* Bremek & Oberm	Alfalfa	*Medicago sativa* L.
		Black lentil	*Lens culinaris* Medikus
Cowpea	*Vigna unguiculata* L. Walp.	Red clover	*Trifolium pratense* L.
Soybean	*Glycine max* L. Merr.	Soybean	*Glycine max* L. Merr.
Cluster bean	*Cyamopsis tetragonoloba*	Faba bean	*Vicia faba* L.
Alfalfa	*Medicago sativa* L.	Crimson clover	*Trifolium incarnatum* L.
Egyptian clover	*Trifoliam alexandrium* L.	Ladino clover	*Trifolium repens* L.
Wild indigo	*Indigofera tinctoria* L.	Subterranean clover	*Trifolium subterraneum* L.
Pigeon pea	*Cajanus cajan* L. Millspaugh		
Mung bean	*Vigna radiata* L. Wilczek	Common vetch	*Vicia sativa* L.
Lablab	*Lablab purpureus* L.	Purple vetch	*Vicia benghalensis* L.
Gray bean	*Mucuna cinerecum* L	Cura clover	*Trifolium ambiguum* Bieb.
Buffalo bean	*Mucuna aterrima* L. Piper & Tracy	Sweet clover	*Melilotus officinalis* L.
		Winter pea	*Pisum sativum* L.
Crotolaria breviflora	*Crotolaria breviflora*	Narrow-leaved vetch	*Vicia angustifolia* L.
White lupin	*Lupinaus albus* L.	Milk vetch	*Artragalus sinicus* L.
Milk vetch	*Astragalus sinicus* L.		
Crotalaria	*Crotalaria striata*		
Zornia	*Zornia latifolia*		
Jack bean	*Canavalia ensiformis* L. DC.		
Tropical kudzu	*Pueraria phaseoloides* (Roxb.) Benth.		
Velvet bean	*Mucuna deeringiana* Bort. Merr.		
Adzuki bean	*Vigna angularis*		
Brazilian stylo	*Stylosanthes guianiensis*		
Jumbie bean	*Leucaena leucocephala* Lam. De Wit		
Desmodium	*Desmodiumovalifolium* Guillemin & Perrottet		
Pueraria	*Pueraria phaseoloides* Roxb.		

Source: Fageria et al. (2005); Fageria (2007a); Fageria and Baligar (2007).

plants are those whose first carbon compound in photosynthesis is composed of a four-carbon chain (Burger, 1984). The C_3 plants have high photorespiration rates compared to C_4 plants. Hence, yield of C_3 plants is low compared to that of C_4 plants.

The C_4 plants are better adapted to adverse environmental conditions (high temperature, limited water supply) compared to C_3 plants. Solar radiation utilization efficiency of C_4 plants is higher than that of C_3 plants. This may lead to a higher photosynthesis rate and consequently a higher yield capacity of C_4 plants compared to C_3 plants (Fageria, 1992). Examples of important C_3 and C_4 plants are presented in Table 1.4 and Table 1.5.

1.8 CROP YIELD

Yield is one of the most important measurements of a crop plant's economic value. Hence, it is important to define yield. Yield is defined as the amount of specific substance produced (e.g., grain, straw, total dry matter) per unit area (Soil Science Society of America, 1997). In the present case, grain yield will be considered for discussion purposes. Grain yield refers to the weight of cleaned and dried grains harvested from a unit area. For cereals or legumes, grain yield is usually expressed either in kilograms per hectare (kg ha^{-1}) or in metric tons per hectare (Mg ha^{-1}) at 13 or 14% moisture. Yield of a crop is determined by management practices, which will maintain the productive capacity of a crop ecosystem. These practices include use of crop genotypes, soil fertility, water management, and control of insects, diseases, and weeds. In addition, preparation of land and plant density influence crop yield.

Potential yield is an estimate of the upper limit of yield increase that can be obtained from a crop plant (Fageria, 1992). Genetic yield potential is defined as the yield of adapted lines in a favorable environment in the absence of agronomic constraints (Reynolds et al., 1999). Evans and Fischer (1999) defined potential yield as the maximum yield that could be reached by a crop or genotype in a given environment, as determined, for example, by simulation models with plausible physiological and agronomic assumptions. These authors further reported that the term *potential yield* is often used synonymously with *yield potential*. However, Evans and Fischer (1999) defined yield potential as the yield of a cultivar when grown in environments to which it is adapted, with nutrients and water nonlimiting and with pests, diseases, and weeds, lodging, and other stresses effectively controlled. Evans and Fischer (1999) further reported that there is no evidence that a ceiling on yield potential has been reached, but should this occur, average yields could still continue to rise as crop management improves and as plant breeders continue to improve resistance to pests, diseases, and environmental stresses.

Long et al. (2006) defined the yield potential of a grain crop as the seed mass per unit ground area obtained under optimum growing conditions without weeds, pests, and diseases. These authors further reported that yield potential is determined by the efficiency of conversion of the intercepted light into biomass and the proportion of biomass partitioned into grain. Hybrid rice between indicas increased yield potential by about 9% under the tropical conditions (Peng et al., 1999). The higher yield potential of indica/indica hybrids compared with indica inbred cultivars was attributed to the greater biomass production rather than harvest index. These authors also reported that new plant type breeding has not yet improved rice yield potential due to poor grain filling and low biomass production. However, work at the International

Rice Research Institute, Philippines, is in progress to remove this yield barrier and increase the yielding potential of rice.

Potential yield is, in a way, the most optimistic estimate of crop yield that is based on present knowledge and available biological material, under ideal management, in an optimum physical environment. Rasmusson and Gengenbach (1984) reported that genetic potential of a plant or genotype is manifested through the interrelationships among genes, enzymes, and plant growth. These authors further reported that a gene contributes to the formation for biosynthesis of an enzyme that functions in a particular metabolic reaction. The combined effect of many genes, through their control of enzymes, results in physiological traits contributing to plant growth, development, and yield (Rasmusson and Gengenbach, 1984). Furthermore, Wallace et al. (1972) reported that yield has long been classified as a characteristic controlled by quantitative genetics (i.e., influenced by many genes with the effects of individual genes normally unidentified). Wallace et al. further reported that a simple way to describe the genetics of yield is to assume that a single gene controls each physiological component. The minimum estimate of gene number controlling yield is then the number of physiological components.

Yield potential is generally determined by calculating photosynthesis during a spikelet-filling period (Murata and Matsushima, 1975). For rice growing in an environment where the daily amount of solar radiation received is 16.7 MJ m^{-2}, assuming an efficiency of 26% in photosynthesis, the net carbohydrate production in a 40-day spikelet-filling period was calculated to be 16.4 Mg ha^{-1} (Austin, 1980; Fageria, 1992). Over the years, rice yields have increased due to advances in breeding and crop management. New rice cultivars have been released that possess yield potential >10 Mg ha^{-1} (Ottis and Talbert, 2005).

1.8.1 YIELD COMPONENTS

Crop yield is determined by yield components, which are (in cereals) the number of panicles/heads, the number of spikelets per panicle/head, the weight of 1000 spikelets, and spikelet sterility or filled spikelet (Fageria et al., 2006; Fageria, 2007b). Therefore, it is very important to understand the formation of yield components during the crop growth cycle and their associations with grain yields and management practices that influence yield components and consequently grain yield.

Yield components are formed during the plant growth cycle. Hence, it is very important to have knowledge of different growth stages during the growth cycle of a crop or plant. Figure 1.4 shows the growth stages of upland rice cultivar having a growth cycle of 130 days from sowing to physiological maturity or 125 days from germination to physiological maturity under Brazilian conditions or tropics. The vegetative growth stage had a duration of 65 days (germination to initiation of panicle primordia), the reproductive growth stage had a duration of 30 days (panicle primordia initiation to flowering), and the spikelet-filling stage also had a duration of 30 days (flowering to physiological maturity). The plant growth stages from panicle initiation to flowering and from flowering to physiological maturity or spikelet filling are important for yield determination because during these stages the potentials for seed number and seed weight components of yield are formed (Fageria et al., 1997).

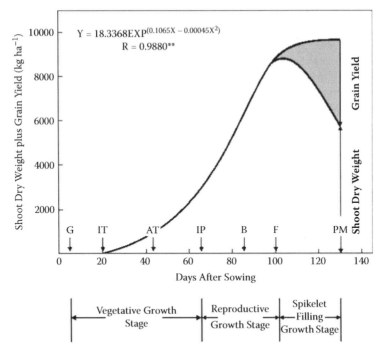

FIGURE 1.4 Shoot dry weight accumulation and grain yield of upland rice during the growth cycle of the crop in central Brazil. G = germination, IT = initiation of tillering, AT = active tillering, IP = initiation of panicle primordia, B = booting, F = flowering, and PM = physiological maturity (Fageria, 2007b).

The number of panicles or heads is determined during the vegetative growth stage, the number of spikelets per panicle/head is determined during the reproductive growth stage, and the weight of spikelets and spikelet sterility are determined during the spikelet-filling or reproductive growth stage. Hence, adequate N supplies throughout the growth cycle of rice plant or cereal crop is one of the main strategies to increase grain yield. Yield of a cereal can be expressed in the form of the following equation by taking into account the yield components (Fageria, 2007b):

$$\text{Grain yield (Mg ha}^{-1}) = \text{number of panicles m}^{-2} \times \text{spikelets per panicle} \times$$
$$\% \text{ filled spikelets} \times 1000 \text{ spikelets weight (g)} \times 10^{-5}$$

Among these yield components, panicles or spikelet per unit area is usually the most variable yield component in rice (Fageria et al., 1997). The number of panicles per unit area is determined during the period up to about 10 days after maximum tiller number is reached (Murata and Matsushima, 1975). The association between panicle number and grain yield of rice is shown in Figure 1.5. Similarly, the association between 1000-grain weight and grain yield and the association between spikelet sterility and grain yield are shown in Figure 1.6 and Figure 1.7, respectively. In addition to these yield components, shoot dry weight and grain harvest index also influenced grain yield of rice significantly (Figure 1.8 and Figure 1.9).

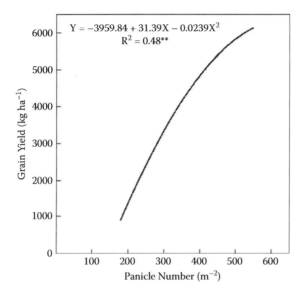

FIGURE 1.5   Relationship between panicle number and grain yield of lowland rice.

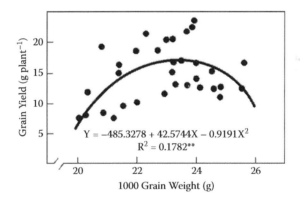

FIGURE 1.6   Association between 1000-grain weight and grain yield of lowland rice.

In legumes, yield is mainly expressed as the product of pods per unit area or per plant, seeds per pod, and seed weight. Grain yield in legumes can be computed by using the following equation:

$$\text{Grain yield (Mg ha}^{-1}) = \text{number of pods} \times \text{number of seeds per pod} \times \text{weight of } 1000 \text{ seeds (g)} \times 10^{-5}$$

In legumes, growth stages (vegetative, reproductive, and pod filling) are not as distinct as in cereals, and reproductive and pod-filling stages overlapped. In the indeterminate type of growth habit, even vegetative growth continues. However,

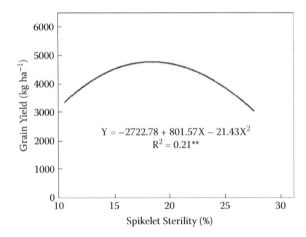

FIGURE 1.7   Relationship between spikelet sterility and grain yield of lowland rice.

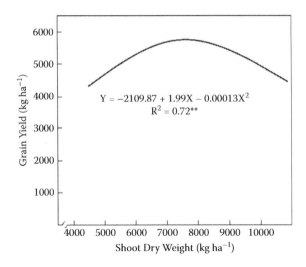

FIGURE 1.8   Relationship between shoot dry weight and grain yield of lowland rice.

the growth cycle of legumes roughly can be divided into vegetative and reproductive growth stages (Figure 1.10). Shoot dry weight of dry bean increased with increasing plant age in an exponential quadratic fashion, and maximum dry weight was produced at 78 days after sowing. From 18 to 60 days' growth periods, shoot dry weight was almost linear.

Number of pods is one of the most important yield components in determining grain yield of dry bean. Figure 1.11 shows the relationship between number of pods and grain yield of dry beans grown on Brazilian Inceptisol. Bean yield increased significantly in a quadratic fashion with increasing number of pods per plant. Variation in grain yield was 67% due to the number of pods per plant. Similarly, shoot dry weight also increased grain yield of dry bean in a quadratic fashion (Figure 1.12).

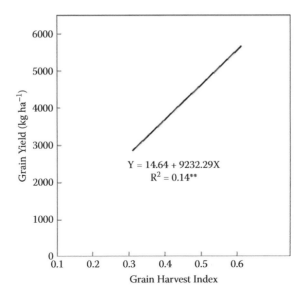

FIGURE 1.9   Association between grain harvest index and grain yield of lowland rice.

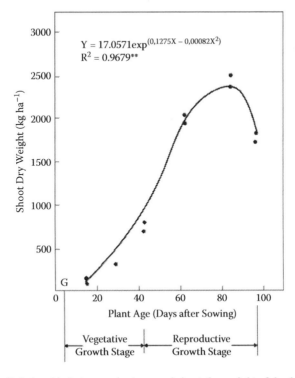

FIGURE 1.10   Relationship between plant age and shoot dry weight of dry beans grown on Brazilian Oxisol (Fageria et al., 2004).

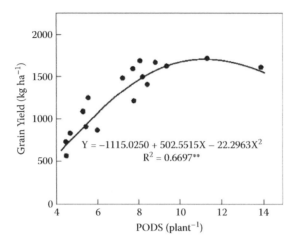

FIGURE 1.11 Relationship between number of pods and grain yield of dry bean grown on Brazilian lowland soil.

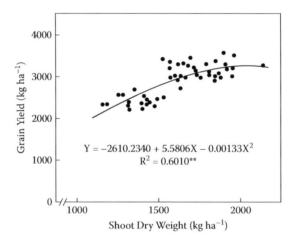

FIGURE 1.12 Association between shoot dry weight and grain yield of dry bean (Fageria et al., 2004).

Management practices that can influence yield components in field crops and consequently yield are use of adequate quantity of essential nutrients, crop species or genotypes within species having high adaptability and yield potential, and control of diseases, insects, and weeds. In acid soils, liming is a very effective practice to improve yield components and grain yield of legumes (Figure 1.13 and Figure 1.14).

1.8.2 Cereal versus Legume Yields

Cereals and legumes account for the majority of mankind's food supply. Hence, it is relevant to compare yield capacity of these two important crop species. Yield of cereals like rice, corn, wheat, barley, and sorghum is much higher compared to that of

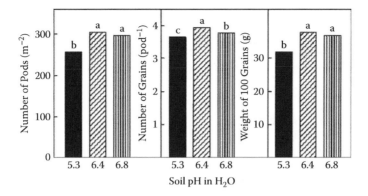

FIGURE 1.13 Influence of soil pH on yield components of dry bean grown on Brazilian Oxisol (Fageria and Santos, 2005).

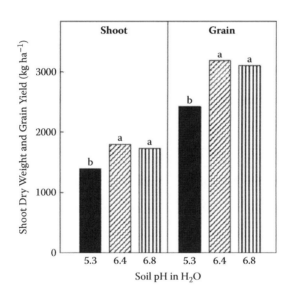

FIGURE 1.14 Influence of soil pH on shoot dry weight and grain yield of dry bean grown on Brazilian Oxisol (Fageria and Santos, 2005).

legumes like soybean, dry bean, peanuts, and cowpea. Data in Table 1.7 show yields of two cereals (rice and corn) and two legumes (dry bean and soybean) grown on Brazilian Oxisols and Inceptisols. Yield of cereals like upland rice, lowland rice, and corn was much higher than that of legumes like dry bean and soybean. Overall, the yield of cereals was 163% higher than that of legumes. The lower yield of legumes compared to cereals is a general trend (Fageria et al., 2006). Several reasons for this have been given in the literature. One reason is that legumes have high photorespiration compared to cereals, and this leads to the lower yield of legumes (Fageria et al., 2006). Another reason cited is the higher protein and lipid content in seeds of legumes compared to seeds of cereals; this leads to higher energy consumption

TABLE 1.7

Grain Yield of Upland Rice, Corn, Dry Bean, and Soybean Grown on Brazilian Oxisols and Inceptisols

Crop Species	Grain Yield (kg ha^{-1})	Reference
Upland rice (Oxisol)	4568	Fageria, 2001
Lowland rice (Inceptisol)	6892	Fageria and Prabhu, 2004
Lowland rice (Inceptisol)	6500	Fageria and Baligar, 2005b
Corn (Oxisol)	8221	Fageria, 2001
Dry bean (Oxisol)	1912	Fageria, 2001
Dry bean (Oxisol)	3086	Fageria, 2006
Dry bean (Oxisol)	2766	Fageria and Stone, 2004
Dry bean (Inceptisol)	1094	Fageria and Santos, 1998
Soybean (Oxisol)	2776	Fageria, unpublished data
Soybean (Oxisol)	3295	Fageria, unpublished data
Average of cereals	6545	
Average of legumes	2490	

and lower yields (Yamaguchi, 1978). Data presented in Table 1.8 show the protein and lipid content in the grain of principal cereal and legume species. Overall, protein content in the legume species was about twofold higher and lipid content about threefold higher than that of cereals. A third reason is that legumes fix atmospheric

TABLE 1.8

Protein and Lipid Content in the Seed of Principal Cereal and Legume Species

Crop Species	Protein (g kg^{-1})	Lipid (g kg^{-1})
Rice	105	22
Wheat	113	17
Barley	132	35
Corn	102	41
Sorghum	88	35
Oat	125	47
Average of cereals	111	33
Soybean	314	201
Dry bean	192	24
Lupin	251	41
Adzuki bean	216	13
Peanut	193	304
Pea	233	15
Chickpea	173	51
Average of legumes	225	93

Source: Adapted from Shinano et al. (1993).

nitrogen, which causes legumes to consume more energy than cereals and consequently lowers grain yields (Fageria et al., 2006). Fageria et al. have also reported that in cereals improved partitioning of dry matter resulted in higher grain yields; similar improvements have not been obtained for grain legumes. Further, dry matter production efficiency per unit N uptake was reported to be lower in legumes than in cereals (Osaki et al., 1992).

1.9. CONCLUSIONS

Increasing crop yields in the 21st century has become an essential component of modern society to keep pace with the increasing world population. In 1970, the U.S. Bureau of the Census estimated the world population at fewer than 4 billion people. This number rose to about 6.5 billion in 2007 and is projected to reach over 9 billion by the year 2030. Global food demand is projected to increase by 50% in the next 30 years. Furthermore, agricultural systems should be economically viable, environmentally sound, and socially acceptable without degrading natural resources. In this context, applications of mineral or organic fertilizers in adequate amount and proportion are vital components of modern agriculture. Sound mineral nutrition programs not only increase crop yields but also are important for sustaining the agroecosystem and maintaining human and animal health. Furthermore, rational use of chemical fertilizers will help reduce land degradation and environmental pollution.

Yield is a complex plant characteristic and is influenced by many yield components and their interactions. The most important yield components and associated plant characteristics are number of panicles or heads, number of spikelets per panicle or head, weight of spikelet, and spikelet sterility in cereals. In legumes, number of pods per unit area, number of seeds per pod, and seed weight are the yield-determining components. In addition to these yield components, shoot dry weight and grain harvest index are also associated with increasing grain yield in cereals and legumes. Experimental results related to the association between yield components and yield of rice and dry bean are presented to make the information as practical as possible. However, higher yield is possible only when there is appropriate balance between these yield-associated plant characteristics. In addition to appropriate plant architecture, the development of crop genotype tolerance to diseases and insects, soil acidity, and increase in N_2 fixation capacity may improve yields. All these yield components and associated plant characteristics are genetically controlled and also influenced by environmental factors. Research findings have proved that crop genotypes differ significantly in the physiological processes that determine yield. Hence, breeding to improve yields components and associated characteristics and adopting appropriate crop management practices are vital for increasing crop yields. Molecular genetics, which can transpose and modify genes, can play a vital role in modifying the yield physiology of field crops in favor of higher yields in the future. In conclusion, the information provided in this chapter can be useful in removing or reducing yield barriers, which would result in improved crop yields.

This introductory chapter provides information on the role of essential nutrients in increasing crop yields, field crops and their classification systems, and how to improve crop yields, which ultimately enhances nutrient use efficiency in crop plants.

REFERENCES

Arnon, D. I. and P. R. Stout. 1939. The essentiality of certain elements in minute quantity for plants with special reference to copper. *Plant Physiol.* 14:371–385.

Austin, R. B. 1980. Physiological limitations to cereal yields and ways of reducing them by breeding. In: *Opportunities for increasing crop yields*, R. G. Hurd, P. V. Biscoe, and C. Dennis, Eds., 3–19. London: Pitman.

Bacon, S. C., L. E. Lanyon, and R. M. Schlander, Jr. 1990. Plant nutrient flow in the management pathways of an intensive dairy farm. *Agron. J.* 82:755–761.

Baligar, V. C., N. K. Fageria, and Z. L. He. 2001. Nutrient use efficiency in plants. *Commun. Soil Sci. Plant Anal.* 32:921–950.

Bray, R. H. 1954. A nutrient mobility concept of soil-plant relationships. *Soil Sci.* 78:9–22.

Bodenheimer, F. S. 1958. *The history of biology: An introduction.* London: Dawson.

Borlaug, N. E. and C. R. Dowswell. 1997. The acid lands: One of the agricultures last frontiers. In: *Plant soil interactions at low pH: Sustainable agriculture and forestry production*, A. C. Moniz, A. M. C. Furlani, R. E. Schaffert, N. K. Fageria, C. A. Rosolem, and H. Cantarella, Eds., 5–15. Campinas, São Paulo: Brazilian Soil Science Society.

Brady, N. C. and R. R. Weil. 2002. *The nature and properties of soils*, 13th edition, Upper Saddle River, New Jersey: Prentice Hall.

Browne, C. A. 1944. *A source book of agricultural chemistry.* Waltham, Massachusetts: Chronica Botanica.

Burger, A. W. 1984. Crop classification. In: *Physiological basis of crop growth and development*, M. B. Tesar, Ed., 1–12. Madison, Wisconsin: American Society of Agronomy.

Cakmak, I. 2001. Plant nutrition research: Priorities to meet human needs for food in sustainable ways. In: *Plant nutrition: Food security and sustainability of agro-ecosystems*, W. J. Horst, M. K. Schenk, A. Burkert, et al., Eds., 4–7. Dordrecht: Kluwer Academic.

Cobley, L. S. 1976. *An introduction to the botany of tropical crops.* New York: Longman.

Daniel, T. C., A. N. Sharpley, and J. L. Leminyon. 1998. Agricultural phosphorus and eutrophication: A symposium overview. *J. Environ. Qual.* 27:251–257.

Epstein, E. 2000. The discovery of the essential elements. In: *Discoveries in plant biology*, Vol. 3, S. D. Kung and S. F. Yang, Eds., 1–16. Singapore: World Scientific.

Epstein, E. and A. J. Bloom. 2005. *Mineral nutrition of plants: Principles and perspectives*, 2nd edition. Sunderland, Massachusetts: Sinauer Associates.

Evans, L. T. 1993. *Crop evolution, adaptation and yield.* New York: Cambridge University Press.

Evans, L. T. and R. A. Fischer. 1999. Yield potentials: Its definition, measurement, and significance. *Crop Sci.* 39:1544–1551.

Fageria, N. K. 1992. *Maximizing crop yields.* New York: Marcel Dekker.

Fageria, N. K. 2001. Response of upland rice, dry bean, corn and soybean to base saturation in cerrado soil. *Rev. Bras. Eng. Agric. Amb.* 5:416–424.

Fageria, N. K. 2005. Soil fertility and plant nutrition research under controlled conditions: Basic principles and methodology. *J. Plant Nutr.* 28:1–25.

Fageria, N. K. 2006. Liming and copper fertilization in dry bean production on an Oxisol in no-tillage system. *J. Plant Nutr.* 29:1219–1228.

Fageria, N. K. 2007a. Green manuring in crop production. *J. Plant Nutr.* 30:691–719.

Fageria, N. K. 2007b. Yield physiology of rice. *J. Plant Nutr.* 30:843–879.

Fageria, N. K. and V. C. Baligar. 1997. Response of common bean, upland rice, corn, wheat, and soybean to soil fertility of an Oxisol. *J. Plant Nutr.* 20:1279–1289.

Fageria, N. K. and V. C. Baligar. 2001. Lowland rice response to nitrogen fertilization. *Commun Soil Sci. Plant Anal.* 32:1405–1429.

Fageria, N. K. and V. C. Baligar. 2005b. Nutrient availability. In: *Encyclopedia of soils in the environment*, D. Hillel, Ed., 63–71. San Diego, California: Elsevier.

Fageria, N. K. and V. C. Baligar. 2007. Agronomy and physiology of tropical cover crops. *J. Plant Nutr.* 30:1287–1339.

Fageria, N. K., V. C. Baligar, and B. A. Bailey. 2005. Role of cover crops in improving soil and row crop productivity. *Commun. Soil Sci. Plant Anal.* 36:2733–2757.

Fageria, N. K., V. C. Baligar, and R. B. Clark. 2002. Micronutrients in crop production. *Adv. Agron.* 77:185–268.

Fageria, N. K., V. C. Baligar, and R. B. Clark. 2006. *Physiology of crop production.* New York: Haworth.

Fageria, N. K., V. C. Baligar, and C. A. Jones. 1997. Growth and mineral nutrition of field crops, 2nd edition. New York: Marcel Dekker.

Fageria, N. K., V. C. Baligar, and Y. C. Li. 2008. The role of nutrient efficient plants in improving crop yields in the twenty first century. *J. Plant Nutr.* 31: (In press).

Fageria, N. K., V. C. Baligar, and R. W. Zobel. 2007. Yield, nutrient uptake, and soil chemical properties as influenced by liming and boron application in common bean in no-tillage system. *Commun. Soil Sci. Plant Analy.* 38:1637–1653.

Fageria, N. K., M. P. Barbosa Filho, L. F. Stone, and C. M. Guimarães. 2004a. Phosphorus nutrition in upland rice production. In: *Phosphorus in Brazilian Agriculture*, T. Yamada and S. R. S. Abdalla, Eds., 401–418. Piracicaba, Brazil: Brazilian Potassium and Phosphate Research Association.

Fageria, N. K., M. P. Barbosa Filho, and L. F. Stone. 2004b. Phosphorus nutrition in dry bean production. In: *Phosphorus in Brazilian Agriculture*, T. Yamada and S. R. S. Abdalla, Eds., 435–455. Piracicaba, Brazil: Brazilian Potassium and Phosphate Research Association.

Fageria, N. K. and A. S. Prabhu. 2004. Blast control and nitrogen management in lowland rice cultivation. *Pesq. Agropec. Bras.* 39:123–129.

Fageria, N. K. and A. B. Santos. 1998. Phosphorus fertilization for bean crop in lowland soil. *Rev. Bras. Eng. Agric. Amb.* 2:124–127.

Fageria, N. K. and A. B. Santos. 2005, October 18–20. Influence of base saturation and micronutrient rates on their concentration in the soil and bean productivity in cerrado soil in no-tillage system. Paper presented at the *VIII National Bean Congress*, Goiânia, Brazil.

Fageria, N. K., N. A. Slaton, and V. C. Baligar. 2003. Nutrient management for improving lowland rice productivity and sustainability. *Adv. Agron.* 80:630152.

Fageria, N. K. and L. F. Stone. 2004. Yield of common bean in no-tillage system with application of lime and zinc. *Pesq. Agropec. Bras.* 39:73–78.

Fageria, N. K. and L. F. Stone. 2006. Physical, chemical and biological changes in rhizosphere and nutrient availability. *J. Plant Nutr.* 29:1327–1356.

FAO. 2000. Fertilizer requirements in 2015 and 2030. Rome, Italy.

Glass, A. D. M. 1989. *Plant nutrition: An introduction to current concepts.* Boston: Jones and Bartlett.

Haneklaus, S., E. Bloem, and E. Schnug. 2007. Sulfur and plant disease. In: *Mineral nutrition and plant disease*, L. E. Datnoff, W. H. Elmer, and D. M. Huber, Eds., 101–118. St. Paul, Minnesota: The American Phytopathological Society.

Heckman, J. R., J. T. Sims, D. B. Beegle, F. J. Coale, S. J. Herbert, T. W. Bruulsema, and W. J. Bamka. 2003. Nutrient removal by corn grain harvest. *Agron. J.* 95:587–591.

Hoagland, D. R. 1922. *Soil analysis and soil plant interrelations.* Agr. Exp. Station Univ. California, Circular No. 235:1–7.

Hoagland, D. R. and T. C. Broyer. 1936. General nature of the process of salt accumulation by roots with description of experimental methods. *Plant Physiol.* 11:471–507.

Hoagland, D. R., J. C. Martins, and G. R. Stewart. 1920. Relation of the soil solution to the soil extract. *J. Agric. Res.* 20:381–395.

Johnston, E. S. and D. R. Hoagland. 1929. Minimum potassium level required by tomato plants grown in water cultures. *Soil Sci.* 27:89–108.

Lal, R. 2006. Soil science in the era of hydrogen economy and 10 billion people. In: *The future of soil science*, A. Hartemink, Ed., 76–79. Wageningen, The Netherlands: International Union of Soil Science.

Long, S. P., X. G. Zhu, S. L. Naidu, and D. R. Ort. 2006. Can improvement in photosynthesis increase crop yields?. *Plant Cell Environ.* 29:315–330.

Loneragan, J. F. 1997. Plant nutrition in the 20th and perspectives for the 21st century. In: *Plant nutrition for sustainable food production and environment*, T. Ando, K. Fujito, T. Mae, et al., Eds., 3–14. Dordrecht: Kluwer Academic.

Lu, R. and Z. Shi. 1998. Effect of long-term fertilization on soil properties. In: *Soil-plant nutrients principles and fertilization*, R. Lu, J. Xie, G. Cai, and Q. Zhu, Eds., 102–110. Beijing: Chemical Industry Press.

Marschner, H. 1995. *Mineral nutrition of higher plants,* 2nd edition. New York: Academic Press.

Murata, Y. and S. Matsushima. 1975. Rice. In: *Crop physiology: Some case histories,* L. T. Evans, Ed., 73–99. London: Cambridge University Press.

Mengel, K., E. A. Kirkby, H. Kosegarten, and T. Appel. 2001. *Principles of plant nutrition*, 5th edition. Dordrecht: Kluwer Academic.

Okajima, H. 2001. *Historical significance of nutrient acquisition in plant nutrition research.* In: *Plant nutrient acquisition: New perspectives*, N. Ae, J. Arihara, K. Okada, A. Srinivasan, Eds., 3–31. Tokyo: Springer.

Osaki, M., T. Shinano, and T. Tadano. 1992. Carbon-nitrogen interaction in field crops production. *Soil Sci. Plant Nutr.* 38:553–564.

Ottis, B. V. and R. E. Talbert. 2005. Rice yield components as affected by cultivar and seedling rate. *Agron. J.* 97:1622–1625.

Peng, S., K. G. Cassman, S. S. Virmani, J. Sheehy, and G. S. Khush. 1999. Yield potential trends of tropical rice since the release of IR8 and the challenge of increasing rice yield potential. *Crop Sci.* 39:1552–1559.

Rasmusson, D. C. and B. G. Gengenbach. 1984. Genetics and use of physiological variability in crop breeding. In: *Physiological basis of crop growth and development,* M. B. Tesar, Ed., 291–321. Madison, Wisconsin: American Society of Agronomy and Crop Science Society of America.

Raun, W. and G. V. Johnson. 1999. Improving nitrogen use efficiency for cereal production. *Agron. J.* 91:357–363.

Reed, H. S. 1942. *A short history of the plant sciences.* New York: Ronald Press.

Reynolds, M. P., S. Rajaram and K. D. Sayre. 1999. Physiological and genetic changes of irrigated wheat in the post-green revolution period and approaches for meeting projected global demand. *Crop Sci.* 39:1611–1621.

Rice, R. W. 2007. The physiological role of minerals in the plant. In: *Mineral nutrition and plant disease*, L. E. Datnoff, W. H. Elmer, and D. M. Huber, Eds., 9–29. St. Paul, Minnesota: The American Phytopathological Society.

Robertson, T. B. 1907. On the normal rate of growth of an individual and its biochemical significance. *Arch. Entwicklungsnech. Organismen* 25:580–614.

Shinano, T., M. Osaki, K. Komatsu, and T. Tadano. 1993. Comparison of production efficiency of the harvesting organs among field crops. I. Growth efficiency of the harvesting organs. *Soil Sci. Plant Nutr.* 39:269–280.

Shinano, T., M. Osaki, S. Yamada, and T. Tadano. 1994. Comparison of root growth and nitrogen absorbing ability between gramineae and leguminosae during the vegetative stage. *Soil Sci. Plant Nutr.* 40:485–495.

Soil Science Society of America. 1997. *Glossary of soil science terms.* Madison, Wisconsin: Soil Science Society of America.

Spillman, W. J. and E. Lang. 1924. *The law of diminishing returns.* Chicago: World Book.

Stewart, W. M., D. W. Dibb, A. E. Johnston, and T. J. Smyth. 2005. The contribution of commercial fertilizer nutrients to food production. *Agron. J.* 97:1–6.

Sumner, M. E. and A. D. Noble. 2003. Soil acidification: The world story. In: *Handbook of soil acidity*, Z. Rengel, Ed., 1–28. New York: Marcel Dekker.

Tisdale, S. L., W. L. Nelson, and J. D. Beaton. 1985. *Soil fertility and fertilizers*, 4th edition. New York: Macmillan.

Tucker, T. C. 1987. Future developments in soil fertility and plant nutrition: An introduction. In: *Future developments in soil science research*, L. L. Boersma, Ed., 169. Madison, Wisconsin: Soil Science Society of America.

Van der Ploeg, R. R., W. Bohm, and M. B. Kirkham. 1999. On the origin of the theory of mineral nutrition of plants and the law of the minimum. *Soil Sci. Soc. Am. J.* 63:1055–1062.

Wallace, D. H., J. L. Ozbun, and H. M. Munger. 1972. Physiological genetics of crop yield. *Adv. Agron.* 24:97–146.

Westerman, R. L. and T. C. Tucker. 1987. Soil fertility concepts: Past, present and future. In: *Future developments in soil science research*, L. L. Boersma, Ed., 171–179. Madison: Soil Science Society of America.

Yamaguchi, J. 1978. Respiration and the growth efficiency in relation to crop productivity. *J Fac. Agric. Hokkaido University*, Japan 59:59–129.

2 Nitrogen

2.1 INTRODUCTION

Nitrogen (N) is one of the most yield-limiting nutrients for crop production in the world. It is also the nutrient element applied in the largest quantity for most annual crops (Huber and Thompson, 2007). Systems of agriculture that rely heavily on soil reserve to meet the N requirements of plants cannot long be effective in producing high yields of crops (Stevenson, 1982). Except for legumes, which have the ability to fix their own N, N must be supplied to plants for growth. It is usually added as a fertilizer and is required for all types of soils (Clark, 1982). To increase crop yields, growers worldwide apply over 80 million metric tons of nitrogen fertilizers per year (Epstein and Bloom, 2005). Use of inorganic N fertilizers has had its most substantial beneficial effect on human health by increasing the yield of field crops and nutritional quality of foods needed to meet dietary requirements and food preferences for growing world populations (Galloway and Cowling, 2002; Galloway et al., 2002). Ridley and Hedlin (1980) concluded that increased use of N fertilizer has had the most dramatic influence on increasing crop yields since the 1950s, in combination with disease-resistant cultivars to a lesser effect. Similarly, Camara et al. (2003) reported that historically, few if any technologies have increased winter wheat yield in the United States more than N fertilization. The main reasons for N deficiency are high-quantity uptake by crop plants compared to other macronutrients (except K in some crops such as rice), including in grains or seeds, and its loss by leaching, denitrification, volatilization, soil erosion, and surface runoff. In addition, N is immobilized by soil microbes and undecomposed plant residues, which may cause temporary deficiency. Nitrogen loss in the form of NH_3 by plant canopy has been reported (Fageria et al., 2005). Furthermore, in intensive cropping systems, where a no-tillage system is adopted, depletion or loss of organic matter has been reported (Johnson et al., 2006), which may result in N deficiency in crop plants.

Use of low rates for high-yielding modern crop cultivars, especially by farmers in developing countries, is another cause of N deficiency (Fageria et al., 2003). In developing countries, intensive agricultural production systems have increased the use of N fertilizer in efforts to produce and sustain high crop yields (Fageria et al., 2003). Consequently, N losses into the environment have also increased (Schmied et al., 2000). Even with the continuing research in N management, average worldwide N use efficiencies (NUE) are reported to be around 50% (Newbould, 1989; Collins et al., 2007), and N recovery efficiency for cereal production (rice, wheat, sorghum, millet, barley [*Hordeum vulgare* L.], maize, oat [*Avena sativa* L.], and rye [*Secale cereale* L.]) is approximately 33% (Raun and Johnson, 1999). This means that large amount of N is lost in the soil–plant system. Bock (1984) reported that in the United Kingdom, 50% of N losses from applied fertilizers are caused by leaching, denitrification of nitrate, and volatilization of ammonia, reducing the efficiency with which they are used as well as exacerbating environmental problems in the hydrosphere, troposphere, and stratosphere. Fageria and Baligar (2001) reported that in Brazilian Inceptisol, flooded rice recovered only 39% of the applied fertilizer N. Similarly, De Datta (1987) reported that in flooded rice, N loss in the tropics is about 56% of the N fertilizer applied. This high N loss not only increases cost of crop production but also creates the problem of environmental pollution.

The increase in crop yields due to N application may be associated with increase in panicles or heads in cereals and number of pods in legumes (Fageria et al., 2006; Fageria, 2007). Nitrogen also improves grain or seed weights in crop plants and reduces grain sterility (Fageria and Baligar, 2001; Fageria, 2007a; Fageria et al., 2006). Nitrogen also increases shoot dry matter, which is positively associated with grain yield in cereals and legumes (Fageria, 2007a; Fageria, 2008). Grain harvest index (grain yield/straw dry weight plus grain yield) and N harvest index (N uptake in the grain/N uptake in grain plus straw) are also reported to be improved by addition of N to crop plants (Fageria and Baligar, 2005a; Fageria et al., 2006). These two plant traits are reported to be positively associated with yield in field crops (Fageria, 2007a; Fageria, 2008).

Nitrogen has been recognized as an essential nutrient for plant growth for more than a century. Significant advances emerged in N fertilizer technology during the last half of the 20th century. Furthermore, the essential role of N in increasing crop production and its dynamic nature and property for N loss from the soil–plant system create a unique and challenging environment for its efficient management (Fageria and Baligar, 2005a). In addition, efficient or optimal management of N in the agro-ecosystem is still a debatable issue. The objectives of this chapter are to discuss the N cycle in the soil–plant system, its functions and deficiency symptoms, and uptake in field crops and suggest management practices to maximize N use efficiency by field crops. The chapter mainly provides a discussion of practical aspects of N management to improve yields of important food crops. This information will increase crop yields, reduce cost of crop production, and minimize environmental pollution.

2.2 CYCLE IN SOIL–PLANT SYSTEMS

The cycle of a nutrient in the soil–plant system can be defined as addition, transformation, and uptake by plants, loss from the soil–plant system, and immobilization. The nitrogen cycle in the soil–plant system is very dynamic and complex due to involvement of climatic, soil, and plant factors. Knowledge of the nutrient cycle in the soil–plant system is an important aspect of understanding the availability of N to plants and adopting management practices to maximize its uptake and use efficiency (Fageria and Baligar, 2005a). A simplified version of the N cycle in the soil–plant system is given in Figure 2.1, which shows that the main input sources of N to the soil–plant system are chemical fertilizers, organic manures, and biological N_2 fixation. Similarly, the main N depletion sources in the soil–plant system are leaching, denitrification, volatilization, surface runoff, and plant uptake. The N

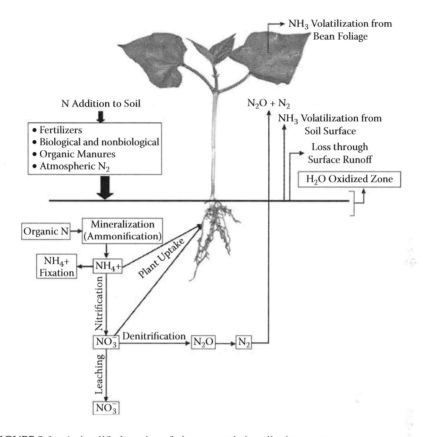

FIGURE 2.1 A simplified version of nitrogen cycle in soil–plant system.

immobilization also has a temporary influence on N uptake to plants. The N immobilization is defined as the transformation of inorganic N compounds (e.g., NH_4^+, NO_3^-, NO_2^-, NH_3) into the organic state. Soil organisms assimilate inorganic N compounds and transform them into organic N constituents of their cells and tissues, the soil biomass (Jansson and Persson, 1982).

In the surface mineral soils, N content ranges from 0.2 to 5.0 g kg^{-1} with a average value of about 1.5 g kg^{-1} (Brady and Weil, 2002). More than 90% of the N in most soils is in the form of organic matter. The organic form of N protects the N from loss; however, it is also not available to crop plants. This organic form of N should be mineralized to NH_4^+ and NO_3^- before its uptake by plants. Mineral N seldom accounts for more than 1 to 2% of the total N in the soil (Brady and Weil, 2002). Mineralization is the conversion of an element from an organic form to an inorganic state as a result of microbial activity (Soil Science Society of America, 1997). According to Stevenson (1982), mineralization is the conversion of organic forms of N to NH_4^+ and NO_3^-. The initial conversion to NH_4^+ is referred to as *ammonification*, and the oxidation of this compound to NO_3^- is termed *nitrification*. The nitrification process, which occurs in two phases in the soil–plant system, can be represented by the following equations:

$$2NH_4^+ + 3O_2 \Leftrightarrow 2NO_2^- + 2H_2O + 4H^+$$

$$2NO_2^- + O_2 \Leftrightarrow 2NO_3^-$$

In the process of nitrification, bacteria known as *Nitrosomonas* are involved in the process of conversion of ammonia into nitrites, and the bacteria that convert nitrites to nitrates are known as *Nitrobacter*. Collectively, the nitrifying organisms are known as *Nitrobacteria*. Under optimal soil temperature and humidity, nitrification occurs at a very fast rate. In addition, nitrification is an oxidation process, and aeration of soil increases nitrification. Plowing and cultivation are recognized means of promoting nitrification. Nitrification results in release of H^+ ions, leading to soil acidification. Furthermore, the enzymatic oxidation of nitrification also releases energy. The utilization of NH_4^+ and NO_3^- by plants and microorganisms constitutes assimilation and immobilization, respectively.

Nitrate ions are negatively charged and easily to leach under heavy rainfall or irrigation. The amount of N leached depends on soil type, source of N fertilizer, crops, and methods of fertilizer application. Also, nitrate is a major factor associated with leaching of such bases as calcium, magnesium, and potassium from the soil. The nitrate and bases move out together. As these bases are removed and replaced by hydrogen, soil becomes more acid. Nitrogen fertilizers containing such strong acid-forming anions as sulfate increase acidity more than other carriers without acidifying anions (Fageria and Gheyi, 1999).

Denitrification is a major loss of N from the soil–plant system, and it is mainly an anaerobic bacterial respiration. The denitrification occurs in the following reductive ways:

$$NO_3^- \text{ (nitrate)} \Rightarrow NO_2^- \text{ (nitrite)} \Rightarrow NO \text{ (nitric oxide)}$$
$$\Rightarrow N_2O \text{ (nitrous oxide)} \Rightarrow N_2 \text{ (dinitrogen)}$$

Most denitrifying bacteria exist in the topsoil (0–30 cm), with the number decreasing exponentially down to 120–150 cm (Parkin and Meisinger, 1989). Denitrification is influenced by several factors like soil pH, temperature, organic C supply, nitrate concentration, aeration, and water status (Aulakh et al., 1992). Due to influence by several physical and chemical factors, exact quantification of N loss due to denitrification is difficult. However, Aulakh et al. (1992) reported that overall, N losses due to denitrification might be about 30% in an agroecosystem. In addition, estimation of global denitrification losses ranges from 83 Tg yr^{-1} (Stevenson, 1982) to 390 Tg yr^{-1} (Hauck and Tanji, 1982; Aulakh et al., 1992). A large part of N is ultimately returned to the atmosphere through biological denitrification, thereby completing the cycle (Stevenson, 1982). Denitrification usually occurs in soil high in organic matter, under extended periods of waterlogged conditions and as temperature rises. Whereas N_2 is an inert gas that poses no known environmental risk, N_2O is one of the greenhouse gases that contribute to destruction of the earth's protective ozone layer (Fageria and Gheyi, 1999). Denitrification was significantly higher in grasslands than in either hardwood or pine forest areas (Lowrance et al., 1995).

The concentration of N_2O has been rising at an increasing rate, particularly since about 1960. Currently N_2O concentrations in the atmosphere are increasing at about 0.3% per year (Seiler, 1986). The atmosphere contains about 1500 Tg (1 Terra gram = 10^{12} g) of N_2O. Perhaps 90% of the emissions are derived from soils through biologically mediated reactions of nitrification and denitrification (Byrnes, 1990). Emission of N_2O may be perceived as a leakage of intermediate products in each of these processes. While soils are a major source of N_2O, they also serve as a sink (Freney et al., 1978; Fageria and Gheyi, 1999).

Ammonia volatilization is an important process of N loss from soil–plant systems when nitrogen fertilizers are surface applied, especially in alkaline soils. Ammonium volatilization can be expressed by the following equation (Bolan and Hedley, 2003):

$$NH_4^+ + OH^- \Leftrightarrow NH_3 + H_2O$$

Ammonia volatilization from flooded rice soils is a major mechanism for N loss and a cause of low fertilizer use efficiency by rice. Reviews on NH_3 volatilization from flooded-rice soils indicate that losses of ammonical-N fertilizer applied directly to floodwater may vary from 10 to 50% of the amount applied (Fillery and Vlex, 1986; Mikkelsen, 1987). Losses, however, are site and soil management specific; thus, disparities may exist in reported rates of volatilization, depending on rate-controlling factors and methods of measurement. Ammonia volatilization under flooded-rice conditions is influenced by five primary factors: NH_4–N concentration, pH, temperature, depth of flooded water, and wind speed (Jayaweera and Mikkelsen, 1990; Fageria and Gheyi, 1999).

Volatilization losses from surface applications of urea-containing N sources can be related to soil and weather conditions following application. Fox and Hoffman (1981) categorized NH_3 volatilization losses in Pennsylvania based on rainfall amount and the length of time between surface application and rainfall. They concluded that 10 mm of rain accruing within 2 days after surface N application resulted in no NH_3 volatilization losses. Losses increased with increased time between application and rain and are substantial (> 30%) if no rain falls within 6 days. Urban et al. (1987) reported that maximum NH_3 losses from surface-applied urea in a growth chamber occurred between 4 to 8 days after application.

Absorption and loss of N through the plant canopy is also an important part of N cycling in soil–plant systems. Controlled as well as field studies showed that plants can absorb NH_3 from the air as well as lose NH_3 to the air by volatilization (Farquhar et al., 1980; Fageria and Baligar, 2005a). Emission of NH_3 has increased considerably in recent decades. Factors influencing NH_3 losses include soil and plant N status and plant growth stage (Sharpe and Harper, 1997). Abundant supply favors NH_3 losses, especially if the supply is in excess of plant requirements (Fageria and Baligar, 2005a). The loss of NH_3 through the plant canopy can occur during the whole growth cycle of a crop (Harper and Sharpe, 1995). However, some scientists have reported that the highest NH_3 volatilization rates for major agricultural crops occur during the reproductive growth stage (Francis et al., 1997). Absorption of atmospheric NH_3 has been associated with low plant N content and with high atmospheric NH_3 concentrations (Harper and Sharpe, 1995).

Fixation of NH_4^+ is an important component of the N cycle in soil–plant systems. Three specific processes are recognized as responsible for the fixation and retention of N applied to soil, namely fixation by clay minerals, NH_3 fixation by soil organic matter, and biological immobilization of NH_4 by heterotrophic microorganisms (Nommik and Vahtras, 1982). Adsorption of NH_4^+ in a nonexchangeable form in the interlayer region of expanding 2:1 layer aluminosilicate clay minerals (fixation) may reduce the fertilizer use efficiency of this nutrient when added to soils in which such minerals predominate (Stehouwer and Johnson, 1991). This type of fixation was first reported early in this century, and many studies have been conducted to investigate this phenomenon with respect to fertilizer use efficiency (Nommik and Vahtras, 1982). Greater fixation of NH_4^+ has been reported with anhydrous NH_3 (AA) than with other forms of NH_4^+-releasing fertilizers (Young and Cattani, 1962). This may be due to the enhanced reaction of NH_3 with more acidic water in 2:1 clay interlayer (Nommik and Vahtras, 1982). Many earlier studies reported that an increase in pH increased fixation of NH_4^+ (Nommik, 1957). This effect is generally attributed to decreased competition with H_3O^+ as pH increased in a lower pH range (2.5–5.5) and to decreasing charge of interlayer hydroxy-Al polycations in a higher pH range (5.5–8.0). Increasing solution concentration of NH_4^+ has been shown to increase the amount of NH_4^+ fixed, while the percentage fixation (amount fixed/amount added) was decreased (Black and Waring, 1972). Removal of organic matter from mineral surfaces has been found to increase NH_4^+ fixation (Hinman, 1966).

Fixed NH_4^+ is involved in the N dynamics of soil and may be an important component of the N fertility status of some agricultural soils. Between 18 and 23% of added $^{15}NH_4$ was fixed (specifically adsorbed) after 15 days' incubation in soils containing relatively high vermiculitic clay (Drury et al., 1989). Ammonium ^{15}N has been observed to be fixed and released in proportion to added $^{15}NH_4$ and vermiculite content (Keerthisinghe et al., 1984). Drury and Beauchamp (1991) reported that immobilization of 5.7% of the added $^{15}NH_4$ occurred in the high-fixing soil and 3.9% in the low-fixing soil. Extractable- and fixed-NH_4 fractions were interrelated pools. When fertilizer NH_4 was added to the soil, a proportional amount was fixed by the clay minerals. When nitrification and immobilization depleted extractable NH_4, fixed NH_4 was released. The fixed-NH_4 pool appeared to be a slow-released reservoir, with the rate of release of fixed NH_4 being slower than the rate of fixation. Juma and Paul (1983) reported that NH_4 fixation was enhanced when nitrification was inhibited with 4-amino-1,3,4-triacole (ATC), a nitrification inhibitor. Neetison et al. (1986) observed that over one-half of added NH_4 disappeared and reappeared after 5 weeks. Since the clay minerals in these soils were not found to fix added NH_4, they postulated that the immobilization occurred as an osmoregulation mechanism, and then NH_4 was remineralized when the microbes died and decomposed. One NH_4 is fixed by clay mica; it is protected against nitrification until it is released from fixation sites (Scherer and Mengel, 1986). It can be concluded that there are a number of ways by which NH_4^+ fixation can be regulated. He et al. (1990) reported that about 16% of the $^{15}NH_3$-N injected was fixed by the soil. Of the N fixed, 48% was accounted for by chemical fixation into the soil organic fraction, most of which was removed by sequential extraction and 52% of which was accounted for as clay-fixed NH_4. These authors also reported that NH_3-N fixed by organic matter occurs in forms that are less stabilized and more biologically available than the native soil N. Nommik and Vahtras (1982) presented an extensive review of the mechanisms and influential factors affecting NH_4 fixation.

A large quantity of nitrogen accumulates in seeds of cereals and legumes; this nitrogen is not recycled and is lost from the soil–plant system. Fageria et al. (1997a) reported that about 50% of the total nitrogen accumulated was in the grains of upland rice cultivars IAC 47 and IR 43. The remaining 50% was in the roots and shoots. In legume crops, grains remove still higher levels of N. This means that in a nitrogen management strategy, the removal of N by a crop should be taken into account.

2.3 FUNCTIONS AND DEFICIENCY SYMPTOMS

Nitrogen has greater influence on growth and yield of crop plants than any other essential plant nutrient. It plays a pivotal role in many physiological and biochemical processes in plants. Nitrogen is a component of many important organic compounds ranging from proteins to nucleic acids. It is a constituent of the chlorophyll molecule, which plays an important role in plant photosynthesis. Many enzymes are proteinaceous; hence, N plays a key role in many metabolic reactions. Nitrogen is also a structural constituent of cell walls. Nitrogen-deficient plants grow slowly, and their leaves are small. Nitrogen deficiency also decreases leaf area index (LAI), lowers radiation use efficiency, and lowers photosynthesis activity in plants (Muchow, 1988; Sinclair and Horie, 1989; Fageria and Baligar, 2005a).

A significant positive association has been reported between the light saturation rate of photosynthesis of a leaf and its nitrogen content (Evans, 1989; Poorter and Evans, 1998). The reason for this strong relationship is the large amount of leaf organic nitrogen (up to 75%) present in the chloroplasts, most of it in the photosynthetic machinery (Evans and Seemann, 1989; Poorter and Evans, 1998). Seeds are small and yields are reduced in cereals and legume crops under N-deficient conditions. The reduction in yield and quality are directly related to the severity of the N deficiency. Data in Table 2.1 show the influence of N on grain yield and yield

TABLE 2.1
Comparison of Yield Components of Lowest and Highest-Yield-Producing Genotypes at Lowest and Highest N Rates (Values Are Averages of Two Years)

Grain Yield and Its Components	0 kg N ha⁻¹		200 kg N ha⁻¹	
	Lowest Yield (BRS Jaburu)	Highest Yield (BRS Alvorada)	Lowest Yield (CNAi 8569)	Highest Yield (BRSGO Guará)
Grain yield (kg ha⁻¹)	1769	2747	4887	6696
Number of panicles (m⁻²)	372	367	433	531
Grain harvest index	0.42	0.50	0.36	0.53
1000 grain weight (g)	25.1	25.4	26.0	25.0
Spikelet sterility (%)	25.9	22.8	29.0	14.2

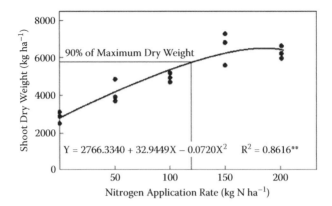

FIGURE 2.2 Influence of N application on shoot dry weight of lowland rice grown on Brazilian Inceptisol. Values are averages of 12 genotypes.

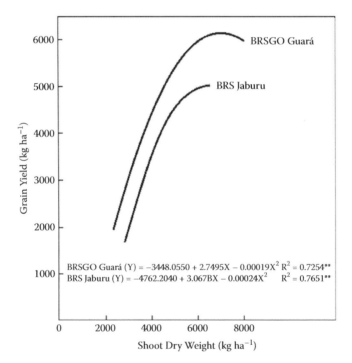

FIGURE 2.3 Relationship between shoot dry weights and grain yield of two lowland rice genotypes.

components of lowland rice genotypes grown on Brazilian Inceptisol. Nitrogen application improved shoot dry weight in crops (Figure 2.2), and shoot dry weight has a significant quadratic association with grain yield (Figure 2.3).

Nitrogen deficiency symptoms are associated with reduced plant height, tillering in cereals, pods in legumes, leaf discoloration, and reduced growth of newly

FIGURE 2.4   Dry bean plants without N application (left) and with N (right).

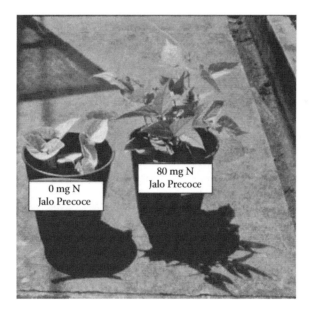

FIGURE 2.5   Dry bean cultivar Jalo Precoce without N (left) and with N (right).

emerging plant parts. Figure 2.4 shows the growth of dry bean plants without and with N. The plants without N showed reduction in growth and yellow color of whole foliage. Similarly, growth of dry bean plants in Figure 2.5 shows reduction in growth and yellow color of leaves much more than in Figure 2.4, because different cultivars were used. This means that there was a difference in genotypes or cultivars of the same species in growth reduction and N deficiency symptoms. The size of the plant canopy is reduced under N-deficient conditions. Nitrogen deficiency also reduces root growth, which negatively influences absorption of water and nutrients by plants. Nitrogen-deficient plants have fewer root hairs compared to plants supplied with adequate amounts of N (Fageria, 1992). Tennant (1976) reported that N fertilization increased seminal and nodal roots in wheat. Nitrogen-deficient plant leaves are yellowish and pale in color due to a loss of chlorophyll. Nitrogen is a mobile nutrient in plants, and its deficiency symptoms first appear in the lower or older leaves. The N deficiency symptoms first start in the tips and margins of the leaves. In cases of severe deficiency, whole leaf becomes yellow and dry. In legumes, N-deficient leaves may fall off.

Observation of nitrogen deficiency symptoms in crop plants is the cheapest method of diagnosis for N disorders. However, this technique requires a lot of experience on the part of the observer, because deficiency symptoms can be confused with problems caused by drought, insect and disease infestation, herbicide damage, soil salinity, soil acidity, and inadequate drainage (Fageria and Baligar, 2005b). Sometimes, a plant may be on the borderline with respect to deficiency and adequacy of a given nutrient. In this situation, there are no visual symptoms, but the plant is not producing at its capacity. This condition is frequently called "hidden hunger." Deficiency symptoms normally occur over an area and not on an individual plant. If a symptom is found on a single plant, it may be due to disease or insect injury or a genetic variation (Fageria and Baligar, 2005b). Mineral-deficient plants are usually more susceptible to diseases, insects, and physical damage (Clark, 1982).

2.4 DEFINITIONS AND ESTIMATION OF N USE EFFICIENCY

Nitrogen use efficiency in crop plants is defined in several ways in the literature (Fageria and Baligar, 2005a). In simple terms, efficiency is ratio of output (economic yield) to input (fertilizers) for a process or complex system (Crop Science Society of America, 1992). Some of the definitions pertaining to N use efficiency reported in the literature are presented and their methods of calculation are given in Table 2.2. Values of different N use efficiency defined and discussed in Table 2.2 are presented in Table 2.3 for five lowland rice genotypes. Nitrogen use efficiency generally has significant positive association with grain yield in crop plants (Table 2.4). This means that improving N use efficiency in crop plants can improve grain yield. Nitrogen use efficiency can be improved by adopting appropriate crop management practices.

2.5 UPTAKE AND PARTITIONING

Uptake of nutrients by crop plants in adequate amount and proportion is very important for producing higher yields. Similarly, distribution of absorbed or accumulated nutrients in shoot and grain (higher N in grain) is associated with yield improvement (Fageria and Baligar, 2005a). Nutrient uptake in crop plants is mainly measured by tissue analysis. Nutrient uptake in crop plants is expressed in concentration or accumulation terms. It is also important how much of the accumulated nutrient in the plant translocated to grain. Nutrient distribution in grain compared to total uptake

TABLE 2.2

Definitions and Methods of Calculating Nitrogen Use Efficiency

Designation of Efficiency	Definitions and Formulas for Calculation
Agronomic efficiency (AE)	The agronomic efficiency is defined as the economic production obtained per unit of nutrient applied. It can be calculated by:
	AE (kg kg^{-1}) = G_f – G_u/N_a, where G_f is the grain yield of the fertilized plot (kg), G_u is the grain yield of the unfertilized plot (kg), and N_a is the quantity of N applied (kg).
Physiological efficiency (PE)	Physiological efficiency is defined as the biological yield obtained per unit of nutrient uptake. It can be calculated by:
	PE (kg kg^{-1}) = BY_f – BY_u/N_f – N_u, where, BY_f is the biological yield (grain plus straw) of the fertilized pot (kg), BY_u is the biological yield of the unfertilized plot (kg), N_f is the nutrient uptake (grain plus straw) of the fertilized plot, and N_u is the N uptake (grain plus straw) of the unfertilized plot (kg).
Agrophysiological efficiency (APE)	Agrophysiological efficiency is defined as the economic production (grain yield in case of annual crops) obtained per unit of nutrient uptake. It can be calculated by:
	APE (kg kg^{-1}) = G_f – G_u/N_f – N_u, where, G_f is the grain yield of fertilized plot (kg), G_u is the grain yield of the unfertilized plot (kg), N_f is the N uptake (grain plus straw) of the fertilized plot (kg), N_u is the N uptake (grain plus straw) of unfertilized plot (kg).
Apparent recovery efficiency (ARE)	Apparent recovery efficiency is defined as the quantity of nutrient uptake per unit of nutrient applied. It can be calculated by:
	ARE (%) = (N_f – N_u/N_a) × 100, where, N_f is the N uptake (grain plus straw) of the fertilized plot (kg), N_u is the N uptake (grain plus straw) of the unfertilized plot (kg), and N_a is the quantity of N applied (kg).
Utilization efficiency (UE)	Nutrient utilization efficiency is the product of physiological and apparent recovery efficiency. It can be calculated by;
	UE (kg kg^{-1}) = PE × ARE

Source: Fageria et al. (1997a), Fageria and Baligar (2003a), Fageria and Baligar (2005a).

in the plant is known as *nutrient harvest index*. It should not be confused with *grain harvest index*, which is the ratio of grain weight to grain plus straw weight.

2.5.1 CONCENTRATION

The concentration is defined as uptake per unit dry weight of plant tissue. Concentration is generally expressed as a percentage, g kg^{-1} or mg kg^{-1}. The concentration is a more stable plant chemical property than uptake due to its expression on a dry weight basis. Nutrient concentration in plant tissue varied with plant tissue analyzed

TABLE 2.3

Nitrogen Use Efficiency in Lowland Rice Genotypes

Genotype	AE (kg kg⁻¹)	PE (kg kg⁻¹)	APE (kg kg⁻¹)	ARE (%)	UE (kg kg⁻¹)
CNAi 8886	23	105	56	37	39
CNAi 8569	17	188	69	29	55
CNAi 9018	21	222	123	29	64
BRS Jaburu	16	114	64	26	30
BRS Biguá	19	145	74	23	33
Average	19	155	77	29	44

Note: AE = agronomical efficiency, PE = physiological efficiency, APE = agrophysiological efficiency, ARE = apparent recovery efficiency, and UE = utilization efficiency.

Source: Adapted from Fageria et al. (2007).

TABLE 2.4

Relationship between N Use Efficiency and Grain Yield in Dry Bean (Values Are Averages of 20 Dry Bean Genotypes Grown on Brazilian Oxisol)

Variable	Regression Equation	R^2
AE (X) vs. grain yield (Y)	$Y = 4.1500 + 1.7022X - 0.0554X^2$	0.5678**
PE (X) vs. grain yield (Y)	$Y = -1.2031 + 0.8366X - 0.0106X^2$	0.2668**
APE (X) vs. grain yield (Y)	$Y = 1.7855 + 1.1349X - 0.0228X^2$	0.3704**
ARE (X) vs. grain yield (Y)	$Y = 0.8856 + 0.4415X - 0.0032X^2$	0.3441**
EU (X) vs. grain yield (Y)	$Y = 3.0899 + 1.2579X - 0.0311X^2$	0.4714**

** Significant at the 1% probability level.

Note: AE = agronomical efficiency, PE = physiological efficiency, APE = agrophysiological efficiency, ARE = apparent recovery efficiency, and UE = utilization efficiency.

(leaves, shoot, or whole plant), plant age, dry matter or grain yield level, crop species or genotypes within species, crop management practices, and environmental factors (Fageria et al., 1997a; Fageria, 1992; Fageria and Baligar, 2005b; Fageria et al., 2006). Although nutrient concentration is influenced by several factors, it is still a more stable and useful parameter than soil analysis for identifying nutritional status of crop plants. Plants have a remarkable ability to regulate nutrient uptake according to their growth demands (Fageria and Baligar, 2005a). Significant variation in nutrient concentrations in the growth medium did produce very small changes in nutrient concentration in plant tissue (Smith, 1986; Fageria and Baligar, 2005a). Hence, it can be concluded that concentrations of most nutrients in plant tissues are restricted to fairly narrow ranges (Fageria and Baligar, 2005a).

Concentration is used to diagnose nutrient deficiency, sufficiency, or toxicity in crop plants. For the interpretation of plant analysis results, a critical nutrient concentration concept was developed from the work of Macy (1936), which was extended by Ulrich (1952). This concept is widely used now in interpretation of plant analysis results for nutritional disorder diagnostic purposes. Critical nutrient concentration is usually designated as a single point within the bend of the crop-yield–nutrient-concentration curve where the plant nutrient status shifts from deficient to adequate. The critical nutrient concentration has been defined in several ways: (1) the concentration that is just deficient for maximum growth (Fageria and Baligar, 2005b); (2) the point where growth is 10% less than the maximum (Ulrich, 1976); (3) the concentration where plant growth begins to decrease (Ulrich, 1952); and (4) the lowest amount of element in the plant accompanying the highest yield (Fageria and Baligar, 2005b). The above definitions are similar but not identical. It is well known that there is a good relationship between concentration of a nutrient and yield of that crop (Fageria et al., 1997a; Fageria and Baligar, 2005b).

Data in Table 2.5 showing N concentration in shoot and grain of one cereal (upland rice) and one legume (dry bean) reveal the change in concentration of this nutrient. Because plant age is one of the most important factors affecting nutrient concentrations in plant tissue, nitrogen concentration in both the crop species decreased significantly in a quadratic fashion with the advancement of plant age. In rice, variability of N concentration was 97% with the advancement of plant age.

TABLE 2.5
Nitrogen Concentration in Shoot and Grain of Upland Rice and Dry Bean during Growth Cycle

Plant Age (Days after Sowing)	N Conc. in Upland Rice Shoot and Grain (g kg^{-1})	Plant Age (Days after Sowing)	N Conc. in Dry Bean Shoot and Grain (g kg^{-1})
19	47.7	15	51.3
43	30.1	29	39.0
69	18.7	43	33.3
90	16.2	62	29.0
102	15.4	84	22.3
130	8.7	96	11.0
130 (grain)	15.4	96 (grain)	40.0

Regression Analysis[a]

Plant age (X) vs. N conc. in upland rice shoot (Y) = $59.6887 - 0.7694X + 0.0030X^2$
$R^2 = 0.9746**$

Plant age (X) vs. N conc. in dry bean shoot (Y) = $56.4079 - 0.5235X + 0.00085X^2$
$R^2 = 0.9512**$

[a] Grain concentration of N was not taken into account for the calculation of regression equation in both the crops.

** Significant at the 1% probability level.

Similarly, in dry bean N concentration, variability was 95% with the advancement of plant age. For example, in upland rice, N concentration at 19 days after sowing was 47.7 g kg^{-1} in the shoot. However, at harvest, the N concentration was 8.7 g kg^{-1}, a decrease of 448%. Similarly, in dry bean shoot, N concentration at 15 days after sowing was 51.3 g kg^{-1} and decreased to 11 g kg^{-1} at harvest, a decrease of 366%.

These results indicate that for nutrient deficiency or sufficiency diagnosis purposes, it is necessary to perform plant tissue analysis at different growth stages. Concentration of N in the grain of both species was higher in the grain than in the shoot at harvest, indicating translocation of N from shoot to grain in cereals as well as in legumes. During grain filling, N content of nongrain tissue generally decreases while grain N content increases (Bauer et al., 1987; Wilhelm et al., 2002). However, shoot dry weight increased with age advancement up to the flowering growth stage and then decreased (Fageria, 2003). Decrease in shoot dry weight at harvest was related to translocation of assimilate to the panicle from flowering to maturity (Black and Siddoway, 1977; Fageria et al., 1997a, 1997b). In rice, 60–90% of the total C accumulated in panicles at the time of harvest was derived from photosynthesis after heading, and the flag leaves are the organs that contribute most to grain filling (Yoshida, 1981).

Adequate concentrations of N in major field crops are given in Table 2.6. Adequate concentration is defined as N concentration ranges in specific plant parts, and changes within this range of concentrations do not increase or decrease growth or production. These N concentration ranges are also termed as intermediate, satisfactory, normal, or sufficient. It is usually considered that fertilizer practices need not change if nutrient concentrations fall within this classification. Another term that is very commonly used for interpretation of plant tissue tests is *critical nutrient range*. Critical nutrient concentration range is defined as the nutrient concentration at which a 10% loss of plant growth occurs. This 10% value has been chosen to agree with significant levels of statistical analysis (Riga and Anza, 2003). On a physiological basis, critical leaf nutrient levels indicate the minimum amount of cell nutrient concentration that allows for maintenance of metabolic functions at nonlimiting growth rates (Riga and Anza, 2003). The 10% yield reduction limit also approximates the economic level of fertilizer addition for annual crops (Fageria et al., 1997a).

2.5.2 UPTAKE

Nutrient uptake or accumulation is defined as the product of concentration and tissue dry weight. The accumulation unit is commonly kg ha^{-1} (macronutrients) and g ha^{-1} (micronutrients) for field trails and g or mg per plant for greenhouse or controlled condition experiments. Data related to the quantity of nutrients accumulated during a cropping season give an idea of fertility depletion and help in managing soil fertility for the succeeding crops. The best time to determine nutrient accumulation is at flowering or at harvest, when accumulation is expected to be at a maximum. Grain as well as shoot or straw should be analyzed and their weights per unit area determined to calculate total accumulation (Fageria and Baligar, 2005b).

TABLE 2.6
Adequate N Concentration in Plant Tissue of Principal Cereal and Legume Crops

Crop Species	Growth Stage	Plant Part	Adequate N Conc. (g kg⁻¹)
Wheat	Tillering	Leaf blade	43–52
Wheat	Shooting	Leaf blade	36–44
Wheat	Heading	Whole tops	21–30
Wheat	Flowering	Leaf blade	27–30
Barley	Tillering	Leaves	47–51
Barley	Shooting	Leaves	45–47
Barley	Heading	Whole tops	20–30
Barley	Flowering	Leaves	29–35
Lowland rice	Initiation of tillering	Whole tops	44–46
Lowland rice	Active tillering	Whole tops	31–35
Lowland rice	Panicle initiation	Whole tops	12–15
Lowland rice	Booting	Whole tops	10–13
Lowland rice	Flowering	Whole tops—grains	9–11
Lowland rice	Physiological maturity	Whole tops—grains	6–7
Corn	30 to 45 DAE¶	Whole tops	35–50
Corn	Before tasseling	Leaf blade below whorl	30–35
Corn	Silking	BOAC§	>32
Sorghum	Seedling	Whole tops	35–51
Sorghum	Early vegetative	Whole tops	30–40
Sorghum	Vegetative	YMB‡	32–42
Sorghum	Bloom	3BBP†	33–40
Soybean	Prior to pod set	UFDT-	45–55
Dry bean	Early flowering	UMB-	52–54
Cowpea	39 DAS-	Whole tops	28–35
Cowpea	Early flowering	PUMB-	11–17
Peanut	Early pegging	Upper stems & leaves	35–45
Cassava	Vegetative	UMB	50–60
Potato	42 DAE	UMB+P-	40–50
Potato	Early flowering	UMB+P	55–65
Potato	Tuber half grown	UMB+P	30–50
Cotton	1st flowering	YMB	38–45

¶ DAE, days after emergence; –DAS, days after sowing; §BOAC, blade opposite & above cob; ‡YMB, youngest (uppermost) mature leaf blade; †3BBP, third blade below panicle; –UFDT, upper fully developed trifoliate; –UMB, uppermost blade; –PUMB, petiole of uppermost mature leaf blade; –UMB+P, uppermost mature leaf blade+petiole.

Source: Piggott (1986), Reuter (1986), Small and Ohlrogge (1973), Fageria et al. (1997a), Fageria (2003), and Fageria and Baligar (2005a).

TABLE 2.7

Nitrogen Uptake in Shoot and Grain of Upland Rice and Dry Bean during Growth Cycle

Plant Age (Days after Sowing)	N Uptake in Upland Rice Shoot and Grain (kg ha⁻¹)	Plant Age (Days after Sowing)	N Uptake in Dry Bean Shoot and Grain (kg ha⁻¹)
19	4	15	6
43	32	29	16
69	65	43	25
90	96	62	57
102	116	84	60
130	56	96	19
130 (grain)	70	96 (grain)	68

Regression Analysis[a]

Plant age (X) vs. N uptake in upland rice shoot (Y) = $0.7406 \, \text{Exp.}(0.1051X - 0.00055X^2)$

$R^2 = 0.9832**$

Plant age (X) vs. N uptake in dry bean shoot (Y) = $1.1579 \, \text{EXP.}(0.1158X - 0.00087X^2)$

$R^2 = 0.9137**$

[a] Grain uptake of N was not taken into account for the calculation of regression equation in both the crops.

** Significant at the 1% probability level.

Data related to nitrogen accumulation in shoot and grain of upland rice and dry bean during the growth cycle are presented in Table 2.7. Nitrogen uptake in both species experienced a significant quadratic increase with the advancement of plant age, as expected. The variation in N accumulation was 98% due to plant age in upland rice, and in dry bean the variation in N accumulation was 91% due to plant age. This means that plant age has a highly significant association with N accumulation in crop plants. Upland rice accumulated higher N in the shoot as well as grain compared to dry bean. This is related to higher yield of cereals compared to legumes (Fageria et al., 2006).

Higher amount of N uptake or accumulation in grain of crop plants is important, because crop yield is significantly and linearly associated with N accumulated in grain (Figure 2.6). Generally, N uptake in grain has positive significant associations with grain yield (Fageria and Baligar, 2001; López-Bellido et al., 2003). Hence, improving N uptake in grain may lead to improved grain yield. Higher N concentration in shoots is also desirable, because if shoots have higher N concentration, during crop growth the N is translocated to grain when plant demand increases, thereby improving yield. Figure 2.7 shows that N uptake in shoots is significantly associated with grain yield in a quadratic fashion. Nitrogen accumulation values in straw and grain of major field crops are given in Table 2.8. Uptake values varied among crop species and were higher in grain than in straw. This indicates that grains are greater sinks for N accumulation compared with other aboveground parts.

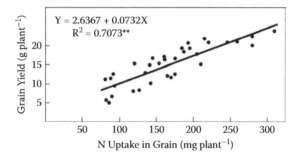

FIGURE 2.6 Relationship between N uptake in grain and grain yield of lowland rice.

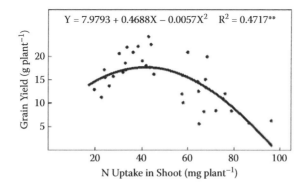

FIGURE 2.7 Relationship between N uptake in shoot and grain yield of lowland rice.

2.5.3 NITROGEN HARVEST INDEX

Nitrogen harvest index (NHI) is defined as the amount of N accumulated in grain divided by the amount of N accumulated in grain plus straw. It is an index and hence has no unit. However, sometimes it is expressed as a percentage. Higher NHI in crop plants or genotypes is desirable because it has positive association with grain yield (Figure 2.8). Figure 2.8 shows that 71% variability in grain yield of lowland rice grown on Brazilian Inceptisol was due to NHI. The NHI values varied among crop species and among genotypes of the same species. Figure 2.9 shows that NHI of five lowland rice genotypes varied from 0.53 to 0.64. Genotype CNA 8569 had the lowest NHI, and genotype BRS Bigua had the highest NHI. Similarly, NHI of 20 dry bean genotypes varied from 0.43 to 0.82 at low N rate (control treatment) and from 0.53 to 0.88 at high N rate (400 mg N kg^{-1} of soil), with an average value of 0.63 at low N rate and 0.75 at high N rate (Table 2.9). Generally, NHI values are higher in legumes than in cereals (Fageria et al., 2006).

2.6 NH$_4^+$ VERSUS NO$_3^-$ UPTAKE

Crop plants absorb N mainly in the form of NH$_4^+$ or NO$_3^-$. In well-drained or oxidized soils, NO$_3^-$ is present in higher concentrations compared to NH$_4^+$. Hence, in

TABLE 2.8

Nitrogen Uptake in Stover and Grain of Major Field Crops

Crop Species	N Uptake in Stover (kg ha⁻¹)	N Uptake in Grain (kg ha⁻¹)	Stover Yield (kg ha⁻¹)	Grain Yield (kg ha⁻¹)	Reference
Upland rice	56	70	6,343	4,568	Fageria (2001)
Upland rice	79	80	6,642	4,794	Fageria et al. (1997a)
Upland rice	35	37	4,341	2,716	Ohno and Marur (1977)
Lowland rice	66	74	9,423	6,389	Fageria and Baligar (2001)
Lowland rice	49	82	8,005	7,093	Fageria and Prabhu (2003)
Lowland rice	75	143	9,000	14,600	De Datta and Mikkelsen (1985)
Corn	72	127	11,873	8,501	Fageria (2001)
Corn	110	150	10,000	9,400	Jacobs (1998)
Dry bean	17	124	2,338	3,859	Fageria et al. (2004)
Wheat	20	55	3,400	2,700	Jacobs (1998)
Wheat	51	104	5,120	3,357	Woolfolk et al. (2002)
Sorghum	73	55	6,700	3,800	Jacobs (1998)
Faba bean	19	89	1,443	2,007	López-Bellido et al. (2003)

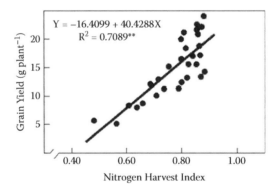

The equation shown in the figure:

$$Y = -16.4099 + 40.4288X$$
$$R^2 = 0.7089^{**}$$

FIGURE 2.8 Relationship between nitrogen harvest index and grain yield of lowland rice.

aerobic soils, NO_3^- is the dominant form of N for plant uptake. In oxygen-deficient or reduced or anaerobic soils (such as flooded rice), the NH_4^+ form of N is the dominant form, and plants take this form of N in higher amounts compared to the NO_3^- form of N. In the literature, there is no consistency about which form of N is taken up in higher amounts or preferred by plants. Fageria et al. (2006) reported that plants supplied with equal proportions of NH_4^+ and NO_3^- grew as well as those supplied with any single amount of N form. These authors also reported that plants can absorb both forms of N equally and that the N form absorbed is mainly determined by what form

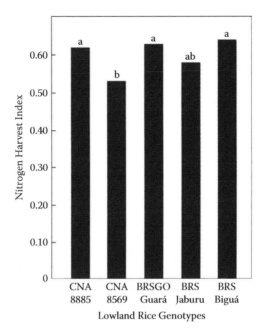

FIGURE 2.9   Nitrogen harvest index of five lowland rice genotypes (Fageria, 2007a).

is abundant and/or accessible at any given time. Mengel et al. (2001) also reported that the uptake rate of NH_4^+ and NO_3^- depends mainly on the availability of these ions in the nutrient medium. In general, one NH_4^+ may substitute for one NO_3^- and vice versa. These authors also reported that uptake rates are determined mainly by the physiological need of the plant and not so much on whether the source is a cation or an anion. Runge (1983) also reported that no particular N form is more readily usable than any other. The more or less favorable effect of NH_4^+ or NO_3^- in individual cases varies among species and is dependent on the concentration, as well as on the pH value, the buffer capacity, and the content of other nutrients in the medium. Runge (1983) also reported that most plant species grow optimally when they are supplied with both N forms simultaneously. If only one N form is available, utilization depends on specific physiological properties of the individual plant species and on interactions with certain soil factors. Marschner (1995) concluded that calcifuges (plants adapted to acid soils and to low redox potential) might prefer NH_4^+ sources, while calcicoles (plants adapted to alkaline soils) may prefer NO_3^- sources.

When NH_4^+ uptake exceeds NO_3^- uptake, soil solution pH decreases, and when NO_3^- uptake exceeds NH_4^+ uptake, soil solution pH increases. Furthermore, nearly 70% of the cations or anions taken up by the plants are ammonium or nitrate (Van Beusichem et al., 1988). The NH_4^+/NO_3^- ion uptake can change rhizosphere pH up to 2 units higher or lower compared with that of bulk soil (Mengel et al., 2001). This change in pH may influence uptake of other essential nutrients from soil solution by plants. In consequence of the lowering of rhizosphere pH by ammonium, increased uptake of phosphate has been reported (Riley and Barber, 1971) and also of micronutrients (Schug, 1985) as compared with nitrate treatment.

TABLE 2.9

Nitrogen Harvest Index (NHI) of 20 Dry Bean Genotypes Grown at Two N Rates

Genotype	0 mg N kg⁻¹ of Soil	400 mg N kg⁻¹ of Soil
Pérola	0.82a	0.73ab
BRS Valente	0.44a	0.53b
CNFM 6911	0.80a	0.79ab
CNFR 7552	0.57a	0.68ab
BRS Radiante	0.67a	0.69ab
Jalo Precoce	0.67a	0.75ab
Diamante Negro	0.65a	0.71ab
CNFP 7624	0.43a	0.78ab
CNFR 7847	0.80a	0.85a
CNFR 7866	0.47a	0.74ab
CNFR 7865	0.71a	0.69ab
CNFM 7875	0.45a	0.85a
CNFM 7886	0.44a	0.78ab
CNFC 7813	0.47a	0.85a
CNFC 7827	0.71a	0.85a
CNFC 7806	0.69a	0.78ab
CNFP 7677	0.71a	0.85a
CNFP`7775	0.70a	0.85a
CNFP 7777	0.79a	0.88ab
CNFP 7792	0.62a	0.72ab
Average	0.63	0.75
F-test		
N rate (N)	**	
Genotype (G)	**	
N X G	**	

** Significant at the 1% probability level. Within same column, means followed by the same letter do not differ significantly at 5% probability level by Tukey's test.

Legumes are exceptions—they acidify the rhizosphere even after the supply of nitrate (Marschner and Romheld, 1983). In legumes, the uptake rate of cations is greater than that of anions, since they acquire nitrogen from the atmosphere through nitrogen fixation rather than by nitrate uptake (Aguilar and van Diest, 1981; Marschner and Romheld, 1983; Jarvis and Hatch, 1985). However, some workers have reported that acidification of legume rhizosphere is associated with specific properties of legumes, which normally acidify the rhizosphere even under nitrate supplied conditions (Hinsinger, 1998; Marschner, 1995). The pH also changes with the excretion of organic acids by roots and by microorganism activity in the rhizosphere. Further, the CO_2 produced by roots and microorganism respiration can dissolve in soil solution, potentially forming carbonic acid and lowering the pH. Soil buffering capacity (clay and organic matter content) and initial pH are the main parameters that determine changes in soil pH.

Absorption of NH_4^+ occurs faster than absorption of NO_3^- (Gaudin and Dupuy, 1999). In addition, assimilation of NO_3^- required energy equivalents up to 20 ATP (adenosine triphosphate) mol^{-1} NO_3^-, whereas NH_4^+ assimilation required only 5 APT mol^{-1} NH_4^+ (Salsac et al., 1987). Similarly, Bloom et al. (1992, 2003) reported that root absorption and assimilation of 1 mol NH_4^+ requires or consumes 0.31 mol O_2, whereas 1 mol of NO_3^- consumes 1.5 mol O_2. This means that NO_3^- uptake consumes about 5 times more energy than NH_4^+ ion uptake. While NH_4^+ can be assimilated directly into amino acids, NO_3^- must first be reduced into NO_2^- and then NH_4^+ via nitrate reductase and nitrite reductase, a process that imparts an additional energetic cost (Hopkins, 1999). Potential energy savings for yield could be obtained if plants were supplied only NH_4^+ (Huffman, 1989). This concept has not been consistently observed, nor is it easy to conduct experiments on this given the nature of the N-cycle dynamic (Raun and Johnson, 1999; Fageria et al., 2006).

The effectiveness of the two N forms on growth and N uptake varies with the type of cultivar and NH_4^+/NO_3^- ratio. For example, solutions containing N entirely as NH_4^+ or NO_3^- have been shown to inhibit plant growth more than solutions containing 25 or 50% of either N form (Nittler and Kenny, 1976). Gashaw and Mugwira (1981) reported that triticale, wheat, and rye produced higher dry matter with 25/75, 50/50, and 75/25 NH_4^+/NO_3^- ratios than with either N source alone. Warncke and Barber (1973) found no differences between the relative rates of NH_3^+ and NO_3^- absorption by corn, but increasing each N form reduced the uptake of the other N form. It has been shown that corn, wheat, and oats prefer 50% of N as NO_3^- for maximum growth (Diest, 1976). Barber et al. (1992) reported that overall, manipulation of soil NH_4^+/NO_3^- ratios had few effects on corn development or yield under field conditions.

Plants grown on NO_3^- maintain a relative homeostasis of their N concentration and internal NO_3^- over a wide range of external concentrations, which suggests an efficient mechanism in control of NO_3^- uptake (Glass et al., 2002). In contrast, NH_4^- uptake seems poorly regulated (Britto and Kronzucker, 2002). The imbalance between NH_4^+ uptake and assimilation rates depends on a variety of factors, including plant species and carbohydrate availability (Schjoerring et al., 2002). Although the mechanisms of toxicity have not been elucidated completely, NH_4^+ accumulation in leaves has been found to depress photosynthesis and leaf growth (Raab and Terry, 1994). Also, NO_3^- is an important osmoticum that intervenes in the expansion of plant cells, and a reduction of its absorption may depress growth (McIntyre, 1997). The reduction of the absorption of cations such as K^+, Ca^{2+}, and Mg^{2+} by NH_4^+ may also limit plant growth (Salsac et al., 1987).

Several investigations have shown that take-all root rot of wheat is less severe when wheat is fertilized with NH_4^+ than when it is fertilized with NO_3^- (Smiley, 1974; Macnish and Speijers, 1982; Taylor et al., 1983). Christensen and Brett (1985) also reported that a critical NH_4^+/NO_3^- ratio of take-all suppression of 3:1 was estimated from data in the literature. Take-all root rot severity in wheat crop was negatively correlated ($R^2 = 0.84$) with the length of time the NH_4^+/NO_3^- ratio remained above the estimated critical ratio. In conclusion, both NO_3^- and NH_4^+ forms are taken up and metabolized by plants; the preferential source has been shown to be dependent on plant species and environmental factors. Ammonium exceeds NO_3^- uptake in ryegrass, wheat, sugar beet, and rice, while NO_3^- exceeds NH_4^+ uptake in corn and tomato (Fageria and Gheyi, 1999).

2.7   INTERACTION WITH OTHER NUTRIENTS

Balanced supply of essential nutrients is one of the most important factors in increasing yields of annual crops. Hence, knowledge of interaction of N with other nutrients is an important factor in improving efficiency of this element and consequently improving crop yields. Nutrient interaction in crop plants is measured in terms of uptake or yield level. Application of a particular nutrient may increase, decrease, or have no effect on uptake of other essential plant nutrients. Similarly, yield level of a crop may increase, decrease, or experience no change with the increase of two nutrient levels in the growth medium. Hence, nutrient interactions may be positive, negative, or neutral. In mineral nutrition, the nutrient interactions are designated as synergistic (positive), antagonistic (negative), or neutral.

Nutrient interaction can occur at the root surface or within the plant. Interactions at the root surface are due to formation of chemical bonds by ions and precipitation or complexes. One example of this type of interaction is the liming of acid soil, which decreases the concentration of almost all micronutrients except molybdenum (Fageria and Zimmermann, 1998; Fageria, 2000). The second type of interaction takes place between ions whose chemical properties are sufficiently similar that they compete for sites of absorption, transport, and function on the plant root surface or within plant tissues. Such interactions are more common between nutrients of similar size, charge, and geometry of coordination and electronic configuration (Robson and Pitman, 1983).

The ratio of the mass of organic carbon to the mass of organic nitrogen in the soil, known as the *C/N ratio*, controls N availability. A C/N ratio >30/1 generally immobilizes N in soil–plant systems and creates the possibility of N deficiency in crop plants (Fixen, 1996). Nitrogen has positive interactions with P and K in crop plants. The increasing N rate increases uptake of P, K, Ca, and Mg in a quadratic fashion in dry bean plants (Table 2.10). Wilkinson et al. (2000) reported that application of N increased uptake of P, K, S, Ca, and Mg, provided that these elements are present in sufficient amounts in the growth medium. The improvement in uptake of macronutrients with the addition of N is reported to be associated with increase in root hairs, chemical changes in the rhizosphere, and physiological changes stimulated by N, which influence transport of these elements (Marschner, 1995; Baligar et al., 2001). Rapid nitrate uptake depends on adequate K in the soil solution. Higher rates of K allowed for the efficient use of more nitrogen, which resulted in better early vegetative growth and higher grain and straw yield as K and N rates increased. In the field, better N uptake and utilization with adequate K means improved N use and higher yields. Crops need more K with higher N rates to take advantage of the extra N (Fageria and Gheyi, 1999).

Nitrogen interactions with micronutrients depend on pH changes in the rhizosphere. If N is absorbed in the form of NH_4^+, rhizosphere pH may decrease, and

TABLE 2.10

Influence of N on Uptake of Macro- and Micronutrients in the Shoot of Dry Bean Plants

N Rate (kg ha⁻¹)	P (kg ha⁻¹)	K (kg ha⁻¹)	Ca (kg ha⁻¹)	Mg (kg ha⁻¹)
0	0.6	4.4	4.1	1.0
40	1.4	9.4	8.2	2.1
80	1.6	17.2	12.8	3.2
120	2.1	16.7	12.9	3.3
160	2.9	28.6	18.4	5.1
200	3.4	33.2	24.9	6.5
F-Test	**	**	**	**

Regression Analysis

N rate (X) vs. P uptake. $(Y) = 0.7906 + 0.0041X + 0.000066X^2$, $R^2 = 0.8058**$

N rate (X) vs. K uptake. $(Y) = 4.7650 + 0.1111X + 0.00016X^2$, $R^2 = 0.7076**$

N rate (X) vs. Ca uptake. $(Y) = 4.8303 + 0.0618X + 0.00017X^2$, $R^2 = 0.7942**$

N rate (X) vs. Mg uptake. $(Y) = 1.1682 + 0.0164X + 0.000048X^2$, $R^2 = 0.7855**$

** Significant at the 1% probability level.

uptake of most micronutrients increases. If N is mainly absorbed as NO_3^-, rhizosphere pH may increase, and uptake of most micronutrients decreases. Other interactions of micronutrients with N may be associated with crop responses to N fertilization. Increase in crop growth with the application of N may increase crop demands for micronutrients, and micronutrient deficiencies may occur (Wilkinson et al., 2000). Zinc deficiency in upland rice and corn in Brazilian Oxisols is very common, and the response of these crop plants to N fertilization is one reason for such deficiency (Fageria and Gheyi, 1999). Chlorine decreases NO_3^- uptake and enhances the uptake of NH_4^+ (Huber and Thompson, 2007). These authors also reported that ammonium increases the uptake of Mn.

2.8 MANAGEMENT PRACTICES TO MAXIMIZING N USE EFFICIENCY

Management of nutrients is an important aspect of improving crop productivity. Nutrient management in crop production means supplying essential plant nutrients to a crop in adequate amount and form to get maximum economic yield in a given agroecological region. The nutrient requirements of a crop vary according to soil, climatic conditions, cultivar planted, and management practices adopted by a farmer. In addition, socioeconomic conditions of the farmer as well as price of the produce in the market also determine fertilizer recommendations. Thus, nutrient management strategies should vary according to type of soil, climatic conditions, crop species, cultivar within species, and socioeconomic considerations. Farmers today cannot afford to apply fertilizers in excess of what their crops require due to economic and ecological reasons. Therefore, fertilizer application rates should coincide with crop requirements to produce economical yields. However, under practical farming, it is not easy to determine the exact fertilizer needs of a crop, because nutrient availability in soil is a dynamic process and changes with moisture content in the soil, soil temperature, type of fertilizer, infestation of diseases, insects, and weeds, and management practices. Further, these factors interact with each other. This means that the determination of optimum nutrients levels for a crop is always an approximation.

Once the mineral deficiency problems have been identified and recognized, amendments or fertilizers can be added to alleviate deficiency or toxicity problems. In addition, one of the current problems in mineral element nutrition is the efficiency with which the elements can be used by plants. It is important that minerals added to soils be used effectively and efficiently (Clark, 1982). Adoption of appropriate management practices may overcome N deficiency and enhance nitrogen use efficiency in crop plants. These practices are associated with soil and crop management and include the liming of acid soils; appropriate source, rate, and timing of N application; supply of adequate soil moisture; crop rotation; conservation or minimum tillage; use of cover crops and animal manures; use of N-efficient crop species or genotypes within species; and control of diseases, insects, and weeds. By adopting these soil and crop management practices, it is possible to improve crop yields and consequently nitrogen use efficiency. These practices are discussed in the succeeding sections.

2.8.1 LIMING ACID SOILS

Soil acidity is one of the most important constraints in crop production in many regions of the world. Acid soils by definition are soils with a pH below 7.0 on a scale of 0 to 14. Generally, soil acidity is measured in terms of H^+ and Al^{3+} ions in the soil solution. However, acid soil toxicity is not caused by a single factor but a complex of factors including toxicities of Al^{3+} and H^+ and deficiencies of many macro- and micronutrients (Fageria and Baligar, 2003b; Fageria, 2006). Liming is an effective and widespread practice for improving crop production on acid soils (Fageria and Baligar, 2003b). Liming acid soils improves soil physical, chemical, and biological activities and consequently improves crop yield and nitrogen use efficiency (Fageria and Baligar, 2003b; Fageria and Baligar, 2005a). Soil acidity indices like pH, base saturation, and aluminum saturation are used as bases for liming acid soils. Furthermore, economic considerations are also important criteria in determining the quantity of lime applied to acid soils. In Brazilian Oxisols, liming to raise pH 6.0 to 6.5 is optimal for the growth of most field crops (Fageria and Baligar, 2003b; Fageria, 2006).

Data in Table 2.11 show that liming in Brazilian Oxisol improves pH, base saturation, and Ca and Mg content; reduces acidity saturation, H^+, and Al; and significantly increases dry bean yield. Regression models were developed relating dry bean grain yield and selected soil chemical properties (Table 2.12). All the soil chemical properties (associated with soil acidity) displayed significant positive and quadratic relationships with bean grain yield. Similar results were reported by Fageria and

TABLE 2.11
Influence of Lime Rate on Soil Chemical Properties and Grain Yield of Dry Beans Grown on Brazilian Oxisol (Values Are Averages of Three Years Determined after Harvesting of Dry Bean Crops)

Soil Chemical Property	Lime Rate (Mg ha^{-1})			F-test
	0	12	24	
PH	5.4c	6.5b	6.8a	**
Base saturation (%)	25.9c	61.6b	73.3a	**
H+Al (cmol$_c$ kg^{-1})	6.6a	2.9b	2.0c	**
Acidity saturation (%)	74.1a	38.4b	26.7c	**
Ca (cmol$_c$ kg^{-1})	1.6c	3.2b	3.8a	**
Mg (cmol$_c$ kg^{-1})	0.5b	1.2a	1.2a	**
Grain yield (kg ha^{-1})	2731.9b	3255.2a	3270.9a	**

** Significant at the 1% probability level. Means followed by the same letter in the same line under different lime treatments are statistically not significant at the 5% probability level by Tukey's test.

Source: Fageria (2006).

TABLE 2.12
Relationship between Soil Chemical Properties and Grain Yield of Dry Bean (Values Are Averages of Three-Years Field Trial)

Soil Chemical Property	Regression Equation	R^2	VMY[a]
PH	$Y = -17887.16 + 6603.7780X - 514.2863X^2$	0.5733**	6.4
Ca (cmol$_c$ kg^{-1})	$Y = -22095.26 + 1195.3480X - 14.0335X^2$	0.6435**	4.2
Mg (cmol$_c$ kg^{-1})	$Y = 1354.74 + 3714.6620X - 1781.2680X^2$	0.6099**	1.0
H+Al (cmol$_c$ kg^{-1})	$Y = 2775.69 + 331.1134X - 51.0827X^2$	0.5769**	3.2
Acidity saturation (%)	$Y = 2484.08 + 41.5085X - 0.5132X^2$	0.5944**	40.4
Base saturation (%)	$Y = 1887.51 + 44.8813X - 0.3557X^2$	0.5677**	63.1
Ca saturation (%)	$Y = 1995.99 + 57.2705X - 0.6277X^2$	0.5677**	45.7
Mg saturation (%)	$Y = 2175.55 + 126.7669X - 3.5108X^2$	0.6069**	18.0
K saturation (%)	$Y = 1090.72 + 1483.4650X - 258.2123X^2$	0.4570**	2.9
Ca/Mg raio	$Y = 2663.77 + 886.2584X - 240.8921X^2$	0.3495**	1.8
Ca/K ratio	$Y = 1765.04 + 140.5983X - 3.1861X^2$	0.5474**	22.1
Mg/K ratio	$Y = 1977.05 + 171.0891X - 5.1843X^2$	0.5551**	16.5

** Significant at the 1% probability levels.

[a] VMY = value for maximum grain yield.

Source: Fageria (2006).

Stone (2004) in relation to soil acidity indices and grain yield of dry bean grown on Brazilian Oxisol.

Activities of most of the beneficial microorganisms that are involved in the process of mineralization of N and biological nitrogen fixation are inhibited under acidic conditions. In general, organic matter decomposes more rapidly in neutral soils than in acid soils (Foy, 1984). Legumes are apparently much more tolerant to low pH when fertilizer N is applied than when grown with only symbiotic N (Andrew, 1978; Munns, 1978). Andrew (1978) used sand culture to determine the effect of low pH on nodulation and growth of several legumes. Alfalfa was the most sensitive species tested. It produces maximum nodules at pH 6.0 (Foy, 1984). Rhizobial strains within species also differ in their abilities to nodulate a given host plant at low pH (Munns, 1978). Liming improves activities of such beneficial microorganisms and improves N use efficiency in crop plants.

2.8.2 USE OF CROP ROTATION

Adopting appropriate crop rotation is one of the most important management strategies for improving N use efficiency in crop plants. Crop rotation is defined as a planned sequence of crops growing in a regularly recurring succession on the same area of land, as contrasted to continuous culture of one crop or growing a variable sequence of crops (Soil Science Society of America, 1997). Scientists and farmers have recognized for centuries the beneficial effects of crop rotation (Varvel and Wilhelm, 2003). Crop rotation was practiced during the Han dynasty of China more than 3000 years ago to maintain soil productivity (MacRae and Mehuys, 1985; Karlen et al., 1994). Planting a legume alternated with a cereal is a simple example of crop rotation. The legume supplies N to the succeeding cereal. Hence, N economy in crop rotations where legumes are part of the cropping sequences is widely reported (Bullock, 1992; Karlen et al., 1994). Legumes may also add large amounts of C to the soil without the inhibitive transportation cost frequently associated with other organic C inputs (Hargrove, 1986; Sharma and Mittra, 1988; Goyal et al., 1992, 1999; Cherr et al., 2006). In addition, crop rotations also control insects, diseases, and weeds, and this may improve N use efficiency in crop plants (Karlen et al., 1994).

Inorganic N fertilizer needs for corn and other crops have been shown to be reduced when these crops are rotated with legumes (Olson et al., 1986). Further, with the current interest in sustainable agriculture systems, the use of legumes in crop rotations to provide N to subsequent crops is increasing. Throughout the past decade, a large research effort has focused on elucidating the N and non-N (rotation) effects of legumes on subsequent nonlegume crops grown in rotation (Fox and Piekielek, 1988; Hesterman, 1988). Still, there is considerable discrepancy regarding the actual quantities of N contributed by legume plant material to subsequent crops. In Australia, Ladd and associates (Ladd et al., 1983; Ladd and Amato, 1986) reported between 11 and 28% recovery of ^{15}N by a subsequent wheat crop from medic (*Medicago littoralis* L.) plant material incorporated in field soil, and an additional 4% recovery by a second wheat crop. Harris and Hesterman (1990) quantified the N contribution from different plant parts to a subsequent corn crop and a second-year spring barley crop in loam and sandy loam soils. Corn recovered 17 and 25% of the alfalfa ^{15}N applied to the loam and sandy loam soils, respectively. Most (96%) of the alfalfa ^{15}N remaining in soil was recovered in the organic fraction, with microbial biomass accounting for 18% of this recovery. Only 1% of the alfalfa ^{15}N from the original application was recovered by a second-year spring barley crop in two soils. Varvel and Peterson (1990) studied N fertilizer recovery by corn in monoculture and rotational systems. They concluded that nitrogen recovery determined by isotopic methods was significantly higher for corn in rotation versus corn in monoculture.

Varvel and Peterson (1990) conducted a long-term field study to determine effects of crop rotation and N fertilizer application on residual inorganic N levels to a depth of 150 cm after 4 years. The study included continuous corn, continuous soybean, continuous grain sorghum, and 2-year rotations of corn/soybean and grain sorghum/ soybean. High N application resulted in greater residual NO_3^- concentration for the continuous corn and grain sorghum systems than for any of the other cropping systems to a depth of 150 cm. Residual NO_3^- concentrations were low (< 4 mg kg^{-1} at depth below 30 cm) at all N application rates in continuous soybean and 2- and 4-year cropping systems. Nitrogen removed by grain accounted for 50% of the applied N in continuous corn and grain sorghum systems at the low N application rate but only 20 to 30% of the applied N at the high rate. Likewise, N removed in the rotation systems at either N application rate accounted for only 20 to 30% of the applied N. Indirect results from this study suggested that immobilization by crop residues and soil organic matter, not leaching, is probably most responsible for apparent N losses in these cropping systems. The study also showed that crop rotations could reduce inorganic N fertilizer needs and at the same time reduce the amount of N available for leaching, both of which are important for efficient crop production.

Many researchers have reported that a previous legume increases the grain yields of nonlegume crops (Fageria, 1989; Fageria and Baligar, 1996). Miller and Dexter (1982) reported that, in North Dakota, fertilized (75 kg N ha^{-1}) hard red spring wheat grain yields were 3 to 15% higher than continuous wheat yields when crops were grown on land previously cropped by barley, flax, or corn. However, yields were 30% higher than those of continuous wheat and at least 15% higher than those of other crops, when grown after soybean. Stoa and Zubriski (1969) reported that wheat yields were nearly 50% higher from land previously cropped to alfalfa for 3 years than from land cropped to a nonlegume without N fertilization, and 10 to 15% higher than that from the nonlegume sequence with 67 kg N ha^{-1}. Meyer (1987) reported that barley grain yields following four to six hayed legumes were increased by 7 to 68% compared with barley following wheat without fertilization, while fertilized (75 kg N ha^{-1}) barley yields were 12 to 15% greater following the hayed legumes than following wheat. Badaruddin and Meyer (1989) reported that nitrogen use efficiency of unfertilized wheat following forage legumes was greater than it was following fallow or wheat. Roder et al. (1989) reported yield increase with crop rotation of soybean with grain sorghum in eastern Nebraska. The yield contribution of soybean in rotation to the succeeding sorghum crop was equivalent to 90 kg N fertilizer ha^{-1}. The relative yield increase (RYI) from rotation was calculated as follows:

$$RYI = [(YR - YC) \times 100]YC$$

where YR = yield rotation and YC = yield continuous. Legume cover crops such as hairy vetch and crimson clover have been reported to have higher potential to contribute significant amounts of biologically fixed N to the subsequent crop. Average estimates of the equivalent amount of fertilizer N replaced by hairy vetch and clover reported by Ebelhar et al. (1984) and Hargrove (1986) were 90 to 100 kg ha^{-1} for corn and 72 kg ha^{-1} for grain sorghum. Touchton et al. (1982) reported that the N provided by crimson clover allowed to reseed was sufficient for maximum grain sorghum yield without the need for supplemental fertilizer N.

Although crop rotation may change soil mineral status, particularly N, there may also be a rotation effect beyond that which can be explained by soil mineral status alone (Copeland and Crookston, 1992). At Urbana, Illinois, high rates of limestone and N, P, and K fertilizers did not substitute for rotation, which increased yield of corn rotated with soybean by 16% over yield of corn grown in monoculture (Welch, 1976). In rotation studies at Lancaster, Wisconsin, in which N was not limiting, the 8-year average yields for continuous corn were less than the average for first-year corn (Higgs et al., 1976). Corn yielded more in a corn–oat rotation than when grown continuously in a tile-drained clay soil that was limed and fertilized (Bolton et al., 1976).

In conclusion, addition of legumes in crop rotation improved N fertilization by the succeeding crop, but the N contribution of legumes varies with the legume species, harvest management, and some environmental factors. The yield improvement in rotation may also be related to improvements in physical as well as chemical soil properties including soil water content. Reduced runoff may also be responsible for improved yield in cereal/legume rotations.

2.8.3 USE OF COVER/GREEN MANURE CROPS

Cover or green manure crops play an important role in improving productivity of subsequent row crops by improving soil's physical, chemical, and biological properties (Fageria et al., 2005; Baligar and Fageria, 2007). The benefit of legume cover crops in crop rotation has long been recognized and is attributed mainly to N contribution to subsequent crops. In recent years, economic and environmental considerations have renewed interest in this old practice for improving crop productivity and soil health and maintaining sustainability of agroecosystems.

Cover crops can be defined as close-growing crops that provide soil protection and soil improvement between periods of normal crop production or between trees in orchards and vines in vineyards (Soil Science Society of America, 1997). Cover crops are not grown for market purposes, but when plowed under and incorporated into the soil, cover crops may be referred to as green manure crops. Cover crops are sometimes called catch crops. Cover crops are usually killed on the soil surface before they are mature by using appropriate herbicides. In most studies, cover crops managed as no-till mulches have been killed with glyphosate (N-[phosphono-methyl]glycine), paraquat (1,1-dimethyl-4-bipyridinium ion; Fageria et al., 2005), or a mixture of nonselective, postemergence, and preemergence herbicides. Because many growers often want to reduce use of chemical inputs, nonchemical methods of killing or suppressing cover crops are desirable. These mechanical methods are

TABLE 2.13
Quantity of Nitrogen Fixed by Legume Cover Crops

Crop Species	N$_2$ Fixed (kg ha^{-1} crop^{-1})	Reference
Peanut (*Arachis hypogaea* L.)	40–80	Brady & Weil (2002)
Cowpea (*Vigna unguiculata* L. Walp.)	30–50	Brady & Weil (2002)
Alfalfa (*Medicago sativa* L.)	78–222	Heichel (1987)
Soybean (*Glycine max* L.)	50–150	Brady & Weil (2002)
Fava bean (*Vicia faba* L.)	177–250	Heichel (1987)
Hairy vetch (*Vicia villosa* Roth.)	50–100	Brady & Weil (2002)
Ladino clover (*Trifolium repens* L)	164–187	Heichel (1987)
Red clover (*Trifolium pratense* L.)	68–113	Heichel (1987)
White lupine (*Lupinus albus* L.)	50–100	Brady & Weil (2002)
Field peas (*Pisum sativum* L.)	174–195	Heichel (1987)
Chickpea (*Cicer arietinum* L.)	24–84	Heichel (1987)
Pigeon pea (*Cajnus cajan* L. Huth.)	150–280	Brady & Weil (2002)
Kudzu (*Pueraria phaseoloides* Roxb. Benth)	100–140	Brady & Weil (2002)
Chickpea (*Cicer arietinum* L.)	24–84	Heichel (1987)
Greengram (*Vigna radiata* L. Wilczek.)	71–112	Chapman & Myers (1987)
Lentil (*Lens culinaris* L.)	57–111	Smith et al. (1987)

mowing, rolling, roll-chopping, undercutting, and partial rototilling. Creamer and Dabney (2002) have reviewed literature on killing cover crops mechanically.

The contribution of N is the most commonly observed primary benefit of leguminous crops. Both legume and nonlegume cover crops affect N fertilizer management (Bauer and Roof, 2004). Legume cover crops fix atmospheric N and reduce N fertilizer needs for succeeding cash crops (Reeves, 1994). The rate of N fixed by cover crops is determined largely by the genetic potential of the legume species and by the amount of plant-available N in the soil. Data in Table 2.13 show quantities of nitrogen fixed by different legume cover crops. Two bacterial species (i.e., *Rhizobium* and *Bradyrhizobium*) are responsible for symbiotic N fixation in legumes. The genus *Rhizobium* contains fast-growing, acid-producing bacteria, while the *Bradyrhizobium* are slow growers that do not produce acid (Fageria et al., 2005). Soil factors such as pH, moisture content, and temperature also determine N fixation capacity of a legume cover crop. In some cases, the amount of N provided by legume cover crops is adequate to produce optimal yields of subsequent nonleguminous crops; however, higher-N-requiring cereals such as corn (*Zea mays* L.) generally need supplemental N fertilizer. In such crops, N fertilizer rates could be lowered appreciably while maintaining optimal economic yields. Table 2.14 shows data relating dry matter production and N uptake by different cover crops. Dry matter yields and N uptake values provide an indication of N uptake capacity of a cover crop from residual soil inorganic N and mineralized soil organic N as well as biological fixed N in the case of a legume cover crop. Data in Table 2.14 show that most legume crops have a high capacity to accumulate N in their dry matter. Similarly, rye (*Secale cereale* L.) as a cereal also accumulated 100 kg N ha^{-1}. Table 2.15 provides data of nitrogen fertilizer

TABLE 2.14
Dry Matter Yield and N Uptake by Cover Crops

Crop Species	Dry Matter Yield (Mg ha^{-1})	N Uptake (kg ha^{-1})
Hairy vetch	5.1	209
Bigflower vetch	1.9	60
Crimson clover	2.4	56
Berseem clover	1.5	45
Australian winter pea	1.6	68
Common vetch	4.3	134
Subterranean clover	4.0	114
Rye	6.3	100
Wheat	1.5	29

Source: Compiled from Frye et al. (1988).

equivalence (NFE) of legume cover crops to succeeding nonlegume crops. The NFE values varied from 12 to 182 kg ha^{-1}. Smith et al. (1987) reported that NFE values range from 40 to 200 kg ha^{-1} but more typically are between 75 to 100 kg ha^{-1}. Interseeding red clover (*Trifolium pretense* L.) into small grains is a common practice in the northeastern United States (Singer and Cox, 1998), and such practice can provide up to 85 kg N ha^{-1} to the subsequent corn crop (Vyn et al., 1999). Researchers in the southeastern United States have estimated that legumes such as hairy vetch can supply well over 100 kg N ha^{-1} to following corn or grain sorghum crops (Hargrove, 1986; Blevins et al., 1990). On prairie soils in Kansas, Sweeney and Moyer (2004) found that grain sorghum following initial kill-down of red clover and hairy vetch yielded as much as 131% more than continuous sorghum with estimated fertilizer N equivalencies exceeding 135 kg ha^{-1}.

Legume cover crops should be inoculated with an appropriate strain of N-fixing bacteria. Perennial legumes fix N during any time of active growth. In annual legumes, N fixation peaks at flowering. With seed formation, it ceases and the nodules fall off the roots. Rhizobia return to the soil environment to await their next encounter with legume roots. These bacteria remain viable in the soil for 3 to 5 years but often at too low a level to provide significant optimal N-fixation capacity when legumes are replanted (Fageria et al., 2005). Figure 2.10 shows growth of gray mucuna (*Mucuna cinereum*) as a cover crop in the central part of Brazil on an Inceptisol.

Green manure crop is defined as any crop grown for the purpose of being turned under while green or soon after maturity for soil improvement (Soil Science Society of America, 1997). Cherr et al. (2006) defined green manure as a crop used primarily as a soil amendment and a nutrient source for future crops. Legumes utilized as green manures may be useful as components of sustainable crop production systems. The positive role of green manuring in crop production has been known since ancient times. The importance of this soil amelioration practice has increased in recent years because of the high cost of chemical fertilizers, increased risk of environmental pollution, and need for sustainable cropping systems (Fageria, 2007b). Leguminous green manure crops may add N to crop systems through biological fixation, and the

TABLE 2.15

Nitrogen Fertilizer Equivalence (NFE) of Legume Cover Crops to Succeeding Nonlegume Crops

Legume/Nonlegume Crop	NFE (kg ha^{-1})
Hairy vetch/cotton	67–101
Hairy vetch + rye/corn	56–112
Hairy vetch/corn	78
Hairy vetch/sorghum	89
Hairy vetch/corn	78
Hairy vetch + wheat/corn	56
Crimson clover/cotton	34–67
Crimson clover/corn	50
Crimson clover/sorghum	19–128
Common vetch/sorghum	30–83
Bigflower vetch/corn	50
Subterranean clover/sorghum	12–103
Sesbania/lowland rice	50
Alfalfa/corn	62
Alfalfa/wheat	20–70
Arachis spps/wheat	28
Subterranean clover/wheat	66
White lupin/wheat	22–182
Arachis spps/corn	60
Pigeon pea/corn	38–49
Sesbania/potato	48
Mungbean/potato	34–148
Chickpea/wheat	15–65

Source: Compiled from Smith et al. (1987) and Kumar and Goh (2000).

slow release of N from decomposing green manure residues may be well timed with plant uptake (Cline and Silvernail, 2002; Cherr et al., 2006). Cherr et al. (2006) reported that sunn hemp (*Crotalaria juncea* L.) produced 12.2 Mg ha^{-1} fresh matter and added 172 kg N ha^{-1} in a 14-week growth period. Decomposition and N release generally occur faster for residues with lower C:N and lignin:N ratios and lower polyphenol concentrations (Andren et al., 1992). Carbon:N and lignin:N ratios are usually lower for legumes compared with nonlegumes and for leaves and flowers compared with stems (Fageria, 2007b). Detailed discussion of the role of green manuring in crop production is given by Fageria (2007b).

2.8.4 Use of Farmyard Manures

Application of farmyard manures in cropping systems can increase organic matter content of the soil, which may enhance N use efficiency for crop plants. The

FIGURE 2.10   Growth of mucuna as a green manure or cover crop in the Brazilian Inceptisol.

importance of soil organic matter in improving soil fertility and productivity is well known (Fageria et al., 2005). Continued crop production potentials of soils are directly related to their organic matter contents. Within limits, crop productivity is positively related to the soil organic matter content (Reicosky and Forcella, 1998). Soil organic matter improves soil's physical, chemical, and biological properties and consequently crop yields (Doran et al., 1998). The primary physical characteristics influenced by soil organic matter are those associated with soil structure, soil aggregation, and aggregate stability (Six et al., 1999). Soil organic matter compounds bind the primary soil particles in the aggregate, physically and chemically, and this, in turn, increases the stability of the aggregates and limits their breakdown during the wetting process (Lado and Bem-Hui, 2004). In turn, aggregates and their stability have tremendous influences on infiltration of water, soil water-holding capacity, and aeration as well as mass bulk density and penetration resistance (Carter, 2002).

Soil organic matter stabilizes soil aggregates, makes soil easier to cultivate, increases aeration, and increases soil water-holding and buffering capacities; soil organic matter breakdown releases available nutrients to plants (Carter and Stewart, 1996). Organic matter content of a soil can be used as an approximate N index. Approximate N value from organic matter can be obtained by the following equation: organic matter (%)/20 = N (%). The soil organic matter mineralization rate is considered to be approximately 2% a year. For example, in a soil with 3% organic matter, there will be 0.15% of N or 3 metric ton ha^{-1} in the arable layer. In this way, mineralization may provide approximately 60 kg N ha^{-1}.

Chemical properties influenced by soil organic matter content are soil pH, nutrient availability and cycling, cation exchange capacity, and buffering capacity (Tisdale et al., 1986). Organic matter can also bind and detoxify potentially toxic cations (e.g., Al, Mn). Although a large proportion of the total soil N remains physically and chemically protected from microbial degradation in the stable soil organic matter pool, and thus is unavailable for immediate plant uptake, more labile fractions of soil organic matter, which are generally much smaller, remain an important source of N (Jenkinson and Parry, 1989).

The complementary use of organic manures and chemical fertilizers has proved to be the best soil fertility management strategy in the tropics (Fageria and Baligar, 2005a). Well-decomposed farmyard manure contains about 12.9 g kg^{-1} total N, 1.0 g kg^{-1} available P, 4.5 g kg^{-1} exchangeable K, 10.8 g kg^{-1} exchangeable Ca, and 0.7 g kg^{-1} exchangeable Mg (Makinde and Agboola, 2002). Organic manure has a greater beneficial residual effect on soils than can be derived from the use of either inorganic fertilizer or organic manure alone. A large proportion (75 to > 90%) of the herbage N and minerals ingested by animals is returned to the soil in the urine and feces (Whitehead, 1970; Wilkinson and Lowry, 1973). But in developing countries, where animal feces or dung is used as a fuel, the situation is different. In rural areas of developing countries, use of cattle dung as a fuel for cooking purposes is a common practice. Under these conditions, the efficiency of N recycling for plant uptake is significantly reduced (Fageria and Gheyi, 1999).

Animal manures can make a substantial contribution to N, P, K, and other nutrients. The potential nutrient contribution is very high in developed as well as undeveloped countries. However, animal manures are often inefficiently used as a result of poor storage and application practices. Runoff, volatilization, and leaching losses of plant nutrients, especially N in the application, handling, and storage of animal manure, may be so high that only a fraction of the original nutrients remain to be applied to cropland (Fageria and Gheyi, 1999). If manure is properly applied and handled, these losses can be reduced significantly, and nutrient use efficiency of crops can be improved through animal wastes. Corn yield increases from the contribution of N in livestock manure have been documented by research in New York (Klausner and Guest, 1981), Vermont (Jokela, 1992), Wisconsin (Montavalli et al., 1989), and Quebec (Xie and Mackenzie, 1986). According to Jokela (1992), manure application rates (9 Mg ha^{-1}) were equivalent, in terms of yield response, for corn to 73 to 122 kg fertilizer N ha^{-1} in individual years, which represented 27 to 44% of the total manure in the year of application.

2.8.5 Adequate Moisture Supply

Adequate moisture supply is fundamental to improving N use efficiency in field crops. Food and Agriculture Organization (FAO) estimates suggest that food production per unit land area is on average 2.5 times greater on land with irrigation than on land without it (Stanhill, 1986). Irrigation has long played a key role in feeding expanding population and is expected to play a still greater role in the future (Hillel and Rosenzweig, 2002). The 13% of cropland that is irrigated in the United States is estimated to yield crops worth 30% of the total production (National Research Council, 1989). Similarly, though irrigated land amounts to only some 17% of the world's cropland, it contributes well over 30% of the total agricultural production (Hillel and Rosenzweig, 2002). That vital contribution is even greater in arid regions, where the supply of water by rainfall is least, even as the demand for water imposed by the bright sun and the dry air is greatest (Hillel and Rosenzweig, 2002). Food production is more stable and less risky with irrigation. Nitrogen movement in soil to reach plant roots and consequent absorption depend on soil moisture content. If water flow does not supply the root requirements, nutrient absorption by the root at or near its surface will reduce its concentration and consequently uptake.

Power (1990) reported that crop responses to applied N depend on water supply. As moisture supply increases, N response to applied N increases in a quadratic fashion. Microbial processes in soils with given structure are adversely affected by both high and low moisture contents (Drury et al., 2003). Grundmann et al. (1995) reported that the lower water limit for net mineralization was –1.5 MPa (wilting point), while the upper limit was 75% of water-filled pore space. In legumes such as soybean, water stress can also reduce N_2 fixation (Serraj and Sinclair, 1996; Fageria and Baligar, 2005a).

2.8.6 ADOPTION OF CONSERVATION/MINIMUM TILLAGE

Conservation tillage is defined as keeping 30% or more cover of crop residue on the soil surface (Soil Science Society of America, 1997). In addition, world minimum tillage is also frequently used in the literature. Minimum tillage can be defined as the minimum use of primary and/or secondary tillage necessary for meeting crop production requirements under the existing soil and climatic conditions, usually resulting in fewer tillage operations than for conventional tillage (Soil Science Society of America, 1997). Adopting conservation or minimum tillage minimizes or reduces loss of soil and water and also improves organic matter content and microbial activities on the soil surface. All these physical, chemical, and biological changes improve N use efficiency for crop plants (Fageria and Baligar, 2005a). The beneficial effects of conservation tillage practice are more pronounced on soils that have low organic matter content and are well drained.

2.8.7 USE OF APPROPRIATE SOURCE, METHOD, RATE, AND TIMING OF N APPLICATION

Use of appropriate source of N is important for improving N use efficiency in crop plants. Such practice not only increases yield but also reduces cost of production and environmental pollution. Urea and ammonium sulfate are the main nitrogen carriers worldwide in annual crop production. However, urea is generally favored by the growers over ammonium sulfate due to lower application cost because urea has a higher N analysis than ammonium sulfate (46% vs. 21% N). In developed countries like the United States, anhydrous NH_3 is an important N source for annual crop production. At normal pressures, NH_3 is a gas and is transported and handled as a liquid under pressure. It is injected into the soil to prevent loss through volatilization. The NH_3 protonates to form NH_4^+ in the soil and becomes XNH_4^+, which is stable (Foth and Ellis, 1988). The major advantages of anhydrous NH_3 are its high N analysis (82% N) and low cost of transportation and handling. However, specific equipment is required for storage, handling, and application. Hence, NH_3 is not a popular N carrier in developing countries. Major N fertilizers available in the world market along with their N contents are presented in Table 2.16.

TABLE 2.16
Major Nitrogen Fertilizers, Their Chemical Formulas, and N Contents

Common Name	Formula	N (%)
Ammonium sulfate	$(NH_4)_2SO_4$	21
Urea	$CO(NH_2)_2$	46
Anhydrous ammonia	NH_3	82
Ammonium chloride	NH_4Cl	26
Ammonium nitrate	NH_4NO_3	35
Potassium nitrate	KNO_3	14
Sodium nitrate	$NaNO_3$	16
Calcium nitrate	$Ca(NO_3)_2$	16
Calcium cyanamide	$CaCN_2$	21
Ammonium nitrate sulfate	$NH_4NO_3(NH_4)_2SO_4$	26
Nitrochalk	$NH_4NO_3 + CaCO_3$	21
Monoammonium phosphate	$NH_4H_2PO_4$	11
Urea ammonium nitrate	$CO(NH_2)_2 + NH_4NO_3$	32
Diammonium phosphate	$(NH_4)_2HPO_4$	18

Source: Foth and Ellis (1988), Fageria (1989), Mengel et al. (2001), Fageria and Baligar (2005a).

Ammonium is less mobile than NO_3 and remains near its original placement in the soil until nitrification (Fenn and Hossner, 1985). Urea anhydrous ammonia or ammonium phosphate may precipitate adsorbed and soluble Ca^{2+} in the fertilizer band, creating a low-Ca^{2+} and a high-NH_4^+ environment (Fenn, 1987). Fenn et al. (1991) reported that addition of Ca^{2+} with urea produced significantly improved plant-use efficiency of nitrogen in onion. Ammonium is monovalent and is similar in size and chemical reactivity to K^+, and absorption appears to be stimulated by increasing soluble Ca^{2+} (Fenn, 1987).

Nitrogen sources studied have included KNO_3 (KN), NH_4NO_3 (AN), $(NH_4)_2SO_4$ (AS), urea (U), urea-NH_4NO_3 (UAN), and anhydrous NH_3 (AA). After 12 years of continuous corn (fall moldboard plow tillage), Nelson and MacGregor (1973) found that AN and U sources of N were equivalent at the 90 and 180 kg N ha^{-1} rates. Also, for conventionally tilled corn, Jung et al. (1972), Stevenson and Baldwin (1969), and Mengel et al. (1982) found no differences between AN, U, and AA; AA and U; and UAN and AA sources of N, respectively. No-till corn that received surface-applied N exhibited differences among N sources, with U generally the least efficient (Bandel et al., 1980; Eckert et al., 1986; Fox et al., 1986; Touchton and Hargrove, 1982).

Author studied the response of lowland rice to ammonium sulfate and urea fertilization applied to a Brazilian Inceptisol. Grain yield significantly ($P < 0.01$) and quadratically increased with increasing N rate from 0 to 400 mg kg^{-1} of soil by ammonium sulfate as well as urea source (Figure 2.11). Maximum grain yield was obtained with the application of 168 mg N kg^{-1} of soil by ammonium sulfate

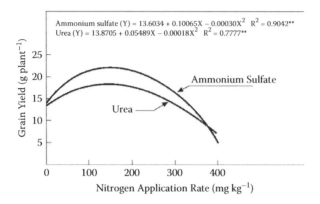

FIGURE 2.11 Relationship between nitrogen rate applied by ammonium sulfate and urea and grain yield of lowland rice.

and 152 mg N kg^{-1} of soil by urea. The variation in grain yield was 5.5 to 22.8 g $plant^{-1}$, with an average yield of 15.9 g $plant^{-1}$, by ammonium sulfate and 8.0 to 19.7 g $plant^{-1}$, with an average value of 14.4 g $plant^{-1}$, by urea fertilization. Ammonium sulfate accounted for 90% of variability in grain yield, whereas urea accounted for 78% of variability in grain yield. Across six N rates, ammonium sulfate produced 10% higher grain yield compared to urea. In addition, averaged across two N sources (160 mg N kg^{-1}), ammonium sulfate produced 22.5 g grain yield per plant and urea produced 18.5 g grain yield per plant. The application of ammonium sulfate at the rate of 160 mg N kg^{-1} produced 22% higher grain yield compared to urea at the same rate of N. This means that ammonium sulfate was superior to urea as a fertilizer for lowland rice grain yield (Figure 2.11). Reddy and Patrick (1978) and Bufogle et al. (1998) reported no differences in straw or grain yield of lowland rice between the two N sources. In a greenhouse study, Phongpan et al. (1988) found no differences in grain and straw yields between addition of urea and ammonium sulfate to an acid sulfate soil at low N rate (160 mg N kg^{-1}), but at higher N rates (320 to 480 mg N kg^{-1}), urea consistently produced higher yields than ammonium sulfate. This means that response of lowland rice to ammonium sulfate and urea depends on soil or climatic conditions (Bufogle et al., 1998).

Author also studied the influence of ammonium sulfate and urea fertilization on grain yield of lowland rice under field conditions. Grain yield expressed in relative yield was significantly increased by urea as well as ammonium sulfate fertilization, and the increase proceeded in a quadratic fashion (Figure 2.12). In fertilizer experiments, 90% of the relative yield is considered as an economic index, and this index was used to calculate adequate N rate (Figure 2.11). Ninety percent of the relative yield (corresponding to 5750 kg grain ha^{-1}) was obtained with the application of 84 kg N ha^{-1} in the case of ammonium sulfate. Similarly, in the case of urea, 90% of the relative grain yield (corresponding to 4811 kg grain ha^{-1}) was obtained with the application of 130 kg N ha^{-1}. This means that ammonium sulfate was a better source of N for lowland rice production under Brazilian conditions. Peterson and Frye (1989) reported that the use of ammonium nitrate in flooded rice enhances the loss of NO_3^- by both denitrification and leaching. Thus, ammonium sulfate is

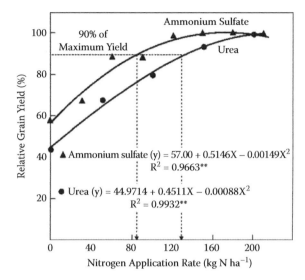

FIGURE 2.12 Relative grain yield of lowland rice as influenced by ammonium sulfate and urea sources.

generally considered to be more effective for flooded rice production than either urea or ammonium nitrate.

Nitrogen can move in the soil by mass flow; hence, it can be applied in the band as well as broadcast. However, it should be incorporated into the soil to avoid volatilization losses. During postemergence, fertilizers may be side-dressed by injecting them into the subsurface and applied as a top dressing. Rao and Dao (1996) reported that subsurface placement of urea under no-till winter wheat conditions had the potential to significantly improve N availability to plants and thereby improve N use efficiency. Mixing fertilizers into soil and injecting them into the subsurface are more efficient methods of N application compared to broadcasting and/or leaving them on the soil surface (Fageria and Baligar, 2005a).

Nelson (1982) reviewed the ammonia losses under field conditions and concluded that the amount of NH_3 volatilized was small when N fertilizers are incorporated into the soil, and NH_3 losses are normally low (15% of applied N). When ammonical fertilizers are surface applied to acidic or neutral soils, however, large amounts of NH_3 may be evolved on addition of N fertilizers, especially urea to the surface of alkaline soils. Surface application of urea fertilizer solutions loses N through NH_3 volatilization. Losses of NH_3 from agricultural soils may reach 55% of applied urea N (Alkanani et al., 1991). The extent of NH_3 volatilization is determined by soil pH, texture, temperature, moisture, exchangeable cations, fertilizer source, and rate of application. Alkanani et al. (1991) reported that ammonia volatilization increased as soil water content increased. Volatilization differences between moist (>–0.038 MPa) and air-dry (<–1.5 MPa) samples were reduced as the clay content of the soil increased. Ammonia volatilization versus water evaporation followed a logarithmic relationship (Alkanani et al., 1991). Deep placement of N-P fertilizer bands in the soil by means of various chisels and knives theoretically should have advantages over a conventional broadcast placement where the surface soil may become dry, thereby reducing water and nutrient uptake. Deeper-placed fertilizer would more likely be in a moist soil zone, available for root exploration.

Janzen et al. (1990) compared the point injection method with the broadcast method for spring application of N to winter wheat. Point injection exhibited significantly higher fertilizer-use efficiency at five out of eight sites. On average, the recovery of fertilizer-derived N by the winter wheat crop was 37% higher in the point-injection treatments than in a broadcast NH_4NO_3 treatment, the next most effective method of N application. Janzen et al. (1991) also tested three methods of N application with time in winter wheat. The experiment included a factorial of three methods (point-injected urea-NH_4NO_3, broadcast-urea, and broadcast NH_4NO_3) and four application times (at seeding, late fall, early spring, and late spring) along with two banding treatments (urea and NH_4NO_3) applied at seeding and a control. For all methods of application, efficiency of fertilizer tended to be highest when applied in early spring. In the point-injection treatments, average recovery of applied N by wheat was 32% for the seeding, 43% for late fall, 56% for early spring, and 29% for late spring application. At the optimum time of application, point injection exhibited significantly higher fertilizer recovery than broadcast applications, probably because of direct placement of N into the rooting zone. Fertilizer recovery from banding treatments applied at seeding was lower than from early-spring point injection, presumably because of overwinter N losses. The results suggest that point injection can significantly enhance fertilizer-use efficiency and that optimum time of injection is in the early spring just prior to the period of maximum N assimilation by the crop (Janzen et al., 1991).

Maddux et al. (1991) evaluated broadcast and subsurface banded urea–ammonium nitrate (UAN) nitrogen applied to corn. Recovery of tagged urea N by all plant components was 61 and 68% for banded UAN in 2 years, respectively, but only 39 and 44% for broadcast-incorporated UAN. The amount of tagged urea N immobilized in the top 90 cm of the soil profile was 58 and 14% greater with the broadcast-incorporated than with the banded treatment for the first and second years, respectively. Tagged urea N that was unaccounted for was 62 and 200% greater, respectively, for the 2 years with the broadcast than with the banded applications. The difference in unaccounted-for tagged N between treatments was attributed to variations in loss by leaching. The differences in uptake of tagged urea N were attributed to a combined effect of greater soil immobilization and loss by leaching with the broadcast-incorporated application (Fageria and Gheyi, 1999).

In a recent review, NH_3 losses across a variety of crops and geographical locations averaged 23% of surface-applied urea N (Scharf and Alley, 1988). Proposed fertilizer additives to reduce NH_3 loss from surface-applied urea include acids such as H_3PO_4 (Keller and Mengel, 1986), Ca or Mg salts (Fenn et al., 1981), and inhibitors of urease, the soil enzyme that catalyzes the hydrolysis of urea to NH_4^+. Fenn and Kissel (1973) reported that, under laboratory conditions, more than 50% of the applied N was lost in 100 hours following the surface application of 550 kg N ha^{-1} as NH_4F, $(NH_4)_2SO_4$, or $(NH_4)_2HPO_4$. They also reported that NH_3 losses from NH_4Cl and NH_4NO_3 were less than 20% of the N applied.

This indicated that NH_4 associated with an anion that forms relatively insoluble Ca compounds in soil is more likely to volatilize as NH_3. Conversely, NH_4 is less likely to volatilize as NH_3 if associated with an anion that forms soluble Ca compounds. Gameh et al. (1990) also reported that during the first 13 days of incubation in the Matapeake soil, 26% of the N applied as urea alone was lost via NH_3 volatilization. From the KCl-coated urea, only 13% of the applied N was lost via volatilization. Those authors suggested that NH_3 volatilization was reduced by KCl because soil pH is decreased in the presence of KCl. In their study, the surface soil pH was lower when KCl was applied in combination with urea than when urea was applied alone. The reduced soil pH obtained with KCl may be due to the replacement of exchangeable Ca followed by the production of NH_4Cl. It is also postulated that the K ion may replace some of the exchangeable Al, which reacts to form hydrous Al oxides, leaving an excess of H ions in solution. This would be especially important in acid soils with substantial amounts of exchangeable Al. All of these mechanisms tend to reduce soil pH, which decreases the NH_4 concentration and thus reduces NH_3 volatilization (Fageria and Gheyi, 1999).

Use of adequate N rates is essential for efficient use of N fertilizer and to maintain the economic sustainability of cropping systems. Excessive use of N fertilizers is economically unfavorable, because incremental increases in yield diminish with increasing amounts of N applied, and it could lead to detrimental effects on the quality of soil and water resources (Fageria and Baligar, 2005a). Nitrogen is a mobile nutrient in soil–plant systems. Hence, creation of crop response curves showing yield versus N rates is the most efficient and effective method for defining the N requirement of a crop. Development of appropriate crop response curves to applied N fertilizer requires optimal environmental conditions during crop growth, and experiments should be repeated over several years to obtain meaningful results.

Nitrogen application timing is also important for improving crop yields and N use efficiency. Application of nutrients when crops need them can improve crop efficiency significantly and consequently improve yields. Nitrogen is important for tillering, to increase the number of grains and to improve seed weights. If we consider yield of a given crop as a function of the number of panicles, heads, or pods, the number of grains, and the grain weight, a crop needs N throughout its growth cycle. The number of panicles, heads, or pods is determined during the vegetative growth stage, the number of grains during the reproductive growth stage, and the grain weight during the ripening or grain-filling growth stage of a crop. This means that to obtain higher yield and higher nitrogen use efficiency, it is necessary to supply N to a crop in three stages (vegetative, reproductive, and ripening or grain filling) of growth and development.

Knowledge of a crop's growth and development is essential for adopting improved management practices including N top-dressing during its growth cycle for higher yields. Additionally, knowledge of occurrence of growth stages can also be used in many physiological studies to identify the critical growth stages during plant growth and development that are sensitive to environmental factors (Fageria, 2007). Growth is defined as the irreversible change in the size of a plant cell or organ (Fageria, 1992). On the other hand, plant development is defined as the sequence of

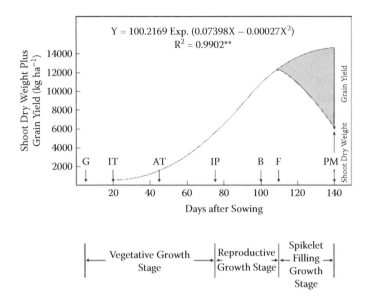

$$Y = 100.2169 \, \text{Exp.} \, (0.07398X - 0.00027X^2)$$
$$R^2 = 0.9902^{**}$$

FIGURE 2.13 Lowland rice shoot dry weight during growth cycle and grain yield. The cultivar planted in the experiment was Metica 1 and experiment was conducted in central part of Brazil in an Inceptisol. G = germination, IT = initiation of tillering, AT = active tillering, IP = initiation of panicle primordia, B = booting, F = flowering, and PM = physiological maturity.

ontogenetic events, involving both growth and differentiation, leading to change in function and morphology. Development is most clearly manifested in changes in the form of organisms, as when crop plants change from the vegetative to the reproductive stage and from the reproductive stage to maturity. The development can be studied through morphological as well as physiological changes (Fageria, 1992).

Different growth stages of lowland rice (cultivar Metica 1 grown in the central part of Brazil) and their occurrence timing are given in Figure 2.13. Lowland rice cultivar Metica 1 has a growth cycle of 140 days from sowing to physiological maturity or 135 days from germination to physiological maturity under Brazilian conditions or in the tropics. The vegetative growth stage comprised 70 days' growth (germination to initiation of panicle primordia), the reproductive growth stage comprised 35 days' growth (panicle primordia initiation to flowering), and the spikelet-filling stage comprised 30 days' growth (flowering to physiological maturity). A study conducted by Fageria and Prabhu (2003) in the Brazilian Inceptisol showed that N fractionated into two or three equal doses produced higher grain yield of lowland rice compared with total applied at sowing (Figure 2.14).

Nitrogen is a mobile nutrient in soil, and its application in large quantities at sowing may result in loss due to leaching or denitrification. Further, if the entire N is applied in the beginning, plant roots are not well developed and cannot absorb all the applied nitrogen. For these reasons, split application of N is generally recommended. First, top-dressing in cereals should be done at the time of panicle initiation, which will determine the number of grains. If economically feasible, a second application may be desirable at about the reduction-division stage, which normally starts

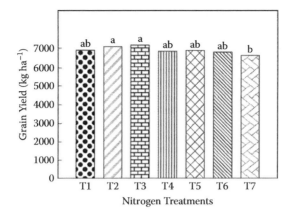

FIGURE 2.14 Grain yield of lowland rice as influenced by N timing treatments. T_1 = all the N applied at sowing, T_2 = 1/3 N applied at sowing + 1/3 N applied at active tillering + 1/3 N applied at the initiation of panicle primordia, T_3 = 1/2 N applied at sowing + 1/2 N applied at active tillering, T_4 = 1/2 N applied at sowing + 1/2 N applied at the initiation of panicle primordia, T_5 = 2/3 N applied at sowing + 1/3 N applied at active tillering, T_6 = 2/3 N applied at sowing + 1/3 N applied at initiation of primordia floral, and T_7 = 1/3 N applied at sowing + 2/3 N applied at 20 days after sowing (Fageria and Prabhu, 2003).

approximately 1 week before flowering. Nitrogen deficiency during this growth stage significantly decreases grain weight and subsequently grain yields (Fageria, 1992).

Legumes fix N through symbiosis by rhizobium and generally need lower N application as compared to cereals. Legumes have been shown to require maximum amounts of N during flowering and pod set for soybean (Brevedan et al., 1978; Israel, 1981). Harper and Gibson (1984) reported that maximum nitrate utilization occurred at full bloom and maximum symbiotic N_2 fixation 3 weeks later during pod fill in soybean. Thus, in legumes, the best time for N top-dressing is a little before flowering. This means that top-dressing of N is related to the growth cycle of a species or cultivar within species. The growth cycle of a cultivar may vary with temperature and nutrient supplies (especially phosphorus). Soil and fertilizer management must be designed to furnish a continuous supply of available N and adequate plant growth factors for producing high yield and quality while withstanding unpredictable plant stress conditions in order to obtain economical, profitable, and reliable production of field crops.

With irrigation, Jung et al. (1972) found that N applied during either the fifth, sixth, seventh, or eighth week after planting led to maximum yield of corn. Russelle et al. (1983) found some yield advantage and greater fertilizer accumulation in grain when N was applied at the 4, 8, or 16 leaf stages than at planting. No-till corn yields (3-year average) were significantly increased by N application at the V5 through V6 growth stage compared to N application at planting or when injected instead of surface broadcast (Fox et al., 1986).

Nitrogen availability during grain fill can be increased by addition of N fertilizer in coordination with irrigation. Nitrogen uptake efficiency was 65% greater for N applied at anthesis than for N applied at planting in wheat (Wuest and Cassman,

1992a, 1992b). These authors reported that mean NHI of labeled fertilizer N applied at anthesis was 0.89 compared to 0.70 for that of fertilizer N applied at planting. From their studies, these authors concluded that both uptake and partitioning of N by irrigated wheat could be manipulated to optimize fertilizer N utilization. Early-season N must be managed to optimize grain yield, but adding excess N at that time reduces overall partitioning efficiency, whereas the late-season N supply can be adjusted to increase grain protein levels without reducing partitioning efficiency.

Field studies under both irrigated and nonirrigated conditions in Nebraska (Rehm and Wiese, 1975), Illinois (Welch et al., 1971), Wisconsin (Bundy et al., 1983), Minnesota (Malzer and Graff, 1983), and New York (Lathwell et al., 1970) have shown increased grain yield and more efficient use of fertilizer N by corn when N application was delayed until several weeks postemergence rather than when N was applied before planting. This was especially true at lower N rates. Fertilizer N efficiency was defined in these studies as either percentage of applied N recovered in the harvested portion, generally the grains, or as grain yield increase due to applied N (Jokela and Randall, 1989). Improved efficiency with delayed N application is consistent with the concept of providing N at the time of maximum uptake, which occurs in a 2- to 3-week period just prior to silking or slightly earlier (Russelle et al., 1983). This increased efficiency is the result of a combination of plant and soil factors. Delay of N application from planting time to the 4-leaf to 12-leaf stage can result in a higher proportion of dry matter and N in the grain portion of the plant (higher harvest index [HI]) and a higher N harvest index (NHI) and therefore more efficient use of N for grain production (Russelle et al., 1983). In addition, later application limits the potential for N loss by leaching or denitrification because the fertilizer N is in the soil for a shorter period of time and because an actively growing root system already exists for uptake of the applied N (Jokela and Randall, 1989). Some uncertainty exists as to whether, if legumes fix atmospheric nitrogen, nitrogen through fertilizers should be added to such crops at planting or at later growth stages. A lot of research data are available to clarify this matter. Early stages of legume establishment in N-poor soils may by retarded due to N stress (Phillips and Dejong, 1984). During seedling development, the supply of N reserves from the seed diminishes and growth of the seedling becomes dependent on available soil N until an effective N_2-fixing symbiosis is established. Addition of mineral N stimulates plant growth and may increase or decrease N_2 fixation (Eaglesham et al., 1983; Morris and Weaver, 1987). Brevedan et al. (1978) showed that increasing N supply to soybean during flowering increased seed yield in the greenhouse (33%) as well as in the field (28–32%). These data demonstrate that high levels of N were necessary during flowering and pod set to maximize seed set. The highest rates of N_2 fixation were observed at the end of flowering or during pod filling (Lawn and Brun, 1974). Most estimates have shown that 25 to 60% of the total N in the mature plant is derived from symbiotic N_2 fixation, the remaining being soil derived (Deibert et al., 1979). In the southeastern coastal plains of the United States, N_2 fixation can account for more than 75% of accumulated N in soybean grown on a loamy sand under irrigated conditions (Matheny and Hunt, 1983).

2.8.8 Use of Efficient Species/Genotypes

Utilization of plant species or genotypes of the same species efficient in absorption and utilization of N is an important strategy in maximizing N use efficiency and sustainable agricultural systems. Nutrient-efficient species or genotypes are defined as those species or genotypes that produce higher yields per unit of nutrient applied or absorbed compared to other standards species/genotypes under similar agroecological conditions. Differences in N uptake and utilization among crop species and cultivars within species for field crops have been reported (Fageria and Baligar, 2005a). In general, C_3 species exhibit higher NO_3^- uptake capacity than C_4 species (Lin et al., 2007). Sehtiya and Goyal (2000) reported a 55 to 91% higher NO_3^- uptake capacity in C_3 plants as compared to C_4 plants. Figure 2.15 shows responses of five lowland rice genotypes to N fertilization. The difference in nutrient uptake and utilization may be associated with better root geometry, plants' ability to take up sufficient nutrients from lower concentrations, plants' ability to solubilize nutrients in the rhizosphere, better transport, distribution, and utilization within plants, and balanced source–sink relationship (Graham, 1984; Baligar et al., 2001; Fageria and Barbosa Filho, 2001; Fageria and Baligar, 2003a).

Nutrient use efficiencies have improved over time. This improvement was associated with increased yield per unit area with better crop management practices and the development of crop genotypes with higher yield potentials. In conclusion,

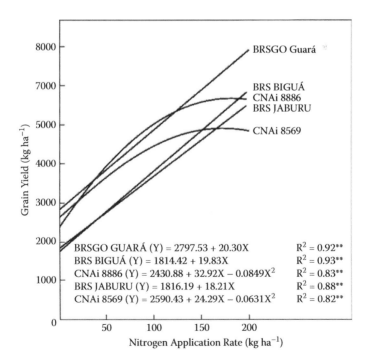

FIGURE 2.15 Response of five lowland rice genotypes to nitrogen fertilization.

efficient use of inorganic fertilizers is essential in today's agricultural practices and will be even more important in years to come. Hence, nutrient-efficient plants will play a vital role in increasing crop yields per unit area and improve the health and quality of life of humans in the 21st century (Fageria et al., 2008).

2.8.9 SLOW-RELEASE FERTILIZERS

According to Gandeza et al. (1991), an ideal fertilizer is one that (1) needs only a single application to supply the amounts of nutrients required for optimum plant growth during the entire growing season, (2) has high maximum percentage recovery in order to achieve higher return to produce input, and (3) has minimum detrimental effects on soil, water, and atmospheric environments. Slow-nitrogen-release fertilizers may satisfy these requirements for N nutrition. A new polyolefin-coated urea (POCU) fertilizer developed in Japan is one example (Fujita, 1989). POCU is a controlled N-release granular fertilizer whose dissolution is primarily temperature dependent. The coating is made of ethylene vinyl acetate, polyolefin, and talc to control N release. Dissolution doubles for every 10°C rise in temperature, a Q_{10} matching that of plants (Gandeza et al., 1991; Shoji et al., 1991). Dissolution rate of POCU under field conditions can be predicted using either soil or air temperature. Cumulative N release gradually increased starting at the time of fertilizer application and reached about 80% of the total N content after about 125 days. Therefore, N uptake by crops can more nearly be optimized with the use of this new type of fertilizer. Since N release is controlled by the moisture permeability of the resin coating, which, in turn, is controlled by temperature, POCU is considered to be a controlled-release fertilizer. In contrast, the release of N from a variety of slow-release fertilizers is largely by chemical or microbiological processes; hence, release is more dependent on soil conditions favorable to microbial growth. For instance, IBDU is dissolved by hydrolytic reaction (Allen, 1984), and urea forms by microbial decomposition of the C band of the urea–polymer complex (Hauck, 1985). Several types of POCU with various N-release rates and duration are already commercially available (Gandeza et al., 1991).

2.8.10 USE OF NITRIFICATION INHIBITOR

Nitrification inhibitors are a group of chemicals capable of retarding the formation of NO_3 from NH_4 in soil by inhibiting the activity of nitrifying microorganisms. Use of nitrification inhibitor (NI) is one of the techniques, which can be incorporated into a best management practice for nitrogen nutrition. Most of the fertilizer N applied to soils is in the form of NH_4 or NH_4-producing compounds such as urea. The NO_3 formed through nitrification of this fertilizer N by soil microorganisms is susceptible to loss by leaching and denitrification and may contribute to NO_3 pollution of groundwater and surface waters (McCarty and Bremner, 1990). Nitrification inhibitors slow the conversion of ammonium N to nitrate N, and numerous compounds have been proposed for this purpose. Most of these compounds are not very effective, however, and only three compounds—nitrapyrin, etridiazole, and dicyandiamide—are among the most commonly used nitrification inhibitors (McCarthy and Bremner, 1990; Jacinthe and Pichtel, 1992). The Dow Chemical Company introduced nitrapyrin under the trade name N-Serve, the Olin Corporation introduced etridiazole under the trade name Dwell, and Showa Denko, Takajo, introduced dicyandiamide (DCD). A recent study by McCarty and Bremner (1990) showed that one of the compounds, 2-ethynylpyridine, is also a potential inhibitor of nitrification and deserves consideration as a fertilizer amendment for retarding nitrification of fertilizer N in soil.

The value of nitrification inhibitors in reducing nitrification of ammonical N under field conditions has been well documented (Hendrickson et al., 1978; Touchton et al., 1979a, 1979b). Touchton et al. (1978) found less movement and increased persistence of nitrapyrin in high-organic-matter soils compared to those low in organic matter. The effectiveness of nitrapyrin has been found to decrease as soil temperature and pH increase (Touchton et al., 1979b). Cooler soil temperature regimes should enhance NI persistence and improve NH_4-N retention in soils. In warmer temperatures, the opposite effect occurs, as increasing soil temperatures hasten NI degradation and disappearance and shorten its period of effectiveness. Sewage sludge and animal manures are alternative sources of nutrients to supplement inorganic fertilizers. However, leaching may result in the loss of a portion of the applied N and/or denitrification after organic N and NH_4-N in the waste is converted to NO_3 by ammonification and nitrification. One way to reduce leaching and denitrification losses is through the use of a nitrification inhibitor.

Nitrapyrin (a nitrification inhibitor) added to NH_4 or NH_4-forming fertilizers reduced N losses from soils and increased yield of agronomic crops. A number of laboratory and field trials have shown that nitrapyrin [2-chloro-6(trichloromethyl) pyridine] markedly reduced the rate of NO_3-N formation in soil following addition of ammonical fertilizers (Meisinger et al., 1980; Schmidt, 1982). Terry et al. (1981) have shown that nitrapyrin added to sewage sludge–amended soil greatly reduced nitrification of sludge-derived NH_4-N. McCormick et al. (1983) found that addition of nitrapyrin delayed nitrification in liquid swine manure bands for > 100 days and markedly decreased loss of N from the manure band. McCormick et al. (1984) studied effects on corn grain yield and plant composition of mixing nitrapyrin with liquid swine manure applied to a silty clay loam soil. The addition of nitrapyrin (at a rate of 50 mg L^{-1}) to liquid swine manure applied at a rate 49 Mg ha^{-1} (154 available N ha^{-1}) significantly increased corn grain yield.

Almost without exception in laboratory, glasshouse, and field studies, heterotrophic immobilization of added ammonium is increased in the presence of a nitrification inhibitor (Hauck, 1990). The nitrification inhibitor chloroacetanilide was found to (1) decrease the uptake of ^{15}N-labeled ammonium by oats, rye, millet, and maize by 25–65%; (2) decrease plant uptake of ammonium sulfate as compared with potassium nitrate; and (3) significantly alter the ratio of soil nitrogen to fertilizer nitrogen taken up by several crop plants (Andreeva and Shcheglova, 1967). These data were consistent with the earlier finding by Jansson et al. (1955) and Jansson (1958) that soil heterotrophic microorganisms preferentially use ammonium rather than nitrate when both sources of nitrogen are present. In a field study with wheat in Canada, Juma and Paul (1983) added ^{15}N-labeled aqueous ammonia or urea to soil with or without the nitrification inhibitor, 4-amino-1,2,4-triazole (ATC). Recovery of applied nitrogen in the soil–plant system was greater for the ATC-amended fertilizers. Additional data could be cited to demonstrate that use of a nitrification inhibitor with ammonium or ammonium-producing fertilizer results in greater immobilization of the fertilizer nitrogen by either soil heterotrophic microorganisms or other soil processes. This effect has been shown using ^{15}N-labeled fertilizers with several nitrification inhibitors, including chloroacetanilide, ATC, dicyandiamide, and nitrapyrin, and has been observed in different cropping systems (Hauck, 1990).

In conclusion, nitrification inhibitor has the potential to increase crop yields where nitrification is followed by conditions for removal of nitrate from the root zone. Nitrification has been shown to depend on the availability of substrate, O_2, CO_2, soil water, and adequate temperature (Schmidt, 1982), while conditions for loss of nitrate are usually dependent on period of high soil moisture where nitrate leaching or denitrification occurs (Tanji, 1982). The efficacy of the inhibitor per se has been related to inhibitor properties such as volatility and solubility, and soil properties such as organic matter content, temperature, moisture, and pH (Gilmour, 1984; Keeney, 1980).

2.8.11 CONTROL OF DISEASES, INSECTS, AND WEEDS

Diseases, insects, and weeds are the most yield-limiting factors for most crop plants grown under most agroecological regions. Evans (1993) in a review reported those losses from pests and diseases before harvest to be about 20% on a global scale, ranging from 9% in the United States to 40–50% in some developing countries. Diseases and insects mostly infect the plants leaf area, which is the site of radiation interception and photosynthesis processes. Owing to reduced photosynthetic activity, utilization of absorbed nutrients is substantially reduced (Fageria, 1992). Controlling these yield-limiting factors or keeping them at threshold levels can improve N use efficiency and lead to higher crop yields (Fageria and Baligar, 2005a). The practices adopted for control of pests may vary from crop to crop and region to region and depend on the socioeconomic conditions of farmers. In modern agriculture, one of the best approaches to reduce risk of pests is using resistant crop species or genotypes within species. This practice not only reduces costs of production but also reduces environmental pollution. However, cultivars that are resistant to many pest stresses are not available, and pest control methods may not be at sufficient or desirable levels. Hence, adopting integrated disease, insect, and weed control methods may be a desirable strategy.

2.8.12 CONCLUSIONS

Nitrogen plays a pivotal role in maintaining soil fertility, environmental quality, and sustainability of agricultural systems. Furthermore, nitrogen occupies a unique position among essential plant nutrients because of the rather large amount required by most field crops, and it limits crop yields in most agroecosystems. Nitrogen deficiency is generally expected in mineral soils with low organic matter content, sandy soils that have been leached by heavy rainfall or irrigation, and intensively cropped soils with low rates of N application. Most of the N in plants (over 90%) is in organic forms and assimilated in carbon-containing compounds. Organic N should be mineralized before plants can absorb it. In addition, N is mainly added to soil by inorganic fertilizers to meet plant requirements.

The nitrogen cycle in soil–plant systems is very dynamic and complex. Continuous turnover of N occurs through mineralization-immobilization, with transfer of biological decay products into stable humus forms. Gains in soil N also occur through fixation of molecular N_2 by microorganisms and from the return of ammonia (NH_3^+) and nitrate (NO_3^-) in rainwater. Nitrogen losses occur through crop removal, leaching, volatilization, soil erosion, loss through plant canopy, and denitrification. Nitrogen is also added in crop residues and animal manures and through chemical fertilizers. Efficient management of N in agroecosystems is important to reduce the cost of crop production and to reduce environmental pollution. Hence, chemical fertilizers require much attention to support their efficient use and avoid environmental pollution.

Nitrogen has many important physiological and biochemical functions in the plant. Overall, uniform yellowing of leaves occurs when a plant first becomes N deficient. Plant growth is reduced, especially tillering in cereals and number of pods in legumes, when N is deficient. Nutrient interaction in crop plants is probably one of the most important factors affecting yields of annual crops. Hence, understanding N interactions with other nutrients is important to improve crop yields and to promote efficient N management. Plant tissue test for N is an important diagnostic technique to determine its deficiency and sufficiency. In addition, plant tissue analysis determines the quantity of N removed by a crop when N concentration is multiplied by dry matter yield. This information is useful in maintaining soil fertility under agroecological conditions for a given crop. However, plant tissue analysis is highly variable with plant age, and defining the physiological crop growth stage is very important in correct analysis and interpretation of tissue analysis results. Due to the dynamic nature of N in soil–plant systems, there is no reliable soil test to identify N deficiency or sufficiency.

Nitrogen fertilization is highly correlated with crop yields and is a major factor in world crop production. Hence, adequate management of this element in crop production is crucial for reducing the cost of crop production and reducing environmental pollution. Principal N management practices are liming acid soils; using crop rotation; using conservation or minimum tillage; supplying adequate moisture; using adequate form, method, rate, and timing of application; planting N-efficient crop species or genotypes within species; and controlling insects, diseases, and weeds. There are several sources of N fertilizers, and choice of fertilizer source depends on economics and the crop situation. Rate and timing of N application are the most important factors from an economic as well as an environmental point of view. Nitrogen nutrition of field crops has been well researched, and adequate technology exists to optimize use, improve yields, and minimize the negative effects of fertilizer N on the environment.

REFERENCES

Aguilar, S. A. and A. van Diest. 1981. Root-phosphate mobilization induced by the alkaline uptake pattern of legume utilizing symbiotically fixed nitrogen. *Plant Soil* 61:27–42.

Allen, S. F. 1984. Slow-release nitrogen fertilizer. In: *Nitrogen in crop production,* R. D. Hauck, Ed., 195–205. Madison, WI: Am Soc. Agron.

Alkanani, T., A. F. Mackenzie, and N. N. Barthakur. 1991. Soil water and ammonia volatilization relationships with surface applied nitrogen fertilizer solutions. *Soil Sci. Soc. Am. J.* 55:1761–1766.

Andreeva, E. A. and G. M. Shcheglova. 1967. Uptake of soil and fertilizer nitrogen by plants as revealed by greenhouse pot experiments using [15]N. *Transactions of the Joint Meeting of Commissions. II and IV Int. Soc. Soil Sci.* 1166:113–124.

Andren, O., E. Steen, and K. Rajkai. 1992. Modelling the effects of moisture on barley straw and root decomposition in the field. *Soil Biol. Biochem.* 24:727–736.

Andrew, C. S. 1978. Mineral characterization of tropical forage legumes. In: *Mineral nutrition of legumes in tropical and subtropical soils*, C. S. Andrew and E. J. Kamprath, Eds., 93–112. East Melbourne: CSIRO.

Aulakh, M. S., J. W. Doran, and A. R. Mosier. 1992. Soil denitrification—Significance, measurement, and effects of management. *Adv. Soil Sci.* 18:1–57.

Badaruddin, M. and D. W. Meyer. 1989. Forage legume effects on soil nitrogen and grain yield, and nitrogen nutrition of wheat. *Agron. J.* 81:419–424.

Baligar, V. C. and N. K. Fageria. 2007. Agronomy and physiology of tropical cover crops. *J. Plant Nutr.* 30:1287–1339.

Baligar, V. C., N. K. Fageria, and Z. L. He. 2001. Nutrient use efficiency in plants: An overview. *Commun. Soil Sci. Plant Anal.* 32:921–950.

Bandel, V. A., S. Dzienia, and G. Stanford. 1980. Comparison of N fertilizers for no-till corn. *Agron. J.* 72:337–341.

Barber, K. L., L. D. Maddux, D. E. Kissel, and G. M. Pierzynski. 1992. Corn responses to ammonium and nitrate-nitrogen fertilization. *Soil Sci. Soc. Am. J.* 56:1166–1171.

Bauer, A., A. B. Frank, and A. L. Black. 1987. Aerial parts of hard red spring wheat: II. Nitrogen and phosphorus concentration and content by plant development stage. *Agron. J.* 79:852–858.

Bauer, P. J. and M. E. Roof. 2004. Nitrogen, aldicarb, and cover crop effects on cotton yield and fiber properties. *Agron. J.* 96:369–376.

Black, A. L. and F. H. Siddoway. 1977. Hard red and durum spring wheat responses to seeding date and NP-fertilization on fallow. *Agron. J.* 69:885–888.

Black, A. S. and S. A. Waring. 1972. Ammonium fixation and availability in some cereal producing soils in Queensland. *Aust. J. Soil Res.* 10:197–207.

Blevins, R. L., J. H. Herbek, and W. W. Frye. 1990. Legume cover crops as nitrogen source for no-till corn and grain sorghum. *Agron. J.* 82:769–772.

Bloom, A. J., P. A. Meyerhoff, A. R. Taylor, and T. L. Rost. 2003. Root development and absorption of ammonium and nitrate from the rhizosphere. *J. Plant Growth Regulation* 21:416–431.

Bloom, A. J., S. S. Sukrapanna, and R. L. Warner. 1992. Root respiration associated with ammonium and nitrate absorption and assimilation by barley. *Plant Physiol.* 99:1294–1301.

Bock, B. R. 1984. Efficient use of nitrogen in cropping systems. In: *Nitrogen in crop production*, R. D. Hauck, Ed., 273–294. Madison, WI: ASA.

Bolan, N. S. and M. J. Hedley. 2003. Role of carbon, nitrogen, and sulfur cycles in soil acidification. In: *Handbook of soil acidity*, Z. Rengel, Ed., 29–56. New York: Marcel Dekker.

Bolton, E. F., V. A. Dirks, and J. W. Aylesworth. 1976. Some effects of alfalfa, fertilizer and lime in corn yield on rotations on clay soil during a range of seasonal moisture conditions. *Can. J. Soil Sci.* 56:21–25.

Brady, N. C. and R. R. Weil. 2002. *The nature and properties of soils*, 13th edition. Prentice Hall, Upper Sadder River, New Jersey.

Brevedan, R. E., D. B. Egli, and J. E. Legget. 1978. Influence of N nutrition on flower and pod abortion and yield of soybeans. *Agron. J.* 70:61–84.

Britto, D. T. and H. J. Kronzucker. 2002. NH_4^+ toxicity in higher plants: a critical review. *J. Plant Physiol.* 159:567–584.

Bruulsema, T. W. and B. R. Christie. 1987. Nitrogen contribution to succeeding corn from alfalfa and red clover. *Agron. J.* 79:96–100.

Bundy, L. G., K. A. Kelling, D. R. Keeney, and R. P. Wolkowski. 1983. Improving nitrogen efficiency on irrigated sand using a nitrification inhibitor p.165 In: Agronomy Abstracts. ASA, Madison, WI.

Bufogle, A. Jr., P. K. Bollich, J. L. Kovar, C. W. Lindau, and R. E. Macchiavellid. 1998. Comparison of ammonium sulfate and urea as nitrogen sources in rice production. *J. Plant Nutr.* 21:1601–1614.

Bullock, D. G. 1992. Crop rotation. *Crit. Rev. Plant Sci.* 11:309–326.

Byrnes, B. H. 1990. Environmental effects of N fertilizer use: An overview. *Fert. Res.* 26:209–215.

Camara, K. M., W. A. Payne, and P. E. Rasmussen. 2003. Long-term effects of tillage, nitrogen, and rainfall on winter wheat yields in the Pacific Northwest. *Agron. J.* 95:828–835.

Carter, M. R. 2002. Soil quality for sustainable land management: Organic matter and aggregation interactions that maintain soil function. *Agron. J.* 94:38–47.

Carter, M. R. and B. Stewart. 1996. *Structure and organic matter storage in agriculture soils.* Boca Raton, FL: CRC Press.

Chapman, A. L. and R. J. K. Myers. 1987. Nitrogen contributions by grain legumes to rice grown in rotation on the cumunurra soils of the ord irrigation area, Western Australia. *Aust. J. Exp. Agric.* 27:155–163.

Cherr, C. M., J. M. S. Scholberg, and R. McSorley. 2006. Green manure as nitrogen source for sweet corn in a warm-temperature environment. *Agron. J.* 98:1173–1180.

Christensen, N. W. and M. Brett. 1985. Chloride and liming effects on soil nitrogen form and take-all of wheat. *Agron. J.* 77:157–163.

Clark, R. B. 1982. Plant response to mineral element toxicity and deficiency. In: *Breeding plants for less favorable environments*, M. N. Christiansen and C. F. Lewis, Eds., 71–142. New York: John Wiley & Sons.

Clark, A. J., J. J. Meisinger, A. M. Decker, and R. F. Mulford. 2007. Effects of a grass-selective herbicide in a vetch-rye cover crop system on nitrogen management. *Agron. J.* 99:36–42.

Cline, G. R. and A. F. Silvernail. 2002. Effects of cover crops, nitrogen and tillage on sweet corn. *Horttechnology* 12:118–125.

Collins, H. P., J. A. Delgado, A. K. Alva, and R. F. Follett. 2007. Use of nitrogen-15 isotopic techniques to estimate nitrogen cycling from a mustard cover crop to potatoes. *Agron. J.* 99:27–35.

Copeland, P. J. and R. K. Crookston. 1992. Crop sequence affects nutrient composition of corn and soybean grown under high fertility. *Agron. J.* 84:503–509.

Creamer, N. G. and S. M. Dabney. 2002. Killing cover crops mechanically: Review of recent literature and assessment of new research results. *Am. J. Alter. Agric.* 17:32–40.

Crop Science Society of America. 1992. *Glossary of crop science terms.* Madison, WI: Crop Science Society of America.

De Datta, S. K. 1987. Advances in soil fertility research and nitrogen fertilizer management for lowland rice. In: *Efficiency of nitrogen for rice*, IRRI, Ed., 27–41. Los Baños, Philippines: IRRI.

De Datta, S. K. and D. S. Mikkelsen. 1985. Potassium nutrition of rice. In: *Potassium in agriculture*, R. D. Munson, Ed., 665–699, Madison, WI: American Society of Agronomy.

Deibert, E. J., M. Bijeriego, and R. A. Olson. 1979. Utilization of ^{15}N fertilizer by nodulating and non-nodulating soybean isolines. *Agron. J.* 71:717–723.

Diest, A. B. 1976. Ammonium and nitrate nutrition of crops. *Stikstof* 7(83/84):389–394.

Doran, J. W., E. T. Elliot, and K. Paustian. 1998. Soil microbial pools as related to fallow tillage management. *Soil Tillage Res.* 49:3–18.

Drury, C. F. and E. G. Beauchamp. 1991. Ammonium fixation, releases nitrification, and immobilization in high and low fixing soils. *Soil Sci. Soc. Am. J.* 55:125–129.

Drury, C. F., E. G. Beauchamp, and L. J. Evans. 1989. Fixation and immobilization of recently applied $^{15}NH_4$ in selected Ontario and Quebec soils. *Can. J. Soil Sci.* 69:391–400.

Drury, C. F., T. Q. Zhang, and D. B. Kay. 2003. The non-limiting and least limiting water ranges for soil nitrogen mineralization. *Soil Sci. Soc. Am. J.* 67:1388–1404.

Eaglesham, A. R. J., S. Hassouna, and R. Seegers. 1983. Fertilizer-N effects on N_2 fixation by cowpeas and soybeans. *Agron. J.* 75:61–66.

Ebelhar, S. A., W. W. Frye, and R. L. Blevins. 1984. Nitrogen from legume cover crops for no-tillage corn. *Agron. J.* 76:51–55.

Eckert, D. J., W. A. Dick, and J. W. Johnson. 1986. Response of no-tillage corn grown in corn and soybean residue to several nitrogen fertilizer sources. *Agron. J.* 78:231–235.

Epstein, M. and A. J. Bloom. 2005. *Mineral nutrition of plants: Principles and perspectives*, 2nd edition. Sunderland, MA: Sinauer Associates.

Evans, J. R. 1989. Photosynthesis and nitrogen relationships in leaves of C_3 plants. *Oecologia* 78:9–19.

Evans, L. T. 1993. Crop yields and world food supply. In: *Crop evolution, adaptation and yield*, L. T. Evans, Ed., 32–61. Cambridge: Cambridge University Press.

Evans, J. R. and J. R. Seemann. 1989. The allocation of protein nitrogen in the photosynthetic apparatus: Costs, consequences, and control. In: *Photosynthesis*, W. R. Briggs, Ed., 183–205. New York: Liss.

Fageria, N. K. 1989. *Tropical soils and physiological aspects of crops*. Brasilia: EMBRAPA/ CNPAF.

Fageria, N. K. 1992. *Maximizing crop yields*. New York: Marcel Dekker.

Fageria, N. K. 2000. Adequate and toxic levels of zinc for rice, common bean, corn, soybean and wheat production in cerrado soil. *Rev. Bras. Eng. Agri. Ambien.* 4:390–395.

Fageria, N. K. 2001. Nutrient management for improving upland rice productivity and sustainability. *Commun. Soil Sci. Plant Anal.* 32:2603–2629.

Fageria, N. K. 2003. Plant tissue test for determination of optimum concentration and uptake of nitrogen at different growth stages in lowland rice. *Commun. Soil Sci. Plant Anal.* 34:259–270.

Fageria, N. K. 2006. Liming and copper fertilization in dry bean production on an Oxisol in no-tillage system. *J. Plant Nutr.* 29:1219–1228.

Fageria, N. K. 2007a. Yield physiology of rice. *J. Plant Nutr.* 30:843–879.

Fageria, N. K. 2007b. Green manuring in crop production. *J. Plant Nutr.* 30:691–719.

Fageria, N. K. 2008. Yield physiology of dry bean. *J. Plant Nutr.* 31 (in press).

Fageria, N. K. and V. C. Baligar. 1996. Response of lowland rice and common bean grown in rotation to soil fertility levels on a varzea soil. *Fert. Res.* 45:13–20.

Fageria, N. K. and V. C. Baligar. 2001. Lowland rice response to nitrogen fertilization. *Commun. Soil Sci. Plant Anal.* 32:1405–1429.

Fageria, N. K. and V. C. Baligar. 2003a. Methodology for evaluation of lowland rice genotypes for nitrogen use efficiency. *J. Plant Nutr.* 26:1315–1333.

Fageria, N. K. and V. C. Baligar. 2003b. Fertility management of tropical acid soils for sustainable crop production. In: *Handbook of soil acidity*, Z. Rengel, Ed., 359–385. New York: Marcel Dekker.

Fageria, N. K. and V. C. Baligar. 2005a. Enhancing nitrogen use efficiency in crop plants. *Adv. Agron.* 88:97–185.

Fageria, N. K. and V. C. Baligar. 2005b. Nutrient availability. In: *Encyclopedia of soils in the environment*, D. Hillel, Ed., 63–71. San Diego, CA: Elsevier.

Fageria, N. K., V. C. Baligar, and Y. C. Li. 2008. The role of nutrient efficient plants in improving crop yields in the twenty first century. *J. Plant Nutr.* 31 (in press).

Fageria, N. K., V. C. Baligar, and B. A. Bailey. 2005. Role of cover crops in improving soil and row crop productivity. *Commun. Soil Sci. Plant Anal.* 36:2733–2757.

Fageria, N. K., V. C. Baligar, and R. B. Clark. 2006. *Physiology of crop production*. New York: The Haworth Press.

Fageria, N. K., V. C. Baligar, and C. A. Jones. 1997a. *Growth and mineral nutrition of field crops*, 2nd edition. New York: Marcel Dekker.

Fageria, N. K. and M. P. Barbosa Filho. 2001. Nitrogen use efficiency in lowland rice genotypes. *Commun. Soil Sci. Plant Anal.* 32:2079–2089.

Fageria, N. K., M. P. Barbosa Filho, and L. F. Stone. 2004. Phosphorus nutrition of common bean. In: *Phosphorus in Brazilian Agriculture*, T. Yamada and S. R. S. Abdalla, Eds., 435–455. Piracicaba, São Paulo, Brazil: Brazilian Potassium and Phosphorus Research Institute.

Fageria, N. K. and H. R. Gheyi. 1999. *Efficient crop production*. Campina Grande, Paraiba, Brazil: University of Paraiba.

Fageria, N. K. and A. S. Prabhu. 2003. Response of lowland rice to nitrogen application and seed treatment with fungicide doses to blast control. *Pesq. Agropec. Bras.* 39:123–129.

Fageria, N. K., N. A. Slaton, and V. C. Baligar. 2003. Nutrient management for improving lowland rice productivity and sustainability. *Adv. Agron.* 80:63–152.

Fageria, N. K., A. B. Santos, and V. C. Baligar. 1997b. Phosphorus soil test calibration for lowland rice on an Inceptisol. *Agron. J.* 89:737–742.

Fageria, N. K., A. B. Santos, and V. A. Cutrim. 2007. Yield and nitrogen use efficiency of lowland rice genotypes as influenced by nitrogen fertilization. *Pesq. Agropec. Bras.* 42:1029–1034.

Fageria, N. K. and L. F. Stone. 2004. Yield of common bean in no-tillage system with application of lime and zinc. *Pesq. Agropec. Bras.* 39:73–78.

Fageria, N. K. and F. J. P. Zimmermann. 1998. Influence of pH on growth and nutrient uptake by crop species in an Oxisol. *Commun. Soil Sci. Plant Anal.* 29:2675–2682.

Farquhar, G. D., P. M. Firth, R. Wetselaar, and B. Weir. 1980. On the gaseous exchange of ammonia between leaves and the environment: Determination of the ammonia compensation point. *Plant Physiol.* 66:710–714.

Fenn, L. B. 1987. Ammonium fertilizer impact on NH_3 loss, the soil solution Ca and Mg, NH_4 and NO_3 content of a calcareous soil. *Biol. Fertil. Soils* 5:171–174.

Fenn, L. B. and L. R. Hossner. 1985. Ammonia volatilization from ammonium forming nitrogen fertilizers. *Adv. Soil Sci.* 1:123–169.

Fenn, L. B. and D. E. Kissel. 1973. Ammonia volatilization from surface applications of ammonium compounds on calcareous soils: I. General theory. *Soil Sci. Soc. Am. Proc.* 37:855–859.

Fenn, L. B., R. M. Taylor, and J. E. Matocha. 1981. Ammonia losses from surface applied nitrogen fertilizer as controlled by soluble calcium and magnesium: General theory. *Soil Sci. Soc. Am. J.* 45:777–781.

Fenn, L. B., R. M. Taylor, M. L. Binzel, and C. M. Burks. 1991. Calcium stimulation of ammonium absorption in onion. *Agron. J.* 83:840–843.

Fillery, I. R. P. and P. L. G. Vlex. 1986. Reappraisal of the significance of ammonia volatilization as an N loss mechanism in flooded rice fields. *Fert. Res.* 9:79–98.

Fixen, P. 1996. Nutrient management following conservation reserve program. *Better Crops* 80:16–19.

Foth, H. D. and B. G. Ellis. 1988. *Soil fertility*. New York: John Wiley & Sons.

Fox, R. H. and L. D. Hoffman. 1981. The effect of N fertilizer source on grain yield, N uptake, soil pH and time requirements in no-till corn. *Agron. J.* 73:891–895.

Fox, R. H., J. M. Kern, and W. P. Piekielek. 1986. Nitrogen fertilizer source, and method and time of application effects on no-till corn yields and nitrogen uptake. *Agron. J.* 78:741–746.

Fox, R. H. and W. P. Piekielek. 1988. Fertilizer N equivalence of alfalfa, birdsfoot trefoil, and red clover for succeeding corn crops. *J. Prod. Agric.* 1:313–317.

Foy, C. D. 1984. Physiological effects of hydrogen, aluminum, and manganese toxicities in acid soil. In: *Soil acidity and liming*, 2nd edition, F. Adams, Ed., 57–97. Madison, WI: ASA, CSSSA and SSSA.

Francis, D. D., J. S. Schepers, and A. L. Sims. 1997. Ammonia exchange from corn foliage during reproductive growth. *Agron. J.* 89:941–946.

Freney, J. R., O. T. Denmead, and J. R. Simpson. 1978. Soil as a source or sink for atmospheric nitrous oxide. *Nature* 273:530–532.

Frye, W. W., R. L. Blevins, M. S. Smith, S. J. Corak, and J. J. Varco. 1988. Role of annual legume cover crops in efficient use of water and nitrogen. In: *Cropping strategies for efficient use of water and nitrogen*, W. L. Hargrove, Ed., 129–154. Madison, WI: American Society of Agronomy.

Fujita, T. 1989. *Invention and development of polyolefin-coated urea*. Ph.D. Thesis. Tohoku Univ., Sendai, Japan.

Galloway, J. N. and E. B. Cowling. 2002. Reactive nitrogen and the world: 200 years of change. *Ambio* 31:64–71.

Galloway, J. N., E. B. Cowling, S. P. Seitzinger, and R. H. Socolow. 2002. Reactive nitrogen: Too much of a good thing? *Ambio* 31:60–63.

Gameh, M. A., J. S. Angle, and J. H. Axley. 1990. Effect of urea potassium chloride and nitrogen transformations on ammonia volatilization from urea. *Soil Sci. Soc. Am. J.* 54:1768–1772.

Gandeza, A. T., S. Shoje, and I. Yamada. 1991. Simulation of crop response to polyolefin-coated urea: I. Field dissolution. *Soil Sci. Soc. Am. J.* 55:1462–1467.

Gashaw, L. and L. M. Mugwira. 1981. Ammonium-N and nitrate-N effects on the growth and mineral composition of triticale, wheat, and rye. *Agron. J.* 73:47–52.

Gaudin, R. and J. Dupuy. 1999. Ammonical nutrition of transplanted rice fertilizer with large urea granules. *Agron. J.* 91:33–36.

Gilmour, J. T. 1984. The effects of soil properties on nitrification and nitrification inhibitor. *Soil Sci. Soc. Am. J.* 48:1262–1266.

Glass, A. D. M., D. T. Britto, B. N. Kaiser, J. R. Kinghorn, H. J. Kronzucker, A. Kumar, M. Okamoto, S. Rawat, M. Y. Siddiqi, E. U. Unkles, and J. J. Vidmar. 2002. The regulations of nitrate and ammonium transport systems in plants. *J. Exp. Botany* 53:855–864.

Goyal, S., M. M. Mishra, I. S. Hooda, and R. Singh. Organic matter-microbial biomass relationships in field experiments under tropical conditions: Effects of inorganic fertilization and organic amendments. *Soil Biol. Biochem.* 24:1081–1084.

Goyal, S., K. Chandler, M. C. Mundra, and K. K. Kapoor. 1999. Influence of inorganic fertilizers and organic amendments on soil organic matter and soil microbial properties under tropical conditions. *Biol. Fertil. Soils* 29:196–200.

Graham, R. D. 1984. Breeding for nutritional characteristics in cereals. In: *Advances in plant nutrition*, Vol. 1, P. B. Tinker and A. Lauchi, Eds., 57–102. New York: Praeger.

Grundmann, G. L., P. Renault, L. Rosso, and R. Bardin. 1995. Differential effects of soil water content and temperature on nitrification and aeration. *Soil Sci. Soc. Am. J.* 59:1342–1349.

Hargrove, W. L. 1986. Winter legumes as a nitrogen source for no-till grain sorghum. *Agron. J.* 78:70–74.

Harris, G. H. and O. B. Hesterman. 1990. Quantifying the nitrogen contribution from alfalfa to soil and two succeeding crops using nitrogen-15. *Agron. J.* 82:129–134.

Harper, J. E. and A. H. Gibson. 1984. Differential nodulation tolerance to nitrate among legume species. *Crop Sci.* 24:797–801.

Harper, L. A. and R. R. Sharpe. 1995. Nitrogen dynamics in irrigated corn: Soil-plant nitrogen and atmospheric ammonia transport. *Agron. J.* 87:669–675.

Hauck, R. D. 1985. Slow-release and bioinhibitor-amended fertilizers. In *Fertilizer technology and use*, 3rd edition, O. P. Inglested, Ed., 293–319. Madison, WI: Soil Sci. Soc. Am.

Hauck, R. D. 1990. Agronomic and public aspects of soil nitrogen research. *Soil Use Management* 6:66–70.

Hauck, R. D. and K. K. Tanji. 1982. Nitrogen transfers and mass balances. In: *Nitrogen in agricultural soils*, F. J. Stevenson, Ed., 891–925. Madison, WI: American Society of Agronomy.

He, X. T., R. L. Mulvaney, F. J. Stevenson, and R. M. Vanden Heuvel. 1990. Characterization of chemically fixed liquid anhydrous ammonia in an Illinois Drummer soil. *Soil Sci. Soc. Am. J.* 54:775–780.

Heichel, G. H. 1987. Legume nitrogen: symbiotic fixation and recovery by subsequent crops. In: *Energy in plant nutrition and pest control*, Z. R. Helsel, Ed., 63–80. Amsterdam: Elsevier Scientific.

Hendrickson, L. L., L. M. Walsh, and D. R. Keeney. 1978. Effectiveness of nitrapyrin in controlling nitrification of fall and spring applied anhydrous ammonia. *Agron. J.* 70:704–708.

Hesterman, O. B. 1988. Exploiting forage legumes for nitrogen contribution strategies for efficient use of water and nitrogen. ASA special publ. 51. Am. Soc. Agron, Madison, WI.

Higgs, R. L., W. H. Paulsen, J. W. Pendleton, A. F. Peterson, J. A. Jackobs, and W. D. Shrader. 1976. Crop rotations and nitrogen: Crop sequence comparisons on soils of the driftless area of southwestern Wisconsin, Univ. of Wisconsin, *College of Agric. and Life Sci. Res. Bull.* R.2761.

Hillel, D. and C. Rosenzweig. 2002. Desertification in relation to climate variability and change. *Adv. Agron.* 77:1–38.

Hinman, W. C. 1966. Ammonium fixation in relation to exchangeable K and organic matter content of two Saskatchewan soils. *Can. J. Soil Sci.* 46:223–225.

Hinsinger, P. 1998. How do plant roots acquire mineral nutrients? Chemical processes involved in the rhizosphere. *Adv. Agron.* 64:225–265.

Hoffland, E., G. R. Findenegg, and J. A. Nelemans. 1989. Solubilization of rock phosphate by rape. I. Evaluation of the role of nutrient uptake pattern. *Plant Soil* 113:155–160.

Hopkins, W. G. 1999. *Introduction to plant physiology*, 2nd edition. New York: John Wiley.

Hoyt, P. B. and R. H. Leitch. 1983. Effects of forage legume species on soil moisture, nitrogen and yield of succeeding barley crops. *Can. J. Soil Sci.* 63:125–136.

Huber, D. M. and I. A. Thompson. 2007. Nitrogen and plant disease. In: *Mineral nutrition and plant disease*, L. E. Datnoff, W. H. Elmer, and D. M. Huber, Eds., 31–44. St. Paul, MN: The American Phytopathological Society.

Huffman, J. R. 1989. Effects of enhanced ammonium nitrogen availability for corn. *J. Agron. Education* 18:325–339.

Israel, D. W. 1981. Cultivar and Rhizobium strain effects on nitrogen fixation and remobilization by soybeans. *Agron. J.* 73:509–516.

Jacinthe, P. A. and J. R. Pichtel. 1992. Interaction of nitrapyrin and dicyandiamide with soil humic compounds. *Soil Sci. Soc. Am. J.* 56:465–470.

Jacobs, L. W. 1998. Nutrient management planning for co-utilization of organic by-products. In: *Beneficial co-utilization of agricultural, municipal and industrial by-products*, S. Brown, J. S. Angle, and L. Jacobs, Eds., 283–287. Dordrecht: Kluwer Academic.

Jansson, S. L. 1958. Tracer studies on nitrogen transformation in soil with special attention to mineralization-immobilization relationships. *Kungliga Lantbruks-Hoegskolan Annaler* 24:101–361.

Jansson, S. L., M. J. Hallam, and W. N. Bartholomeu. 1955. Preferential utilization of ammonium over nitrate by microorganisms in the decomposition of oat straw. *Plant Soil* 6:382–390.

Jansson, S. L. and J. Persson. 1982. Mineralization and immobilization of soil nitrogen. In: *Nitrogen in agricultural soils*, F. J. Stevenson, Ed., 229–252. Madison, WI: ASA, CSSA, and SSSA.

Janzen, H. H., C. W. Lindwall, and C. J. Roppel. 1990. Relative efficiency of point-injection and broadcast methods for N fertilization of winter wheat. *Can. J. Soil Sci.* 70:189–201.

Janzen, H. H., T. L. Roberts, and C. W. Lindwall. 1991. Uptake of point-injected nitrogen by winter wheat as influenced by time of application. *Soil Sci. Soc. Am. J.* 55:259–264.

Jarvis, S. C. and D. J. Hatch. 1985. Rates of hydrogen ion efflux by nodulated legumes grown in flowing solution culture with continuous pH monitoring and adjustment. *Anna. Bot.* 55:41–51.

Jayaweera, G. R. and D. S. Mikkelsen. 1990. Ammonia volatilization from flooded soil systems: A computer model I. Theoretical aspects. *Soil Sci. Soc. Am. J.* 54:1447–1455.

Jenkinson, D. S. and L. C. Parry. 1989. The nitrogen cycle in the Broadbalk wheat experiment: A model for the turnover of nitrogen through microbial biomass. *Soil Biol. Biochem.* 21:535–541.

Johnson, J. M. F., R. R. Allmaras, and D. C. Reicosky. 2006. Estimating source carbon from crop residues, roots and rhizodeposits using the national grain-yield database. *Agron. J.* 98:622–636.

Jokela, W. E. 1992. Nitrogen fertilizer and dairy manures effects on corn yield and soil nitrate. *Soil Sci. Soc. Am. J.* 56:148–154.

Jokela, W. E. and G. W. Randall. 1989. Corn yield and residual soil nitrate as affected by time and rate of nitrogen application. *Agron. J.* 81:720–726.

Juma, N. G. and E. A. Paul. 1983. Effect of a nitrification inhibitor on N immobilization and release of ^{15}N from nonexchangeable ammonium and microbial biomass. *Can. J. Soil Sci.* 63:167–175.

Jung, P. E., Jr., L. A. Peterson, and L. E. Schraeder. 1972. Response of irrigated corn to time, rate, and source of applied N on sandy soils. *Agron. J.* 64:668–670.

Karlen, D. L., G. E. Varvel, D. G. Bullock, and R. M. Cruse. 1994. Crop rotations for the 21st century. *Adv. Agron.* 53:1–45.

Keeney, D. R. 1980. Factors affecting the persistence and bioactivity of nitrification inhibitors. In: *Nitrification inhibitors-potentials and limitations,* J. J. Meinsinger, Ed., 33–46. Madison, WI: Am. Soc. Agron.

Keerthisinghe, G., K. Mengel, and S. K. De Datta. 1984. The release of nonexchangeable ammonium (^{15}N labeled) in Wetland rice soils. *Soil Sci. Soc. Am. J.* 48:291–294.

Keller, O. D. and D. B. Mengel. 1986. Ammonia volatilization from nitrogen fertilizers applied to no-till corn. *Soil Sci. Soc. Am. J.* 50:1060–1063.

Klausner, S. D. and R. W. Guest. 1981. Influence of NH_3 conservation from dairy manure on the yield of corn. *Agron. J.* 73:720–723.

Kumar, K. and K. M. Goh. 2000. Crop residues and management practices: Effects on soil quality, soil nitrogen dynamics, crop yield, and nitrogen recovery. *Adv. Agron.* 68:197–319.

Ladd, J. N. and M. Amato. 1986. The fate of nitrogen from legume and fertilizer sources in soils successively cropped with wheat under field conditions. *Soil Biol. Biochem.* 18:417–425.

Ladd, J. N., M. Amato, R. B. Jackson, and J. H. A. Butler. 1983. Utilization by wheat crops of nitrogen from legume residues decomposing in the field. *Soil Biol. Biochem.* 15:231–238.

Lado, M. and M. Bem-Hui. 2004. Organic matter and aggregate size interactions in infiltration, seal formation and soil loss. *Soil Sci. Soc. Am. J.* 68:935–942.

Lathwell, D. J., D. R. Boudin, and W. S. Reid. 1970. Effects of nitrogen fertilizer application in agriculture. In: *Relationship of agriculture to soil and water pollution*, 192–206. Proc. Cornell Univ. Conf. on Agric. Waste Management, Syracuse, NY, January 19–21, 1970, Cornell University, Ithaca.

Lawn, R. J. and W. A. Brun. 1974. Symbiotic nitrogen fixation in soybeans. I. Effect of photosynthetic source-sink manipulations. *Crop Sci.* 14:11–16.

Lin, C. H., R. N. Lerch, H. E. Garrett, D. Jordan, and M. F. George. 2007. Ability of forage grasses exposed to atrazine and isoxaflutole to reduce nutrient levels in soils and shallow groundwater. *Commun. Soil Sci. Plant Anal.* 38:1119–1136.

López-Bellido, R. J., L. López-Bellido, F. J. López-Bellido, and J. E. Castillo. 2003. Faba bean (*Vicia faba* L.) response to tillage and soil residual nitrogen in a continuous rotation with wheat (*Triticum aestivum* L.) under rainfed Mediterranean conditions. *Agron. J.* 95:1253–1261.

Lowrance, R., G. Vellidis, and R. K. Hubbard. 1995. Denitrification in a restored riparion forest wetland. *J. Environ. Qual.* 24:808–815.

Macnish, G. C. and J. Speijers. 1982. The use of ammonium fertilizers to reduce the severity of take-all (*Gaeumannomyces graminis var. tritici*) on wheat in Western Australia. *Ann. Appl. Biol.* 100:83–90.

MacRae, R. J. and G. R. Mehuys. 1985. The effect of green manuring on the physical properties of temperate-area soils. *Adv. Soil Sci.* 3:71–94.

Macy, P. 1936. The quantitative mineral nutrient requirements of plants. *Plant Physiol.* 11:749–764.

Maddux, L. D., C. W. Raczkowski, D. E. Kissel, and P. L. Barnes. 1991. Broadcast and sub-surface-banded urea nitrogen in urea ammonium nitrate applied to corn. *Soil Sci. Soc. Am. J.* 55:264–267.

Makinde, E. A. and A. A. Agboola. 2002. Soil nutrient changes with fertilizer type in cassava based cropping system. *J. Plant Nutr.* 25:2303–2313.

Malzer, G. L. and T. Graff. 1983. Influence of nitrogen rate, timing of nitrogen application and use of nitrification inhibitors for irrigated corn production. In: *A report on field res. in soil*, 14–19. Minnesota Agric. Exp. Stn. Misc. Publ. 2.

Marschner, H. 1995. *Mineral nutrition of higher plants*, 2nd edition. New York: Academic Press.

Marschner, H. and V. Romheld. 1983. In vivo measurement of root-induced pH changes at soil-root interface. Effect of plant species and nitrogen source. *Z. Pflanzenphysiol.* 111:241–251.

Matheny, T. A. and P. G. Hunt. 1983. Effects of irrigation on accumulation of soil and symbiotically fixed N by soybean grown as Norfolk loamy sand. *Agron. J.* 75:719–722.

McCarty, G. W. and J. M. Bremner. 1990. Evaluation of 2-ethynylpyridine as a soil nitrification inhibitor. *Soil Sci. Soc. Am. J.* 54:1017–1021.

McCormick, R. A., D. W. Nelson, A. L. Sutton, and D. M. Huber. 1983. Effect of nitrapyrin on nitrogen transformation in soil treated with liquid swine manure. *Agron. J.* 75:947–950.

McCormick, R. A., D. W. Nelson, A. L. Sutton, and D. M. Huber. 1984. Increased N efficiency from nitrapyrin added to liquid swine manure used as a fertilizer for corn. *Agron. J.* 76:1010–1014.

McIntyre, G. I. 1997. The role of nitrate in the osmotic and nutritional control of plant development. *Aust. J. Plant Physiol.* 24:103–118.

Meisinger, J. J., G. W. Randall, and M. L. Vitosh. 1980. *Nitrification inhibitors: Potentials and limitations.* Am. Soc. Agron. Sp. Publ. 38 ASA, Madison, WI.

Mengel, K., A. Kirkby, H. Kosegarten, and T. Appel. 2001. *Principles of plant nutrition*, 5th edition. Dordrecht: Kluwer Academic.

Mengel, D. B., D. W. Nelson, and D. M. Huber. 1982. Placement of nitrogen fertilizers for no-tilled conventional corn. *Agron. J.* 74:515–518.

Meyer, R. J. K. 1987. Influence of green manured, hayed, or grain legumes on grain yield and availability of the following barley crop in the northern Great Plains. In: *The role of legumes in conservation tillage systems*, J. F. Power, Ed., 94–95. Proc. of Natl. Conf. Athens, GA, April 27–29, 1987. Ankeny, IA: Soil Conserv. Soc. Am.

Mikkelsen, D. S. 1987. Nitrogen budgets in flooded soils used for rice production. *Plant Soil* 100:71–97.

Miller, S. D. and A. G. Dexter. 1982. No-till crop production in the Red River Valley. ND *Farm Res.* 40:3–5.

Montavalli, P. P., K. A. Kelling, and J. C. Conserve. 1989. First year nutrient availability from injected dairy manure. *J. Environ. Qual.* 18:180–185.

Morris, D. R. and R. W. Weaver. 1987. Competition for nitrogen 15 depleted ammonium nitrate and nitrogen fixation in arrowleaf clover-gulf ryegrass mixtures. *Soil Sci. Soc. Am. J.* 51:115–119.

Muchow, R. C. 1988. Effect of nitrogen supply on the comparative productivity of maize and sorghum in semi-arid tropical environment: I. Leaf growth and leaf nitrogen. *Field Crops Res.* 18:1–16.

Munns, D. N. 1978. Legume-rhizobium relations. In: *Mineral nutrition of legumes in tropical and subtropical soils*, C. S. Andrew and E. J. Kamprath, Eds., 247–263. East Melbourne: CSIRO.

National Research Council. 1989. *Alternative agriculture.* Washington DC: National Academic Science.

Neetison, J. J., D. J. Greenwood, and E. J. M. H. Habets. 1986. Dependence of soil mineral N on N-fertilizer application. *Plant Soil* 91:417–420.

Nelson, D. W. 1982. Gaseous losses of nitrogen other than through denitrification. In: *Nitrogen in agricultural soils*, F. J. Stevenson, Ed., 327–363. Madison, WI: Am. Soc. Agron. Monograph 22.

Nelson, W. W. and J. M. MacGregor. 1973. Twelve years of continuous corn fertilization with ammonium nitrate or urea nitrogen. *Soil Sci. Soc. Proc.* 37:583–586.

Newbould, P. 1989. The use of nitrogen fertilizer in agriculture. Where do we go practically and ecologically? *Plant Soil* 115:297–311.

Nittler, L. W. and J. J. Kenny. 1976. Effect of ammonium to nitrate ratio on growth and anthocyanin development of perennial ryegrass cultivars. *Agron. J.* 68:680–682.

Nommik, H. 1957. Fixation and defixation of ammonium in soils. *Acta Agriculture Scandinavian* 7:395–439.

Nommik, H. and K. Vahtras. 1982. Retention and fixation of ammonium and ammonia in soils. In: *Nitrogen in agricultural soils*, F. J. Stevenson, Ed., 123–171. Madison, WI: ASA, CSSA, and SSSA.

Ohno, Y. and Marur, C. J. 1977. Physiological analysis of factors limiting growth and yield of upland rice. Annual Report of Ecophysiological Study of Rice. IAPAR, Londrina, Brazil.

Olson, R. A., W. R. Raun, Y. S. Chun, and J. Skopp. 1986. Nitrogen management and interseeding effects on irrigated corn and sorghum on soil strength. *Agron. J.* 78:856–862.

Parkin, T. B. and J. J. Meisinger. 1989. Denitrification below the crop rooting zone as influenced by surface tillage. *J. Environ. Qual.* 18:12–16.

Peterson, G. A. and W. W. Frye. 1989. Fertilizer nitrogen management. In: *Nitrogen management and ground water protection*, R. F. Follett, Ed., 183–219. New York: Elsevier.

Phillips, D. A. and T. M. Dejong. 1984. Dinitrogen fixation in leguminous crop plants. In: *Nitrogen in crop production*, R. D. Hauck, Ed., 121–132. Madison, WI: ASA.

Phongpan, S., S. Vacharotayan, and K. Kumazawa. 1988. Efficiency of urea and ammonium sulfate for wetland rice grown on an acid sulfate soil as affected by rate and time of application. *Fert. Res.* 15:237–246.

Piggott, T. J. 1986. Vegetable crops. In: *Plant analysis: An interpretation manual,* D. J. Reuter and J. B. Robinson, Eds., 148–187. Melbourne: Inkata Press.

Poorter, H. and J. R. Evans. 1998. Photosynthetic nitrogen-use efficiency of species that differ inherently in specific leaf area. *Oecologia* 116:26–37.

Power, J. F. 1990. Fertility management and nutrient cycling. *Adv. Soil Sci.* 13:131–149.

Raab, T. K. and N. Terry. 1994. Nitrogen source regulation of growth and photosynthesis in *Beta vulgaris* L. *Plant Physiol.* 105:1159–1166.

Rao, S. C. and T. H. Dao. 1996. Nitrogen placement and tillage effects on dry matter and nitrogen accumulation and redistribution in winter wheat. *Agron. J.* 88:365–371.

Raun, W. and G. V. Johnson. 1999. Improving nitrogen use efficiency for cereal production. *Agron. J.* 91:357–363.

Reeves, D. W. 1994. Cover crops and rotations. In: *Crop residue management*, J. T. Hatifield and B. A. Stewart, Eds., 125–172. Boca Raton, FL: Lewis Publishers.

Rehm, G. W. and R. A. Wiese. 1975. Effect of method of N application on corn grown on irrigated sandy soils. *Soil Sci. Soc. Am. Proc.* 39:1217–1220.

Reddy, K. R. and W. H. Patrick, Jr. 1978. Utilization of labeled urea and ammonium sulfate by lowland rice. *Agron. J.* 70:465–467.

Reicosky, D. C. and F. Forcella. 1998. Cover crop and soil quality interaction in agroecosystems. *J. Soil Sci. Water Conserv.* 53:224–229.

Reuter, D. J. 1986. Temperate and sub-tropical crops. In: *Plant Analysis: An interpretation manual*, D. J. Reuter and J. B. Robinson, Eds., 38–99. Melbourne: Inkata Press.

Ridley, A. O. and R. A. Hedlin. 1980. Crop yields and soil management on the Canadian Prairies: Past and present. *Can. J. Soil Sci.* 60:393–402.

Riga, P. and M. Anza 2003. Effect of magnesium deficiency on pepper growth parameters: Implications for determination of magnesium-critical value. *J. Plant Nutr.* 26:1581–1593.

Riley, D. and S. A. Barber. 1971. Effect of ammonium and nitrate fertilization on phosphorus uptake as related to root induced pH changes at the root-soil interface. *Soil Sci. Soc. Am. Proc.* 35:301–306.

Robson, A.D. and J. B. Pitman. 1983. Interactions between nutrients in higher plants. In: *Inorganic plant nutrition: Encyclopedia of plant physiology*, Vol. 1A, A. Lauchli and R. L. Bieleski, Eds., 147–180. New York: Springer-Verlag.

Roder, W., S. C. Mason, M. D. Clegg, and K. R. Kniep. 1989. Yield-soil-water relationships in sorghum-soybean cropping systems with different fertilizer regimes. *Agron. J.* 81:470–475.

Runge, M. 1983. Physiology and ecology of nitrogen nutrition. In: *Encyclopedia of plant physiology*, New Series, Vol. 12C, O. L. Lange, P. S. Nobel, C. B. Osmond, and H. Ziegler, Eds., 163–200. Berlin: Springer-Verlag.

Russelle, M. P., R. D. Hauck, and R. A. Olson. 1983. Nitrogen accumulation rates for irrigated maize. *Agron. J.* 75:593–598.

Salsac, L., S. Chaillou, J. F. Morot-Gaudry, C. Lesaint, and E. Jolivoe. 1987. Nitrate and ammonium nutrition in plants. *Plant Physiol. Biochem.* 25:805–812.

Scharf, P. C. and M. M. Alley. 1988. Nitrogen loss pathways and nitrogen loss inhibitors: A review. *J. Fert. Issues* 5:109–125.

Schjoerring, J. K., S. Husted, G. Mack, and M. Mattson. 2002. The regulation of ammonium translocation in plants. *J. Exp. Bot.* 53:883–890.

Schmidt, E. L. 1982. Nitrification in soil. In: *Nitrogen in agricultural soils*, F. J. Stevenson, Ed., 253–288. Madison, WI: ASA.

Schmied, B., K. Abbaspour, and R. Schulin. 2000. Inverse estimation of parameters in a nitrogen model using field data. *Soil Sci. Soc. Am. J.* 64:533–542.

Scherer, H. W. and K. Mengel. 1986. Importance of soil type on the release of non exchangeable NH_4^+ and availability of fertilizer NH_4^+ and fertilizer NO_3^-. *Fert. Res.* 8:249–258.

Schug, E. 1985. Mikronahrstoff-mangel-ein stussfaktor im ertragreichen pflanzenbau. *Kali-Briefe* 17:419–430.

Sehtiya, H. L. and S. S. Goyal. 2000. Comparative uptake of nitrate by intact seedlings of C_3 (barley) and C_4 (corn) plants: Effect of light and exogenously supplied sucrose. *Plant Soil* 227:185–190.

Seiler, W. 1986. Other greenhouse gases and aerosol; nitrous oxide. In: *The greenhouse effect, climatic change and ecosystem,* B. Bolin, B. R. Doos, J. Jager, and R. A. Warrick, Eds., 170–174. New York: Wiley and Sons.

Serraj, R. and T. R. Sinclair. 1996. Process contributing to N fixation insensitivity to drought in the soybean cultivar Jackson. *Crop Sci.* 36:961–968.

Sharma, A. R. and B. N. Mittra. 1988. Effect of green manuring and mineral fertilizer on growth and yield of crops in rice-based cropping system on acid lateritic soil. *J. Agric. Sci.* 110:605–608.

Sinclair, T. R. and T. Horie. 1989. Leaf nitrogen, photosynthesis, and crop radiation use efficiency: A review. *Crop Sci.* 29:90–98.

Singer, J. W. and W. J. Cox. 1998. Agronomic of corn production under different crop rotations. *J. Prod. Agric.* 11:462–468.

Six, J., E. T. Elliott, and K. Paustian. 1999. Aggregate and soil organic matter dynamics under conventional and no-tillage systems. *Soil Sci. Soc. Am. J.* 63:1350–1358.

Sharpe, R. R. and L. A. Harper. 1997. Apparent atmospheric nitrogen loss from hydroponically grown corn. *Agron. J.* 89:605–609.

Shoji, S., A. T. Gandeza, and K. Kimura. 1991. Simulation of crop response to polyolefincoated urea: II. Nitrogen uptake by corn. *Soil Sci. Soc. Am. J.* 55:1468–1473.

Small, H. G. and A. J. Ohlrogge. 1973. Plant analysis as an aid in fertilizing soybeans and peanuts. In: *Soil testing and plant analysis,* L. M. Walsh and J. D. Beaton, Eds., 315–327. Madison, WI: Soil Science Society of America.

Smiley, R. W. 1974. Take-all of wheat as influenced by organic amendments and nitrogen fertilizers. *Phytopathology* 64:822–825.

Smith, F. W. 1986. Interpretation of plant analysis: Concepts and principles. In: *Plant analysis: An interpretation manual,* D. J. Reuter and J. B. Robinson, Eds., 1–12. Melbourne: Inkata Press.

Smith, M. S., W. W. Frye, and J. J. Varco. 1987. Legume winter cover crops. *Advances in Soil Sci.* 7:95–139.

Soil Science Society of America. 1997. *Glossary of soil science terms.* Soil Science Society of America, Madison, WI.

Stanhill, G. 1986. Irrigation in arid lands. *Phil. Trans. R. Soc. London.* A316:261–273.

Stehouwer, R. C. and J. W. Johnson. 1991. Soil adsorption interactions of band-injected anhydrous ammonia and potassium chloride fertilizers. *Soil Sci. Soc. Am. J.* 55:1374–1381.

Stevenson, F. J. 1982. Origin and distribution of nitrogen in soil. In: *Nitrogen in agricultural soils,* F. J. Stevenson, Ed., 1–42. Madison, WI: ASA, CSSA, and SSSA.

Stevenson, C. K. and C. S. Baldwin. 1969. Effect of time and method of nitrogen application and source of nitrogen on the yield and nitrogen content of corn. *Agron. J.* 61:381–384.

Stoa, T. E. and J. C. Zubriski. 1969. Crop rotation, crop management and soil fertility studies on Fargo Clay. North Dakota Res. Rep.20.

Sweeney, D. W. and J. L. Moyer. 2004. In-season nitrogen uptake by grain sorghum following legume green manures in conservation tillage systems. *Agron. J.* 96:510–515.

Tanji, K. K. 1982. Modeling of the soil nitrogen cycle. In: *Nitrogen in agricultural soils,* F. J. Stevenson, Ed., 721–772. Am. Soc Agron. Monograph 22, ASA, Madison, WI.

Taylor, R. G., T. L. Jackson, R. L. Powelson, and N. W. Christensen. 1983. Chloride, nitrogen form, lime, and planting date effects on take-all root rot of winter wheat. *Plant Dis.* 67:1116–1120.

Tennant, D. 1976. Root growth of wheat. I. Early patterns of multiplication and extension of wheat roots including effects of levels of nitrogen, phosphorus and potassium. *Aust. J. Agric. Res.* 27:183–196.

Terry, R. E., D. W. Nelson, and L. E. Sommers. 1981. Nitrogen transformations in sewage sludge-amended soils as affected by soil environmental factors. *Soil Sci. Soc. Am. J.* 45:506–512.

Tisdale, J. M., W. L. Nelson, and J. D. Beaton. 1986. *Soil fertility and fertilizers*, 4th edition. New York: Macmillan.

Touchton, J. T., R. G. Hoeft, and L. F. Welch. 1978. Nitrapyrin degradation and movement in soil. *Agron. J.* 70:811–816.

Touchton, J. T., R. G. Hoeft, and L. F. Welch. 1979a. Effect of nitrapyrin on nitrification of broadcast-applied urea, plant nutrient compositions and corn yield. *Agron. J.* 71:787–791.

Touchton, J. T., R. G. Hoeft, L. F. Welch, and W. L. Argyilan. 1979b. Loss of nitrapyrin from soils as affected by pH and temperature. *Agron. J.* 71:865–869.

Touchton, J. T., W. A. Gardner, W. L. Hargrove, and R. R. Duncan. 1982. Reseeding crimson clover as a N source for no-tillage grain sorghum production. *Agron. J.* 74:283–287.

Touchton, J. T. and W. L. Hargrove. 1982. Nitrogen sources and methods of application for no-tillage corn production. *Agron. J.* 74:823–826.

Ulrich, A. 1952. Physiological basis for assessing the nutritional requirements of plants. *Annu. Rev. Plant Physiol.* 3:207–228.

Ulrich, A. 1976. Plant analysis as a guide in fertilizing crops. In: *Soil and plant tissues testing in California*, Division of Agriculture Science, Ed., 1–4. Davis: University of California Bull., 1879.

Urban, W. J., W. L. Hargrove, B. R. Bock, and R. A. Raunikar. 1987. Evaluation of urea-urea phosphate as a nitrogen source for no-tillage production. *Soil Sci. Soc. Am. J.* 51:242–246.

Van Beusichem, M. L., E. A. Kirkby, and R. Baas. 1988. Influence of nitrate and ammonium nutrition and the uptake, assimilation, and distribution of nutrients in *Ricinus communis*. *Plant Physiol.* 86:914–921.

Varvel, G. E. and T. A. Peterson. 1990. Nitrogen fertilizer recovery by corn in monoculture and rotation systems. *Agron. J.* 82:935–938.

Varvel, G. E. and W. W. Wilhelm. 2003. Soybean nitrogen contribution to corn and sorghum in western corn belt rotations. *Agron. J.* 95:1220–1225.

Vyn, T. J., K. L. Janovicek, M. H. Miller, and E. G. Beauchamp. 1999. Soil nitrate accumulation and corn response to preceding small-grain fertilization and cover crops. *Agron. J.* 91:17–24.

Warncke, D. D. and S. A. Barber. 1973. Ammonium and nitrate uptake by corn as influenced by nitrogen concentration and NH_4^+/No_3^- ratio. *Agron. J.* 65:950–953.

Welch, L. F. 1976. The Murrow plots—Hundred years of research. *Ann. Agron.* 27:881–890.

Welch, L. F., D. L. Mulvaney, M. G. Oldham, L. V. Boone, and J. W. Pendleton. 1971. Corn yields with fall, spring and sidedress nitrogen. *Agron. J.* 63:119–123.

Wilhelm, W. W., G. S. McMaster, and D. M. Harrell. 2002. Nitrogen and dry matter distribution by culm and leaf position at two stages of vegetative growth in winter wheat. *Agron. J.* 94:1078–1086.

Wilkinson, S. R., D. L. Grunes, and M. E. Sumner. 2000. Nutrient interactions in soil and plant nutrition. In: *Handbook of soil science*, M. E. Sumner, Ed., 89–112. Boca Raton, FL: CRC Press.

Whitehead, D. C. 1970. *The role of nitrogen in grassland productivity*. Comm. on Wealth Agric. Bureaux Bull.98. Hurley, England.

Wilkinson, S. R. and R. W. Lowry. 1973. Cycling of mineral nutrients in pasture ecosystems. In: *Chemistry and biochemistry of herbage*, Vol. 2, G. W. Butler and R. W. Bailey, Eds., 247–315. New York: Academic Press.

Woolfolk, C. W., W. R. Raun, G. V. Johnson, W. E. Thomason, R. W. Mullen, K. J. Wynn, and K. W. Freeman. 2002. Influence of late season foliar nitrogen application on yield and grain nitrogen in winter wheat. *Agron. J.* 94:429–434.

Wuest, S. B. and K. G. Cassman. 1992a. Fertilizer-nitrogen use efficiency of irrigated wheat. I. Uptake efficiency of preplant versus late-season application. *Agron. J.* 84:682–688.

Wuest, S. B. and K. G. Cassman. 1992b. Fertilizer-nitrogen use efficiency of irrigated wheat. II. Partitioning efficiency of preplant versus late-season application. *Agron. J.* 84:689–694.

Xie, R. J. and A. F. Mackenzie. 1986. Urea and manure effects on soil nitrogen and corn dry matter yields. *Soil Sci. Soc. Am. J.* 50:1504–1509.

Yoshida, S. 1981. *Fundamentals of rice crop science*. Los Baños, Philippines: IRRI.

Young, J. L. and R. A. Cattani. 1962. Mineral fixation of anhydrous ammonia by air-dry soils. *Soil Sci. Soc. Am. Proc.* 26:147–152.

3 Phosphorus

3.1 INTRODUCTION

In modern agriculture, maximizing and sustaining crop yields are the main objectives. One of the major problems constraining the development of an economically successful agriculture is nutrient deficiency (Fageria and Baligar, 2005). After nitrogen, phosphorus (P) has more widespread influence on both natural and agricultural ecosystems than any other essential plant element (Brady and Weil, 2002). Phosphorus is an essential nutrient for both plants and animals. It is estimated that some 30 to 50% of the increase in world food production since the 1950s is attributable to fertilizer use, including P use (Higgs et al., 2000). Phosphorus deficiency in crop plants is a widespread problem in various parts of the world, especially in highly weathered acidic soils (Fageria and Baligar, 1997; Fageria and Baligar, 2001; Faye et al., 2006). Worldwide applications of phosphate fertilizers now exceed over 30 million metric tons annually (Epstein and Bloom, 2005). The deficiency of this element is related to several factors. These factors are low natural level in some soils, high immobile or fixation capacity of acidic soils, uptake of modern crop cultivars in large amount, loss by soil erosion, and use of low rate by farmers in developing countries. Biotic stresses such as crop infestation of insects, diseases, and weeds also reduce P use efficiency in crop plants.

Soil acidity is one of the key causes of reduced P use efficiency in crop plants. It is estimated that acid soils occupy about 3.95 billion ha of land area, i.e., about 30% of the ice-free land area of the world (Von Uexkull and Mutert, 1995). The highly weathered soils of the humid tropics cover large areas in South America and central Africa and are mainly classified as Oxisols and Ultisols (Thurman et al., 1980). According to Sanchez et al. (1983), the 580 million ha of potentially arable tropical lands, with udic moisture regimes and supporting rain forest vegetation, represent the largest available area for expanding the world's agriculture frontier. Preliminary surveys indicate that at least 50×10^6 ha of the cerrado region of Brazil have potential for intensive mechanized agriculture (Goedert, 1983). These lands have no major temperature or moisture limitations and generally have topography suitable for year-round crop production. However, crop production has been limited by soil-related constraints. Most of these soils are acidic and have very low native fertility, especially in terms of phosphorus (Fageria et al., 2004; Fageria and Breseghello, 2004). When P fertilizers are applied to replenish soil fertility of these soils, about 70–90% of the P fertilizer is adsorbed and becomes locked in various soil P compounds of low solubility without giving any immediate contribution to crop production (Holford, 1977). Because of the importance of phosphorus in agriculture and because of the limited supplies in most soils, enhancing P use efficiency in crop plants is fundamental for obtaining maximum economic yields. Hence, the objective of this chapter is to review the P cycle in soil–plant systems, functions and deficiency symptoms of P in crop plants, P uptake and use efficiency, and management options for enhancing P use by annual crops.

3.2 PHOSPHATE FERTILIZER–RELATED TERMINOLOGY

Many phosphorus fertilizer–associated terms are used in the literature. Familiarity with these terms is important to understanding the agricultural science of mineral nutrition.

1. **Phosphate rock (PR):** If an apatite-bearing rock is high enough in P content to be used directly to make fertilizer or as a furnace charge to make elemental P, it is called phosphate rock. The term is also used to designate a beneficiated apatite concentrate (Cathcart, 1980).
2. **Phosphorite:** It is a rock term for sediment in which a phosphate mineral is a major constituent.
3. **Ore or matrix:** It can be defined as material in the ground that can be mined and processed at a profit. *Matrix* is synonymous with *ore*.
4. **Absolute citrate solubility (ACS):** As the use of other solvents became more common, the absolute solubility was expressed in more general terms to include any type of solvent (McClellan and Gremillion, 1980). The term *absolute solubility index* (ASI) then was defined as follows: ASI = (solvent – soluble P_2O_5, %/theoretical P_2O_5 concentration of apatite, %. The three solvents most commonly used throughout the world in solubility tests are neutral ammonium citrate, 2% citric acid, and 2% formic acid (McClellan and Gremillion, 1980).
5. **Acidulation processes:** Acidulation involves treating phosphatic raw materials with mineral acids to prepare water- and citrate-soluble P compounds. The most widely used processes involve treatment of apatitic PR with sulfuric acid to prepare normal superphosphate or phosphoric acid; triple or concentrated superphosphate is prepared from PR and phosphoric acid.
6. **Phosphorus immobilization:** Soluble phosphorus compounds, when added to soil, become chemically or biologically attached to the solid phase of soil so as not to be recovered by extracting the soil with specific extracting under specific conditions. Immobilized phosphorus generally is not absorbed by plants during the first cropping year.
7. **Chemisorbed phosphorus:** Phosphorus adsorbed or precipitated on the surface of clay minerals or other crystalline material as a result of the attractive forces between the phosphate ion and constituents in the surface of the solid phase.
8. **Water-soluble phosphorus:** Water-soluble, citrate-soluble, citrate-insoluble, available, and total phosphorus—these are the terms frequently used to describe phosphate contained in fertilizers. Tisdale et al. (1985) described all these terms as follows. (1) *Water-soluble:* A small sample of the material to be analyzed is first extracted with water for a prescribed period of

time. The slurry is then filtered and the amount of phosphorus contained in the filtrate is determined. Expressed by a percentage by weight of the sample, it represents the fraction of the sample that is water-soluble. (2) *Citrate-soluble*: The residue from the leaching process is added to a solution of neutral 1N ammonium citrate. It is extracted for a prescribed period of time by shaking, and the suspension is filtered. The P content of the filtrate is determined, and the amount determined is expressed as a percentage of the total weight of the sample; this is termed the citrate-soluble phosphorus. (3) *Citrate-insoluble*: The residue remaining from the water and citrate extraction is analyzed. The amount of phosphorus found is termed citrate-insoluble. (4) *Available phosphorus:* The sum of water-soluble and citrate-soluble P is termed available P. (5) *Total phosphorus:* The sum of available and citrate-insoluble P is termed total phosphorus.

9. **Labile phosphorus:** From the plant nutrition point of view, three soil phosphate fractions are important. These soil fractions are known as solution P, labile P, and nonlabile P. The phosphorus dissolved in the soil solution is known as solution phosphorus. The fraction of phosphorus that is held on the solid surface is in rapid equilibrium with soil solution phosphate. It can be determined by means of isotopic exchange and is called labile phosphate. The third fraction is the insoluble phosphate. The phosphate in this fraction can be released only very slowly into the labile pool and is known as nonlabile phosphorus. Isotopically exchangeable P, sampled by a growing plant over the span of a growing season, is called the L-value. It is customary to describe soil P in terms of the following relationship: soil solution P $\Leftrightarrow$ labile soil P $\Leftrightarrow$ nonlabile P, where equilibrium is rapidly established between labile soil solution P; true equilibrium is seldom, if ever, established between the labile and nonlabile pools of soil P (Olsen, 1980).

10. **Phosphorus buffer power of soils:** The term *buffer power* can be defined as the total amount of diffusible ion (solution plus sorbed) per unit of volume of soil required to increase the solution concentration by one unit. Buffer power, however, has also been described as the relationship between the concentration of ions adsorbed on the solid phase and the concentration of ions in solution.

11. **Phosphate beneficiation:** Almost all phosphate rock is mined by strip-mining. It usually contains about 6.55% P (15% P_2O_5) and must be upgraded for use as fertilizer. Upgrading removes much of the clay and other impurities. This process is called *beneficiation*. It raises P to 13.11–15.29% (30–35% P_2O_5). Following beneficiation, the rock phosphate is finely ground and treated to make the P more soluble. Fertilizer phosphates are classified as either *acid-treated or thermal-processed*. Acid-treated P is by far the most important. Sulfuric and phosphoric acids are basic in producing acid-treated phosphate fertilizers. Sulfuric acid is produced from elemental sulfur or from sulfur dioxide. More than 60% of this industrial acid is used to produce fertilizers. Treating rock phosphate with sulfuric acid produces a mixture of phosphoric acid and gypsum. Filtration removes the gypsum to leave green or wet-process phosphoric acid (Potash and Phosphate Institute, 1979).

12. **Polyphosphates:** Most liquid P sources start with wet-process phosphoric acid. But wet-process acid can be further concentrated to form superphosphoric acid. In this process, water is driven off and molecules with two or more P atoms are formed. Such molecules are called *polyphosphates.*

13. **Phosphorus longevity:** Longevity is defined as the time when the P concentration in the center of the applied band is 5 times the original water-soluble soil-P concentration (Eghball et al., 1990).

14. **Phosphorus-efficient plant:** Plants able to absorb, translocate, and utilize P effectively in growth.

15. **Phosphoric acid:** In commercial fertilizer manufacturing, it is used to designate orthophosphoric acid, H_3PO_4. In fertilizer labeling, it is the common term used to represent the phosphate concentration in terms of available P, expressed as percent P_2O_5 (Soil Science Society of America, 1997).

16. **Phosphogypsum:** *Phosphogypsum* is the term used for the gypsum by-product of wet-acid production of phosphoric acid from rock phosphate. It is essentially hydrated $CaSO_4 \cdot 2H_2O$ with small proportions of P, F, Si, Fe, and Al, several minor elements, heavy metals, and radionuclides as impurities (Alcordo and Rechcigl, 1993).

3.3   CYCLE IN SOIL–PLANT SYSTEMS

Knowledge of nutrient cycling in soil–plant systems is fundamental to an understanding of nutrient availability and the loss of nutrient balance to crop plants, beginning with sources of P in soils. Unlike nitrogen, which can be returned to the soil by fixation from the air, phosphorus cannot be replenished except from external sources once it leaves the soil in agricultural products or by erosion. Origins of P in soils include residual soil minerals or inputs of P from commercial fertilizers and organic manures. Hence, both organic and inorganic sources of P are found in soil–plant systems, and both are important P sources for plants. However, only a small fraction of the total phosphorus is in a form available to plants. Inorganic forms of phosphorus are mainly calcium and iron or aluminum bounded. Calcium-bounded P predominates in alkaline soils, and iron- and aluminum-bound P predominates in acidic soils. The organic fraction of P varies from soil to soil and may constitute 20 to 80% of the total phosphorus of the surface soil horizons (Brady and Weil, 2002). Organic P should be mineralized before it is taken up by plants. Mineralization is defined as the process of conversion of P from organic form to an inorganic form by microbial activity.

Main features of the phosphorus cycle involve its addition in soil–plant systems, transformation, losses, and uptake by plants (Figure 3.1). Solubilization and immobilization are the main transformation processes of P in soil–plant systems that control its availability to plants and potential losses. Immobilization or fixation is defined as the strong adsorption or precipitation of P ions on Al and Fe hydroxides. In most soils, soil solution P ranges between < 0.01 to 1 mg L^{-1}, and a value of 0.2 mg P L^{-1} is commonly accepted as the solution P concentration needed to meet the nutritional needs of most agronomic crops (Wood, 1998). Soil erosion, surface and subsurface

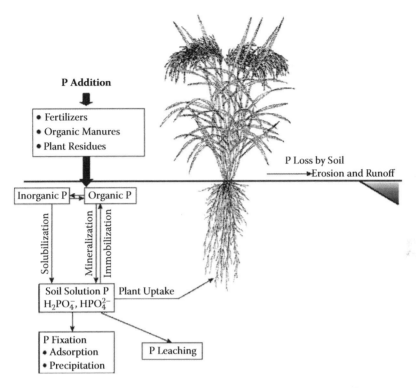

FIGURE 3.1 Simplified version of P cycle in soil–plant system.

runoff, leaching, and uptake by plants are the main avenues of P loss in soil–plant systems (Figure 3.1).

Recovery efficiency of applied soluble P fertilizers by annual crops during their growth cycle is less than 20% in most of the acid soils (Baligar et al., 2001; Fageria et al., 2003; Fageria and Barbosa Filho, 2007). Phosphorus uptake by plants mainly occurs in the form of $H_2PO_4^-$ ion in acid soils and in the form of HPO_4^{2-} ion in basic or alkaline soils. The proportion of these two ions in the soil solution is governed by pH. At pH 5, most of P is in the form of $H_2PO_4^-$, and at pH 7, both of these ions are present more or less in equal amounts (Mengel et al., 2001). Foth and Ellis (1988) reported that some studies with excised roots suggested that plants preferred the $H_2PO_4^-$ ion over the HPO_4^{2-} ion by about 10 to 1. But since conversion between the two species in solution is very rapid, this preference is probably of little importance for soils in the pH range of 4 to 8 (Foth and Ellis, 1988). The uptake of P by plants is governed by the ability of a soil to supply P to plant roots and by the desorption characteristics of the soil (Fageria et al., 2003). In acid soils, phosphorus is mainly immobilized or fixed by Al and Fe ions by following reactions:

$$Al^{3+} + H_2PO_4^- \text{(soluble)} + 2H_2O \Leftrightarrow Al(OH)_2H_2PO_4\text{(insoluble)} + 2H^+$$

$$Fe^{3+} + H_2PO_4^- \text{(soluble)} + 2H_2O \Leftrightarrow Fe(OH)_2H_2PO_4\text{(insoluble)} + 2H^+$$

In basic or alkaline soils, phosphorus immobilization or fixation takes place by the following reaction:

$$Ca(H_2PO_4)_2(\text{soluble}) + 2Ca^{2+} \Leftrightarrow Ca_3(PO_4)_2(\text{insoluble}) + 4H^+$$

Phosphorus immobilization is high in soils containing higher amount of amorphous Fe and Al hydroxides and allophane. This complexing reduces ion mobility and renders a large proportion of the total inorganic P insoluble and unavailable to plants. Thus, P acquisition is not a problem of total supply but of unavailability caused by the extreme insolubility of P at both acidic and alkaline pH. As a result, concentrations of P in soil solution are often low for adequate plant nutrition (Jayachandran et al., 1989). Reaction of P with soil may involve both adsorption and precipitation, which are thought to result from the same chemical force (Lin et al., 1983; Bolan et al., 1999). Many believe that adsorption mechanisms prevail at low P concentrations and precipitation mechanisms at high P concentrations (Lin et al., 1983).

It is also reported by Van Riemsdijk and De Haan (1981) that the reaction of P with soils that are initially free of sorbed P and with metal oxides/hydroxides is a very fast reaction at the beginning, slowing down substantially during the course of reaction. Phosphorus recovery increased with increasing levels of P. Fageria and Barbosa Filho (1987) reported that P immobilization capacity of Brazilian Oxisol varied from 77 to 90% depending on the level of soluble P applied. The high P fixation capacity of Brazilian Oxisol is related to low pH and high iron oxide contents, Al saturation, and clay fractions that are commonly composed primarily of kaolinite, gibbsite, and iron oxides (Smyth and Sanchez, 1982). The P immobilization is higher in soils containing high clay content compared with coarse-textured soils.

Fageria and Gheyi (1999) reported that in Oxisols and Ultisols of the tropics, P immobilization capacity is high. These authors recommend that in these soils, soluble P should not be applied in anticipation of sowing crops. Because most crops need P throughout their growth cycle, if P is applied in advance, a large part of it may be fixed in the beginning and crops may suffer from P deficiency.

The P fixation discussed above is related to well-drained soils. What happens to P in flooded soils? Large areas around the world are planted under flooded rice. Sah and Mikkelsen (1986) reported that flooding and subsequent draining of soil affect P transformations, increased amorphous Fe levels and P sorption, and induced P deficiency in crops grown after flooded rice. In California, wheat, safflower, corn, and sorghum have shown P deficiency following flooded rice (Brandon and Mikkelsen, 1979). Similar observations were also reported in Australia (Willet and Higgins, 1980). The process leading to increased immobilization of P in flooded-drained soils perhaps starts during the flooding periods of soil. The decrease in redox potential of flooded soils causes transformation of several chemical species. Reversal of these processes after soil drainage leads to increased chemical reactivity of soil minerals with P (Sah and Mikkelsen, 1986), which may immobilize P for several years (Willet and Higgins, 1980). A flooding period as short as 2 to 4 days increased P sorptivity in flooded-drained soils (Willet, 1982). Factors such as added organic matter and favorable soil temperatures that accelerate soil anaerobiosis may also enhance P sorption. Under aerobic soil conditions, addition of organic matter has been reported to decrease P sorption and increase P desorption (Kuo, 1983; Singh and Jones, 1976). Application of organic matter to a flooded soil intensifies the soil reduction processes, increases the transformation of soil Fe and P minerals, and leads to high P sorption in flooded-drained soil (Sah and Mikkelsen, 1989; Sah et al., 1989a, 1989b). Sah et al. (1989a) reported that increased soil P sorption in flooded-drained soil is related to an increase in amorphous Fe oxides in these environments, and they concluded that the sorptivity in flooded-drained soils is correlated with the Fe transformation.

3.4 FUNCTIONS AND DEFICIENCY SYMPTOMS

Phosphorus is essential for plant growth and reproduction and is a major nutrient along with nitrogen and potassium. Its functions cannot be performed by any other nutrients. Without adequate supply of P, a plant cannot reach its maximum yield potential (Fageria and Gheyi, 1999). Phosphorus plays an important role in energy storage and transfer in crop plants. Adenosine diphosphate (ADP) and adenosine triphosphate (ATP), summarized through both respiration and photosynthesis, are compounds with high-energy phosphate groups that drive most physiological processes in plants including photosynthesis, respiration, protein nucleic acid synthesis, and ion transport across cell membranes (Eastin and Sullivan, 1984; Wood, 1998). Phosphorus is an essential part of the structure of triphosphopyridine nucleotide (DPN and TPN). The DPN and TPN act as carriers of electrons or hydrogen between sites of oxidation and reduction reactions, which occur in respiration, fermentation, and photosynthesis (Fageria and Gheyi, 1999).

In cereals, P increases tillering; in legumes, branches increased with P fertilization. Figure 3.2 shows growth of upland rice with the application of N, P, and K fertilization and with the omission of N, P, and K fertilization in a Brazilian Oxisol. Plant growth and tillering were minimized in the pot that did not receive P fertilization compared to the pot that received N and P fertilization. Similarly, Figure 3.3

FIGURE 3.2 Growth of upland rice with the fertilization of N + P + K and omission of N, P, and K in Brazilian Oxisol.

FIGURE 3.3   Dry bean growth without P (left) and with P (right) in a Brazilian Oxisol.

shows the growth of dry bean with and without P application. Dry bean growth was significantly reduced in the pot that did not receive P compared to pot that received 200 mg P kg⁻¹ of soil. Growth of corn was also significantly improved with the addition of P (Figure 3.4). Root development increases with P fertilization in cereals as well as legume crops (Baligar et al., 1998). Figure 3.5 shows the influence of P fertilization on the root growth of upland rice grown on a Brazilian Oxisol. Similarly, root growth of corn was also increased with the application of P (Figure 3.6). Phosphorus strengthens culm strength in plants and hence prevents lodging. Crop quality in forages and vegetables is also improved with P fertilization. The formation of seeds and fruits is especially depressed in plants suffering from P deficiency. Thus, not only yields but also poor-quality seeds and fruits are obtained from P-deficient

FIGURE 3.4   Growth of corn plants at the 0 mg P kg⁻¹ soil (left), 50 mg P kg⁻¹ (middle), and 175 mg P kg⁻¹ (right) grown on a Brazilian Oxisol.

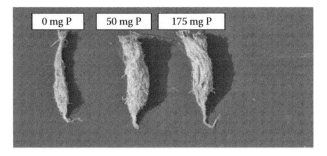

FIGURE 3.5 Upland rice root growth at 0, 50, and 175 mg P kg⁻¹ of soil.

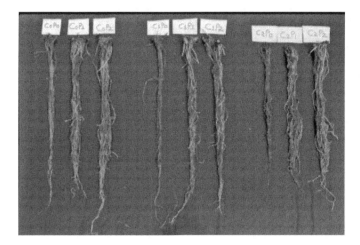

FIGURE 3.6 Corn root growth at 0 lime level kg⁻¹ soil with 0, 50, and 175 mg P kg⁻¹ soil (left); 2 g lime level kg⁻¹ soil with 0, 50, and 175 mg P kg⁻¹ soil (middle); and 4 g lime level kg⁻¹ soil with 0, 50, and 175 mg P kg⁻¹ soil (right) applied to a Brazilian Oxisol.

plants (Mengel et al., 2001). Phosphorus is needed in especially large amounts in meristematic tissue, where cells are rapidly dividing and enlarging (Brady and Weil, 2002). Phosphorus is also a key component of Phytin, a seed component that is essential to inducing germination. Phosphorus deficiency can reduce seed size, seed numbers, and viability. Phosphorus nutrition has been related to all facets of N_2 fixation in leguminous plants, probably due to the relationship between P and energy transfer mechanisms. Phosphorus helps in improving crop quality.

Leaf area, leaf numbers, and leaf expansion decreased under P stress (Lynch et al., 1991; Qui and Israel, 1992; Halsted and Lynch, 1996). Low P decreased shoot branching in bean (Lynch et al., 1991), and an increase in root-to-shoot ratio was often observed (Smith et al., 1990; Lynch et al., 1991). Phosphorus stress has also been reported to decrease carbon fixation and assimilation (Fredeen et al., 1990; Usuda and Shimogawara, 1991; Halsted and Lynch, 1996). Suggested mechanisms for decrease in carbon fixation and assimilation include the limitation of the triose-phosphate translocator function (Heber and Heldt, 1981), limitations of phosphate recycling (Stitt, 1986), and a decrease in ribulose bisphosphate (RuBP) regeneration (Abadia et al., 1987).

The best parameter for evaluating nutrient deficiency in a given soil is crop response to applied nutrient. According to this criterion, acid as well as alkaline soils around the world are extremely deficient in phosphorus for crop production. Phosphorus is a mobile nutrient in the plant; hence, P deficiency symptoms first appear on the older leaves. The visual symptoms of a P shortage, apart from stunted growth and reduced yields, are purple or reddish color on the older leaves. Phosphorus is not a constituent of chlorophyll; thus, in P-deficient plants, the concentration of chlorophyll in a leaf becomes comparatively high and the color of leaves, especially younger ones, changes to dark green (Fageria and Gheyi, 1999).

Phosphorus deficiency symptoms in plants include severe stunting, thin stems, and erect and dark green leaves. Phosphorus deficiency reduces seedling height, tiller number, stem diameter, leaf size, and leaf duration (Fageria et al., 2003). When P is deficient, cell and leaf expansions are retarded more than chlorophyll formation. Thus, the chlorophyll content per unit leaf area increases, but the photosynthetic efficiency per unit of chlorophyll decreases (Marschner, 1995). Phosphorus is not a constituent of chlorophyll; hence, the concentration of chlorophyll in P-deficient plants becomes comparatively high and the leaf color changes from green to dark green. If a P deficiency persists, the older leaves may turn an orange color and desiccate from the leaf tip back toward the base. Rice maturity can be delayed by as much as 10–12 days by P deficiency (Fageria et al., 2003). Phosphorus deficiency symptoms of many crops include a reddish or purple tint on leaves due to the accumulation of anthocyanins (Hewitt, 1963). Phosphorus-deficient roots were long and spindly, with little fibrousness, and turned dark (reddish purple to dark red) with increased severity of deficiency (Clark, 1982).

3.5   DEFINITIONS AND ESTIMATION OF P USE EFFICIENCY IN CROP PLANTS

The P use efficiency (PUE) can be defined as the maximum economic yield produced per unit of P applied or absorbed by the plant. Nutrient use efficiency has been defined in several ways in the literature, although most definitions denote the ability of a system to convert inputs into outputs. Definitions of nutrient use efficiencies have been classified as agronomic efficiency, physiological efficiency, agrophysiological efficiency, apparent recovery efficiency, and utilization efficiency (Table 3.1).

The five P use efficiencies mentioned in Table 3.1 were calculated for lowland rice and are presented in Table 3.2. On average, all P use efficiencies were higher at lower P rates and decreased at higher P rates. This indicated that rice plants were unable to absorb P when P was applied in excess, perhaps because their absorption mechanisms had been saturated (Fageria et al., 2003).

3.6   CONCENTRATION IN PLANT TISSUE

Tissue analysis is an important technique for diagnosing nutritional disorders in crop plants. Nutrient concentration is defined as content per unit of dry weight and is

TABLE 3.1

Definitions and Methods of Calculating Phosphorus Use Efficiency

Nutrient Efficiency	Definitions and Formulas for Calculation
Agronomic efficiency (AE)	The agronomic efficiency is defined as the economic production obtained per unit of nutrient applied. It can be calculated by: AE (kg kg^{-1}) = $G_f - G_u/N_a$, where G_f is the grain yield of the fertilized plot (kg), G_u is the grain yield of the unfertilized plot (kg), and N_a is the quantity of P applied (kg).
Physiological efficiency (PE)	Physiological efficiency is defined as the biological yield obtained per unit of nutrient uptake. It can be calculated by: PE (kg kg^{-1}) = $BY_f - BY_u/N_f - N_u$, where, BY_f is the biological yield (grain plus straw) of the fertilized plot (kg), BY_u is the biological yield of the unfertilized plot (kg), N_f is the P uptake (grain plus straw) of the fertilized plot (kg), and N_u is the P uptake (grain plus straw) of the unfertilized plot (kg).
Agrophysiological efficiency (APE)	Agrophysiological efficiency is defined as the economic production (grain yield in case of annual crops) obtained per unit of nutrient uptake. It can be calculated by: APE (kg kg^{-1}) = $G_f - G_u/N_{uf} - N_{uu}$, where, G_f is the grain yield of fertilized plot (kg), G_u is the grain yield of the unfertilized plot (kg), N_{uf} is the P uptake (grain plus straw) of the fertilized plot (kg), N_{uu} is the P uptake (grain plus straw) of unfertilized plot (kg).
Apparent recovery efficiency (ARE)	Apparent recovery efficiency is defined as the quantity of nutrient uptake per unit of nutrient applied. It can be calculated by: ARE (%) = $(N_f - N_u/N_a) \times 100$, where, N_f is the P uptake (grain plus straw) of the fertilized plot (kg), N_u is the P uptake (grain plus straw) of the unfertilized plot (kg), and N_a is the quantity of P applied (kg).
Utilization Efficiency (EU)	Nutrient utilization efficiency is the product of physiological and apparent recovery efficiency. It can be calculated by: EU (kg kg^{-1}) = PE × ARE

Source: Fageria and Barbosa Filho (2007).

expressed in g kg^{-1} or %. Nutrient concentration in plant tissue varied with plant age. As a general rule, nutrient concentration decreased with the advancement of plant age. The decrease in nutrient concentration with the advancement of plant age is associated with increased dry matter yield. The decrease in nutrient concentration with advancement of plant age is known as the *dilution effect* in mineral nutrition of plants. Data in Table 3.3 show that P concentration decreased significantly in a quadratic fashion in upland rice and dry bean with the advancement of plant age. The variability in P concentration in upland rice was about 90% and in dry bean it was about 80% with the advancement of plant age. Adequate concentrations of P in tissue of principal field crops are presented in Table 3.4.

TABLE 3.2

Phosphorus Use Efficiency in Lowland Rice under Different P Rates (Values Are Averaged across Two Years; Phosphorus Source Was Thermophosphate and Applied as Broadcast)

P Rate (kg ha⁻¹)	AE (kg kg⁻¹)	PE (kg kg⁻¹)	APE (kg kg⁻¹)	ARE (%)	UE (kg kg⁻¹)
131	15.5	604.4	300.9	6.3	39.6
262	12.7	536.8	269.5	5.1	27.1
393	10.5	521.8	477.4	3.8	19.7
524	6.8	443.6	277.8	3.4	14.8
655	6.2	439.3	296.6	2.7	10.9
Average	10.3	509.2	324.4	4.3	22.4

Regression Analysis

P Rate (X) vs. agronomical efficiency (Y) = $17.66 - 0.0178X$, $R^2 = 0.52**$

P Rate (X) vs. physiological efficiency (Y) = $636.22 - 0.3232X$, $R^2 = 0.23^{NS}$

P Rate (X) vs. agrophysiological efficiency (Y) = $324.70 - 0.00025X$, $R^2 = 0.02^{NS}$

P Rate (X) vs. recovery efficiency (Y) = $6.93 - 0.0067X$, $R^2 = 0.43**$

P Rate (X) vs. utilization efficiency (Y) = $43.27 - 0.0529X$, $R^2 = 0.50**$

*, **, NS Significant at the 5 and 1% probability levels and nonsignificant, respectively.

AE = agronomical efficiency, PE = physiological efficiency, APE = agrophysiological efficiency, RE = recovery efficiency, and UE = utilization efficiency.

TABLE 3.3

Phosphorus Concentration (g kg⁻¹) in Shoot of Upland Rice and Dry Bean during Growth Cycle (Data Are Averages of Two Years Field Trial Conducted on Brazilian Oxisols)

Plant Age (Days after Sowing)	Upland Rice	Plant Age (Days after Sowing)	Dry Bean
19	3.6	15	3.0
43	1.7	29	2.2
68	1.1	43	2.1
90	1.1	62	1.4
102	1.1	84	1.8
130	0.5	96	0.6

Regression Analysis

Age vs. P conc. in upland rice shoot (Y) = $4.4638 - 0.0661X + 0.00029X^2$, $R^2 = 0.8995*$

Age vs. P conc. in dry bean shoot (Y) = $3.2976 - 0.0323X + 0.000084X^2$, $R^2 = 0.7996*$

* Significant at the 5% probability level.

TABLE 3.4
Adequate P Concentration in Plant Tissue of Principal Cereal and Legume Crops

Crop Species	Growth Stage	Plant Part	Adequate P Conc. (g kg⁻¹)
Wheat	Tillering	Leaf blade	3.5–4.9
Wheat	Flowering	Leaf blade	2.5–3.4
Barley	Tillering	Leaves	5.0–6.8
Barley	Whole tops	Heading	2.0–5.0
Lowland rice	75DAS	Whole tops	2.5–4.8
Lowland rice	At harvest	Whole tops	1.6–2.0
Corn	35 to 45 DAE	Whole tops	4.0–8.0
Corn	Before tasseling	LB below	2.5–4.5
Corn	Silking	Ear LB	2.5–4.0
Sorghum	Seedling	Whole tops	3.0–6.0
Sorghum	Early vegetative	Whole tops	2.1–5.0
Sorghum	Bloom	3BBP	1.5–2.5
Soybean	Prior to pod set	UFD trifoliate	2.6–5.0
Dr bean	Early flowering	UMB	4.0–6.0
Cowpea	56DAS	Whole tops	3.0–3.5
Peanut	Early pegging	Whole tops	2.0–3.5

Note: DAS = days after sowing, DAE = days after emergence, LB = leaf blade, 3BBP = third blade below panicle, UFD = upper fully developed, UMB = uppermost blade.

Source: Compiled from Reuter (1986); Fageria et al. (1997b).

3.7 UPTAKE AND P HARVEST INDEX

Plants do not require as much phosphorus as they do nitrogen and potassium. But phosphorus is just as essential. Phosphorus uptake by plants is influenced by climatic, plant, and soil factors. The important climatic factors that control P availability to plants are soil temperature, moisture content, and solar radiation. The important soil factors controlling P availability are concentration of P in the soil solution, soil texture, organic matter content, soil pH, presence of other essential nutrients in quantity and proportion, and microbial activities. Plant species and genotypes within species also influence P uptake.

Quantity of P removed by a crop depends on yield level, fertility level, and other management practices. However, after fixation, a large part of the phosphorus is removed from the soil–plant system in harvesting parts of the plants. Of course, soil erosion is considered at a minimal expected level. According to McCollum (1991), phosphorus removal in harvested products averaged 16kg P ha⁻¹ per year (corn = 17 kg; soybean = 15 kg) during the first 8 years of the experiment on Portsmouth soil in North Carolina; while crop yields varied among years, neither yield nor P concentration in harvested products in any given year was markedly altered by P treatment (average P concentration in dry seed: corn = 0.29%; soybean = 0.65%).

TABLE 3.5

Phosphorus Uptake by Annual Crop Species Grown on Brazilian Oxisols

Crop Species	Shoot Weight (kg ha⁻¹)	Grain Yield (kg ha⁻¹)	P Uptake in Shoot (kg ha⁻¹)	P Uptake in Grain (kg ha⁻¹)	Total Uptake (kg ha⁻¹)
Upland rice	6104	4413	2.9	9.5	12.4
Corn	11006	8221	4.5	16.7	21.2
Dry bean	1807	3182	2.3	13.9	16.2
Soybean	3518	4003	1.8	14.3	16.1

TABLE 3.6

Grain Harvest Index and Phosphorus Harvest Index of Four Crop Species

Crop Species	Grain Harvest Index[a]	P Harvest Index[b]
Upland rice	0.42	0.77
Corn	0.43	0.79
Dry bean	0.64	0.86
Soybean	0.53	0.89

[a] Grain harvest index = grain yield/grain yield plus straw yield.

[b] P harvest index = P uptake in grain/P uptake in grain plus straw.

Phosphorus uptake by upland rice, corn, dry bean, and soybean grown in Brazilian Oxisols is presented in Table 3.5. In all four crops, P uptake was higher in the grain compared to shoot. Hence, it can be concluded that P requirements are higher for grain compared to shoot in the cereals as well as in the legumes. The P harvest index (P uptake in grain/P uptake in grain plus straw) values were 0.77 for upland rice, 0.79 for corn, 0.86 for dry bean, and 0.89 for soybean (Table 3.6). This means that grain harvest index and P harvest index were higher for legumes compared to cereals. This means that P requirements are higher for legumes compared to cereals.

3.8   INTERACTION WITH OTHER NUTRIENTS

Nutrient interactions are important factors in improving the yield of field crops. Nutrient interaction is influenced by factors such as concentration of nutrient, temperature, light intensity, soil aeration, soil moisture, soil pH, architecture of root, the rate of plant transpiration and respiration, plant age and growth rate, plant species, and internal nutrient concentration of plants. Generally, P has positive significant interaction with N, K, and Mg. The positive interaction of P with macronutrients may be associated with improvement in growth and yield of crop plants with the P fertilization. Increased growth and yield required more nutrients compared to low growth and yield. Wilkinson et al. (2000) also reported that increased growth requires more nutrients to maintain tissue composition within acceptable limits; mutually synergistic effects for N and P promote growth even more. Most P–lime interactions are intimately associated with toxic soil Al that limits root growth and proliferation and nutrient uptake (Sumner and Farina, 1986).

Among micronutrients, P–Zn interaction is widely reported in the literature (Sumner and Farina, 1986; Wilkinson et al., 2000). In Brazilian Oxisols, repose of upland rice to P application is very common, and this generally induces Zn deficiency (Fageria, 1989). The Zn deficiency in this situation is associated with rapid growth of plants, and soil-available Zn cannot fulfill demands of rapidly growing plants. Zinc deficiency induced P toxicity, a phenomenon that has been well described in the literature (Loneragan and Webb, 1993).

3.9 PHOSPHORUS VERSUS ENVIRONMENT

Without phosphorus in the environment, no living organisms could exist. Phosphorus is present in all plant and animal tissue. It is necessary for such life processes as photosynthesis, the synthesis and breakdown of carbohydrates, and the transfer of energy within the plant. The plant takes up phosphorus from the soil. In intensive cropping systems, application of chemical fertilizers is indispensable to maintain and/or increase crop yields; in some cases, nutrients may accumulate in higher levels in the agricultural lands. In these situations, there is an increased risk of nutrient losses including P to surface water and groundwater. Nonpoint source pollution of surface waters by agricultural P is a major environmental concern in many areas of the United States (Sharpley et al., 2000; McDowell and Sharpley, 2001; Shober and Sims, 2003). Additionally, nonpoint P pollution may also occur in areas of intensive animal production (Shober and Sims, 2003). Because of concerns regarding the influence of P on water quality in the United States, many state and federal agencies now recommend or require P-based nutrient management plans for animal manures (Shober and Sims, 2003). Peltonen-Sainio et al. (2006) reported that use of P fertilizers is restricted in Finland if soil P values are high because, with N, it is a major contributor of pollution in Finnish lakes and in the Baltic Sea. Turner et al. (2002) also reported that in past decades, high import of P in animal feeds and generous applications of P fertilizers have resulted in accumulation of P in European soils to an extent that high P fertilizer dose poses a particular threat to the environment due to runoff losses, which can result in alga blooms. The loss of P in dissolved and particulate forms is a function of topography, soil type, soil test phosphorus concentration, and soil hydrology (McDowell and Sharpley, 2001). Additionally, the amount of P released from soil to water is dependent on the soil P quantity–intensity relationship and the kinetics of P desorption. Phosphorus threshold soil test P values for agronomic and environmental purposes are given in Table 3.7.

Phosphorus is an immobile nutrient in soil, and its leaching downward in the soil profile is possible when its accumulation is very high. In developing countries,

TABLE 3.7

Threshold Soil Test P Values for Agronomic and Environmental Purposes

P Extraction Method	Agronomic (mg kg⁻¹)[a]	Environmental (mg kg⁻¹)[b]
Mehlich-1	13–25	>55
Mehlich-3	25–50	>50
Bray-1	20–40	>75
Olsen	12	>50

[a] Threshold value is defined as the soil test concentration above which the soil test level is considered optimum for plant growth and responses to the addition of the nutrient are unlikely to occur. Threshold values cited in this table are approximate and can be affected by soil type, crop, and management practices. [b] Environmental-threshold value is defined as the soil test concentration above which risk of environmental contamination is very high.

Source: Compiled from Sharpley et al. (1996); Fageria et al. (1997a); Shober and Sims (2003).

farmers use very low rates of P, and P leaching seldom occurs. However, in intensive agricultural areas of Europe and North America, much more phosphorus has been applied during the past few decades than has been removed by crops. As a result, soil in these areas has accumulated rather high levels of total and available phosphorus (Daniel et al., 1998; Sharpley et al., 1996; Brady and Weil, 2002; Zhang et al., 2005). As a result of soil erosion, runoff, and leaching, some of the phosphorus from these areas has contaminated reservoirs, lakes, and ponds, triggering the process of eutrophication (growth of unwanted algae and aquatic weeds; Brady and Weil, 2002). *Eutrophication* is a Greek word (*eutruphos*) that means "well nourished."

Eutrophication restricts water use for fisheries, recreation, industry, and drinking due to the increased growth of undesirable algae and aquatic weeds and also due to oxygen shortages caused by their death and decomposition (Fageria et al., 2003). Phosphorus concentrations that cause eutrophication can range from 0.01 to 0.03 mg L⁻¹ (Sharpley et al., 1996). To determine the threshold level of soil P accumulation, Dutch regulators have set a critical limit of 0.10 mg L⁻¹ as dissolved P tolerance in groundwater at a given soil depth (mean highest water level; Daniel et al., 1998). Pote et al. (1996) found that a Mehlich 3 P concentration of 50 mg kg⁻¹ (optimum for many crops) gave a dissolved reactive phosphorus (DRP) concentration in overland flow from grassland in Arkansas of 0.5 mg L⁻¹. Thus, soils with soil test phosphorus (STP) concentrations similar to those recommended for optimum crop growth may sustain DRP concentrations in surface runoff above levels accelerating eutrophication in surface water bodies (McDowell and Sharpley, 2001).

Hence, appropriate management of P is an important aspect not only for higher crop yields but also for environmental protection. Measures to stop soil erosion can significantly decrease particulate and dissolved forms of P loss (Withers and Jarvis, 1998). Multispecies riparian buffer systems have been recognized as one of the most cost-effective bioremediation approaches to alleviate nonpoint sources of agricultural pollutants on adjacent croplands (Schultz et al., 1995; Lin et al., 2007). Sediment retention and increased infiltration are the initial mechanisms that reduce nutrient loads in surface runoff. It has been generally observed that 50 to 90% of the sediments and the nutrients attached to the sediment are trapped by vegetative buffers (Mendez et al., 1999; Schmitt et al., 1999). In a runoff simulation study, 6-m grass buffers removed from 38 to 47% of NO_3-N and 39 to 46% of orthophosphate (PO_4-P) from surface runoff (Lee et al., 1997). Blanco-Canqui et al. (2004) also reported a comparable effectiveness in reducing N and P losses between tall fescue (*Festuca arundinacea* Schreb) buffers and a buffer composed of a switchgrass (*Panicum virgatum* L.) barrier in combination with native grass and forb species. Lin et al. (2007) reported that switchgrass decreased PO_4-P leaching by 60 to 74% compared to bare soil treatment.

3.10 MANAGEMENT PRACTICES TO MAXIMIZE P USE EFFICIENCY

Phosphorus use efficiency in crop plants can be improved by adopting appropriate management practices. These management practices include liming of acid soils; using appropriate source, timing, method, and rate of P fertilization; using balanced nutrition; using P-efficient crop species or genotypes within species; supplying adequate soil moisture; improving soil organic matter content; improving activities of beneficial microorganisms in the rhizosphere; controlling soil erosion; and controlling diseases, insects, and weeds.

3.10.1 LIMING ACID SOILS

Soil acidity affects large land area in various parts of the world. Liming acid soil influences physical, chemical, and biological properties in favor of higher crop yields and consequently higher P use efficiency. Liming supplies Ca^{2+} and Mg^{2+} and neutralizes Al^{3+} and H^+ ion toxicity. Soil pH is improved by liming, which produces many beneficial effects for plant growth and nutrient uptake. Data in Table 3.8 show that liming Brazilian Oxisol improves pH, improves Ca^{2+} and Mg^{2+} content, and reduces soil acidity ($H^+ + Al^{3+}$) significantly. The variation in soil pH was 97%, variation in Ca^{2+} was 95%, variation in Mg^{2+} was 92%, and variation in $H^+ + Al^{3+}$ was 96% with lime application (Table 3.8). The improvement in Ca and Mg content and reduction in soil acidity with liming can be explained by the following equation:

$$CaMg(CO_3)_2 + 2H^+ \leftrightarrow 2HCO_3^- + Ca^{2+} + Mg^{2+}$$

$$2HCO_3^- + 2H^+ \leftrightarrow 2CO_2 + 2H_2O$$

$$CaMg(CO_3)_2 + 4H^+ \leftrightarrow Ca^{2+} + Mg^{2+} + 2CO_2 + 2H_2O$$

TABLE 3.8
**Influence of Liming on Soil pH, Ca^{2+}, Mg^{2+}, and Soil
Acidity in Brazilian Oxisol after Harvesting of Two
Soybean Crops**

Lime Rate (Mg ha^{-1})	pH (in H$_2$O)	Ca^{2+} (cmol$_c$ kg^{-1})	Mg^{2+} (cmol$_c$ kg^{-1})	$H^+ + Al^{3+}$ (cmol$_c$ kg^{-1})
0	5.2	0.48	0.19	3.88
3	5.8	1.41	0.78	2.64
6	6.4	2.24	0.96	1.92
12	6.9	2.93	1.18	1.23
18	7.3	3.45	1.25	0.65
R^2	0.97**	0.95**	0.92**	0.96**

** Significant at the 1% probability level.

The above equations show that the acidity-neutralizing reactions of lime occur in two steps. In the first step, Ca and Mg react with H to replace these ions with Ca^{2+} and Mg^{2+} on the exchange sites (negatively charged particles of clay or organic matter), forming HCO_3^-. In the second step, HCO_3^- reacts with H^+ to form CO_2 and H_2O to increase pH. Soil moisture and temperature and quantity and quality of liming material mainly determine the liming reaction rate. To get maximum benefits from liming to improve crop yields, liming materials should be applied in advance of crop sowing and thoroughly mixed into the soil.

Oxisols are naturally deficient in phosphorus and have high P immobilization capacity due to either precipitation of P as insoluble Fe/Al-phosphates or chemisorption to Fe/Al-oxides and clay minerals (Nurlaeny et al., 1996). Smyth and Cravo (1992) reported that Oxisols are notorious for P immobilization because they have higher iron oxide contents in their surface horizons than any other kind of soils. The P fixation capacity in Oxisols is directly related to the surface area and clay contents of the soil material and inversely related to SiO_2/R_2O_3 ratios (Curi and Camargo, 1988).

Bolan et al. (1999) reported that in variable-charge soils, a decrease in pH increases the anion exchange capacity, increasing the retention of P. Hence, improving crop yields on these soils require high P rates (Sanchez and Salinas, 1981; Fageria, 1989). There are conflicting reports on the effects of liming on P availability in highly weathered acid soils (Friesen et al., 1980a; Haynes, 1984). Liming can increase, decrease, or have no effect on P availability (Fageria, 1984; Mahler and McDole, 1985; Anjos and Rowell, 1987). However, in a recent study, Fageria and Santos (2005) reported that there was a linear increase in Mehlich-1 extractable P with increasing soil pH in the range of 5.3 to 6.9 (average of 0–10 and 10–20 cm soil depth) in the Brazilian Oxisol (Figure 3.7). Fageria (1984) also reported that there was a quadratic increase in the Mehlich-1 extractable P in the pH range of 5 to 6.5, and thereafter it was decreased in the Brazilian Oxisol. This author further reported that the P increase in the pH range of 5 to 6.5 was associated with release of P ions from Al and Fe oxides, which were responsible for P fixation. At higher pH (>6.5) the reduction of extractable P was associated with fixation by Ca ions. This increase in extractable P

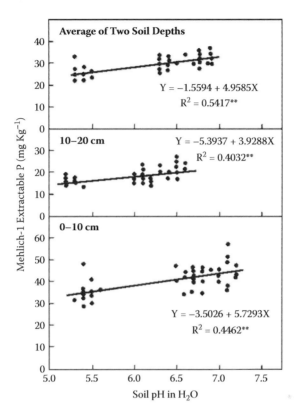

FIGURE 3.7 Relationship between soil pH and Mehlich-1 extractable soil phosphorus in Brazilian Oxisol (Fageria and Santos, 2005).

or liberation of this element in the pH range of 5 to 6.5 and reduction in the higher pH range (>6.5) can be explained with the help of the following equations:

$$AlPO_4 \text{ (P fixed)} + 3OH^- \Leftrightarrow Al(OH)_3 + PO_4^{3-} \text{ (P released)}$$

$$Ca(H_2PO_4)_2 \text{ (soluble P)} + 2Ca^{2+} \Leftrightarrow Ca_3(PO_4)_2 \text{ (insoluble P)} + 4H^+$$

The liming of acid soils results in the release of P for plant uptake; this effect is often referred to as the "P spring effect" of lime (Bolan et al., 2003). Bolan et al. (2003) reported that in soils high in exchangeable and soluble Al, liming might increase plant P uptake by decreasing Al, rather than by increasing P availability per se. This may be due to improved root growth where Al toxicity is alleviated, allowing a greater volume of soil to be explored (Friesen et al., 1980b).

3.10.2 USE OF APPROPRIATE SOURCE, TIMING, METHOD, AND RATE OF P FERTILIZATION

The use of proper source, method, and P rate is an important management practice to improve P use efficiency in crop plants. There are several chemical fertilizer sources

TABLE 3.9
Major Phosphorus Fertilizers, Their P Content, and Their Solubility

Common Name	Chemical Composition	P_2O_5 Content (%)	Solubility
Simple superphosphate	$Ca(H_2PO_4)_2 + CaSO_4$	18–22	Water soluble
Triple superphosphate	$Ca(H_2PO_4)_2$	46–47	Water soluble
Monoammonium phosphate	$NH_4H_2PO_4$	48–50	Water soluble
Diammonium phosphate	$(NH_4)_2HPO_4$	54	Water soluble
Phosphoric acid	H_3PO_4	55	Water soluble
Thermophosphate (Yoorin)	$[3MgO.CaO.P_2O_5 + 3(CaO.SiO_2)]$	17–18	Citric acid soluble
Rock phosphates	Apatites	24–40	Citric acid soluble
Basic slag	$Ca_3P_2O_8.Cao+CaO.SiO_2$	10–22	Citric acid soluble

of P (Table 3.9). Among these sources, water-soluble sources are more effective for annual crops. Water-insoluble or citric acid–soluble sources of P are effective for acid soils. Since liming is an essential and effective practice for improving yield of annual crops on acid soils, as discussed earlier, rock phosphates are not effective on limed acid soils or soils having higher pH (>6.0) (Foth and Ellis, 1988). Dissolution of phosphate rocks in soils requires an adequate supply of acid (H^+) as shown in the following equation (Bolan et al., 1997):

$$Ca_{10}(PO_4)_6F_2 + 12H^+ \leftrightarrow 10Ca^{2+} + 6H_2PO_4^- + 2F^-$$

Bolland and Gilkes (1990) reported that efficiency of rock phosphates was much lower than that of superphosphate tested under the arid conditions of Australia. Tisdale et al. (1985) reported that water-insoluble phosphates and those of low water solubility may give equally good results under some conditions; however, they are not as suitable as the water-soluble materials. These authors also reported that the thermal process phosphates, when finely ground, are satisfactory sources of phosphorus for most crops on acid soils, but they have failed generally to give favorable results on neutral and alkaline soils.

The immobile nature of P requires that the P source be water soluble to be effective for crop production. Several different views exist regarding the proportion of water-soluble P necessary in commercial phosphate fertilizers. The European Economic Community (EEC) requires 93% of the available P in commercial normal and triple superphosphate fertilizers to be in the water-soluble form (Fageria and Gheyi, 1999). Yet investigations conclude that there would probably be no significant yield increase from commercial phosphate fertilizers containing > 80% of their available P in the water-soluble form (Mullins, 1988; Mullins and Sikora, 1990). Bartos et al. (1992) tested five monoammonium phosphate (MAP) fertilizers, representing the major U.S. sources of PR (Florida, North Carolina, and Idaho) using sorghum as a test crop. They concluded that 57 to 68% water-soluble P is necessary to obtain 90% of the maximum yield.

Phosphorus should be applied at the time of sowing of annual crops in bands in low-P soils. This reduces the fixation of the fertilizer P to a minimum as it allows the crop the best opportunity to compete with the soil P utilization (Mengel et al., 2001). If soil P content is higher or in the adequate range for crop growth, band and broadcast applications will be equally effective. However, in P-fixing acid soils or low-P soils, band application is more effective, and in such soils, broadcast application may require three to four times more P compared to band application. The band application in soils may concentrate in a limited soil volume, saturating the soil phosphate adsorption capacity and increasing the soluble phosphate in the soil (Mengel et al., 2001). Early-season growth effects may also be obtained by band placement of fertilizer, even in soils that have considerable available P (Foth and Ellis, 1988). The band P usually increased early crop growth more than the broadcast placement because of increased uptake (Eghball et al., 1990; Zhang and Barber, 1992; Barber, 1995).

Smyth and Cravo (1990) reported that banded P provided maximum yields of corn and cowpea crops at lower rates than when applied broadcast in a Brazilian Amazon Oxisol. Banded P rates to each crop of 22 kg ha^{-1} for corn and 44 kg ha^{-1} for cowpea would sustain near maximum yields during 5 years of continuous crop production. Fiedler et al. (1989) reported that medium and low levels of soil P seed application resulted in 2 to 4 times more profit from fertilizer P application than broadcast P in winter wheat. Banding P fertilizer increases fertilizer effectiveness for two primary reasons. First, banding reduces soil fertilizer contact, which affects the degree of fertilizer P adsorption or precipitation and reduces what is often generally referred to as immobilization or fixation. In addition, banding relatively close to the plant results in a greater probability of root contact with the fertilizer.

Optimal P rates for crops are determined on the basis of soil test P calibration studies or crop response curves to applied P rates. The soil test calibration is defined as the process of determining the crop nutrient requirement at different soil test values (Soil Science Society of America, 1997). The use of an adequate level of P for maximum economic yield is possible if P soil test calibration data are available for a given crop and soil type (Fageria et al., 1997a). In order to achieve better soil P availability indices for any given soil type and crop species, detailed field calibrations are needed in relation to the P extracted by any particular extracting and the economic crop yield response to various soil P fertilization levels (Fageria et al., 1997a). Various extractants (e.g., Bray-1-P, Mehlich-1 and Mehlich-3, Olsen-P) have been extensively used in different parts of the world to assess the plant-available soil P status (Cope and Evans, 1985; Sharpley et al., 1994b).

When soil testing programs began around the middle of the 20th century, collection of soil test field calibration data was a major research activity (Hanna and Flannery, 1960; Hanway, 1963). Today, it is not considered as a priority research program. This may in part be due to the fact that soil test calibration research is perceived academically as lacking originality and as low priority, and grant funds to carry it out are limited (Heckman et al., 2006). However, it is widely recognized that soil test information and fertilizer recommendations require continual updating and reevaluation (Peck and Soltanpour, 1990; Heckman et al., 2006). In addition, such

TABLE 3.10

Mehlich-1 Soil Test P Availability Indices and P Recommendations for Common Bean

Soil P Test (mg kg⁻¹)	P Test Interpretation	Relative Grain Yield (%)	Band P Application (kg ha⁻¹)
0–5.3	Very low	0–70	66
5.3–7.1	Low	70–90	44
7.1–9.0	Medium	90–100	44
>9.0	High	100	22

Source: Adapted from Fageria and Santos (1998).

research is important for growers as well as for protecting the environment (Sharpley et al., 1994a; Heckman et al., 2006).

Effectiveness of soil P extraction as a soil P availability index is known to be influenced by soil type, soil properties (pH, type and amount of clay minerals, hydrated oxides of Al and Fe, exchangeable Ca and $CaCO_3$), and crop species involved (Kamprath and Watson, 1980; Sharpley et al., 1994). The differences in soil clay content are known to have an influence on critical levels of P needed for good soybean yields in Brazil (Lins et al., 1985). Mehlich-1 extractable P and clay content showed that, in high-clay soil, critical P levels were lower and needed higher P application per unit of soil test P (Lins et al., 1985; Cox, 1994).

Mehlich-1 soil test P availability indices and P fertilizer recommendations for common bean grown on Brazilian Inceptisol are presented in Table 3.10. The lowland rice response curve to applied thermophosphate P fertilization (Yoorin) is presented in Figure 3.8. The response curves are significant and quadratic in fashion in 2 years and average of 2 years. Based on Mehlich-1 and Bray-1 extractable soil P and relative grain yield, four categories were established for the P soil test: very low (VL), low (L), medium (M), and high (H) in relation to grain yield response zones (where VL = 0 to 70% relative grain yield zone, L = 70 to 90%, M = 90 to 100%, and H = > 100%; Figure 3.9 and Figure 3.10). The sufficiency P level is generally defined at 90% relative yield and here coincides with the low limit of the medium or optimum range. The 90% relative yield in fertilizer experiments is also considered as an economical rate (Fageria et al., 1997b). This is a standard convention for soil test calibration research. The soil P tests availability indices, calculated on the basis of Figure 3.7 and Figure 3.8, are presented in Table 3.11. These P index values can be a useful reference guide for interpreting soil P test results for lowland rice grown on an Inceptisol and making P fertilizer recommendations. Soil P test values of Mehlich-1 extracting solution were slightly higher compared with Bray-1 extracting solution, especially at higher P rates. However, regression analysis was conducted relating Mehlich-1 and Bray-1 extracting P (X) with grain yield (Y) to verify correlation of two extractants with grain yield. (Mehlich-1 extractable P (X) vs. grain yield (Y) = $1541.48 + 163.92X - 1.71X^2$, $R^2 = 0.74**$, Bray-1 extractable P (X) vs. grain yield (Y) = $875.79 + 211.51X - 2.21X^2$, $R^2 = 0.76**$). Phosphorus extracted by both the extractants had a highly significant association with grain yield, and coefficients of

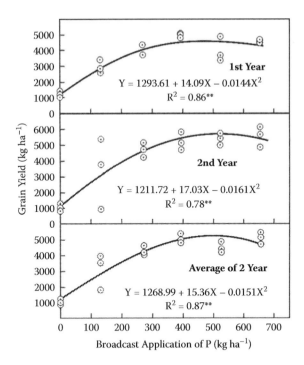

FIGURE 3.8 Relationship between broadcast application rate of thermophosphate (Yoorin) and grain yield of lowland rice. *Source:* Fageria and Santos (2008).

determination of both the extractants were very close to each other. This means that these extractants are equally effective in defining P availability indices for lowland rice grown on an Inceptisol.

Beegle and Oravec (1990) determined the critical P level for corn in the state of Pennsylvania of 19 mg kg^{-1} for Bray-Kurtz-P1 and 20 mg kg^{-1} for the Mehlich-3 extracting solution. Similarly, Mallarino and Blackmer (1992) in Iowa determined the critical P level of 13 mg kg^{-1} for Bray-P1 and 12 mg kg^{-1} for Mehlich-3 extracting for corn. Fageria et al. (1997a) determined the critical P level of 13 mg kg^{-1} for lowland rice by Mehlich-1 extracting solution.

3.10.3 Use of Balanced Nutrition

Use of balanced nutrition means supplying other essential plant nutrients in adequate amount and proportion along with P. At balanced nutrition, crop yields are maximized and P use efficiency improves. Supply of N and K in adequate amounts and ratios along with P is a classic example to promote optimum growth and yield of crops and consequently higher P use efficiency. Data in Table 3.12 show the importance of adequate amount of N, P, and K nutrition for upland rice and common bean grown on Brazilian Oxisol. Grain yield of rice as well as bean was significantly higher at NPK treatment compared with control and PK and NK treatments. Yield reduction of both the crops was higher when N and P were omitted from the growth medium compared to K omission. This means that N and P were the most

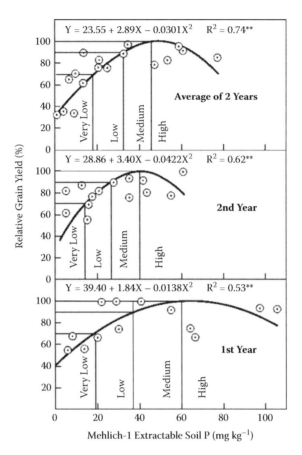

FIGURE 3.9   Relationship between Mehlich-1 extractable P and relative grain yield of lowland rice grown on an Inceptisol. *Source:* Fageria and Santos (2008).

yield-limiting nutrients in these soils and that K was the least yield-limiting nutrient. Tisdale et al. (1985) reported that greater effectiveness of fertilizer P has been found in many areas of the United States and Canada when modern fertilizer application systems place P in close association with ammonical nitrogen sources.

3.10.4   USE OF P EFFICIENT CROP SPECIES OR GENOTYPES WITHIN SPECIES

The cost of P fertilizers has dramatically increased in recent decades. This has given rise to interest in fitting the plants to nutrient-deficient soils rather than modifying the soils by adding fertilizers. Significant differences have been reported among crop species and genotypes of same species in absorption and utilization of nutrients including P (Clark, 1990; Marschner, 1995; Fageria et al., 1997b; Baligar et al., 2001; Mengel et al., 2001; Epstein and Bloom, 2005; Fageria et al., 2006). Clark (1990) reported differences in the abilities of sorghum to take up P and utilize it efficiently, including its remobilization within the plant over time. Plant properties and characteristics relating to efficient uptake and utilization of mineral elements are

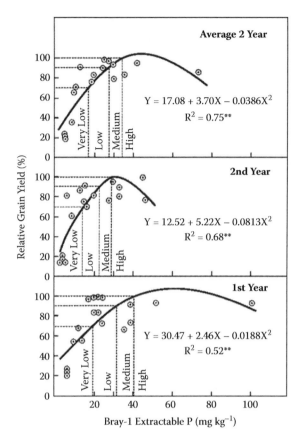

FIGURE 3.10 Relationship between Bray-1 extractable P and relative grain yield of lowland rice grown on an Inceptisol. *Source:* Fageria and Santos (2008).

heritable and will, no doubt, be exploited more vigorously as fertilizer costs continue to rise (Eastin and Sullivan, 1984). Inter- and intraspecific variation for plant growth and mineral nutrient use efficiency are known to be under genetic and physiological control and are modified by plant interactions with environmental variables (Baligar et al., 2001). Baligar et al. (2001) also reported that identification of traits such as nutrient absorption, transport, utilization, and mobilization in plant cultivars should greatly enhance fertilizer use efficiency.

Many investigators have reported that plants display a wide array of adaptive responses to low P availability to enhance P mobility in the soil and increase its uptake (Fageria et al., 1988; Fageria et al., 1997a; Kochian et al., 2004; Faye et al., 2006). A well-known adaptive response is the alteration of root morphology and architecture to increase P acquisition from the soil at minimum metabolic cost (Neumann et al., 1999; Jonathan and Kathleen, 2001; Yong et al., 2003; Faye et al., 2006). In this regard, morphological and genotypic variations are well documented for many crop plants (Gorny and Patyna, 1984; Aina and Fapohunda, 1986). Bruck et al. (2002) and Faye et al. (2006) reported genotype differences in shoot and rooting parameters of pearl millet (*Pennisetum glaucum* L. R. Br.) in fields with P deficiency.

TABLE 3.11
Mehlich-1 and Bray-1 Soil Test P Availability
Indices for Lowland Rice in Brazilian Inceptisol

Soil P Test (mg kg^{-1})	P Test Interpretation	Relative Yield (%)
Mehlich-1 (1st Year)		
0–20	Very low	0–70
20–36	Low	70–90
36–60	Medium	90–100
>60	High	100
Mehlich-1 (2nd Year)		
0–14	Very low	0–70
14–26	Low	70–90
26–40	Medium	90–100
>40	High	100
Mehlich-1 (Averages of 2 Years)		
0–17	Very low	0–70
17–32	Low	70–90
32–45	Medium	90–100
>45	High	100
Bray-1 (1st Year)		
0–19	Very low	0–70
19–32	Low	70–90
32–41	Medium	90–100
>41	High	100
Bray-1 (2nd Year)		
0–14	Very low	0–70
14–23	Low	70–90
23–30	Medium	90–100
>30	High	100
Bray-1 (Averages of 2 Years)		
0–17	Very low	0–70
17–28	Low	70–90
28–35	Medium	90–100
>35	High	100

Source: Fageria and Santos (2008).

Formation of cluster roots is one of the mechanisms that enhance the capacity of plants to acquire sparingly soluble P from soil (Shu et al., 2007). Cluster roots comprise a number of tightly grouped, determinate rootlets that undergo initiation, growth, and arrest in a synchronized manner (Skene, 2001). The development and functional synchrony within the cluster roots leads to a concentrated change in soil

TABLE 3.12

Response of Upland Rice and Common Bean to N, P, and K Fertilization

Nutrient Treatment	Upland Rice Grain Yield (g/pot)	Common Bean Grain Yield (g/pot)
Control	18.40b	15.15b
NPK	52.92a	29.50a
PK	29.93b	16.15b
NK	25.53b	15.20b
NP	50.56a	19.48b

Note: Means in the same column followed by the same letter are not significantly different at the 5% probability level by Tukey's studentized test.

Source: Adapted from Fageria and Baligar (2004).

chemistry around the cluster roots and is thought to mobilize phosphate, iron, and other elements in the rhizosphere (Dinkelaker et al., 1995; Watt and Evans, 1999; Vance et al., 2003). White lupin (*Lupinas albus* L.) is considered a model plant species for studying cluster root formation and root exudation (Johnson et al., 1996; Neumann et al., 1999; Watt and Evans, 1999). Formation of cluster roots by white lupin is induced under P deficiency (Dinkelaker et al., 1995; Keerthisinghe et al., 1998; Watt and Evans, 1999).

Data in Table 3.13 show grain yield of six upland rice genotypes at low and high P levels. Genotypes CNA 4361 and CNA 4261 were the lowest-yielding genotypes at low as well as high P levels, whereas genotypes CNA 4180 and CNA 4166 were the highest-yielding genotypes at low as well as high P levels. Hence, these

TABLE 3.13

Grain Yield and Grain Yield Efficiency Index (GYEI) of Six Upland Rice Genotypes at Two P Levels

Genotype	Grain Yield (kg ha^{-1})		GYEI
	Low P (3 mg kg^{-1})	High P (4.7 mg kg^{-1})	
CNA 4361	991	1370	0.80
CNA 4209	1190	1590	1.12
CNA 4617	1570	1910	1.77
CNA 4640	1620	1740	1.67
CNA 4180	2490	2720	4.02
CNA 4166	2230	2560	3.48

Note: GYEI = (yield at low P level/exp. mean yield at low P level) × (yield at high P level/exp. mean yield at high P level).

Source: Adapted from Fageria et al. (1988).

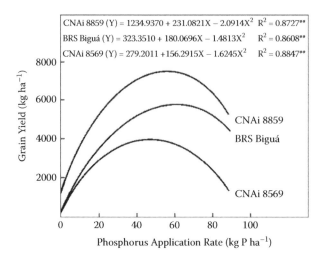

$CNAi\ 8859\ (Y) = 1234.9370 + 231.0821X - 2.0914X^2\quad R^2 = 0.8727^{**}$

$BRS\ Biguá\ (Y) = 323.3510 + 180.0696X - 1.4813X^2\quad R^2 = 0.8608^{**}$

$CNAi\ 8569\ (Y) = 279.2011 + 156.2915X - 1.6245X^2\quad R^2 = 0.8847^{**}$

FIGURE 3.11 Relationship between P rate and grain yield of lowland rice genotypes.

two genotypes were the most P efficient. The higher efficiency of these two geno-types was also proved by their higher grain yield efficiency indices compared with those of the other four genotypes. Lowland rice genotypes also showed yield dif-ferences with increasing P rate (Figure 3.11). Although three genotypes displayed similar responses to applied P, the magnitude of responses varied from genotype to genotype. For example, at 50 kg P ha^{-1}, genotype CNAi 8569 produced 4000 kg ha^{-1} grain yield and genotype CNAi 8859 produced 7500 kg ha^{-1} grain yield. Similarly, genotype BRS Biguá produced 5600 kg ha^{-1} grain yield at 50 kg P ha^{-1}. Hence, among these three genotypes, CNAi 8859 was most P efficient and CNAi 8569 was least P efficient. Similarly, there was a difference in growth of two dry bean cultivars at low as well as high P levels (Figure 3.12). The cultivar Novo Jalo had better growth at low as well as high P levels compared with cultivar Carioca. The difference in P utilization among rice and bean cultivars may be associated with better root geom-etry, capacity of plants to uptake at lower concentration, plants' ability to solubilize nutrients at the rhizosphere, better distribution and utilization within plants, and balanced source–sink relationship (Graham, 1984; Clark and Duncan, 1991; Baligar et al., 2001). Root exudation of organic acids has been found to efficiently enhance the availability of inorganic soil P, while enhanced secretion of acid phosphatase into the soil rhizosphere provides P from organic soil sources (Marschner, 1995). Ryan et al. (2001) reported that many organic acids such as citrate, malate, oxalate, and fumarate are exuded from roots and improve P nutrition of plants. Studies have shown that roots can also induce high activity of rhizosphere acid phosphatase, which improves P availability to plants (Asmar et al., 1995; Kandeler et al., 2002; Gahoo-nia and Nielsen, 2004). Hinsinger (1998), Richardson (2001), and Fageria and Stone (2006) have reviewed the various root-induced chemical and biological processes in relation to plants' ability to mobilize nutrients.

Phosphorus use efficiency is also different among C$_3$ and C$_4$ plants and mono-cot and dicot plants (Halsted and Lynch, 1996). An example in this respect is the

0 mg P
Carioca

0 mg P
Novo Jalo

200 mg P
Carioca

200 mg P
Novo Jalo

FIGURE 3.12 Growth of two dry bean cultivars without application of P and with application of P in a Brazilian Oxisol. The first two pots from left to right contain cultivar carioca and Novo Jalo without P; the next two pots contain the same cultivar with the application of 200 mg P kg^{-1} of soil.

legume-based tropical pasture systems, where low-P soils predominate; in these areas, the aggressive C_4 grasses outcompete C_3 leguminous species (Sanchez, 1976). In the temperate zone, Morris et al. (1982) observed that warm-season C_4 grasses had higher yield with a lower tissue P concentration than cool-season C_3 grasses on a low-P soil. However, Halsted and Lynch (1996) reported that C_4 species are not inherently more P efficient than C_3 species but that monocots are more P efficient than dicots, because of contrasting P and biomass allocation under stress.

3.10.5 Supply of Adequate Moisture

Phosphorus is an immobile nutrient in soil; thus, maintaining an adequate supply of moisture in the soil–plant system is essential for its movement to the root vicinity and its uptake. Adequate soil moisture also favors dissolution and diffusion of phosphatic fertilizers in the soil and their consequent uptake by plants (Mengel et al., 2001). As soil water content decreases, the radii of water-filled pores decrease and tortuosity increases; therefore, P mobility also decreases. Nye and Tinker (1977) reported that lower water availability causes a reduction in P availability and, therefore, in its absorption. The interaction of P and the water supply has been reported on pearl millet growth and development under Sahel-like conditions (Payne et al., 1991; Faye et al., 2006).

Under prolonged drought conditions during the reproductive growth stage (initiation of panicle primordial to 1 week before flowering), upland rice yield of plots that received higher rates of P (>40 kg P ha^{-1}) fertilizer was lower than that of plots that received a control treatment (Fageria, 1980). Tisdale et al. (1985) reported that with soil water content at field capacity, 50 to 80% of the water-soluble phosphorus can be expected to move out of the fertilizer granule within a 24-hour period. These authors also reported that under wet conditions, the response to granular phosphates of high water solubility is superior to the response to powdered materials. Under dry conditions, however, powdered materials are likely to give better results.

3.10.6 IMPROVING ORGANIC MATTER CONTENT OF THE SOIL

Recycling of organic matter through use of animal manures, green manuring, and crop residues can improve P uptake by plants. Organic materials can release P by microbial activities, and crop plants can enhance its availability. Also, organic materials can reduce P fixation by masking the fixation sites on the soil colloids and by forming organic complexes or chelates with Al, Fe, and Mn ions, thereby improving P uptake efficiency of crop plants (Brady and Weil, 2002). Soil organic matter can be improved by applying organic manures, adopting appropriate crop rotation, using of conservation tillage, and recycling crop residues.

3.10.7 IMPROVING ACTIVITIES OF BENEFICIAL MICROORGANISMS IN THE RHIZOSPHERE

Microorganisms play an important role in the acquisition and transfer of nutrients in soil (Richardson, 2001). For P, soil microorganisms are involved in a range of processes that affect P transformation and thus influence the subsequent availability of P to plant roots. In particular, microorganisms can solubilize and mineralize P from inorganic and organic pools of total soil P. Also, the microbial biomass itself contains a large pool of immobilized P that potentially is available to plants (Richardson, 2001). In particular, the release of root exudates, which either directly influence P availability or support the growth of microbial populations within the rhizosphere, and the production of phosphatase enzymes are important, as are mycorrhizal associations (Richardson, 2001). Various types of microorganisms and fungi that are able to solubilize precipitated P have been reported (Kucey et al., 1989; Whitelaw, 2000). Predominant among these organisms are *Bacillus, Pseudomonas, Penicillium,* and *Aspergillus* spp. On the basis of laboratory screening assays, it has been shown that P-solubilizing microorganisms may constitute up to 40% of the cultivable population of soil microorganisms and that a significant proportion of them can be isolated from rhizosphere soil (Kucey, 1987; Richardson, 2001).

Beneficial microorganisms such as arbuscular mycorrhizal fungi (AMF) form beneficial symbioses with roots of many plants to allow them to maintain themselves and grow well under relatively harsh mineral stress conditions (Clark and Zeto, 2000). A common endomycorrhizal association is produced by phycomycetous fungi of the family *Endogonaceae*. The host range includes most agricultural and horticultural crops (Soil Science Society of America, 1997). The AMF fungi especially benefit plants grown in soils where P limits plant growth. This may be explained mainly by the increased soil volume explored by AMF hyphae relative to that of root hairs of non-AMF plants (Bolan, 1991; Jakobsen, 1995; Clark and Zeto, 2000). Reduced effectiveness of AMF colonization of roots often occurs when soluble soil P levels increase (Bolan, 1991; Marschner, 1995). Furthermore, AMF hyphae normally transport P located at greater distance from the root than do non-AMF roots (Tinker et al., 1992).

Solubilization of sparingly soluble soil P compounds might occur from increased activity of exocellular enzymes such as phosphatase/phytase in the rhizosphere/ hyphosphere from AMF roots or AMF hyphae (Clark and Zeto, 2000). Rhizosphere soil often has higher activity of phosphatase than bulk soil (Adams and Pate, 1992). Even though external root surfaces have extensive acid/alkaline phosphatase activity, particularly under low or P deficiency conditions, these enzymes could potentially solubilize organic P (Boddington and Dodd, 1998; Fries et al., 1997; Clark and Zeto, 2000).

3.10.8 CONTROL OF SOIL EROSION

Soil erosion by water and wind is a serious problem in several parts of the world. The erosion process washes away productive soil, silts up waterways and water reservoirs, and increases the danger of flooding of large areas. Consequently, the economic damage from erosion is very high. If uncontrolled, high rate of erosion combined with organic matter depletion and nutrient mining reduced yields to unprofitable levels within a short period after land clearing and cultivation. The release and migration of nutrients in runoff from agricultural land is both an economic loss and a threat to the quality of surface water and groundwater. A substantial proportion of P (7–30%) and other nutrients may also be in solution form (Schuman et al., 1973; Menzel et al., 1978; McDowell et al., 1984). Therefore, control of soil erosion helps to preserve topsoil layers and consequently prevents loss of nutrients.

Topsoil loss by erosion often decreases productivity and usually decreases the nutrient supply. Unfavorable physical conditions resulting from erosion can reduce infiltration, induce crusting, and reduce the effective rootzone (Fageria et al., 1997a). Plots desurfaced 20 cm and fertilized with twice the recommended rate usually yielded less than the control plots of Manitoba, Canada (Ives and Shaykewich, 1987). Mielke and Schepers (1986) reported that adequately fertilized topsoil yielded 25% more than did fertilized subsoil alone in Nebraska. Massee (1990) also reported that wheat yield on eroded soil could be partially alleviated by applying appropriate fertilizers in silt loam soil near Albion, Idaho. From the foregoing review, it can be concluded that certain beneficial characteristics of topsoil cannot be replaced by fertilization and that erosion control is the best solution for sustainable agricultural production on arable lands.

Use of appropriate soil conservation practices that control or minimize soil erosion can reduce P loss by surface runoff and increase its availability to crop plants. Use of cover crops and crop residues can improve water infiltration into the soil and reduces soil erosion. In most soils, the P content of surface horizons is greater than that of the subsoil due to sorption of added P, greater biological activity, cycling of P from roots to aboveground plant biomass, and more organic material in surface layers (Sharpley et al., 1999). Hence, adopting proper soil conservation practices can minimize soil loss or P loss from surface soil by erosion.

Conservation tillage is one of the most important practices to reduce soil erosion. Conservation tillage is practiced on 45 million ha worldwide, predominantly in North and South America, but its use is also increasing in South Africa, Asia, and Australia (Holland, 2004). It is primarily used to control erosion, reduce compaction of soil, conserve moisture, and reduce the cost of crop production. Under conservation tillage, a richer soil biota develops that can improve nutrient recycling.

3.10.9 CONTROL OF DISEASES, INSECTS, AND WEEDS

Diseases, insects, and weeds are the important biotic stresses in crop production around the world. Yield losses caused by cereal diseases are between 5 and 15% in developed agriculture systems (Jenkins and Lescar, 1980), although losses in individual crops may be greater than 40% (Wright and Gaunt, 1992). A factor often limiting a yield response to applied N is the severity of diseases (Boquet and Johnson, 1987; Howard et al., 1994). Fageria (1989) reported that the average loss at the global level of production potential of annual crops caused by diseases is about 10%; by insects, 12%; and by weeds, 11%.

Diseases, insects, and weeds have adverse effects on leaf area, photosynthesis, and metabolism and consequently lower nutrient use efficiency. Similarly, diseases, insects, and weeds can also adversely affect root development, which reduces absorption of water and nutrients. Weeds compete for light, nutrients, and water with crop plants and adversely affect growth and development and consequently absorption and utilization of nutrients. Hence, controlling diseases, insects, and weeds at harmful threshold levels can improve availability of nutrients, including P, to crop plants.

3.11 CONCLUSIONS

Phosphorus deficiency is a global constraint for crop production. The main role of phosphorus in plants is its participation in energy-transfer processes. Plant morphological characteristics such as leaves, height, and stem diameter are reduced by P deficiency. Dark green or blue-green foliage is one of the first symptoms of P deficiency in crop species. The main causes of P deficiency are low total and available P content in highly weathered acidic soils; immobilization of fertilizer-applied P; loss by soil erosion, by surface runoff, and in drainage water; and uptake in large amount by modern crop cultivars. Plants absorb less than 20% of the total fertilizer P applied during growing season due to immobilization. However, in some European countries and the United States, accumulation of P in excess in intensive agriculture has been reported to be a major source of contamination of aquatic resources.

Important management practices to improve P use efficiency are liming acid soils; using appropriate method, timing, form, and rate of application; using organic manures; recycling crop residues; sowing P-efficient plant species or genotypes within species; reducing soil erosion; improving water use efficiency; providing a balanced nutrient supply; improving plant root association with beneficial microorganisms; and controlling plant diseases, insects, and weeds. Genetic variations in crop species and genotypes within species in P uptake and utilization provide an opportunity to save nonrenewable P resources through selection and breeding of P-efficient cultivars. Phosphorus recovery is low; however, P use efficiency (grain weight/unit of P uptake) is much higher than that of N and K. A larger amount of P is translocated to grain compared to straw or shoot, indicating the greater importance of this element in improving grain yield of crops. Thus, it can be concluded that enhancing P use efficiency by crop plants is vital for sustaining or increasing crop production, reducing cost of crop production, and reducing environmental pollution.

REFERENCES

Abadia, J., I. M. Rao, and N. Terry. 1987. Changes in leaf phosphate status have only small effects on the photochemical apparatus of sugar beet leaves. *Plant Science* 50:49–55.

Adams, M. A. and J. S. Pate. 1992. Availability of organic and inorganic forms of phosphorus to lupins (*lupinus sp.*) *Plant Soil* 145:107–113.

Aina, P. O. and H. O. Fapohunda. 1986. Root distribution and water uptake patterns of maize cultivars field grown under differential irrigation. *Plant Soil* 94:257–265.

Alcordo, I. S. and J. E. Rechcigl. 1993. Phosphogypsum in agriculture: A review. *Adv. Agron.* 49:55–118.

Anjos, J. T. and D. L. Rowell. 1987. The effect of lime on phosphorus adsorption and barley growth in three acid soils. *Plant Soil* 103:75–82.

Asmar, F. 1977. Variation in activity of root extracellular phytase between genotypes of barley. *Plant Soil* 195:61–64.

Baligar, V. C., N. K. Fageria, and M. A. Elrashidi. 1998. Toxicity and nutrient constraints on root growth. *HortScience* 33:960–965.

Baligar, V. C., N. K. Fageria, and Z. L. He. 2001. Nutrient use efficiency in crop plants. *Commun. Soil Sci. Plant Anal.* 32:921–950.

Barber, S. A. 1995. *Soil nutrient bioavailability: A mechanistic approach*, 2nd edition. New York: John Wiley & Sons.

Bartos, J. M., G. L. Mullins, J. C. Williams, F. J. Sikora, and J. P. Copeland. 1992. Water-insoluble impurity effects on phosphorus availability in monoammonium phosphate fertilizers. *Soil Sci. Soc. Am. J.* 56:972–976.

Beegle, D. B. and T. C. Oravec. 1990. Comparison of field calibrations for Mehlich 3 P and K with Bray-Kurtz P1 and ammonium acetate K for corn. *Commun. Soil Sci. Plant Anal.* 21:1025–1036.

Blanco-Canqui, H., C. J. Gantzer, S. H. Anderson, E. E. Alberts, and A. L. Thompson. 2004. Grass barrier and vegetative filter strip effectiveness in reducing runoff, sediment, nitrogen, and phosphorus loss. *Soil Sci. Soc. Am. J.* 68:1670–1678.

Boddington, C. L. and J. C. Dodd. 1998. A comparison of the development and metabolic activity of mycorrhizas formed by arbuscular mycorrhizal fungi from different genera on two tropical forage legumes. *Mycorrhiza* 8:149–157.

Bolan, N. S. 1991. A critical review on the role of mycorrhizal fungi in the uptake of phosphorus by plants. *Plant Soil* 134:187–207.

Bolan, N. S., D. C. Adriano, and D. Curtin. 2003. Soil acidification and liming interactions with nutrient and heavy metal transformation and bioavailability. *Adv. Agron.* 78:215–272.

Bolan, N. S., J. Elliott, P. E. H. Gregg, and S. Weil. 1997. Enhanced dissolution of phosphate rocks in the rhizosphere. *Biology of Fertile Soils* 24:169–174.

Bolan, N. S., R. Naidu, J. K. Syers, and R. W. Tillman. 1999. Surface charge and solute interactions in soils. *Adv. Agron.* 67:88–141.

Bolland, M. D. A. and R. J. Gilkes. 1990. Rock phosphates are not effective fertilizers in Western Australia soils: A review. *Fert. Res.* 22:79–85.

Boquet, D. J. and C. C. Johnson. 1987. Fertilizer effects on yield, grain composition, and foliar disease of double crop soft red winter wheat. *Agron. J.* 79:135–141.

Brandon, D. M. and D. S. Mikkelsen. 1979. Phosphorus transformations in alternately flood California soils. I. Cause of plant phosphorus deficiency in rice rotation crops and correctional methods. *Soil Sci. Soc. Am. J.* 43:989–994.

Brady, N. C. and R. R. Weil. 2002. *The nature and properties of soils*, 13th edition. Upper Saddle River, NJ: Prentice Hall.

Bruck, H., B. Sattelmacher, and W. A. Payne. 2002. Varietal differences in shoot and rooting parameters of pearl millet on sandy soils in Niger. *Plant Soil* 25:175–185.

Cathcart, J. B. 1980. World phosphate reserves and resources. In: *The role of phosphorus in agriculture*, F. E. Khasawneh, E. C. Sample, and E. J. Kamprath, Eds., 1–18. Madison, WI: Am. Soc. Agron. Madison.

Clark, R. B. 1982. Plant response to mineral element toxicity and deficiency. In: *Breeding plants for less favorable environments*, M. N. Christiansen and C. F. Lewis, Eds., 71–142. New York: John Wiley & Sons.

Clark, R. B. 1990. Physiology of cereals for mineral nutrient uptake, use, and efficiency. In: *Crops as enhancers of nutrient use*, V. C. Baligar and R. R. Duncan, Eds., 131–209. San Diego: Academic Press.

Clark, R. B. and R. R. Duncan. 1991. Improvement of plant mineral nutrition through breeding. *Field Crops Res.* 27:219–240.

Clark, R. B. and S. K. Zeto. 2000. Mineral acquisition by arbuscular mycorrhizal plants. *J. Plant Nutr.* 23:867–902.

Cope, J. T. and C. E. Evans. 1985. Soil testing. *Adv. Soil Sci.* 1:201–228.

Cox, F. R. 1994. Current phosphorus availability indices: Characteristics and shortcomings. In: *Soil testing: Prospects for improving nutrient recommendations*, L. P. Wilding, Ed., 101–113. Madison, WI: Soil Science Society of America.

Curi, N. and O. A. Camargo. 1988. Phosphorus adsorption characteristics of Brazilian Oxisols. In: *Proceedings of the 8th International Soil Classification Workshop*, Part 1, F. H. Beinroth, M. N. Camargo, and H. Eswarn, Eds., 56–63. Washington, DC: Soil Management Support Services, U.S. Department of Agriculture.

Daniel, T. C., A. N. Sharpley, and J. L. Lemunyon. 1998. Agricultural phosphorus and eutrophication: A symposium overview. *J. Environ. Qual.* 27:251–257.

Dinkelaker, B., C. Hengler, and H. Marschner. 1995. Distribution and function of proteoid roots and other root clusters. *Botanica Acta* 108:183–200.

Eastin, J. D. and C. Y. Sullivan. 1984. Environmental stress influences on plant persistence, physiology, and production. In: *Physiological basis of crop growth and development*, M. B. Tesar, Ed., 201–236. Madison, WI: American Society of Agronomy.

Eghball, B., D. H. Sander, and J. Skopp. 1990. Diffusion, adsorption and predicted longevity of banded phosphorus fertilizer in three soils. *Soil Sci. Soc. Am. J.* 54:1161–1165.

Epstein, E. and A. J. Bloom. 2005. *Mineral nutrition of plants: Principles and perspectives.* Sunderland, Massachusetts: Sinauer Associates, Inc. Publishers.

Fageria, N. K. 1980. Rice in cerrado soils with water deficiency and its response to phosphorus. *Pesq. Agropec. Bras.* 15:259–265.

Fageria, N. K. 1984. Response of rice cultivars to liming in cerrado soil. *Pesq. Agropec. Bras.* 19:883–889.

Fageria, N. K. 1989. *Tropical soils and physiological aspects of crops.* Brasilia, Goiânia, Brazil: EMBRAPA-CNPAF.

Fageria, N. K. and V. C. Baligar. 1997. Response of common bean, upland rice, corn, wheat, and soybean to fertility of an Oxisol. *J. Plant Nutr.* 20:1279–1289.

Fageria, N. K. and V. C. Baligar. 2001. Improving nutrient use efficiency of annual crops in Brazilian acid soils for sustainable crop production. *Commun. Soil Sci. Plant Anal.* 32:1303–1319.

Fageria, N. K. and V. C. Baligar. 2004. Properties of termite mound soils and response of rice and bean to nitrogen, phosphorus, and potassium fertilization on such soil. *Commu. Soil Sci. Plant Anal.* 35:2097–2109.

Fageria, N. K. and V. C. Baligar. 2005. Nutrient availability. In: *Encyclopedia of soils in the environment,* D. Hillel, Ed., 63–71. San Diego, CA: Elsevier.

Fageria, N. K., V. C. Baligar, and R. B. Clark. 2006. *Physiology of crop production*. New York: The Haworth Press.

Fageria, N. K., V. C. Baligar, and C. A. Jones. 1997b. *Growth and mineral nutrition of field crops*, 2nd edition. New York: Marcel Dekker.

Fageria, N. K. and M. P. Barbosa Filho. 1987. Phosphorus fixation in Oxisol of Central Brazil. *Fertilizers and Agriculture* 94:33–37.

Fageria, N. K. and M. P. Barbosa Filho. 2007. Dry matter and grain yield, nutrient uptake, and phosphorus use efficiency of lowland rice as influenced by phosphorus fertilization. *Commun. Soil Sci. Plant Anal.* 38:1289–1297.

Fageria, N. K., M. P. Barbosa Filho, L. F. Stone, and C. M. Guimarães. 2004. Phosphorus nutrition in upland rice production. In: *Phosphorus in Brazilian agriculture*, T. Yamada and S. R. S. Abdalla, Eds., 401–418. Piracicaba, Brazil: Brazilian Potassium and Phosphate Institute.

Fageria, N. K. and F. Breseghello. 2004. Nutritional diagnostic in upland rice production in some municipalities of State of Mato Grosso, Brazil. *J. Plant Nutr.* 27:15–28.

Fageria, N. K. and H. R. Gheyi. 1999. *Efficient crop production*. Campina Grande, Paraiba, Brazil. Federal University of Paraiba.

Fageria, N. K., O. P. Morais, V. C. Baligar, and R. J. Wright. 1988. Response of rice cultivars to phosphorus supply on an Oxisol. *Fert. Res.* 16:195–206.

Fageria, N. K. and A. B. Santos. 1998. Phosphorus fertilization for bean crop in lowland soil. *Rev. Bras. Eng. Agric. Amb.* 2:124–127.

Fageria, N. K. and A. B. Santos. 2005. Influence of base saturation and micronutrient rates on their concentration in the soil and bean productivity in cerrado soil in no-tillage system. Paper presented at the *VIII National Bean Congress*, Goiânia, Brazil, 18 to 20 October, 2005.

Fageria, N. K. and A. B. Santos. 2008. Lowland rice response to thermophosphate fertilization. *Commun. Soil Sci. Plant Anal.* 39:873–889.

Fageria, N. K., A. B. Santos, and V. C. Baligar. 1997a. Phosphorus soil test calibration for lowland rice on an Inceptisol. *Agron. J.* 89:737–742.

Fageria, N. K., N. A. Slaton, and V. C. Baligar. 2003. Nutrient management for improving lowland rice productivity and sustainability. *Adv. Agron.* 80:63–152.

Fageria, N. K. and L. F. Stone. 2006. Physical, chemical, and biological changes in the rhizosphere and nutrient availability. *J. Plant Nutr.* 29:1327–1356.

Faye, I., O. Diouf, A. Guisse, M. Sene, and N. Diallo. 2006. Characterizing root responses to low phosphorus in pearl millet (*Pennisetum glaucum* L. R. Br.). *Agron. J.* 98:1187–1194.

Fiedler, R. J., D. H. Sander, and G. A. Peterson. 1989. Fertilizer phosphorus recommendations for winter wheat in terms of method of phosphorus application, soil pH, and yield goal. *Soil Sci. Soc. Am. J.* 53:1282–1287.

Fredeen, A. L., T. K. Raab, I. M. Rao, and N. Terry. 1990. Effects of phosphorus nutrition on photosynthesis in *Glyne max* L. Merr. *Planta* 181:399–405.

Fries, L. L. M., R. S. Pacovsky, G. R. Safir, and J. O. Siqueira. 1997. Plant growth and arbuscular mycorrhizal fungal colonization affected by exogenously applied phenolic compounds. *J. Chem. Ecology* 23:1755–1767.

Friesen, D. K., A. S. R. Juo, and M. H. Miller. 1980a. Liming and lime phosphorus-zinc interactions in two Nigerian Ultisols. I. Interactions in the soil. *Soil Sci. Soc. Am. J.* 44:1221–1226.

Friesen, D. K., M. H. Miller, and A. S. R. Juo. 1980b. Lime and lime-phosphate-zinc interactions in two Nigerian Ultisols. II. Effects on maize root and shoot growth. *Soil Sci. Soc. Am. J.* 44:1227–1232.

Foth, H. D. and B. G. Ellis. 1988. *Soil fertility*. New York: John Wiley & Sons.

Gahoonia, T. S. and N. E. Nielsen. 2004. Root traits as tools for creating phosphorus efficient crop varieties. *Plant Soil* 260:47–57.

Goedert, W. J. 1983. Management of the Cerrado soils of Brazil: A review. *J. Soil Sci.* 34:405–428.

Gorny, A. G. and H. Patyna. 1984. The development of root system in seven spring barley varieties under high and low soil irrigation levels. *J. Agron. Crop Sci.* 153:264–273.

Graham, R. D. 1984. Breeding for nutritional characteristics in cereals. In: *Advances in plant nutrition*, Vol. 1, P. B. Tinker and A. Lauchi, Eds., 57–102. New York: Praeger.

Halsted, M. and J. Lynch. 1996. Phosphorus responses of C_3 and C_4 species. *J. Exp. Bot.* 47:497–505.

Hanna, W. J. and R. L. Flannery. 1960. Current New Jersey research in chemical soil testing. *Agric. Food Chem.* 8:92–94.

Hanway, J. J. 1963. Improving soil test calibrations. *Better Crops Plant Food* (July–August): 33–40.

Haynes, R. J. 1984. Lime and phosphate in the soil-plant system. *Adv. Agron.* 37:249–315.

Heber, U. and H. W. Heldt. 1981. The chloroplast envelope. *Annu. Rev. Plant Physiol.* 33:73–96.

Heckman, J. R., W. Jokela, T. Morris, D. B. Beegle, J. T. Soms, F. J. Coale, S. Herbert, T. Griffin, B. Hoskins, J. Jemison, W. M. Sullivan, D. Bhumbla, G. Estes, and W. S. Reid. 2006. Soil test calibration for predicting corn response to phosphorus in the Northeast USA. *Agron. J.* 98:280–288.

Hewitt, E. J. 1963. The essential nutrients: requirements and interactions in plants. In: *Plant physiology*, F. C. Steward, Ed., 137–360. New York: Academic Press.

Higgs, B., A. E. Johnston, J. L. Salter, and C. J. Dawson. 2000. Some aspects of achieving sustainable phosphorus use in agriculture. *J. Environ. Qual.* 29:80–87.

Hinsinger, P. 1998. How do plant roots acquire mineral nutrients? Chemical processes involved in the rhizosphere. *Adv. Agron.* 64:225–265.

Holland, J. M. 2004. The environmental consequences of adopting conservation tillage in Europe: reviewing the evidence. *Agriculture, Ecosystems and Environment* 103:1–25.

Holford, I. C. P. 1977. Soil phosphorus, its measurements and its uptake by plants. *Aust. J. Soil Res.* 35:227–239.

Howard, D. D., A. Y. Chambers, and J. Logan. 1994. Nitrogen and fungicide effects on yield components and disease severity in wheat. *J. Prod. Agric.* 7:448–454.

Ives, R., M. and C. F. Shaykewich. 1987. Effect of simulated soil erosion on wheat yields on the humid Canadian Prairie. *J. Soil Water Conserv.* 42:205–208.

Jakobsen, I. 1995. Transport of phosphorus and carbon VA mycorrhizas. In: *Mycorrhiza: Structure, function, molecular biology and biotechnology*, A. Varma and B. Hock, Eds., 297–324. Berlin: Springer-Verlag.

Jayachandran, K., A. P. Schwab, and B. A. D. Hetrick. 1989. Micorrhizal mediation of phosphorus availability: Synthetic iron chelates effects on phosphorus solubilization. *Soil Sci. Soc. Am. J.* 53:1701–1706.

Jenkins, J. E. E. and L. Lescar. 1980. Use of foliar fungicides on cereals in eastern Europe. *Plant Disease* 64:987–994.

Johnson, J. F., D. L. Allan, and C. P. Vance. 1996. Phosphorus deficiency in Lupinus albus: Altered lateral root development and enhanced expression of phosphoenolpyruvate carboxylase. *Plant Physiol.* 112:31–41.

Jonathan, P. L. and M. B. Kathleen. 2001. Topsoil foraging: An architectural adaptation of plants to low p availability. *Plant Soil* 237:225–237.

Kamprath, E. J. and M. E. Watson. 1980. Conventional soil and tissue tests for assessing the phosphorus status of soils. In: *The role of phosphorus in agriculture*, F. E. Khasawneh, Ed., 433–469. Madison, WI: ASA, CSSA and SSSA.

Kandeler, E., P. Marschner, D. Tscherko, T. S. Gahoonia, and N. E. Nielsen. 2002. Microbial community composition and functional diversity in the rhizosphere of maize. *Plant Soil* 238:301–312.

Keerthisinghe, G., P. J. Hocking, P. R. Ryan, and E. Delhaize. 1998. Effects of phosphorus supply on the formation and function of proteoid roots of white lupin (*Lupinus albus* L.). *Plant Cell Environ.* 21:467–478.

Kochian, L. V., O. A. Hoekenga, and M. A. Pineros. 2004. How do crop plants tolerate acid soils? Mechanisms of aluminum tolerance and phosphorus efficiency. *Annu. Rev. Plant Biol.* 55:459–493.

Kucey, R. M. N. 1987. Increased phosphorus uptake by wheat and field beans inoculated with a phosphate-solubilizing penicillium bilaji strain with vesicular-arbuscular mycorrhizal fungi. *Applied Environ. Microbiol.* 53:2699–2703.

Kucey, R. M. N., H. H. Janzen, and M. E. Leggett. 1989. Microbially mediated increases in plant-available phosphorus. *Adv. Agron.* 42:199–228.

Kuo, S. 1983. Effects of organic residues and nitrogen transformations on phosphorus sorption desorption by soil. *Agronomy abstracts*, ASA, Madison, Wisconsin.

Lee, K. H., T. M. Isenhart, R. C. Schultz, and S. K. Mickelson. 1997. Nutrient and sediment removal by switchgrass and cool-season grass filter strips. In: *Exploring the opportunity for agroforestry in changing rural landscapes: Proceedings of the fifth conference on agroforestry in North America*, L. E. Buck and J. P. Lassoie, Eds., 68–73. Ithaca, NY: Department of Natural Resources, College of Agriculture and Life Sciences.

Lin, C., W. J. Busscher, and L. A. Douglas. 1983. Multifactor Kinetics of phosphate reactions with minerals in acidic soils. I. Modeling and simulation. *Soil Sci. Soc. Am. J.* 47:1097–1103.

Lin, C. H., R. N. Lerch, H. E. Garrett, D. Jordan, and M. F. George. 2007. Ability of forage grasses exposed to atrazine and isoxaflutole to reduce nutrient levels in soils and shallow groundwater. *Commun. Soil Sci. Plant Anal.* 38:1119–1136.

Lins, I. D. G., F. R. Fox, and J. J. Nicholaides III. 1985. Optimizing phosphorus fertilization rates for soybean grown on Oxisols and associated Entisols. *Soil Sci. Soc. Am. J.* 49:1457–1460.

Loneragan, J. and M. J. Webb. 1993. Interactions between zinc and other nutrients affecting the growth of plants. In: *Zinc in soils and plants*, A. D. Robson, Ed., 119–134. Dordrecht: Kluwer Academic Publishers.

Lynch, J., E. Epstein, A. Lauchli, and G. I. Wiegt. 1990. An automated greenhouse sand culture system suitable for studies of P nutrition. *Plant Cell Environ.* 13:547–554.

Lynch, J., A. Lauchli, and E. Epstein. 1991. Vegetative growth of the common bean in response to phosphorus nutrition. *Crop Sci.* 31:380–387.

Mallarino, A. P. and A. M. Blackmer. 1992. Comparison of methods for determining critical concentrations of soil test phosphorus for corn. *Agron. J.* 84:850–856.

Marschner, H. 1995. *Mineral nutrition of higher plants*, 2nd edition. New York: Academic Press.

Mahler, R. L. and R. E. McDole. 1985. The influence of lime and phosphorus on crop production in northern Idaho. *Commun. Soil Sci. Plant Anal.* 16:485–499.

Massee, T. W. 1990. Simulated erosion and fertilizer effects on winter wheat cropping in an intermountain dryland area. *Soil Sci. Soc. Am. J.* 54:1720–1725.

McDowell, L. L., G. H. Willis, and C. E. Murphree. 1984. Plant nutrient yields in runoff from a Mississippi delta watershed. *Trans. ASAE* 27:1059–1066.

McClellan, G. H. and L. R. Gremillion. 1980. Evaluation of phosphate raw materials. In: *The role of phosphorus in agriculture*, F. E. Khasawneh, E. C. Sample, and E. J. Kamprath, Eds., 43–80. Madison, WI: Am. Soc. Agron.

McDowell, R. W. and A. N. Sharpley. 2001. Approximating phosphorus release from soils to surface runoff and subsurface drainage. *J. Environ. Qual.* 30:508–520.

McCollum, R. E. 1991. Buildup and decline in soil phosphorus: 30-year trends on a typic umprabuult. *Agron. J.* 83:77–85.

Mendez, A., T. A. Dillaha, and S. Mostaghimi. 1999. Sediment and nitrogen transport in grass filter strips. *J. Am. Water Resource Association* 35:867–875.

Mengel, K., A. Kirkby, H. Kosegarten, and T. Appel. 2001. *Principles of plant nutrition,* 5th edition. Dordrecht, The Netherlands: Kluwer Academics.

Menzel, R. G., E. D. Rhoades, A. E. Olness, and S.J. Smith. 1978. Variability of annual nutrient and sediment discharges in runoff from Oklahoma cropland and rangeland. *J. Environ. Qual.* 7:401–406.

Mielke, L. N. and J. S. Schepers. 1986. Plant response to topsoil thickness on an eroded loess soil. *J. Soil Water Conserv.* 41:59–63.

Morris, R. J., R. H. Fox, and G. A. Jung. 1982. Growth, P uptake and quality of warm and cool season grasses on a low available P soil. *Agron. J.* 74:125–129.

Mullins, G. L. 1988. Plant availability of P in commercial superphosphate fertilizers. *Commun. Soil Sci. Plant Anal.* 19:1509–1525.

Mullins, G. L. and F. J. Sikora. 1990. Field evolution of commercial monoammonium phosphate fertilizers. *Soil Sci. Soc. Am. J.* 54:1469–1472.

Neumann, G., A. Massonneau, E. Martonoia, and V. Romheld. 1999. Physiological adaptation to P deficiency during proteoid root development in white lupin. *Planta* 208:373–382.

Neumann, G. and V. Romheld. 1999. Root excretion of carboxylic acids and protons in phosphorus-deficient plants. *Plant Soil* 211:121–130.

Nurlaeny, N., H. Marschner, and E. George. 1996. Effects of liming and mycorrhizal colonization on soil phosphate depletion and phosphate uptake by maize (*Zea mays* L.) and soybean (*Glycine max* L.) grown in two tropical acid soils. *Plant Soil* 181:275–285.

Nye, P. H. and P. B. Tinker. 1977. Solute exchange between solid, liquid and gas phases in the soil. In: *Solute movement in the soil-root system*, 33–68. Oxford, UK: Stud. Ecol. Blackwell Sci.

Olsen, S. R. 1980. Use and limitations of physical-chemical criteria for assessing the status of phosphorus in soils. In: *The role of phosphorus in agricultural*, F. E. Khasawneh, E. C. Sample, and E. J. Kamprath, Eds., 361–410. Madison, WI: ASA, CSSA, SSSA.

Payne, W. A., R. J. Lascano, L. R. Hossner, C. W. Wendt, and A. B. Onken. 1991. Pearl millet growth as affected by P and water. *Agron. J.* 83:942–948.

Peck, T. R. and R. N. Soltanpour. 1990. The principles of soil testing. In: *Soil testing and plant analysis*, R. L. Westerman, Ed., 1–9. Madison, WI: ASA, CSSSA, and SSSA.

Peltonen-Sainio, P., M. Konturi, and J. Peltonen. 2006. Phosphorus seed coating enhancement on early growth and yield components in oat. *Agron. J.* 98:206–211.

Potash and Phosphate Institute. 1979. *Soil fertility manual,* 2nd edition. Atlanta, Georgia: PPI.

Pote, D. H., T. C. Daniel, A. N. Sharpley, P. A. Moore, Jr., D. R. Edwards, and D. J. Nichols. 1996. Relating extractable soil phosphorus to phosphorus losses in runoff. *J. Environ. Qual.* 60:855–859.

Qui, J. and D. W. Israel. 1992. Diurnal starch accumulation and utilization in phosphorus-deficient soybean plants. *Plant Physiol.* 98:316–323.

Reuter, D. J. 1986. Temperate and sub-tropical crops. In: *Plant analysis: An interpretation manual*, D. J. Reuter and J. B. Robinson, Eds., Melbourne: Inkata Press.

Richardson, A. E. 2001. Prospects of using soil microorganisms to improve the acquisition of phosphorus by plants. *Aust. J. Plant Physiol.* 28:897–906.

Ryan, P. R., E. Delhaize, and D. L. Jones. 2001. Function and mechanism of organic anion exudation from plant roots. *Annu. Rev. Plant Physiol.* 52:527–560.

Sah, R. N. and D. S. Mikkelsen. 1986. Transformations of inorganic phosphorus during flooding and draining periods of soil. *Soil Sci. Soc. Am. J.* 50:62–67.

Sah, R. N. and D. S. Mikkelsen. 1989. Phosphorus behavior in flooded-drained soils. I. Effects on phosphorus sorption. *Soil Sci. Soc. Am. J.* 53:1718–1722.

Sah, R. N., D. S. Mikkelsen, and A. A. Hafez. 1989a. Phosphorus behavior in flooded-drained soils. I. Iron transformation and phosphorus sorption. *Soil Sci. Soc. Am. J.* 53:1723–1729.

Sah, R. N., D. S. Mikkelsen, and A. A. Hafez. 1989b. Phosphorus behavior in flooded-drained soils. III. Phosphorus desorption and availability. *Soil Sci. Soc. Am. J.* 53:1729–1732.

Sanchez, P. A. 1976. *Properties and management of soils in the tropics.* New York: John Wiley & Sons.

Sanchez, P. A. and J. G. Salinas. 1981. Low-input technology for managing Oxisols and Ultisols in tropical America. *Adv. Agron.* 34:279–406.

Sanchez, P. A., J. H. Villachica, and D. E. Bandy. 1983. Soil fertility after clearing a tropical rainforest in Peru. *Soil Sci. Soc. Am. J.* 47:1171–1178.

Schmitt, T. J., M. G. Dosskey, and K. D. Hoagland. 1999. Filter strip performance and processes for different vegetation, widths, and contaminants. *J. Environ. Qual.* 28:1479–1489.

Schultz, R. C., J. P. Colletti, T. M. Isenhart, W. W. Simpkins, C. W. Mize, and M. L. Thompson. 1995. Design and placement of a multi-species riparian buffer strip system. *Agroforestry Systems.* 29:201–226.

Schultz, R. C., J. P. Colletti, C. Miza, A. Skadberg, M. Christian, W. W. Simpkins, M. L. Thompson, and B. Menzel. 1991. Sustainable tree-shrubs-grass buffer strips along midwestern-waterway. In: *Proceedings of the second conference on agroforestry in North America*, H. E. Garrett, Ed., 312–326, Columbia, MO: University of Missouri.

Schuman, G. E., R. G. Spomer, and R. F. Piest. 1973. Phosphorus losses from four agricultural watersheds in Missouri Valley loess. *Soil Sci. Soc. Am. Proc.* 37:424–427.

Sharpley, A. N., S. C. Chapra, R. Wedepohl, J. T. Sims, T. C. Daniel, and K. R. Reddy. 1994a. Managing agricultural phosphorus for the protection of surface waters: Issues and options. *J. Environ. Qual.* 23:437–451.

Sharpley, A. N., T. C. Daniel, J. T. Sims, J. Lemunyon, R. Stevens, and R. Parry. 1999. Agricultural phosphorus and eutrophication. University Park, Pennsylvania: USDA-ARS.

Sharpley, A. N., T. C. Daniel, J. T. Sims, and D. H. Pote. 1996. Determining environmentally sound soil phosphorus levels. *J. Soil Water Conser.* 51:160–166.

Sharpley, A. N., B. Foy, and P. Withers. 2000. Practical and innovative measures for the control of agricultural phosphorus losses to water: An overview. *J. Environ. Qual.* 29:1–9.

Sharpley, A. N., J. T. Sims, and G. M. Pierzynski. 1994b. Innovative soil phosphorus availability indices: Assessing inorganic phosphorus. In: *Soil testing: Prospects for improving nutrient recommendations*, L. P. Wilding, Ed., 115–142. Madison, WI: Soil Science Society of America.

Shu, L., J. Shen, Z. Rengel, C. Tang, and F. Zhang. 2007. Cluster root formation by *Lupinus Albus* is modified by stratified application of phosphorus in a split-root system. *J. Plant Nutr.* 30:271–288.

Singh, B. B. and J. P. Jones. 1976. Phosphorus sorption and desorption characteristics of soil as affected by organic residues. *Soil Sci. Soc. Am. J.* 40:389–394.

Shober, A. L. and J. T. Sims. 2003. Phosphorus restrictions for land application of biosolids: Current status and future trends. *J. Environ. Qual.* 32:1955–1964.

Skene, K. R. 2001. Cluster root: model experimental tools for key biological problems. *J. Exp. Bot.* 52:479–485.

Smith, F. W., W. A. Jackson, and P. J. V. Berg. 1990. Internal phosphorus flows during development of phosphorus stress in *Stylosanthes hamata*. *Aust. J. Plant Physiol.* 17:451–464.

Smyth, J. T. and M. S. Cravo. 1990. Phosphorus management for continuous corn-cowpea production in a Brazilian Amazon oxisol. *Agron. J.* 82:305–309.

Smyth, T. J. and M. S. Cravo. 1992. Aluminum and calcium constraints to continuous crop production in a Brazilian Oxisol. *Agron. J.* 84:843–850.

Smyth, T. T. and P. A. Sanchez. 1982. Phosphate rock dissolution and availability in cerrado soils as affected by phosphorus sorption capacity. *Soil Sci. Soc. Am. J.* 46:339–345.

Soil Science Society of America. 1997. *Glossary of soil science terms.* Madison, Wisconsin: Soil Science Society of America.

Stitt, M. 1986. Limitation of photosynthesis by carbon metabolism. I. Evidence for excess electron transport capacity in leaves carrying out photosynthesis in saturating light and CO_2. *Plant Physiol.* 81:1115–1122.

Sumner, M. E. and M. P. W. Farina. 1986. Phosphorus interactions with other nutrients and lime in field cropping systems. *Adv. Soil Sci.* 5:201–236.

Thurman, L. G., K. D. Ritchey, and G. C. Naderman, Jr. 1980. Nitrogen fertilization on an Oxisol of the cerrado of Brazil. *Agron. J.* 72:261–265.

Tinker, P. B., M. D. Jones, and D. M. Durall. 1992. A functional comparison of ecto and endo mycorrhizas. In: *Mycorrhizas in ecosystems*, D. J. Reid, D. H. Lewis, A. H. Fitter, and I. J. Alexander, Eds., 303–310. Wellingford, UK: CAB International.

Tisdale, S. L., W. L. Nelson, and J. D. Beaton. 1985. *Soil fertility and fertilizers*, 4th edition. New York: Macmillan.

Turner, B. L., M. J. Paphazy, P. M. Haygarth, and I. D. McKelvie. 2002. Inositol phosphates in the environment. *Philosophy Transactions of Royal Society of London Series B—Biology Science* 357:449–469.

Usuda, H. and K. Shimogawara. 1991. Phosphate deficiency in maize II. Enzyme activities. *Plant Cell Physiol.* 32:1313–1317.

Vance, C. P., C. Uhde-Stone, and D. L. Allan. 2003. Phosphorus acquisition and use: critical adaptations by plants for securing a nonrenewable resource. *New Phytologist* 157:423–447.

Van Riemsdijk, W. H. and De Haan, F. A. M. 1981. Reaction of orthphosphate with a sandy soil at constant supersaturation. *Soil Sci. Soc. Am. J.* 45:261–266.

Von Uexkull, H. R. and E. Mutert. 1995. Global extent, development and economic impact of acid soil. In: *Plant-soil interactions at low pH: Principles and management*, R. A. Date, N. J. Grundon, G. E. Rayment, and M. D. Probert, Eds., 5–19. Dordrecht, The Netherlands: Kluwer Academic.

Watt, M. and J. R. Evans. 1999. Linking development and determinancy with organic acid efflux from proteoid roots of white lupin grown with low phosphorus and ambient or elevated atmosphere CO_2 concentration. *Plant Physiol.* 120:705–716.

Whitelaw, M. A. 2000. Growth promotion of plants inoculated with phosphate-solubilizing fungi. *Adv. Agron.* 69:99–151.

Willet, I. R. 1982. Phosphorus availability in soils subject to short period of flooding and drying. *Aust. J. Soil Res.* 20:131–138.

Willet, I. R. and M. L. Higgins. 1980. Phosphorus sorption and extractable iron in soils during irrigated rice-upland crop rotation. *Aust. J. Exp. Agric. Anim. Husb.* 20:346–353.

Wilkinson, S. R., D. L. Grunes, and M. E. Sumner. 2000. Nutrient interactions in soil and plant nutrition. In: *Handbook of soil science*, M. E. Sumner, Ed., 89–112. Boca Raton, FL: CRC Press.

Withers, P. J. A. and S. C. Jarvis. 1998. Mitigation option for diffuse phosphorus loss to water. *Soil Use Management* 10:348–354.

Wood, C. W. 1998. Agriculture phosphorus and water quality: An overview. In: *Soil testing for phosphorus: Environmental uses and implications*, J. Thomas, Ed., 5–12. Newark, DE: University of Delaware.

Wright, A. C. and R. E. Gaunt. 1992. Disease-yield relationship in barley. I. Yield, dry matter accumulation and yield loss models. *Plant Pathology* 41:676–687.

Yong, H., L. Hong, and Y. Xiaolong. 2003. Localized supply of P induces root morphological and architectural changes of rice in split and stratified soil cultures. *Plant Soil* 248:247–256.

Zhang, J. and S. A. Barber. 1992. Maize root distribution between phosphorus fertilized and unfertilized soil. *Soil Sci. Soc. Am. J.* 56:819–822.

Zhang, H., J. L. Schroder, J. K. Fuhrman, N. T. Basta, D. E. Storm, and M. E. Payton. 2005. Path and multiple regression analyses of phosphorus sorption capacity. *Soil Sci. Soc. Am. J.* 69:96–106.

4 Potassium

4.1 INTRODUCTION

Potassium (K) is an essential element for all life. It is abundant in nature and occurs in considerable total amounts in most soils. Forms of potassium available for plant uptake, however, are often deficient in soils. Potassium, nitrogen, and phosphorus are considered the major nutrient elements for plants because their use as fertilizers is more widespread and in greater amounts than other elements. Potassium deficiency in crop plants under different agroecosystems is not as common as N and P deficiencies. Furthermore, K^+ deficiency is not as easily identified as N and P deficiencies, which are accompanied by major changes in leaf color and tillering. Figure 4.1 provides a clear example of this. Figure 4.1 shows the response of dry bean to K^+ fertilization in Brazilian Oxisol. There was a reduction in growth of bean plants in the pot that did not receive potassium fertilization compared to the pot that received potassium fertilization. However, plants without K^+ fertilization experienced reduced growth but did not show any foliar symptoms of K deficiency. The reason for this is that nonavailable or fixed K^+ might replenish available K^+ when deficiency of this element occurs under high K^+ demands by crop plants. However, K^+ uptake by field crops is as high as nitrogen uptake. In lowland and upland rice, uptake of K^+ is higher than N uptake (Fageria et al., 1997a; Fageria, 2001; Fageria and Baligar, 2001; Fageria et al., 2003).

Crop yields have significantly increased in the last few decades in developed as well as developing countries through the introduction of modern production technologies; as a result, supplies of K^+ in the soils rapidly depleted. Pretty and Stangel (1985) reported that 17% of the total land area in Africa, 21% of the total land area in Asia, and 29% of the total land area in Latin America are K^+ deficient. Most of the K^+-deficient soils on these three continents are acid savanna soils. Buol et al. (1975) estimated that one-fourth of the soils in the tropics and subtropics have a low K^+ status. Fageria et al. (1990a) reported that 50% of the Amazon basin was characterized by soils with low K^+ reserves. Fageria (1989) also reported that many soils of the tropical and temperate regions are unable to supply sufficient K^+ to field crops. Hence, application of this element in adequate amount is essential for obtaining optimal crop yields and maintaining soil fertility for sustainable crop production. Slaton et al. (1995) and Williams and Smith (2001) reported that prior to the early 1990s, K^+ deficiency of rice was rare in the rice-producing areas of the United States. However, K deficiency is now recognized as an annual problem on many soils as rice and rotation crop yields have increased, soils have been mined of K^+, and production practices have changed.

FIGURE 4.1   Response of dry bean to potassium fertilization. Plants without K (left) and with 200 mg K kg^{-1} soil (right).

On soils where present K levels are considered adequate, the increased use of other fertilizers and production inputs can quickly shift the soil K status from one of adequacy to one of deficiency. Soils that have recently been brought under cultivation may contain relatively high levels of available K$^+$; however, the low level of K$^+$ reserves and poor retention capacity, especially on soils with low cation exchange capacity (CEC) and a preponderance of 1:1 clay minerals, combined with higher crop removal, can result in depletion of available supplies after only a few cropping seasons (Pretty and Stangel, 1985). Furthermore, in many crops, maximum amount of K$^+$ is retained in the straw (Fageria et al., 1990a). Where this is removed for fodder, fuel, or other uses, the depletion of soil K$^+$ will be much too rapid. Hence, supply of adequate K$^+$ rate for field crop production is essential not only to increase productivity but also to reduce the cost of crop production, reduce environmental pollution, and maximize efficiency of K$^+$ use. To achieve these objectives, a better understanding and better management of the dynamics of potassium in soils and plants are required. Hence, the objectives of this chapter are to discuss the latest advances in K nutrition of crop plants.

4.2   CYCLE IN SOIL–PLANT SYSTEMS

Potassium salts were first mined commercially in Germany in 1861 after Liebig's doctrine of mineral nutrition had demonstrated their value as a fertilizer (Sheldrick, 1985). A list of principal K$^+$ minerals, their composition, and their K$^+$ concentration is given in Table 4.1. Canada, the Soviet Union, and Europe have large deposits of K ores. Deposits of potassium ores are also found in some countries of Africa, Asia, and Latin America. Most of the K$^+$ fertilizer in the world is produced in the form of potassium chloride (KCl) or muriate of potash. Potassium, like N, is not easily lost from soil–plant systems and, unlike P, is not immobilized in the soil. The main K$^+$ addition sources for plant growth are chemical fertilizers, crop residues, organic

TABLE 4.1
Principal Potassium Minerals and Ores

Mineral	Composition	K Content (g kg⁻¹)
Sylvite	KCl	524
Niter	KNO_3	386
Glaserite	$3K_2SO_4 \cdot Na_2SO_4$	353
Syngenite	$K_2SO_4 \cdot CaSO_4 \cdot H_2O$	238
Leonite	$K_2SO_4 \cdot MgSO_4 \cdot 4H_2O$	213
Schoenite	$K_2SO_4 \cdot MgSO_4 \cdot 6H_2O$	194
Langbeinite	$K_2SO_4 \cdot 2MgSO_4$	188
Kainite	$4KCl \cdot 4MgSO_4 \cdot 11H_2O$	160
Carnallite	$KCl \cdot MgCl_2 \cdot 6H_2O$	141
Polyhalite	$K_2SO_4 \cdot 2MgSO_4 \cdot 2CaSO_4 \cdot H_2O$	130
Alunite	$K_2 \cdot Al_6 \cdot (OH)_{12} \cdot (SO_4)_4$	95
Krugite	$K_2SO_4 \cdot MgSO_4 \cdot 4CaSO_4 \cdot 2H_2O$	89
Kalinite	$K_2SO_4 \cdot Al_2(SO_4)_3 \cdot 24H_2O$	82
Hanksite	$KCl \cdot 9Na_2SO_4 \cdot 2Na_2CO_3$	25

Source: Adapted from Sheldrick (1985).

manures, and K^+-bearing minerals. The main K^+ depletion sources in the soil are removal by crops, leaching losses, erosion and runoff losses, and immobilization by microorganisms or soil colloids. Potassium fertilizer losses by leaching are minor, except in very sandy soils. A major part of K^+ depletion in soil–plant systems is due to losses by erosion and runoff and uptake by crops. Figure 4.2 shows a simplified K cycle in the soil–plant system.

The majority of K^+ moves to plant roots by diffusion. This diffusion process is dependent on several factors including the soil water content, tortuosity of the diffusion path, temperature, diffusion coefficient of K^+ in water, and the K^+ concentration gradient, particularly if a depletion zone exists around the absorbing roots (Bertsch and Thomas, 1985). The potassium concentration in most soil solution is very low (0.1 to 0.2%) relative to exchangeable K^+ (1 to 2%) because of strong K^+ adsorption by many 2:1 layer silicate minerals. The nonexchangeable K^+ is in the range of 1 to 10%, and 90 to 98% of K^+ occurs as mineral K (Figure 4.2). Hence, nonexchangeable K^+ and mineral K^+ are the major K^+ forms in the soil–plant system. Potassium ions move from one category to another whenever the removal or addition of K^+ disturbs the equilibrium within this soil K^+ pool (Fageria et al., 2003). The ability of a soil to replenish solution K^+ is dependent on the transformations between the various labile K^+ forms and the nature of their respective equilibrium with the soil solution (Bertsch and Thomas, 1985).

Most of the soil K^+ occurs in the crystal lattice structure of silicate minerals, especially feldspars ($MAlSi_3O_8$; M represents combinations of the cations K^+, Na^{2+}, and Ca^{2+}) and micas. Silicate minerals release K^+ slowly by means of the weathering process. Secondary clay minerals, especially the 1:1 clay minerals (kaolinite), are important sources, yielding K^+ more easily than the 2:1 clay minerals

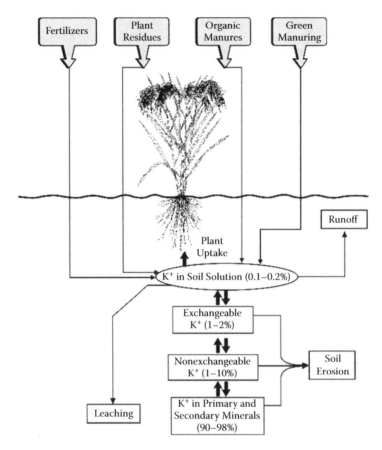

FIGURE 4.2 A simplified K cycle in soil–plant system.

(vermiculite, montmorillonite, micas, hydrous mica, and beidellite; De Datta and Mikkelsen, 1985). The equilibrium between soil solution K^+, exchangeable K^+, and nonexchangeable K^+ is reversible, and soils containing large amounts of vermiculite or K^+-depleted micas can fix large amounts of applied potassium. Both low-charge montmorillonite and kaolinite do not exhibit the K^+ fixation mechanisms (Malavolta, 1985). In the field, with adequate drainage, leaching of soil K^+ may be considerable. Studies in Taiwan summarized by Chang (1971) suggest that K^+ loss by leaching could be substantial. Potassium fixation by clay decreases leaching susceptibility and luxury uptake by crop plants. The leaching of K^+ is determined by the amount of rainfall and soil types. Light-textured soils are more prone to K^+ leaching compared to heavy-textured soils. This may be due to the low cation exchange capacity of light-textured or sandy soil. Malavolta (1985) reported that in Brazilian clayey Oxisol, leaching of K^+ occurred when the potassium fertilizer rate was >125 kg ha^{-1} for corn crop. Fageria et al. (1990b) also studied leaching of K^+ in a Brazilian Inceptisol cultivated with flooded rice (Table 4.2). Data in Table 4.2 show that the extractable K^+ level was higher in the 20–40 cm soil depth at the higher K^+ rates (126 and 168 kg K ha^{-1}) compared to the 20–40 cm soil depth at control treatment (0 kg K ha^{-1}).

TABLE 4.2
Influence of Potassium Fertilizer Application Rates on Exchangeable Soil K⁺ (mg kg⁻¹) at Different Soil Depths

K Applied (kg ha⁻¹)	Soil Depth (cm)	1st Crop	2nd Crop
0	0–20	38	26
0	20–40	27	27
42	0–20	39	26
42	20–40	31	24
84	0–20	44	33
84	20–40	35	32
126	0–20	50	38
126	20–40	39	36
168	0–20	56	44
168	20–40	36	36

Source: Adapted from Fageria et al. (1990b).

Potassium uptake during plant growth is a dynamic process with periods of K⁺ depletion in the root zones and release of nonexchangeable K⁺ to exchangeable and solution factions by K⁺-bearing soil minerals. This process does not result in significant disintegration of the mineral matrix, which can occur with some laboratory methods used to determine nonexchangeable K⁺ supply (Havlin et al., 1985). Havlin et al. (1985) have shown that rate and amount of nonexchangeable K⁺ released during intensive greenhouse cultivation of alfalfa was directly related to clay content of several soils from Colorado, Kansas, and Nebraska. These authors also reported that clay soils had a long-term supply of plant-available K⁺ while light-textured soil did not. After 16 cuttings of alfalfa, the exchangeable K⁺ levels in the clay, loam, and sand-textured soils declined 44, 33, and 58%, respectively. Reddy (1976) reported that medium- and fine-textured Colorado soils were capable of supplying sufficient K⁺ to a 12-metric-ton ha⁻¹ annual alfalfa crop for a period of 3 years while the K⁺ supply in coarse-textured soils was sufficient for only 1 year. Soil chemical properties that influence the phytoavailability of K+ include K⁺ activity in soil solution and K⁺ diffusion rate, buffering capacity, and exchange equilibrium. Activity of K⁺ in the soil solution is reported to represent readily available K⁺.

4.3 FUNCTIONS AND DEFICIENCY SYMPTOMS

Potassium plays many vital roles in crop plants, and the essentiality of K⁺ has been recognized since the work of von Leibig, a German scientist published in 1840. Functions of potassium to increase crop yields can be summarized as follows (Fageria and Gheyi, 1999):

1. K⁺ increases root growth and improves water and nutrient uptake.
2. K⁺ builds cellulose and reduces lodging.
3. K⁺ is required to activate at least 60 different enzymes involved in plant growth.

4. K⁺ reduces respiration, preventing energy losses.
5. K⁺ aids in photosynthesis and food formation.
6. K⁺ helps translocation of sugars and starch.
7. K⁺ produces grain rich in starch.
8. K⁺ increases the protein content of plants.
9. K⁺ maintains turgor and reduces water loss and wilting.
10. K⁺ helps retard crop diseases.
11. Although K⁺ is not a constituent of chlorophyll, a characteristic symptom of K⁺ deficiency is the destruction of chlorophyll. This means it is suspected that part of the function of K⁺ is related to the formation of chlorophyll precursor or to the prevention of the decomposition of chlorophyll.
12. K⁺ deficiency can inhibit N fixation by reducing plant growth in legumes.
13. K⁺ neutralizes acids produced during metabolism of carbohydrates in the plant cell.
14. K⁺ is intimately involved in the opening and closing of stomata.
15. K⁺ is implicated in increased uptake and transport of Fe in both monocotyledonous and dicotyledonous plants.
16. Not only can K⁺ increase the resistance of plant tissues, but it may also reduce fungal populations in the soil, reduce their pathogenicity, and promote more rapid healing of injuries (Huber and Arny, 1985).

Potassium, like N and P, is highly mobile in plant tissues. Hence, K⁺ deficiency symptoms first appear in the older leaves. Potassium deficiency symptoms show up as scorching along leaf margins of older leaves (Figure 4.3). Potassium-deficient plants grow slowly. They have poorly developed root systems. Stalks are weak, and lodging is common. Seed and fruit are small and shriveled, and plants possess low resistance to disease. Plants under stress from short K⁺ supplies are very susceptible

FIGURE 4.3 Growth of rice plants without and with K in nutrient solution.

to unfavorable weather. Although it cannot be detected as it is happening, stand loss in forage grasses and legumes is a direct result of K^+ deficiency. In grass/legume pastures, the grass crowds out the legume when the K^+ runs short, because the grass has a greater capacity to absorb K and the legume is starved out.

4.4 CONCENTRATION AND UPTAKE

Plant tissue analysis is a useful diagnostic technique for determining the adequate level of a nutrient during a particular growth stage. Plant analysis is based on the concept that the concentration of an essential element in a plant or part of a plant indicates the soil's ability to supply that nutrient. This means it is directly related to the quantity in the soil that is available to the plant. For annual crops, the primary objective of plant analysis is to identify nutritional problems or to determine or monitor the nutrient status during the growing season. Like soil analysis, plant analysis also involves plant sampling, plant tissue preparation, analysis, and interpretation of analytical results. All these steps are important for a meaningful plant analysis program. Many factors such as soil, climate, plants, and their interactions affect absorption of nutrients by growing plants. However, the concentrations of the essential nutrients are maintained within rather narrow limits in plant tissues. Such consistency is thought to arise from the operation of delicate feedback systems, which enable plants to respond in a homeostatic fashion to environmental fluctuations (Fageria and Baligar, 2005).

For the interpretation of plant analysis results, a critical nutrient concentration concept was developed. This concept is widely used now in the interpretation of plant analysis results for the diagnosis of nutritional disorders. Critical nutrient concentration is usually designated as a single point within the bend of the crop-yield–nutrient-concentration curve where the plant nutrient status shifts from deficient to adequate. The critical nutrient concentration has been defined in several ways (Fageria and Baligar, 2005): (1) the concentration that is just deficient for maximum growth; (2) the point where growth is 10% less than the maximum; (3) the concentration where plant growth begins to decrease; and (4) the lowest amount of element in the plant accompanying the highest yield. Nutrient concentration in plant tissue decreased with the advancement of plant age. Figure 4.4 and Figure 4.5 show that K^+ concentration in the shoot of upland rice and dry bean decreased with the advancement of plant age. Hence, for nutrient sufficiency or deficiency diagnosis purposes, plant tissue analysis should be done at different growth stages. Data in Table 4.3 show adequate concentration of K^+ in plant tissues of principal crops.

Data related to the quantity of nutrients accumulated during a cropping season give an idea of fertility depletion and help in managing soil fertility for sustainable crop production. Grains as well as straw should be analyzed and their weights per unit area should be determined to calculate total accumulation. Uptake of K^+ in the shoot of upland rice and common bean during the crop growth cycle of these two crops is presented in Figure 4.4 and Figure 4.5. In both crops, potassium accumulation responded quadratically with the advancement of plant age. In upland rice, variation in K^+ uptake was 98% due to plant age, and in dry bean, the variability in K^+ uptake was 95% due to plant age. This means that plant age plays a significant role

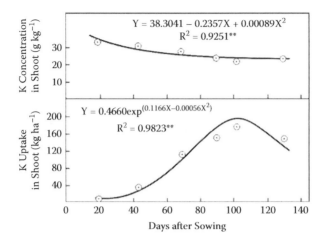

FIGURE 4.4   Potassium concentration and uptake in upland rice as a function of plant age.

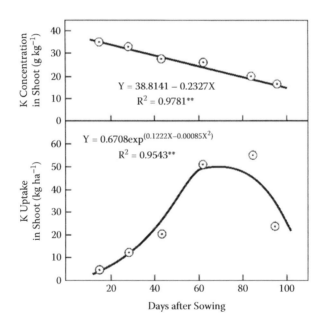

FIGURE 4.5   Potassium concentration and uptake in dry bean as a function of plant age.

in accumulation of K^+ in crop plants. Uptake of K^+ in the shoot of these two crops followed shoot dry weight during their growth cycles (Figure 4.6). The variation in shoot weight of upland rice was 99% due to plant age. Similarly, the variability in shoot weight of dry bean was 96% due to plant age.

Quantity of K^+ accumulated in shoot and grain of principal crops is presented in Table 4.4. From these data, it is clear that the majority of K^+ is accumulated in the shoot of cereals like rice and corn and a small amount is translocated to grain.

TABLE 4.3

Adequate Level of K⁺ in Plant Tissue of Principal Field Crops

Crop Species	Growth Stage	Plant Part	Adequate K Level (g kg⁻¹)
Barley	Tillering	Leaves	42–47
Barley	Heading	Whole tops	15–30
Wheat	Tillering	Leaf blade	34–42
Wheat	Heading	Whole tops	15–30
Rice	75 days after sowing	Whole tops	15–40
Rice	Flowering	Whole tops	12–30
Corn	30 to 45 days after emergence	Whole tops	30–50
Corn	Before tasseling	Leaf blade below	20–25
Sorghum	Seedling	Whole tops	30–45
Sorghum	Early vegetative	Whole tops	25–40
Sorghum	Bloom	3rd blade below panicle	10–15
Soybean	Prior to pod set	Upper fully developed trifoliate	17–25
Dry bean	Early flowering	Uppermost blade	15–35
Cowpea	Early flowering	PUMB	1.2–6.0
Peanut	Early pegging	Upper stems and leaves	17–30

Note: PUMB = Petiole of uppermost mature leaf blade.

Source: Piggott (1986); Reuter (1986); Fageria et al. (1997a).

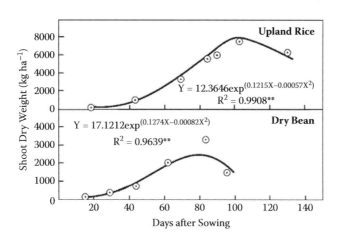

FIGURE 4.6 Shoot dry weight of upland rice and dry bean as influenced by plant age.

Hence, incorporation of cereal straw into the soil may have a significant effect on recycling of K⁺ in soil–plant systems. In dry bean, about 60% of the total K⁺ accumulated was translocated to grain and 40% remained in the shoot. In soybean, 50% of the accumulated K⁺ was translocated to grain and 50% remained in the shoot. Overall, cereals accumulated more K⁺ in shoot plus grain compared to legumes. This

TABLE 4.4
Shoot Dry Weight, Grain Yield, and Uptake of K+ by Principal Field Crops

Crop Species	Shoot Dry Wt. (kg ha⁻¹)	Grain Yield (kg ha⁻¹)	K+ Uptake in Shoot (kg ha⁻¹)	K Uptake in Grain (kg ha⁻¹)	Total (kg ha⁻¹)
Upland rice	6343	4568	150	56	206
Lowland rice	8093	5000	138	14	152
Lowland rice	9423	6389	156	20	176
Corn	11873	8501	153	34	187
Dry bean	2200	3409	41	64	105
Dry bean	1481	1912	25	36	61
Soybean	2225	1441	30	28	58
Soybean	3244	3102	76	80	156

Source: Fageria (2001); Fageria (2004); Fageria and Santos (2008); Fageria et al. (2007).

was associated with higher shoot yield of cereals compared to legumes. The higher K+ content of grain legumes compared to cereals mirrors the especial importance of legumes in human nutrition.

4.5   GRAIN HARVEST INDEX AND K HARVEST INDEX

Grain harvest index (GHI = grain yield/grain plus straw yield) and potassium harvest index (KHI = potassium uptake in grain/potassium uptake in grain plus straw) are important indices in the determination of crop yields and potassium distribution in plants, respectively. These two indices are influenced by potassium fertilization in dry bean crop (Table 4.5). Data in Table 4.5 show that GHI and KHI had significant genotype × K level interactions, suggesting variation in these two indices with increasing K+ levels. The GHI varied from 0.30 to 0.55 at low level of K+ and 0.42 to 0.57 at higher K+ level. Across the 10 genotypes, GHI was 0.46 at low K+ level and 0.51 at high K+ level. This means that an adequate K+ level improves GHI in dry bean. Snyder and Carlson (1984) reported that GHI of 23 cultivars of dry bean varied from 0.39 to 0.58. Results of grain GHI presented in Table 4.5 fall in this range. Efficiency of grain production in crop plants is frequently expressed as GHI. Sinclair (1998) and Hay (1995) have stated that GHI is an important trait associated with the dramatic increase in crop yields during the 20th century. The GHI reflects the partitioning of photosynthate between the grain and the vegetative plant part, and improvements in GHI emphasize the importance of carbon allocation in grain production (Fageria, 1992). Snyder and Carlson (1984) reviewed the relation of GHI to grain yield in legumes and cereals and reported that GHI correlated positively with grain yield. Fageria et al. (2001) also found significant and positive correlation between grain yield and grain harvest index in dry bean (Table 4.6). Similarly, GHI displays a significant quadratic relationship with grain yield of upland rice (Figure 4.7).

TABLE 4.5
Grain Harvest Index and Potassium Harvest Index at Two K⁺ Levels of 10 Dry Bean Genotypes

Genotype	Grain Harvest		K Harvest Index	
	0 mg K kg⁻¹	200 mg K kg⁻¹	0 mg K kg⁻¹	200 mg K kg⁻¹
Iraí	0.51a	0.52ab	0.48ab	0.50ab
Jalo precoce	0.44ab	0.43b	0.62ab	0.42ab
Novo jalo	0.51a	0.51ab	0.63ab	0.53ab
L 93300166	0.30b	0.51ab	0.50ab	0.51ab
L 93300176	0.51a	0.57a	0.53ab	0.37ab
Carioca	0.52a	0.50ab	0.67a	0.57a
Diamante negro	0.41ab	0.48ab	0.31b	0.51ab
Pérola	0.47a	0.57a	0.59ab	0.46ab
Rosinha G-2	0.55a	0.57a	0.68a	0.53ab
Xamego	0.42ab	0.42b	0.55ab	0.33b
Average	0.46	0.51	0.56	0.47
F-test				
K rate (K)	**		**	
Genotype (G)	**		**	
K X G	**		*	
CV(%)	9		19	

Note: Means in the same column followed by the same letter are not significantly different at the 5% probability levels by Tukey's test.

*,** Significant at the 5 and 1% probability levels, respectively.

Source: Fageria et al. (2001).

TABLE 4.6
Correlation (r) between Dry Bean Grain Yield and Growth, Yield Components, and Potassium Harvest Index

Plant Parameter	Grain Yield
Dry matter yield	0.78**
Number of pod	0.62**
Number of grain	0.82**
Grain harvest index	0.39**
100 grain weight	−0.03[ns]
K harvest index	−0.21[ns]

**,[ns] Significant at the 1% probability level and not significant, respectively.

Source: Fageria et al. (2001).

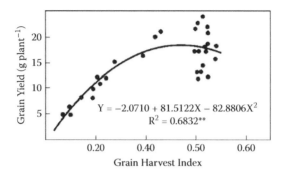

FIGURE 4.7 Relationship between grain harvest index and grain yield of lowland rice.

KHI varied from 0.31 to 0.68 at low K level and 0.33 to 0.57 at higher K level (Table 4.5). Across the 10 genotypes, KHI values were 0.56 at low K level and 0.47 at high K level. This means that improving KHI did not improve grain yield. This is also confirmed by correlation analysis as shown in Table 4.6.

4.6 USE EFFICIENCY

Nutrient absorption and nutrient use efficiency of a crop species or variety are important aspects for the improvement of crop yield and the reduction of production costs (Fageria et al., 2003). Nutrient use efficiency has been expressed in several ways, as discussed in chapters 2 and 3. Efficiency in a broader sense is output divided by input: the higher the value of a quotient, the higher the efficiency of the applied input. Potassium use efficiency of upland rice, corn, dry bean, and soybean, which is calculated by dividing the shoot dry weight or grain yield by K^+ accumulated in the shoot or grain, is presented in Table 4.7. Regression analysis showed that K use efficiency in the shoot of upland rice, corn, and dry bean as a function of plant age increased in a quadratic fashion. In soybean shoot weight production, K use efficiency increased in a linear fashion with the advancement of plant age. At harvest, K use efficiency was higher in corn, followed by soybean, dry bean, and upland rice. In grain production, K use efficiency was higher in cereals (upland rice and corn) compared to legumes (dry bean and soybean). The higher K use efficiency in cereals was associated with lower accumulation of K^+ in the grain of cereals compared to legumes (Table 4.7).

Potassium efficiency expressed as apparent recovery efficiency has especial significance in defining K^+ balance and mineral nutrition of crops. Apparent recovery efficiency (ARE) can be calculated by using the following formula (Fageria, 1992):

$$ARE\ (\%) = \left(\frac{\begin{array}{c}\text{Potassium uptake in grain plus straw at applied K rate} -\\ \text{Potassium uptake in grain plus straw at zero K rate}\end{array}}{\text{K rate applied}} \right) \times 100$$

The AREs of 15 upland rice genotypes grown on Brazilian Oxisol are presented in Table 4.8. The AREs varied from 44 to 88% depending on genotype, with an average

TABLE 4.7

Potassium Use Efficiency in Upland Rice, Corn, Dry Bean, and Soybean during Growth Cycle Grown on Brazilian Oxisols

Age in Days after Sowing	Upland Rice (kg kg^{-1})	Age in Days after Sowing	Corn (kg kg^{-1})	Age in Days after Sowing	Dry Bean (kg kg^{-1})	Age in Days after Sowing	Soybean (kg kg^{-1})
19	30	18	29	15	32	27	36
43	34	35	26	29	34	41	44
69	36	53	35	43	45	62	51
90	43	69	51	62	41	82	56
102	46	84	69	84	57	120	59
130	43	119	94	96	71	158	73
130 (grain)	239	119 (grain)	306	96 (grain)	54	158 (grain)	52
Regression	Quadratic		Quadratic		Quadratic		Linear
R^2	0.78*		0.97**		0.92*		0.95**

Note: Potassium use efficiency = (shoot or grain yield/K accumulated in shoot or grain).

*, ** Significant at the 5 and 1% probability level, respectively.

value of 66%. Fageria et al. (2003) reported that K$^+$ recovery efficiency in lowland rice genotypes ranged from 51 to 81% and varied among lowland rice genotypes. This means that K$^+$ recovery efficiency in crop plants is much higher compared to N and P use efficiencies.

In addition to recovery efficiency, agronomic efficiency (AE), defined as grain yield obtained with the application of per unit nutrient, is an important index in determining crop yields. AE can be calculated with the help of the following formula (Fageria, 1992):

$$AE \ (kg \ kg^{-1}) = \frac{Grain \ yield \ at \ applied \ K \ level \ in \ kg - Grain \ yield \ at \ zero \ K \ level \ in \ kg}{K \ level \ applied \ in \ kg}$$

Fageria et al. (2003) reported that AE in lowland rice varied from 44 to 80 kg kg^{-1}, with an average value of 66 kg kg^{-1}. Generally, AE is positively associated with grain yield in crop plants. Data in Figure 4.8 show that AE has significant positive linear association with grain yield in lowland rice. Fageria (2000a) reported significant positive association of AE and ARE with grain yield of upland rice genotypes. This means that improving potassium use efficiency in crop plants can improve their yields.

4.7 INTERACTION WITH OTHER NUTRIENTS

Ionic interactions may occur as cation–cation interactions, anion–anion interactions, or cation–anion interactions. Greater knowledge of this subject is warranted, due to the intensification of agriculture, in devising efficient fertilization systems (Fageria, 1983). Potassium interaction with other nutrients is an important aspect in improving

TABLE 4.8

Apparent Recovery Efficiency of 15 Upland Rice Genotypes Grown on Brazilian Oxisol

Genotype	Apparent Recovery Efficiency (%)
Rio Paranaíba	88a
CNA6975-2	76ab
CNA7690	64abc
L141	64abc
CNA7460	79ab
CNA6843-1	76ab
Guarani	44c
CNA7127	62abc
CNA6187	68abc
CNA7911	61abc
CNA7645	81ab
CNA7875	57bc
CNA7680	54bc
CNA6724-1	59bc
CNA7890	58bc
Average	66

Note: Means followed by the same letter in the same column do not differ significantly at the 5% probability level by Duncan multiple range test.

Source: Adapted from Fageria (2000a).

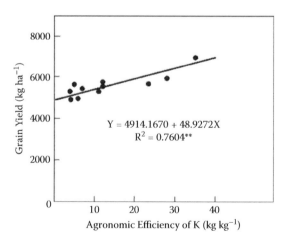

$$Y = 4914.1670 + 48.9272X$$
$$R^2 = 0.7604**$$

FIGURE 4.8 Relationship between agronomic efficiency of potassium and grain yield of lowland rice.

crop yields. Potassium plays several roles in plant metabolism, and to perform these roles positively, it should interact positively with other essential nutrients. Positive interactions of K with N and P have been reported (Dibb and Thomson, 1985). Dibb and Welch (1976) reported that the increased K allowed for rapid assimilation of absorbed NH_4^+ ions in the plant, maintaining a low, nontoxic level of NH_3. Increased yield of crops with the addition of N and P requires higher level of K in the soil (Dibb and Thompson, 1985; Fageria et al., 1997a, 1997b).

Antagonistic interaction between K^+ and Mg^{2+} and Ca^{2+} uptake has been widely reported (Johnson et al., 1968; Fageria, 1983; Dibb and Thompson, 1985). Fageria (1983) reported that reduction in Ca uptake with increasing K concentration in the growth medium was closely associated with increased uptake of K, indicating that there may have been a competitive effect. A competition between K and Ca^{2+} and Mg^{2+} due to physiological properties of these ions has been reported (Johnson et al., 1968; Fageria, 1983).

Potassium and micronutrient interactions have been observed with many crop plants. Hill and Morrill (1975) and Gupta (1979) reported that high K rates reduced B uptake and intensified B deficiency in crop plants. Dibb and Thompson (1985) reviewed interaction between K^+ and Cu in crop plants and reported that Cu^{2+} uptake increased with the addition of K^+. Matocha and Thomas (1969) reported that Fe application with K^+ increased sorghum yield. Fageria (1984) reported that Fe^{2+} toxicity in flooded rice reduced with the addition of adequate rate of K^+ in the soil. These authors also reported that plants with adequate K^+ have more metabolic activity in the roots and high levels of Fe^{2+}-excluding power. Dibb and Thompson (1985) in a review reported that K^+ improved Mn^{2+} uptake when this element is in low concentration in the growth medium and decreased its uptake when it was present in higher concentration that might be toxic. Beneficial effect of K on the uptake of Zn^{2+} has been reported (Dibb and Thompson, 1985).

4.8 MANAGEMENT PRACTICES TO MAXIMIZE K USE EFFICIENCY

Appropriate management practices can improve potassium use efficiency in crop plants and result in efficient crop production. The following important practices can be adopted in this respect: (1) liming acid soils, (2) applying at adequate rate, (3) applying at appropriate time, (4) using appropriate method of application, (5) using efficient crop species and genotypes within species, (6) incorporating crop residues, (7) supplying adequate moisture, (8) using farmyard manures, and (9) maintaining adequate K^+ saturation ratio. These management practices are briefly discussed in the succeeding section. In addition to these, other practices such as rotating crops, adopting conservation tillage, improving organic matter content of soil, and controlling diseases, insects, weeds, and erosion, which are discussed for N and P nutrients, can be employed to promote efficient use of potassium in crop production.

4.8.1 LIMING ACID SOILS

Liming is the best way to improve fertilizer use efficiency in acid soils. Liming acid soils not only decreases soil Al but also can increase retention of applied K^+ and

TABLE 4.9

Influence of Liming on Chemical Properties of an Oxisol

Lime Rate (g kg⁻¹ soil)	pH in H₂O	Ca (cmol$_c$ kg⁻¹)	Mg (cmol$_c$ kg⁻¹)	Al³⁺ (cmol$_c$ kg⁻¹)	Base Saturation (%)
0	4.6	0.9	1.0	0.7	22
6	5.7	5.6	2.8	0.0	68
12	6.2	7.4	2.8	0.0	83
18	6.4	7.8	2.3	0.0	90
24	6.6	8.0	2.2	0.0	90
36	6.8	9.2	2.1	0.0	94
R^2	0.99**	0.99**	0.96**	0.98**	0.99**

** Significant at the 1% probability level.
Source: Fageria (2000b).

TABLE 4.10

Relationship between Soil pH (X) and Shoot Dry Weight (Y), Grain Yield (Y), and Panicle Number of Upland Rice (Y)

Plant Parameter	Regression Equation	R^2	Adequate pH
Shoot dry weight	$Y = -73.64 + 35.68X - 3.53X^2$	0.99**	5.1
Grain yield	$Y = -94.26 + 39.98X - 3.72X^2$	0.91*	5.4
Panicle number	$Y = -8.19 + 5.34X - 0.54X^2$	0.94*	5.0

*, ** Significant at the 5 and 1% probability level, respectively.
Source: Fageria (2000b).

thereby decrease K^+ leaching. Liming increases K^+ retention in soils by replacing Al^{3+} on the exchange sites with Ca^{2+}, allowing K^+ to compete better for exchange sites and increasing the effective cation exchange capacity. Data in Table 4.9 show that liming a Brazilian Oxisol increases pH, Ca^{2+}, and Mg^+ concentrations and base saturation and decreases Al^{3+} concentration. Improvements in these chemical properties improve upland rice yield and lead to higher K use efficiency (Fageria, 2000b). Fageria (1989) also reported that improvements in these properties increase grain yield of many crops in acid soils. Data in Table 4.10 show that improvement in soil pH with liming increases shoot dry weight grain yield and number of panicles in upland rice. An increase in yields means an improvement in K^+ use efficiency in crop plants.

Liming also helps prevent K^+ leaching from soil–plant systems. Nemeth (1982a, 1982b) analyzed Malaysian and Brazilian soils before and after lime application. In both cases, it was found that desorption, that is, soil solution K^+, decreased when lime was applied. The occupation of selective adsorption sites by K^+ prevents its removal by leaching. Mielniczuk (1977) also reported that K^+ buffering capacity of Brazilian soils from Rio Grande do Sul increased with liming. The increase in

K^+ buffering capacity by liming indicates a rise in K^+ adsorption; thus, K^+ becomes less subject to loss by leaching (Malavolta, 1985). Liming can also improve root development in the limed soil depth by neutralizing Al^{3+} and improving pH and Ca^{2+} and Mg^{2+} contents, which can improve uptake of water and nutrients. Muzilli (1982) studied the response of soybean to levels of potassium fertilization in the presence and absence of lime in a Brazilian Oxisol. Soybean yield response to K was higher in limed plots compared to unlimed plots. Leaf analysis for K also showed a significant increase in K content in the limed plots compared to unlimed plots.

4.8.2 APPROPRIATE SOURCE

Potassium chloride is the major source of single K^+ fertilization; potassium sulfate and potassium-magnesium sulfate are minor sources of K^+ to field crops. Potassium chloride is widely used because of its lower cost of production and high analysis. Use of potassium sulfate may be useful in areas where S deficiency is reported. However, it should not be used in flooded rice where sulfate reduction and hydrogen sulfide toxicity are problems (De Datta and Mikkelsen, 1985). Similarly, use of KCl is discouraged in saline soils, because it may increase the salt concentration in the rhizosphere. Potassium is also used in the form of formulated fertilizer mixtures such as $N-P_2O-K_2O$. In Brazil, fertilizer mixture of 4-30-16 or 4-20-20 is very common and used for fertilization of annual crops. The nutrient contents of these fertilizer mixtures are expressed as percentages. Principal K^+ fertilizers are presented in Table 4.11. All the potassium fertilizers are highly water soluble, except potassium metaphosphate (Fageria, 1989). In general, crop yields are comparable whether one or the other source is used, though this may depend on the soil, the crop, or the manner of application (Stewart, 1985). However, the fertilizer source selected by farmers mainly depends on cost of transportation and handling convenience in application.

4.8.3 ADEQUATE RATE OF APPLICATION

Due to the introduction of high-yielding cultivars and the increase of multiple cropping indices, yield per unit area has increased around the world, resulting in the removal of considerable quantities of K^+ from soils. Due to this fact, since the early

TABLE 4.11
Principal Potassium Fertilizers, Their Potassium Content, and Their Solubility

Common Name	Formula	K_2O (%)	Solubility
Potassium chloride	KCl	60	Water soluble
Potassium sulfate	K_2SO_4	50	Water soluble
Potassium-magnesium sulfate	$K_2SO_4 \cdot MgSO_4$	23	Water soluble
Potassium nitrate	KNO_3	44	Water soluble
Kainit	$MgSO_4 + KCl + NaCl$	12	Water soluble
Potassium metaphosphate	KPO_3	40	Low water solubility

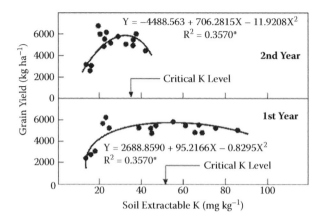

FIGURE 4.9 Relationship between soil extractable potassium and grain yield of lowland rice (Fageria and Barbosa Filho, 2007).

1970s, a very rapid increase in global K^+ consumption has occurred, and in spite of periodic market adjustments, this increase will continue in the long run. Adequate fertility management requires knowledge of crop response to K fertilizer. Adequate rate of K application varied from crop species to crop species and within genotypes of the same species. Other factors that determine K rates are yield level, environmental factors, and crop management practices. Hence, supplying an adequate rate of K^+ for crops is fundamental to improve yields and potassium use efficiency.

The most important criteria to achieve an adequate K application rate is the crop response to applied K and the corresponding soil analysis values of a given nutrient determined by a soil test calibration study. Dahnke and Olson (1990) defined soil test calibration as the process of ascertaining the degree of limitation to crop growth or the probability of getting a growth response to an applied nutrient at any soil test level. The amount of extractable nutrient is usually expressed as low, medium, or high or as a range of critical concentration. The final step is to develop fertilizer recommendations (Dahnke and Olson, 1990). Figure 4.9 shows the relationship between soil extractable K^+ and grain yield of lowland rices grown on Brazilian Inceptisol. Similarly, soil test values for making dry bean potassium recommendations in various states of Brazil are presented in Table 4.12. If soil test calibration data are not available, potassium fertilizer recommendations can also be made on the basis of crop response curves to applied potassium. In this case, maximum economic rate of K^+ is determined or defined on the basis of 90% maximum yield. Sometimes, absolute yield is transformed into relative yield and a relationship between potassium fertilizer rates and relative yield is determined. In this way, it is easy to determine economic K^+ rate at 90% relative yield. Relative yield can be calculated by using the following formula (Fageria et al., 1997a):

$$\text{Relative yield } \% = \left(\frac{\text{Yield of control or fertilized plot}}{\text{Maximum yield of fertilized plot}} \right) \times 100$$

TABLE 4.12

Soil Test Values for Potassium Recommendations for Dry Bean in Brazil

State	Exchangeable K Level in the Soil (mg kg^{-1})		
	Low	Medium	High
Rio Grande do Sul & Santa Catarina	4–60	61–80	81–120
Paraná	<39	40–117	118–235
São Paulo	<27	48–59	60–117
Espírito Santo	<30	30–60	>60
Minas Gerais	<30	31–60	>60
Pernambuco	<15	16–30	31–45
Goiás	<25	26–50	>50

Source: Moraes (1988) and Fageria (2002).

According to Fageria et al. (1997a), a wide scattering of absolute yields may occur as a result of factors other than soil fertility. This scattering of absolute yields does not necessarily mean that there is poor correlation, but a better relationship may be obtained by using a relative yield to eliminate some of the climate and site influences.

The absolute minimum level of exchangeable K$^+$ for tropical agriculture is considered to be close to 0.10 cmol$_c$ kg^{-1} (39 mg kg^{-1}) but may vary from 0.07 to 0.20 cmol$_c$ kg^{-1} (27 to 78 mg kg^{-1}) depending on the kinds of soil and crop species involved (Fageria and Gheyi, 1999). A level below 0.13 cmol$_c$ K kg^{-1} is usually inadequate to support normal plant growth of most crop species (Fageria and Gheyi, 1999). Muns (1982) reviewed the suggested critical concentrations of exchangeable K$^+$ in a number of tropical and subtropical soils. Deficiencies were reported to occur in soil having concentrations ranging from 0.15 to 0.45 cmol$_c$ kg^{-1}, depending on soil characteristics and crop requirements. According to Mahapatra and Prasad (1970), an exchangeable K$^+$ content of about 0.2 cmol$_c$ kg^{-1} is considered a satisfactory level for flooded rice. This value has been suggested as appropriate to upland rice in Latin America (De Datta and Mikkelsen, 1985).

4.8.4 Appropriate Time of Application

Efficient use of K$^+$ fertilizer for crop production is essential to maximize economic return to the grower. Timing of fertilizer K$^+$ application is an important management tool in this effort. Maximum efficiency is obtained when K$^+$ is applied so that it is available for uptake by the plants as needed. Generally, K$^+$ fertilizer is applied as a basal application at the time of sowing because of its relative immobility in clay soils. In contrast to N, few studies have been conducted to define the best time for K$^+$ application. However, the large losses of potassium are attributable to leaching and runoff. Oxisols and Ultisols, two predominant tropical soil groups, have a very low

TABLE 4.13

Influence of Potassium Fertilizer Application Rates on Extractable K (mg kg⁻¹) at Four Soil Depths in an Oxisol of Central Brazil (Soil Analysis Was Done after Harvest of Each Upland Rice Crop)

K Applied (kg ha⁻¹)	Soil Depth (cm)	1st Crop	2nd Crop	3rd Crop
0	0–20	30	26	31
0	20–40	19	20	22
0	40–60	16	15	15
0	60–80	14	14	13
42	0–20	31	29	45
42	20–40	21	24	27
42	40–60	17	18	17
42	60–80	18	16	16
84	0–20	34	31	50
84	20–40	24	28	33
84	40–60	20	18	21
84	60–80	19	19	21
126	0–20	37	34	55
126	20–40	29	37	40
126	40–60	24	23	25
126	60–80	23	24	25
168	0–20	43	40	59
168	20–40	29	42	46
168	40–60	23	25	30
168	60–80	24	28	30
F-test				
K rate	**			
Soil depth (D)	**			
K X D	**			

** Significant at the 1% probability level.

Source: Adapted from Fageria et al. (1990c).

cation exchange capacity (CEC). They do not contain K-fixing minerals, so there is little chance of large amounts of K being retained in these soils. Table 4.13 shows extractable K^+ values in an Oxisol of central Brazil after harvest of each upland rice crop under field conditions. Significant differences ($P < 0.01$) in extractable K existed with K application rate and depth. Extractable K increased with K application rate and decreased with soil depth (Table 4.13). Applied K increased extractable K in both the surface and the subsoil. Leaching of K^+ to lower depths was especially noticeable at the higher K rates. In this situation, split application of K^+ may be an appropriate management practice to reduce K^+ losses by leaching and improving K^+ utilization efficiency by crop plants.

Split application of K^+ is also recommended when high rates are required and fertilizer is applied in the furrow. High rate (>500 kg ha^{-1} fertilizer mixture of 4-30-16) may create high salt index of the fertilizer placed too close to the seed or to the root system. Generally, split application of K^+ is done along with top-dressing of N in annual crops. For example, in lowland rice in Brazil, a beneficial effect of top-dressing of K^+ (half of the recommended rate is applied at sowing and the remaining half is applied at 45 days after emergence along with N) has been observed (Fageria, 1991a).

Malavolta (1985) reported beneficial effects of split application of K^+ in several annual crops in Brazil. Similarly, Haque et al. (1982) in Bangladesh and Ismunadji et al. (1982) in Indonesia also reported yield increases in lowland rice with the split application of K^+. Kim and Park (1973) reported that split application of K^+ significantly increased rice yield on heavy sulfate soils in South Korea. From experiments conducted in Japan and Taiwan, Su (1976) reported the beneficial effect to rice of split application of K^+. Similar benefits of K top-dressing in rice have been reported in India (Singh and Kumar, 1981).

4.8.5 Appropriate Method of Application

Generally, fertilizers are applied as broadcast, in a furrow, or in a band below or alongside the planted seed row. As a relatively immobile ion in soil, K^+ supply to roots depends mostly on diffusion (Barber, 1995). This means that in K^+-deficient soil, band application of K^+ may be more efficient than broadcast application, because the mean level of K^+ within developing roots and diffusion rates are increased. Maintaining a high level of K^+ intensity in a portion of the root system may ensure that the K^+-supplying power of the soil does not limit the rate of K^+ accumulation. Welch et al. (1966) found that less K^+ fertilizer was required to obtain a given corn yield when the K^+ was banded rather than broadcast. Randall and Hoeft (1988), in a review on fertilizer placement studies, concluded that, in many situations, localized placement of fertilizer can improve fertilizer efficiency and economic return. Heckman and Kamprath (1992) reported that band placement of K^+ (either alone or in combination with broadcast) yielded little benefit when corn was grown with intensive production practices on sandy soils testing relatively high in K^+. De Datta and Mikkelsen (1985) reported that the placement of K^+ fertilizers in upland soils often enhances their effectiveness, since the ion is not very mobile. Placement becomes less critical, however, as soil-available K levels increase (De Datta and Mikkelsen, 1985).

Plant accumulation and stover yield was, however, increased in a dry year when a band was included with broadcast placement. According to Fageria (1982), the application of moderate amounts of K^+ (30–45 kg ha^{-1}) in the planting furrow of upland rice produced the same yield as twice as much K^+ broadcast and incorporated. Lopes (1983) recommended that K^+ fertilization of the Brazilian cerrado soils be made in two complementary ways: (1) broadcasting and incorporation to raise K^+ saturation to 3%, and (2) application of maintenance fertilizer in bands 15 to 20 cm wide. Localized placement has little advantage over broadcast if the soil is well supplied with K^+ (Malavolta, 1985).

In addition to application on the band or broadcast, uniform application of fertilizer is usually considered essential for maximum yield and high K^+ utilization efficiency. Nonuniform fertilizer application can occur because of faulty machinery, faulty machine operation, or fertilizer properties that adversely affect the performance of the machine. Hence, uniform application of fertilizers is very important for uptake of nutrients by plant roots and improving or maximizing fertilizer use efficiency.

4.8.6 USE OF EFFICIENT CROP SPECIES/CULTIVARS

Use of nutrient-efficient crop species or genotypes within species is an important management strategy for maximizing potassium use efficiency in crop plants. In the literature, nutrient-efficient plants are defined in several ways. Some of these definitions are presented in Table 4.14. From a practical standpoint, crop nutrient use efficiency reflects the ability to produce a higher yield under nutrient-limited growth conditions. But to classify plant genotypes within a crop species as nutrient use efficient or inefficient, Gerloff (1976) proposed that inefficient genotypes must yield approximately the same amount as efficient strains under optimum supplies of the (limiting) element.

This method avoids comparison of genotypes with widely differing growth potential and nutrient demand. Cassman et al. (1989) evaluated cultivar differences

TABLE 4.14
Definitions of Nutrient-Efficient Plants

Definition	Reference
A nutrient-efficient plant is defined as a plant that absorbs, translocates, or utilizes more of a specific nutrient than another plant under conditions of relatively low nutrient availability in the soil or growth media.	Soil Science Society of America (1997)
The nutrient efficiency of a genotype (for each element separately) is defined as the ability to produce a high yield in a soil that is limiting in that element for a standard genotype.	Graham (1984)
Nutrient efficiency of a genotype/cultivar is defined as the ability to acquire nutrients from growth medium and/or to incorporate or utilize them in the production of shoot and root biomass or utilizable plant material (grain).	Blair (1993)
An efficient genotype is one that absorbs relatively high amounts of nutrients from soil and fertilizer, produces a high grain yield per unit of absorbed nutrient, and stores relatively few nutrients in the straw.	Isfan (1993)
Efficient plants are defined as those that produce more dry matter or have a greater increase in harvested portion per unit time, area, or applied nutrient, have fewer deficiency symptoms, or have greater increment increases and higher concentrations of mineral nutrients than other plants grown under similar conditions or compared to a standard genotype.	Clark (1990)
An efficient germplasm requires less nutrient than an inefficient one for normal metabolic processes.	Gourley et al. (1994)
An efficient plant is defined as that plant which produces higher economic yield with a determined quantity of applied or absorbed nutrient compared to other or standard plants under similar growing conditions.	Fageria et al. (2008)

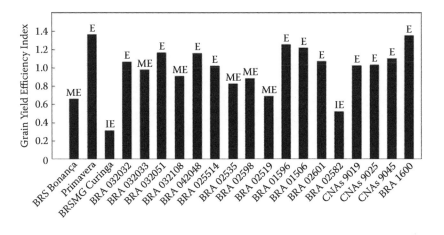

FIGURE 4.10 Classification of upland rice genotypes for K use efficiency. Letters on bar: E = efficient, ME = moderately efficient, and IE = inefficient.

in K+ use efficiency (defined as higher yield with a limited K+ supply) in relation to K+ uptake, K+ partitioning, and critical internal and external K+ requirements for cotton. Without K+ addition, yield was 29% and 35% greater in 2 years of experimentation in the K-use-efficient cultivar when K+ supply was not limited; cultivar yields were similar. Yield of both cultivars was closely associated with leaf K+ concentration and soil K+ availability, but response curves indicated a lower leaf and soil K+ requirement for the K+-use-efficient cultivar. The K+-use-efficient cultivar had a higher K+ uptake rate during fruit development and greater total K+ accumulation, particularly at low soil K+ levels.

Author evaluated 20 upland rice genotypes for potassium use efficiency in Brazilian Oxisol. Based on grain yield efficiency index (GYEI), genotypes were classified into efficient, moderately efficient, and inefficient (Figure 4.10). The GYEI was calculated by using the following formula:

$$\text{Grain yield efficiency index} = \left(\frac{\text{Grain yield at low K level}}{\text{Average grain yield of 20 genotypes at low K level}} \right)$$
$$\times \left(\frac{\text{Grain yield at high K level}}{\text{Average grain yield of 20 genotypes at high K level}} \right)$$

The GYEI is useful in separating high-yielding, stable, nutrient-efficient genotypes from low-yielding, unstable, nutrient-inefficient genotypes (Fageria et al., 1988). Genotypes having GYEI higher than 1 were considered K efficient, inefficient genotypes were in the range of 0 to 0.5 GYEI, and genotypes in between these two limits were considered intermediate in K use efficiency. The genotypes Primavera and BRA 1600 were the most efficient in K use, and genotypes BRSMG Curinga and BRA 02582 were the most inefficient in K use.

Cultivar differences in K+ uptake have been reported in a number of crop species (Glass and Perley, 1980; Clark, 1990; Fageria et al., 1990a; 1990b; Clark and

Duncan, 1991; Fageria et al., 1997a; Epstein and Bloom, 2005; Fageria et al., 2006). The difference in nutrient uptake and utilization may be associated with better root geometry, ability of plants to take up sufficient nutrients from lower concentrations, plants' ability to solubilize nutrients in the rhizosphere, better transport, distribution, and utilization within plants, and balanced source–sink relationship (Graham, 1984; Baligar et al., 2001; Fageria et al., 2006). In addition, soil and plant mechanisms and processes and other factors that influence the genotype differences in plant nutrient use efficiency are summarized in Table 4.15. Root system is one of the main features that affect uptake of nutrients. Figure 4.11 and Figure 4.12 show differences in root systems at low as well as high K levels in the dry bean cultivars. However, no attempt is made here to discuss these mechanisms or processes in detail. For extensive reviews related to nutrient flux at the soil–root interface and across roots and shoot and mechanisms of uptake and utilization in the soil–plant system, see Fageria et al. (1997), Barber (1995), Marschner (1995), Baligar et al. (2001), and Mengel et al. (2001).

There are also differences in K^+ utilization among crop species. Generally sugar beet, sugarcane, potatoes, tomatoes, and celery have a very high demand, whereas cotton and wheat take up much smaller amounts of K^+ (Mengel et al., 2001). Pretty and Stangel (1985) reported that roots and sugar crops, fiber crops, and forages are among those with a high K^+ requirement. Oilseeds and grains such as soybean and corn have intermediate needs, whereas the cereal grains wheat and rice require relatively small amounts of K^+. Romero et al. (1988) also reported that alfalfa cultivers differed in yield and in herbage K^+ concentration responses to increasing K^+ concentrations. These cultivers suggested that it is possible to improve K^+ utilization and forage:mineral composition of the species through use of efficient cultivers.

4.8.7 INCORPORATION OF CROP RESIDUES

The incorporation of crop residues in the soil after harvest enables a substantial amount of plant K to be recycled. Fageria et al. (1990c) and Fageria and Gheyi (1999) reported that approximately 70 to 80% of the total K^+ content remains in the vegetative shoot of cereals such as wheat and rice, while about 40 to 50% remains in the shoot of legumes such as cowpea and beans. Fageria et al. (1982) also reported that most of the K^+ is retained in the straw of upland rice at harvest, and hence it is advisable to incorporate rice straw in the soil to contribute K^+ for the succeeding crop. Fageria (1991b) determined K^+ distribution in different plant parts of cowpea and concluded that 5% was in the roots, 49% was in the tops, 36% was in the seeds, and 10% was in pod husks. Similarly, Fageria (1991c) reported distribution of K^+ in upland rice plants as 3% in the roots, 86% in the tops, and 10% in the grain. Fageria et al. (1997b) reported that in lowland rice at the time of harvest, 15% of total plant K^+ was in the grain and the remaining 85% was in the shoot.

De Datta and Mikkelsen (1985) reported that recycling K^+ occurs to a considerable extent when crop residues are left in the field and incorporated into the soil. These authors also reported that recycling rice straw produces up to about 28 kg of K^+ per metric ton of rice straw. Returning crop residues was found to be essential for prolonging the residual effects from K applications under high-input systems in acid,

TABLE 4.15

Soil and Plant Mechanisms and Processes and Other Factors That Influence Genotypic Differences in Nutrient Use Efficiency

Nutrient Acquisition

1. Diffusion and mass flow (buffer capacity, ionic concentration, ionic properties, tortuosity, soil moisture, bulk density, and temperature)

2. Root morphological factors (number, length, root hair density, root extension, and root density)

3. Physiological [root:shoot, root microorganisms such as mycorrhizal fungi, nutrient status, water uptake, nutrient influx and efflux, rate of nutrient transport in roots and shoots, affinity to uptake (K_m), threshold concentration (C_{min})]

4. Biochemical (enzyme secretion as phosphate, chelating compounds, and phytosiderophore), proton exudes, organic acid production such as citric, transaconitic, and malic acid exudes

Nutrient Movement in Root

1. Transfer across endodermis and transport within root

2. Compartmentation/binding within roots

3. Rate of nutrient release to xylem

Nutrient Accumulation and Remobilization in Shoot

1. Demand at cellular level and storage in vacuoles

2. Retransport from older to younger leaves and from vegetative to reproductive parts

3. Rate of chelates in xylem transport

Nutrient Utilization and Growth

1. Metabolism at reduced tissue concentration of nutrient

2. Lower element concentration in supporting structure, particularly the stem

3. Elemental substitution, e.g., Na for K function

Other Factors

1. Soil factors

   a. Soil solution (ionic equilibrium, solubility precipitation, competing ions, organic ions, pH, and phytotoxic ions)

   b. Physicochemical properties of soil (organic matter, pH, aeration, structure, texture, compaction, soil moisture)

2. Environmental effects

   a. Intensity and quality of solar radiation

   b. Temperature

   c. Moisture supply

   d. Plant diseases, insects, weeds, and allelopathy

Source: Fageria et al. (1997a); Baligar et al. (2001); Fageria and Baligar (2003).

low-cation exchange capacity (CEC) Oxisols (Da Silva and Ritchey, 1982). The same authors reported that one application of 150 kg K ha^{-1} provided enough K$^+$ for five crops when stover was returned. In another study established after the clearing of a virgin forest from a clayey Oxisol in the Brazilian Amazon basin, a seven-crop rotation was adequately supplied with K$^+$ when crop stover was returned (Smyth et al., 1987; Cox and Uribe, 1992).

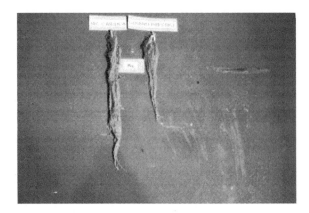

FIGURE 4.11 Root growth of two dry bean cultivars at low level of K in Brazilian Oxisol: (left) IAC Carioca; (right) Goiana Precoce. K level was 0 mg kg^{-1} soil.

FIGURE 4.12 Root development of three dry bean cultivars at 0 and 200 mg k kg^{-1} of soil. Cultivars were Apore (left), Perola (middle), and IAC Carioca (right). Left side of each cultivar is zero level of K and right side 200 mg K kg^{-1} of soil.

4.8.8 ADEQUATE MOISTURE SUPPLY

Soil moisture is an important factor affecting diffusion of K$^+$ in the rhizosphere and its uptake. Low soil water content reduces K$^+$ uptake by plant roots because K$^+$ diffusion rates decrease as soil moisture content decreases. Crop response to fertilizers also is affected by the interaction between fertilizer placement and soil moisture content. Placement of fertilizers in soil zones that are susceptible to drying may limit plant uptake. Kaspar et al. (1989) reported that drying of the fertilized soil layer even though water was available at greater depth decreased soybean growth and K$^+$ utilization efficiency. Changes in soil moisture content have a greater effect on K$^+$ diffusion rate than changes in soil temperature. Schaff and Skogley (1982) found that

K^+ diffusion rates increased, on average, 2.8-fold as soil moisture was raised from 10 to 28% (w/w). This compared with only a 1.6- to 1.7-fold increase in K^+ diffusion as soil temperature was raised from 5 to 30°C. Bertsch and Thomas (1985) reported that as the soil moisture is reduced, the concentration of divalent ions like Ca^{2+} and Mg^{2+} increases faster than the concentration of K^+ in the soil solution. This results in a decreasing K^+ concentration with increasing soil moisture tension and consequently decreased K^+ uptake.

4.8.9 USE OF FARMYARD MANURES

Livestock manure may be a complimentary source of N, P, and K when applied at rates that account for the composition of the manure and the nutrient-supplying potential of the soil (Chang et al., 1993; Fageria and Baligar, 2003) for crops grown on tropical soils for sustainable crop production. Integration of manure with inorganic fertilizers may result in the benefits of greater residual effects of organic than inorganic sources and advantages of manure in addition to a supply of N, P, and K (e.g., improved soil physical properties and supply of bases; Fageria and Baligar, 2003).

In addition, use of farmyard manures improves soil organic matter content. Soil organic matter is the storehouse of nutrients and water, improves soil structure, improves activities of beneficial microorganisms, reduces elemental toxicity, and makes soil less susceptible to erosion. Thus, improving and/or maintaining soil organic matter is one of the most important management practices to maintain soil fertility of tropical soils for sustainable crop production (Fageria and Baligar, 2003).

4.8.10 OPTIMUM K SATURATION IN SOIL SOLUTION

Optimum K^+ saturation in soil solution is an important index for improving crop yields and potassium use efficiency in crop plants. There was a linear increase in dry bean yield when K^+ saturation increased from 1 to 3% across two soils layers (Figure 4.13). This indicates that K^+ saturation level was not sufficient for achieving optimum dry bean yield in the soil under investigation. The optimum K^+ saturation value reported by Eckert (1987) from various studies is in the range of 2 to 5%. CIAT (1992) reported that about 15% of the bean-growing area in Latin America and 20% in Africa might be affected by soil K deficiency. Thung and Rao (1999) reported that yield losses can reach 20% in K-deficient soils, but this deficiency is seldom reported in bean literature. Fageria et al. (2001) reported response of bean genotypes grown on an Oxisol to K fertilization. However, response varied from genotype to genotype.

4.9 CONCLUSIONS

Fertilizers, including potassium, have contributed significantly to crop production in both developed and developing countries in the 20th century. Potassium plays an important role in many biochemical and physiological processes in crop plants. Among these processes, K^+ plays an important role in plants to ensure pH stabilization, osmoregulation, membrane transport processes, and enzyme activation.

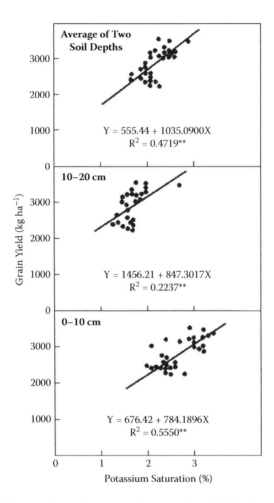

FIGURE 4.13 Influence of potassium saturation on grain yield of dry bean (Fageria, 2008).

Deficiency of potassium may lead to lodging, increased water stress, reduced photo-synthetic rates, and decline in quality of economic products of crops. It accumulates in large quantity by field crops. Potassium, along with N and P, has contributed significantly to increases in crop yields; still, there are many constraints to the use of this element in adequate amounts, particularly in developing countries. Most of the constraints are economic, social, and political in nature.

The soil mineralogy has a profound influence on the potassium cycle in soil–plant systems. Soils with high amounts of vermiculite can fix enormous amounts of K^+. Exchangeable K^+ concentration, determined by a given extracting solution and its association with grain yield, is a principal index of potassium fertilizer recommendations for field crops. In addition to use of adequate rate, maximizing efficiency of the K^+ is also important from an economic and environmental point of view. Important management practices to improve K use efficiency are soil, plant, and environmental manipulation. These practices are discussed in this chapter.

REFERENCES

Baligar, V. C., N. K. Fageria, and Z. L. He. 2001. Nutrient use efficiency in plants. *Commun. Soil Sci. Plant Anal.* 32:921–950.

Barber, S. A. 1995. *Soil nutrient bioavailability: A Mechanistic approach.* New York: John Wiley & Sons Inc.

Bertsch, P. M. and G. W. Thomas. 1985. Potassium status of temperate region soils. In: *Potassium in agriculture*, R. D. Munson, Ed., 131–162. Madison, WI: ASA, CSSA, and SSSA.

Blair, G. 1993. Nutrient efficiency—What do we really mean? In: *Genetic aspects of mineral nutrition,* P. J. Randall, E. Delhaize, R. A. Richards, and R. Munns, Eds., 205–213. Dordrecht: Kluwer Academic Publishers.

Brady, N. C. and R. R. Weil. 2002. *The nature and properties of soils,* 13th edition, Upper Saddle River, NJ: Prentice Hall.

Buol, S. W., P. A. Sanchez, R. B. Cote, Jr., and M. A. Granger. 1975. Soil fertility capability classification. In: *Soil management in tropical America.* Proceedings of CIAT Seminar, February 1974, 126–141. Raleigh: North Carolina State University.

Cassman, K. G., T. A. Kerby, B. A. Roberts, D. C. Bryant, and S. M. Brouder. 1989. Differential response of two cotton cultivars to fertilizer and soil potassium. *Agron. J.* 81:870–876.

Chang, S. C. 1971. *Chemistry of paddy soils.* Ext. Bull. 7. ASPAC Food and Fertilizer Technology Center, Taipei, Taiwan.

Chang, C., T. G. Sommerfield, and T. Entz. 1993. Barley performance under heavy applications of cattle feedlot manure. *Agron. J.* 85:1013–1018.

CIAT (Centro Internacional de Agricultura Tropical). 1992. *Constraints to and opportunities for improving bean production: A planning document 1993–98 and achieving document 1987–92,* Cali, Colombia: CIAT.

Clark, R. B. 1990. Physiology of cereals for mineral nutrient uptake, use, and efficiency. In: *Crops as enhancers of nutrient use,* V. C. Baligar and R. R. Duncan, Eds., 131–209. San Diego: Academic Press.

Clark, R. B. and R. R. Duncan. 1991. Improvement of plant mineral nutrition through breeding. *Field Crops Res.* 27:219–240.

Cox, F. R. and E. Uribe. 1992. Management and dynamics of potassium in a humid tropical Ultisol under a rice-cowpea rotation. *Agron. J.* 84:655–660.

Da Silva, J. E. and K. D. Ritchey. 1982. Potassium fertilization of cerrado soils. In: *Potassium in Brazilian agriculture*, T. Yamada, Ed., 323–338. Piracicaba, São Paulo: Brazilian Potassium and Phosphate Association.

Dahnke, W. C. and R. A. Olson. 1990. Soil test correlation, calibration, and recommendation. In: *Soil testing and plant analysis*, R. L. Westerman, Ed., 45–71. Madison, WI: SSSA.

De Datta, S. K. and D. S. Mikkelsen. 1985. Potassium nutrition of rice. In: *Potassium in agriculture*, D. Munson, Ed., 665–699. Madison, WI: ASA, CSSA, and SSSA.

Dibb, D. W. and W. R. Thompson, Jr. 1985. Interaction of potassium with other nutrients. In: *Potassium in agriculture*, R. D. Munson, Ed., 515–533. Madison, WI: ASA, CSSA, and SSSA.

Dibb, D. W. and L. F. Welch. 1976. Corn growth as affected by ammonium vs. nitrate absorbed from soil. *Agron. J.* 68:89–94.

Eckert, D. J. 1987. Soil test interpretations: basic cation saturation ratios and sufficiency levels. In: *Soil testing: sampling, correlation, calibration, and interpretation*, J. R. Brown, Ed., 53–64. Madison, WI: SSSA.

Epstein, E. and A. J. Bloom. 2005. *Mineral nutrition of plants: Principles and perspectives*, 2nd edition, Sunderland, Massachusetts: Sinauer Associations, Inc.

Fageria, N. K. 1982. Fertilization and potassium nutrition of rice in Brazil. In: *Potassium in Brazilian agriculture,* T. T. Yamada, Ed., 421–436. Piracicaba, São Paulo: Brazilian Potassium and Phosphate Association.

Fageria, N. K. 1983. Ionic interactions in rice plants from dilute solutions. *Plant Soil* 70:309–316.

Fageria, N. K. 1984. *Fertilization and mineral nutrition of rice*. Campus, Rio de Janeiro/ CNPAF, Goiania, Brazil.

Fageria, N. K. 1989. *Tropical soils and physiological aspects of crops*. Brasilia, Goiânia, Brazil: EMBRAPA/CNPAF.

Fageria, N. K. 1991a. Response of rice to fractional applied potassium in Brazil. *Better Crops International* 7:19.

Fageria, N. K. 1991b. Response of cowpea to phosphorus on an Oxisol with special reference to dry matter production and mineral ion contents. *Trop. Agric.* 68:384–388.

Fageria, N. K. 1991c. Response of rice cultivars to phosphorus fertilization on a dark red latosol of central Brazil. *Rev. Bras. Ci. Solo* 15:63–67.

Fageria, N. K. 1992. *Maximizing crop yields*. New York: Marcel Dekker.

Fageria, N. K. 2000a. Potassium use efficiency of upland rice genotypes. *Pesq. Agropec. Bras.* 35:2115–2120.

Fageria, N. K. 2000b. Upland rice response to soil acidity in cerrado soil. *Pesq. Agroec. Bras.* 35:2303–2307.

Fageria, N. K. 2001. Response of upland rice, dry bean, corn and soybean to base saturation in cerrado soil. *Rev. Bras. Eng. Agric. Ambien.* 5:416–424.

Fageria, N. K. 2002. Nutrient management for sustainable dry bean production in the tropics. *Commun. Soil Sci. Plant Anal.* 33:1537–1575.

Fageria, N. K. 2004. Dry matter yield and nutrient uptake by lowland rice. *J. Plant Nutr.* 6:947–958.

Fageria, N. K. 2008. Optimum soil acidity indices for dry bean production on an Oxisol in no-tillage system. *Commun. Soil Sci. Plant Anal.* 39:845–857.

Fageria, N. K. and V. C. Baligar. 2003. Fertility management of tropical acid soil for sustainable crop production. In: *Handbook of soil acidity*, Z. Rengel, Ed., 359–385. New York: Marcel Dekker.

Fageria, N. K. and V. C. Baligar. 2001. Lowland rice response to nitrogen fertilization. *Commun. Soil Sci. Plant Anal.* 32:1405–1429.

Fageria, N. K. and V. C. Baligar. 2005. Nutrient availability. In: *Encyclopedia of soils in the environment*, D. Hillel, Ed., 63–71. San Diego, CA: Elsevier.

Fageria, N. K., V. C. Baligar, and R. B. Clark. 2006. *Physiology of crop production*. New York: The Haworth Press.

Fageria, N. K., V. C. Baligar, and D. G. Edward. 1990a. Soil-plant nutrient relationships at low pH stress. In: *Crops as enhancers of nutrient use*, V. C. Baligar and R. R. Duncan, Eds., 475–507. New York: Academic Press.

Fageria, N. K., V. C. Baligar, and C. A. Jones. 1997a. *Growth and mineral nutrition of field crops*, 2nd edition. New York: Marcel Dekker.

Fageria, N. K., V. C. Baligar, and Y. Li. 2008. The role of nutrient efficient plants in improving crop yields in the 21st century. *J. Plant Nutrition* 31 (in press).

Fageria, N. K., V. C. Baligar, R. J. Wright, and J. R. P. Carvalho. 1990b. Lowland rice response to potassium fertilization and its effect on N and P uptake. *Fert. Res.* 21:157–162.

Fageria, N. K., V. C. Baligar, and R. W. Zobel. 2007. Yield, nutrient uptake and soil chemical properties as influenced by liming and boron application in common bean in no-tillage system. *Commun. Soil Sci. Plant Anal.* 38:1637–1653.

Fageria, N. K. and M. P. Barbosa Filho. 2007. Response of irrigated rice to potassium fertilization in the State of Tocantins. Paper presented at the 31st Brazilian Soil Science Congress, August 5–10, 2007, Gramado, Rio Grande do Sul: Brazilian Soil Science Society.

Fageria, N. K., M. P. Barbosa Filho, and J. R. P. Carvalho. 1982. Response of upland rice to phosphorus fertilization on an Oxisol of central Brazil. *Agron. J.* 74:51–56.

Fageria, N. K., M. P. Barbosa Filho, and J. G. C. da Costa. 2001. Potassium use efficiency in common bean genotypes. *J. Plant Nutr.* 24:1937–1945.

Fageria, N. K. and H. R. Gheyi. 1999. *Efficient crop production.* Campina Grande, Brazil: Federal University of Paraiba.

Fageria, N. K., O. P. Morais, V. C. Baligar, and R. J. Wright. 1988. Response of rice cultivars to phosphorus supply on an Oxisol. *Fert. Res.* 16:195–206.

Fageria, N. K., A. B. Santos, and V. C. Baligar. 1997b. Phosphorus soil test calibration for lowland rice on an Inceptisol. *Agron. J.* 89:737–742.

Fageria, N. K. and A. B. Santos. 2008. Lowland rice response to thermophosphate fertilization. *Commun. Soil Sci. Plant Analy.* 39:873–889.

Fageria, N. K., N. A. Slaton, and V. C. Baligar. 2003. Nutrient management for improving lowland rice productivity and sustainability. *Adv. Agron.* 80:63–152.

Fageria, N. K., R. J. Wright, V. C. Baligar, and J. R. P. Carvalho. 1990c. Upland rice response to potassium fertilization on an Oxisol. *Fert. Res.* 21:141–147.

Gerloff, G. C. 1976. Plant efficiencies in the use of nitrogen, phosphorus, and potassium. In: *Plant adaptation to mineral stress in problem soils,* M. J. Wright, Ed., 161–173. Ithaca, NY: Cornell Univ. Agric. Exp. Stn.

Glass, A. D. M. and J. E. Perley. 1980. Varietal differences in potassium uptake by barley. *Plant Physiol.* 65:160–164.

Graham, R. D. 1984. Breeding for nutritional characteristics in cereals. In: *Advances in plant nutrition,* Vol. 1, P. B. Tinker and A. Lauchi, Eds., 57–102. New York: Praeger Publisher.

Gourley, C. J. P., D. L. Allan, and M. P. Russelle. 1994. Plant nutrient efficiency: A comparison of definitions and suggested improvement. *Plant Soil* 158:29–37.

Gupta, U. C. 1979. Boron nutrition of crops. *Adv. Agron.* 31:273–307.

Haque, S. A., Z. H. Bhuya, A. K. M. Idris Ali, F. A. Choudhury, M. Jahiruddin, and M. M. Rahman. 1982. Response of HVY paddy to potash fertilization in different regions of Bangladesh. In: *Phosphorus and potassium in the tropics*, E. Pushparajah and E. H. A. Hamid, Eds., 425–430. Proceedings of International Conference on Phosphorus and Potassium in the Tropics, Kuala Lumpur, Malaysia, August 1981, Malaysian Society of Soil Science, Kula Lumpur.

Havlin, J. L. and D. G. Westfall. 1985. Potassium release kinetics and plant response in calcareous soil. *Soil Sci. Soc. Am. J.* 49:366–370.

Havlin, J. L., D. G. Westfall, and S. R. Olsen. 1985. Mathematical models for potassium release kinetics in calcareous soils. *Soil Sci. Soc. Am. J.* 49:371–376.

Hay, R. K. M. 1995. Harvest index: A review of its use in plant breeding and crop physiology. *Ann. Appl. Biol.* 126:197–216.

Heckman, J. R. and E. J. Kamprath. 1992. Potassium accumulation and corn yield related to potassium fertilizer rate and placement. *Soil Sci. Soc. Am. J.* 56:141–148.

Hill, W. E. and L. G. Morrill. 1975. Boron, calcium and potassium interactions in Spanish peanuts. *Soil Sci. Soc. Am. Proc.* 39:80–83.

Huber, D. M. and D. C. Arny. 1985. Interactions of potassium with plant diseases. In: *Potassium in agriculture*, R. D. Munson, Ed., 467–488. Madison, WI: ASA, CSSA, and SSA.

Isfan, D. 1993. Genotypic variability for physiological efficiency index of nitrogen in oats. *Plant Soil* 154:53–59.

Ismunadji, M., I. Nasution, and S. Parthohardjono. 1982. Potassium nutrition of lowland rice. In: *Phosphorus and potassium in the tropics*, E. Pushparajah and E. H. A. Hamid, Eds., 315–321. Proceedings of International Conference on Phosphorus and Potassium in the Tropics, Kuala Lumpur, Malaysia, August 1981, Malaysian Society of Soil Science, Kula Lumpur.

Johnson, C., D. G. Edwards, and J. F. Loneragan. 1968. Interactions between potassium and calcium in their absorption by intact barley plants. I. Effects of potassium on calcium absorption. *Plant Physiol.* 43:1717–1721.

Kaspar, T. C., J. B. Zahler, and D. R. Timmons. 1989. Soybean response to phosphorus and potassium fertilizers as affected by soil drying. *Soil. Sci. Soc. Am. J.* 53:1448–1454.

Kim, Y. S. and S. C. Park. 1973. Effect of split potassium application a paddy rice grown on acid sulfate soils. *Potash Rev.* 7:1–9.

Lopes, A. S. 1983. *Soil under cerrado.* Piracicaba, São Paulo: Potassium and Phosphate Institute, Brazil.

Mahapatra, I. C. and R. Prasad. 1970. Response of rice to potassium in relation to its transformation and availability under waterlogged condition. *Fert. News* 15:34–41.

Malavolta, E. 1985. Potassium status of tropical and subtropical region soils. In: *Potassium in agriculture*, R. D. Munson, Ed., 163–200. Madison, WI: ASA, CSSA, and SSSA.

Marschner, H. 1995. *Mineral nutrition of higher plants*, 2nd edition. New York: Academic Press.

Matocha, J. E. and G. W. Thomas. 1969. Potassium and organic nitrogen content of grain sorghum as affected by iron. *Agron. J.* 61:425–428.

Mengel, K., E. A. Kirkbay, H. Kosegarten, and T. Appel. 2001. *Principles of plant nutrition,* 5th edition. Noewell, MA: Kluwer Academic Publisher.

Mielniczuk, J. 1977. Potassium forms in Brazilian soils. *Rev. Bras. Ci. Solo* 1:55–56.

Moraes, J. F. V. 1988. Liming and fertilization. In: *Bean crop: Factors affecting productivity*, M. J. Zimmermann, M. Rocha, and T. Yamada, Eds., 260–301. Piracicaba, São Paulo: Brazilian Potash and Phosphate Institute.

Muzilli, O. 1982. Fertilization and nutrition of soybean in Brazil. In *Potassium in Brazilian agriculture*, T. Yamada, Ed., 373–392. Piracicaba, São Paulo: Brazilian Potassium and Phosphate Institute.

Muns, R. D. 1982. *Potassium, calcium and magnesium in the tropics and subtropics*, J. C. Brosheer, Ed., Tech. Bull. No. IFDC-T-23. Muscle Shoals, AL: International Fertilizer Development Center.

Nemeth, K. 1982a. Potassium analysis methods in soil and their interpretation. In: *Potassium in Brazilian agriculture*, T. Yamada, Ed., 77–94. Piracicaba, São Paulo: Brazilian Potassium and Phosphate Institute.

Nemeth, K. 1982b. Nutrient dynamics in some humid tropical soils as determined by electro-ultrafiltration, In: *Phosphorus and potassium in the tropics*, E. Pushparajah and E. H. A. Hamid, Eds., 3–14. Proceedings of International Conference on Phosphorus and Potassium in the Tropics, Kuala Lumpur, Malaysia, August 1981, Malaysian Society of Soil Science, Kula Lumpur.

Piggott, T. J. 1986. Vegetable crops. In: *Plant analysis: An interpretation manual*, D. J. Reuter and J. B. Robinson, Eds., 148–187. Melbourne: Inkata Press.

Pretty, K. M. and P. J. Stangel. 1985. Current and future use of world potassium. In: *Potassium in agriculture*, R. D. Munson, Ed., 99–128. Madison, WI: ASA, CSSA, and SSSA.

Randall, G. W. and R. G. Hoeft. 1988. Placement methods for improved efficiency of P and K fertilizers: A review. *J. Prod. Agric.* 1:70–79.

Reddy, S. V. 1976. *Availability of potassium in different Colorado soils*. PhD dissertation, Ft. Collins, CO: Colorado State Univ. (Diss. Abstr. 37-3195B).

Reuter, D. J. 1986. Temperate and sub-tropical crops. In: *Plant analysis: An interpretation manual*, D. J. Reuter and J. B. Robinson, Eds., 38–99. Melbourne: Inkata Press.

Romero, W., J. Augustin, and G. Schilling. 1988. The relationship between phosphate absorption and root length in nine wheat cultivars. *Plant Soil* 111:199–201.

Schaff, B. E. and E. O. Skogley. 1982. Diffusion of potassium, calcium, and magnesium in Bozeman silt loam as influenced by temperature and moisture. *Soil Sci. Soc. Am. J.* 46:521–524.

Sheldrick, W. F. 1985. World potassium reserves. In: *Potassium in agriculture*, D. Munson, Ed., 3–28. Madison, WI: ASA, CSSA, and SSSA.

Sinclair, T. R. 1998. Historical changes in harvest index and crop nitrogen accumulation. *Crop Sci.* 38:638–643.

Singh, R. P. and A. Kumar. 1981. Effect of levels and times of potassium application on upland rice. *Indian Potash J.* 6:12–15.

Slaton, N. A., C. D. Cartwright, and C. E. Wilson, Jr. 1995. Potassium deficiency and plant diseases observed in rice fields. *Better Crops* 82:10–12.

Smyth, J. T., M. Cravo, and J. B. Bastos. 1987. Soil nutrient dynamics and fertility management for sustained crop production on Oxisols in the Brazilian Amazon. In: *Tropsoils Tech-Report 1985–1986,* 88–91. Raleigh, NC: Dept. Soil Sci., North Carolina State Univ.

Snyder, F. W. and G. E. Carlson. 1984. Selecting for partitioning of photosynthetic products in crops. *Adv. Agron.* 37:47–72.

Stewart, J. A. 1985. Potassium sources, use, and potential. In: *Potassium in agriculture,* R. D. Munson, Ed., 83–98. Madison, WI: ASA, CSSA, and SSSA.

Soil Science Society of America. 1997. *Glossary of soil science terms.* Madison, WI: Soil Science Society of America.

Su, N. R. 1976. Potassium fertilization of rice. In: *The fertility of paddy soils and fertilizer application for rice,* Asian Pacific Food and Fertilizer Technology Center, Ed., 117–148. Taipei: Taiwan.

Thung, M. and I. M. Rao. 1999. Integrated management of abiotic stresses. In: *Common bean improvement in the twenty-first century,* S. P. Singh, Ed., 331–370. Dordrecht, The Netherlands: Kluwer Academic Publishers.

Welch, L. F., P. E. Johnson, G. E. Mckibben, L. V. Boone, and J. W. Pendleton. 1966. Relative efficiency of broadcast versus banded potassium for corn. *Agron. J.* 58:618–621.

Williams, J. and S. G. Smith. 2001. Correcting potassium deficiency can reduce rice stem diseases. *Better Crops* 85:7–9.

5 Calcium

5.1 INTRODUCTION

Calcium (Ca^{2+}) is a divalent alkaline cation and plays many important roles in plant growth and development. When the bioavailability of Ca^{2+} is low, crop yield and quality can be suboptimal, depending on crop species and environmental conditions. Calcium is abundant in neutral and alkaline soils. However, calcium deficiency is most common in highly weathered acid soils, like Oxisols and Ultisols. Large areas of Oxisols and Ultisols are found in the tropics, covering about 43% of the tropics, including large areas in South America and central Africa (Thurman et al., 1980; Fageria and Baligar, 2003). According to Sanchez et al. (1983), the 580 million ha of potentially arable tropical land, with udic moisture regimes and supporting rainforest vegetation, present the largest available area for expanding the world's agriculture frontier. These lands have no major temperature or moisture limitations and generally have topography suitable for year-round crop production. However, these soils are acidic in reaction and have low fertility (Yost et al., 1979; Goedert, 1989; Fageria and Baligar, 1997, 2003).

Brennan et al. (2007) reported responses of wheat and oilseed rape (*Brassica napus* L.) grown on sandy acidic soil of Western Australia. Fageria and Breseghello (2004) conducted a survey with the objective to evaluate soil fertility and nutritional status of Oxisols covering 43 sites and 33 rural properties in three municipalities of the Chapada dos Parcecis Region, State of Mato Grosso, Brazil. From soil analysis data, it was concluded that Oxisols of the region contained low levels of P, K^+, base saturation, Ca^{2+} saturation, and Mg^{2+} saturation, and high levels of Al^{3+}. Hence, most of these soils need liming for crop production, especially for legumes like dry bean and soybean. Similarly, Fageria and Baligar (2003) reported that most of the central part of Brazil is occupied by a tropical savanna, locally known as the cerrado, which covers about 205 million ha or 23% of the national territory. Goedert (1983) reported that at least 50 million ha of the cerrado region have potential for intensive mechanized agriculture. Most soils in this region are highly weathered Oxisols (46%), Ultisols (15%), and Entisols (15%), with low natural soil fertility, high aluminum saturation, and high P fixation capacity.

Calcium-deficient soils have low cation exchange capacity (CEC) and have high leaching capacity. In such soils, Al^{3+} content is high and sometimes toxic to plants. Calcium usually has to be added to acid soils, and the most effective and dominant practice is liming to supply the necessary Ca^{2+} for plant growth. Calcium content of the acid soils can also be increased by application of gypsum ($CaSO_4 \cdot 2H_2O$). However, gypsum cannot substitute for lime, which brings many chemical and biological

TABLE 5.1
Influence of Soil pH on Root Growth of Wheat and Dry Bean Grown on Brazilian Oxisol

Soil pH in H_2O	Root Dry Weight of Wheat (g plant^{-1})	Root Dry Weight of Dry Bean (g plant^{-1})
4.1	1.21	0.48
4.7	2.51	0.89
5.3	2.95	1.17
5.9	2.97	1.35
6.6	2.75	1.18
7.0	2.77	0.95

Regression Analysis

Soil pH (X) vs. root dry wt. of wheat (Y) = $-14.5996 + 5.9145X - 0.4937X^2$, $R^2 = 0.9323*$

Soil pH (X) vs. root dry wt. of dry bean (Y) = $-7.8220 + 3.0959X - 0.2627X^2$, $R^2 = 0.9856**$

*, **Significant at the 5 and 1% probability level, respectively.

Source: Adapted from Fageria and Zimmermann (1998).

changes in favor of higher crop yields. Mixing of gypsum with lime (25% gypsum and 75% lime) is considered a very effective practice to reduce soil acidity, improve Ca^+ and Mg^+ content, and reduce Al toxicity (Fageria, 1989). Mixing gypsum with lime may leach Ca^{2+} with SO_4^{2-} ion in the subsoil and improve growth of root systems to a greater extent.

Bruce et al. (1988) have shown that the primary limitations to root elongation in acid soils (mainly subsoils) from southeast Queensland, Australia, was Ca^{2+} deficiency and not Al^{3+} toxicity, in spite of high Al^{3+} saturation and relatively low soil pH. Similarly, Fageria and Zimmermann (1998) evaluated influence of increasing pH with liming on root growth of wheat and dry bean in Brazilian Oxisol (Table 5.1). Root weight of both the crops was significantly increased with increasing soil pH from 4.1 to 7.0. The variability in root dry weight of wheat due to the rise in pH was 93%, and maximum root weight was obtained at pH 6.0. In the case of dry bean, the variability in root dry weight due to change in soil pH was 98%, and maximum root dry weight was obtained at pH 5.8 (Table 5.1). Hence, it can be concluded that liming is an essential practice for improving root growth of field crops in acid soils.

Calcium is called a secondary nutrient. This does not mean that this nutrient plays a secondary role in plant growth and development. Calcium is as essential as N, P, K, and other nutrients for healthy plant growth (Fageria and Gheyi, 1999). Calcium is a constituent of calcium pectate, which is found in the middle lamella of the cell wall. Once calcium enters the middle lamella of the cell wall, it has left its metabolic site. This movement is irreversible; consequently, if a plant suffers from a shortage of calcium at any stage, the newly growing parts cannot receive a supply of calcium from older tissues (Ishizuka, 1978). The objective of this chapter is to discuss mineral nutrition aspects of calcium as well as management practices, which can maximize use efficiency of this element in crop production.

5.2 CYCLE IN SOIL–PLANT SYSTEMS

The processes of removal and addition regulate the calcium cycle in soil–plant systems. Calcium is mostly removed or lost from soil–plant systems by leaching, soil erosion, and crop removal. It is also adsorbed on soil colloids under specific conditions. Similarly, addition of Ca^{2+} is through liming (21–32% Ca), application of gypsum (23% Ca), fertilizers containing calcium such as calcium nitrate (19% Ca), superphosphates (14 to 20% Ca), and farmyard manures (0.5 to 2.3%). Under acid conditions, Ca^{2+} ion is displaced by Al^{3+} and H^+ ions from the exchange complex, and leaching of this element may occur.

The Ca^{2+} content of soil depends on its parent material, degree of weathering, and whether liming and fertilizers have added calcium. A large amount of Ca^{2+} in the soil is present as exchangeable Ca^{2+}, which depends on the cation exchange capacity of a soil. Exchangeable Ca^{2+} is in equilibrium with soil solution calcium. The concentration of soil solution Ca^{2+} is determined by CEC, the nature of the bonding with exchange sites, pH, type of soil, and the level of anions in solution. Exchangeable Ca^{2+} is more tightly held on soil colloids than either K^+ or Mg^{2+}. Calcium may also be present in soil minerals having varying degrees of solubility. Solution and exchangeable calcium are the main forms that can move to the plant root and be absorbed (Barber, 1995). Gypsum and calcium carbonate are soil minerals having greater solubility and higher calcium content. Low-calcium-content minerals are plagioclase feldspars, augite, hornblende, and epidote (Barber, 1995). Mass flow is generally the primary mechanism for supplying calcium to plant roots for absorption.

Calcium concentration in soil solutions varies extensively. Adams (1984) reported Ca^{2+} concentrations in soil solutions of 1.7 to 19.4 mM. Barber (1995) points out that Ca^{2+} is one element that can be supplied to plants in sufficient quantity by mass flow. However, many tropical agricultural soils in West Africa and South America are acid and low in plant-available Ca^{2+}. Frequently, these soils require lime, which raises the soil pH along with supplying other nutrients, such as Ca and Mg, for optimum crop production (Kamprath, 1984). Moore and Patrick (1989) also reported that deficiency of Ca has been a limiting factor for rice production in acid sulfate soils of the central plains region of Thailand.

As the soil pH increases, adsorption of Ca and Mg increases, especially in soils rich in iron and aluminum oxides. The pH in these situations acts in two ways in determining the adsorption of Ca and Mg by soils with variable charge. One way is by increasing the negative charge of the soil and providing an increased number of exchangeable sites with a higher affinity for divalent cations. The other way is to increase the population of hydrolyzed divalent cations as $Ca^{2+} + OH^- \Leftrightarrow CaOH^+$, which can then become specifically adsorbed into the stem layer (James and Healy, 1972). Since these two effects occur simultaneously as the pH is increased, a rapid increase in the divalent cations adsorbed takes place (Chan et al., 1979). At pH < 6, most of the adsorbed Ca remains exchangeable, but as the pH is increased, more divalent cations become specifically adsorbed and therefore are no longer exchangeable (Chan et al., 1979).

5.3   FUNCTIONS AND DEFICIENCY SYMPTOMS

Calcium is involved in cell division and cell elongation and plays a major role in the maintenance of cell membrane integrity (Fageria et al., 1997). In addition, calcium plays an important role in maintaining nutrient balance in plant tissues and also ameliorates toxicity of heavy metals. Plant cells without Ca^{2+} tend to lose their semipermeability and collapse. Calcium has been found to catalyze some enzymes involved in the hydrolysis of adenosine triphosphate (ATP) and phospholipids and to partially replace Mg^{2+} in some reactions. Calcium is required in the fruits to alleviate *bitter bit* or *blossom end rot* symptoms (Clark, 1982).

Root growth is severely restricted in Ca^{2+}-deficient plants, and the roots become prone to infection by bacteria and fungi. Calcium protects the plasma membrane from the deleterious effects of H^+ ions at lower pH and also reduces harmful effects of Na^+ in salt-affected soils (Epstein and Bloom, 2005). Many biotic and abiotic stresses are reduced by the presence of adequate amounts of Ca in the rhizosphere (Reddy and Reddy, 2002). The role of Ca^{2+} in regulation of the stomata aperture is also elucidated by Schroeder et al. (2001) and Epstein and Bloom (2005). Calcium acts as a regulator ion in the translocation of carbohydrates through its effect on cells and cell walls (Bennett, 1993). It plays a role in mitosis. Calcium is cited for its beneficial effect on plant vigor and stiffness of straw and also on grain and seed formation (Follett et al., 1981). It is a part of certain structural compounds such as calcium oxalate and calcium pectate (Bould et al., 1984). Calcium is also involved in the metabolism of nitrogen. Calcium promotes ion uptake and the formation of root mitochondria and membrane permeability. Low supply of Ca inhibits the nodulation, growth, and nitrogen fixation of bacteria associated with the root of legumes (Pan, 2000). Calcium is considered to be indispensable for the germination of pollen and growth of the pollen tube in plants from numerous families (Fageria and Gheyi, 1999).

Calcium mediates ion transport across plasmalemma. The following possibilities have been proposed and reviewed by Fageria (1983): (1) calcium enhanced the formation or turnover of the carrier; (2) calcium functioned as a cofactor in the formation of the carrier; (3) calcium increased the rate of breakdown of the ion carrier complex; (4) calcium enhanced the affinity between carrier and ion; (5) in the presence of Ca, the stability of the structure of the protoplasm increased, and consequently the absorbed ion can bind in a more stable form; (6) calcium enhanced the formation of mitochondria, which are considered to play an important role in ion accumulation; (7) calcium protected RNA, which is important in the membrane for the transport of ions, against breakdown by endogenous enzymes; and (8) calcium influenced the permselectivity of a barrier situated at the cell surface, through which the ions must migrate to reach absorption sites. The stimulating effect of Ca was considered to block the entrance of interfering ions.

A few days after germination, seedlings depend on the seed reserve for their mineral elements; upon mineral depletion of this reserve, plants depend on the external medium for minerals. If these required nutrients are not available in the growth medium in sufficient amounts, deficiency symptoms appear. Very little Ca^{2+} accumulates in the seeds of most crop plants. Hence, Ca^{2+} requirements of crop plants are met by addition of this element in the soil–plant system. Calcium is immobile in plants, and Ca^{2+} deficiency appears first in the newly emerging leaves or forming tissues. Even with Ca^{2+} deficiency of new leaves, the older or lower leaves may contain sufficient amounts of Ca^{2+}. Calcium deficiency is characterized by dead and tightly curled leaf tips that are usually bent over and sticky or gummy to the touch (Clark, 1982). Plants with severe Ca^+ deficiency are stunted since internodes fail to elongate, and the new growth grows in a rosette form. Leaf tips fail to unfold, are deformed, and form swordlike projections. Leaves become brittle, frequently coalesce, turn brown, and form sticky vehicles at or near the margins. Shoot–root ratios usually decrease because shoots are affected more extensively than roots (Clark, 1993). Root tips turn brownish, root extension is inhibited, lateral branching is reduced, and taproots are small in diameter (Cassman, 1993).

Flowering and maturity are delayed by mild calcium deficiency, but ear or head size is often unaffected in cereals. However, when calcium deficiency is severe, the head may not form or it may be partly barren (Grundon et al., 1987). New tissue needs Ca^{2+} pectate for cell wall formation. So Ca^{2+} deficiencies cause gelatinous leaf tips and growing point. Even though Ca^{2+} is considered a macronutrient and accumulates in plants at relatively high concentrations, considerable evidence indicates that many Ca^{2+} functions occur at very low Ca^{2+} concentrations (Hanson, 1984).

5.4 CONCENTRATION AND UPTAKE

Calcium concentration (uptake per unit dry weight) in plant tissues is an important parameter for diagnosis of deficiency or sufficiency level in crop plants. The Ca^{2+} concentration in plant tissues varied with soil, climatic, and plants factors. Among plant factors, age and part of tissue analyzed are the most important in determining Ca concentration. Figure 5.1 and Figure 5.2 show Ca^{2+} concentrations in the shoot of upland rice and dry bean grown on Brazilian Oxisols. In upland rice, Ca^{2+}

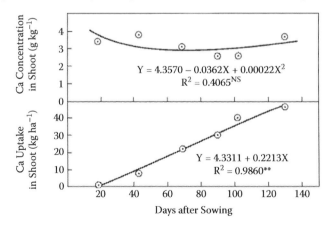

FIGURE 5.1 Relationship between plant age and calcium concentration and uptake in upland rice shoot.

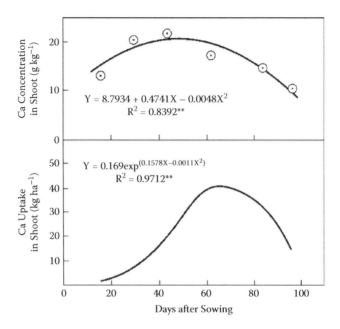

FIGURE 5.2   Relationship between plant age and calcium concentration and uptake in dry bean shoot.

concentration in the shoot decreased quadratically with the advancement of plant age. However, in dry bean, Ca^{2+} concentration in shoot increased in a quadratic fashion. Concentration of Ca in the shoot of dry bean was much higher compared with that of dry bean. Higher Ca^{2+} requirements of legumes compared to cereals have been widely reported (Loneragan et al., 1968; Fageria et al., 1997; Barber, 1995). Adequate or sufficient levels of Ca^{2+} in plant tissue of principal crop plants are presented in Table 5.2.

Calcium uptake (concentration × shoot or grain dry weight) is an important index for measuring soil fertility depletion with the cultivation of crop plants. The uptake of Ca^{2+} depends on dry weight of shoot and grain yield and varied from crop species to crop species (Table 5.3). The Ca^{2+} uptake was higher in shoot compared to grain in cereals as well as legume crops. In addition, it was higher in the grain of legumes compared to the grain of cereals. To produce 1 metric ton grain of upland rice, Ca^{2+} accumulated in the grain plus straw was 6 kg. Similarly, to produce 1 metric ton grain of lowland rice and corn, it is necessary to accumulate 5 kg Ca^{2+} in the grain plus straw. To produce 1 metric ton grain of dry bean, 9 kg Ca^{2+} should be accumulated in the grain and straw. To produce 1 metric ton grain of soybean, 17 kg Ca^{2+} is required in the grain plus straw. Hence, higher quantity of Ca is required by legumes compared to cereals to produce similar amounts of economic yield.

Calcium uptake in upland rice and dry bean during the growth cycle is presented in Figure 5.1 and Figure 5.2. In upland rice, Ca^{2+} accumulation in the shoot was linear with the advancement of plant age. The variability in Ca uptake in shoot was about 99% due to plant age. In dry bean, Ca^{2+} uptake increased quadratically with the advancement of plant age. The variability in Ca^{2+} uptake due to plant age was 97%.

TABLE 5.2
Adequate Calcium Concentration in Tissues of Principal Field Crops

Crop Species	Plant Part and Growth Stage	Adequate Ca Conc. (g kg⁻¹)
Wheat	Whole tops at heading	2–5
Barley	Whole tops at heading	3–12
Rice	Whole tops 100 days after sowing	2.5–4
Corn	Whole tops at 30 to 45 days after emergence	9–16
Corn	Ear leaf blade at silking	2.1–5
Sorghum	Whole tops at seedling	9–13
Sorghum	Whole tops at early vegetative	1.5–9
Sorghum	3ʳᵈ blade below panicle at bloom	2–6
Soybean	Upper fully developed trifoliate prior to pod set	3.6–20
Dry bean	Upper fully developed trifoliate at early flowering	15–25
Cowpea	Petiole of uppermost mature leaf blade at early flowering	7.2–10

Source: Adapted from Fageria et al. (1997).

TABLE 5.3
Shoot Dry Weight, Grain Yield, and Uptake of Ca²⁺ by Principal Field Crops

Crop Species	Shoot Dry Wt. (kg ha⁻¹)	Grain Yield (kg ha⁻¹)	Ca²⁺ Uptake in Shoot (kg ha⁻¹)	Ca²⁺ Uptake in Grain (kg ha⁻¹)	Total (kg ha⁻¹)
Upland rice	6642	4794	27	2	29
Lowland rice	8093	5000	20	4	24
Lowland rice	9423	6389	26	5	31
Corn	13670	8148	34	8	42
Dry bean	2200	3409	22	9	31
Dry bean	1773	1674	21	5	26
Soybean	2901	1323	48	5	53
Soybean	3244	3102	41	12	53

Source: Fageria and Baligar (2001); Fageria (2004); Fageria and Santos (2008); Fageria et al. (2007).

Hence, it can be concluded that plant age has a significant influence on Ca accumulation in field crops.

5.5 USE EFFICIENCY AND Ca HARVEST INDEX

Calcium use efficiency, which is defined as straw or grain yield per unit of Ca uptake in straw and grain, varies with crop species (Table 5.4). It was higher for the production of grain of cereals as well as legumes compared to shoot of cereals or legumes. The higher Ca efficiency for grain production of cereals and legumes compared to shoot was associated with lower amounts of Ca accumulation in the grain, as

TABLE 5.4
Calcium Use Efficiency in Upland Rice, Corn, Dry Bean, and Soybean during Growth Cycle Grown on Brazilian Oxisols (Values Are Averages of Two-Year Field Trial)

Age in Days after Sowing	Upland Rice (kg kg⁻¹)	Age in Days after Sowing	Corn (kg kg⁻¹)	Age in Days after Sowing	Dry Bean (kg kg⁻¹)	Age in Days after Sowing	Soybean (kg kg⁻¹)
19	299	18	176	15	107	27	86
43	266	35	191	29	49	41	59
69	326	53	226	43	47	62	60
90	406	69	272	62	59	82	71
102	392	84	297	84	72	120	85
130	281	119	379	96	99	158	74
130 (grain)	1324	119 (grain)	1001	96 (grain)	345	158 (grain)	245
Regression	Quadratic		Linear		Quadratic		Quadratic
R^2	0.42^{NS}		0.99**		0.82*		0.11^{NS}

Note: Calcium use efficiency = (shoot or grain yield/Ca accumulated in shoot or grain).

*, **, NS Significant at the 5 and 1% probability level and not significant, respectively.

TABLE 5.5
Calcium Harvest Index in Principal Field Crops

Crop Species	Ca Harvest Index (%)
Upland rice	7
Lowland rice	17
Lowland rice	16
Corn	19
Dry bean	29
Dry bean	19
Soybean	23

Note: Calcium harvest index (%) = (Ca uptake in grain/Ca uptake in grain plus straw) × 100.

indicated by the Ca harvest index values of crops presented in Table 5.5. In upland rice, only 7% of the total Ca accumulated (grain plus straw) was translocated to the grain. In the case of lowland rice, Ca translocation to grain was about 17%; for corn, it was 19%. In dry bean, the calcium harvest index was 19–29%; in soybean, 23%. This means that legume seeds have higher Ca content compared to cereals, and this has especial significance for human nutrition.

5.6   INTERACTION WITH OTHER NUTRIENTS

Many factors affect Ca uptake and activity. Extensive interactions of other mineral elements with Ca have been reported (Clark, 1984; Hanson, 1984). The ions that strongly inhibit Ca uptake are H^+, K^+, Na^+, Mg^{2+}, Al^{3+}, and NH_4^+. The inhibitory role of NH_4^+ on calcium uptake is of special interest since NH^{4+} fertilization of plants is a common practice. Nutrient solutions containing some NH_4^+ become low in pH (near 4.0 or below) when sorghum and corn are grown in them (Bernardo et al., 1984; Clark, 1982). The uptake of cations at low pH (high H^+ concentration) is less than that at higher pH value, and Ca uptake is depressed more than uptake of the other cations (Marschner, 1995). Ammonium is absorbed more readily than is Ca, and this can inhibit growth. Growth inhibition of sorghum by NH_4^+ was closely related to decreases in nutrient solution pH imposed by NH_4^+ uptake. Growth inhibition by NH^{4+} may be a result of Ca deficiencies that occur under these conditions.

Interaction between lime and P was investigated for the production of upland rice, dry bean, and corn in Brazilian Oxisol (Table 5.6). Interaction here is defined as comparison of yield increase of each crop with the addition of lime rate (2 and 4 g kg^{-1}) as compared to control and increasing P levels. In the case of upland rice, shoot dry weight decreased or displayed no change with increasing lime rate. This means that for upland rice, the interaction between lime and P was negative with increasing lime and P rate for shoot dry weight production. The negative interaction between lime and P was also observed in root growth of upland rice (Figure 5.3). In the case of dry bean and corn, it was positive with increasing lime and P rates. The negative interaction between lime and P in upland rice production may be associated with acidity tolerance of upland rice (Fageria et al., 2006; Fageria and Zimmermann, 1998). Fageria et al. (2004) determined correlation coefficients between grain yield and soil chemical properties associated with soil acidity (Table 5.7). Data in

---

TABLE 5.6
Shoot Dry Weight of Upland Rice Dry Bean and Corn under Different Lime and P Rates

Lime Rate (g kg^{-1})	P Rate (mg kg^{-1})	Upland Rice Shoot Dry Wt. (g pot^{-1})	Dry Bean Shoot Dry Wt. (g pot^{-1})	Corn Shoot Dry Wt. (g pot^{-1})
0	0	0.72	1.25	1.10
0	50	15.08	8.30	4.93
0	175	18.63	10.60	8.73
2	0	0.73	1.30	1.43
2	50	15.23	9.00	7.13
2	175	13.23	10.60	11.47
4	0	0.33	1.70	1.10
4	50	10.20	10.50	6.60
4	175	13.20	12.00	9.93

Source: Adapted from Fageria et al. (1995).

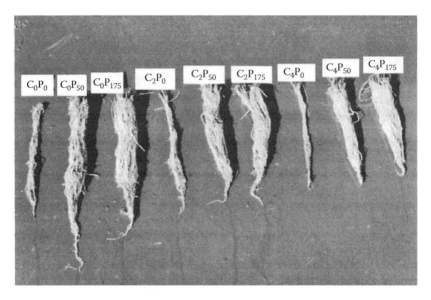

FIGURE 5.3 Root growth of rice at three lime rates and three P levels. Lime rates were 0, 2, and 4 g kg^{-1} of soil and P levels were 0, 50, and 175 mg kg^{-1} of soil.

TABLE 5.7
Correlation Coefficient (r) between Grain Yield and Soil Acidity Indices across Two Soil Acidity Levels (pH 4.5 and 6.4) and 20 Upland Rice Genotypes

Soil Acidity Index	Grain Yield (g pot^{-1})
PH	–0.232*
Ca (mmol$_c$ dm^{-3})	–0.168
Al (mmol$_c$ dm^{-3})	0.189*
H+Al (mmol$_c$ dm^{-3})	0.231*
Base saturation (%)	–0.207*

* Significant at the 5% probability level.
Source: Adapted from Fageria et al. (2004).

Table 5.7 show that grain yield of upland rice genotypes was significantly and positively related with Al and H+Al and possessed significant negative correlations with pH, Ca saturation, and base saturation. Fageria et al. (2004) reported that Brazilian upland rice cultivars are most tolerant to soil acidity and are used worldwide in breeding program to develop modern, high-yielding cultivars for tropical acid soils. An increasing response to applied P with increasing rates of added lime in the case of dry bean might be attributable to either an improved rate of supply of P by the soil or an improved ability of the plant to absorb P when Al toxicity has been eliminated (Friesen et al., 1980).

5.7 MANAGEMENT PRACTICES TO MAXIMIZE Ca²⁺ USE EFFICIENCY

To improve Ca^{2+} use efficiency by crop plants, adoption of some soil and crop management practices is essential. These practices include liming acid soils, applying at an optimum rate, using an appropriate source, using appropriate Ca/Mg and Ca/K ratios, and using efficient crop species or genotypes within species. In addition, some practices like the use of farmyard manures, conservation tillage and crop rotation, and control of diseases, insects, and weeds, which are discussed in chapters 2 and 3, are also valid for adequate calcium management for efficient crop production.

5.7.1 LIMING ACID SOILS

Liming acid soils is one of the oldest practices in agriculture. Kamprath (1970, 1971; Foy, 1992; Fageria and Baligar, 2003) suggested that elimination of toxic quantities of Al^{3+} and Mn^{2+} and addition of adequate quantities of Ca^{2+} and Mg^{2+} are the goals of liming, especially in highly leached and weathered soils. Data in Table 5.8 show that liming of Brazilian Oxisol significantly improved pH, increased Ca^{2+} and Mg^{2+}

TABLE 5.8
Soil pH, Ca, Mg, and Al Contents as Influenced by Liming in Brazilian Oxisol

Lime Rate (Mg ha⁻¹)	pH in H₂O	Ca²⁺ (cmol_c kg⁻¹)	Mg²⁺ (cmol_c kg⁻¹)	Al³⁺ (cmol_c kg⁻¹)
18 Days after Sowing of Upland Rice				
0	5.0	1.07	0.38	0.62
3	5.3	1.33	0.72	0.28
6	5.6	1.90	0.99	0.13
9	5.9	2.18	1.32	0.10
12	6.0	2.51	1.49	0.08
67 Days after Sowing of Upland Rice				
0	5.0	1.17	0.33	0.52
3	5.4	1.60	1.09	0.26
6	5.6	2.09	1.53	0.12
9	5.8	2.52	1.78	0.07
12	6.0	2.91	2.35	0.04

Regression Analysis across Two Sampling Dates

Lime rate (X) vs. pH (Y) = $5.0085 + 0.1309X - 0.0039X^2$, $R^2 = 0.9920**$
Lime rate (X) vs. Ca (Y) = $1.1180 + 0.1353X$, $R^2 = 0.9940**$
Lime rate (X) vs. Mg (Y) = $0.4480 + 0.1253X$, $R^2 = 0.9860**$
Lime rate (X) vs. Al (Y) = $0.554 - 0.1009X + 0.0051X^2$, $0.9870**$

* Significant at the 1% probability level.
Source: Adapted from Fageria et al. (1991).

content, and decreased Al^{3+} content. The variability in these soil chemical properties was about 99% due to liming. Substantial increases in exchangeable Ca^{2+} and Mg^{2+} between the 18- and 67-day sampling periods give evidence of the continued dissolution of the dolomitic lime.

5.7.2 APPLICATION OF OPTIMUM RATE

Application of the optimum rate is essential for improving Ca^{2+} use efficiency by crop plants. This rate may vary with crop species, cultivar within species, and soil type and climatic conditions. All these factors should be taken into account when defining the optimum rate of Ca^{2+} for a given crop. Response of soybean to applied lime in a Brazilian Oxisol is shown in Figure 5.4. Grain yield was significantly increased with increasing lime rate in the range of 0 to 18 Mg ha^{-1}. The increase in grain yields followed a quadratic pattern, and variability in grain yield due to liming was 86% (Figure 5.4). The increase in grain yield was 35% at 3 Mg lime ha^{-1} compared to control or without lime treatment. Similarly, grain yield increase was 53% with the application of 6 Mg lime ha^{-1} compared with control treatment.

Based on a regression equation (Figure 5.4), maximum grain yield was achieved at 12.6 Mg lime ha^{-1}. However, 90% of the maximum grain yield, considered as an economic rate for grain yield in annual crops (Fageria, 2001a), was achieved at about 6 Mg lime ha^{-1}. Increase in grain yield of soybean by liming in Brazilian Oxisols has been reported by Fageria (2001a) and Barbosa Filho et al. (2005). Raij and Quaggio (1996) reported that the economic rate of lime for soybean in cerrado soil (Oxisol) is 5.5 Mg ha^{-1}.

The grain yield increase was associated with an increase in shoot dry weight, an increase in number of pods per plant, and an increase in grain harvest index with the addition of lime (Table 5.9). The increase in shoot dry weight was 42% at 6 Mg lime ha^{-1} compared to control treatment. Similarly, overall increase in pod number per

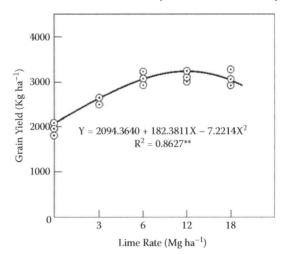

FIGURE 5.4 Relationship between lime rate and grain yield of soybean grown on Brazilian Oxisol. Values are averages of 3-year field trial.

TABLE 5.9

Association between Grain Yield (Y) and Plant Growth and Yield Components (X) of Soybean Grown on Brazilian Oxisol

Plant Parameter	Regression Equation	R^2
Shoot dry wt. vs. grain yield	$Y = 2107.0430 + 0.3693X$	$0.1936*$
Number of pods per plant vs. grain yield	$Y = -3.6860 + 0.0182X$	$0.4556**$
Number of grain per pod vs. grain yield	$Y = 263.2958 - 190.7731X + 43.4871X^2$	0.0136^{NS}
Weight. of 100 grain vs. grain yield	$Y = -3049.9900 + 492.8400X - 19.5433X^2$	0.0731^{NS}
Grain harvest index vs. grain yield	$Y = -959.5784 + 4124.8170X - 4177.7350X^2$	$0.2458*$

$*, **, NS$ Significant at the 5 and 1% probability levels and nonsignificant, respectively.

plant was 28% at 6 Mg lime ha^{-1} compared to control treatment. Association between grain yield and growth and yield components showed that pod number per plant contributed 46% variation in grain yield, shoot dry matter contributed 19% variation in grain yield, and grain harvest index contributed 25% variation in grain yield. This means that pods per plant is the most important yield component in increasing grain yield of soybean compared to other growth and yield components.

In addition to the crop response curve to added lime, soil pH, base saturation, calcium saturation, exchangeable calcium, and Al^{3+} saturation are important indices of soil acidity in defining lime rates. Figure 5.5 shows the relationship between soil pH and dry bean grain yield at 0–10 and 10–20 cm depths. Soil pH significantly influenced grain yield of bean at 0–10 and 10–20 cm soil depths (Figure 5.5). Maximum bean yield was achieved at pH 6.7 at 0–10 cm soil depth. At 10–20 cm soil depth, maximum yield was obtained with a soil pH of 6.3. Across two soil depths, maximum grain yield was achieved at a pH of 6.5. At 0–10 cm soil depth, at the lowest pH value of 5.3, the grain yield was about 2270 kg ha^{-1}. However, when pH at this depth (0–10 cm) was increased to 6.7, the grain yield was 3330 kg ha^{-1}. The grain yield increase was 47% when soil pH was raised from 5.3 to 6.7. A similar increase in grain yield was obtained at the 10–20 cm soil depth with the increase in pH of similar range. Fageria (2001a) reported that maximum grain yield of dry bean in Brazilian Oxisol was achieved at soil pH of about 6.2. Dry bean root growth was also increased when soil pH was raised from 4.9 to 5.9 and then decreased (Figure 5.6). Lathwell and Reid (1984) also indicated that soil pH values for optimum crop production should be between pH 6.0 and 7.0. Similarly, Alley and Zelazny (1987) reported that soils in crop production should be limed to at least pH 6.5 to provide a generally more productive environment. In the present study, maximum grain yield of bean was achieved more or less in the same pH range. The increase in grain yield with increase in soil pH may be associated with improvement in the content of Ca and Mg and reduction in the Al concentration in soil solution (Fageria, 2002).

Bean yield had a significant quadratic response in relation to base saturation (Figure 5.7). Maximum yield was obtained with base saturation of 73% at 0–10 cm soil depth, with base saturation of 62% at 10–20 cm soil depth and at 67% base saturation when averaged across two soil depths. Hence, higher base saturation was

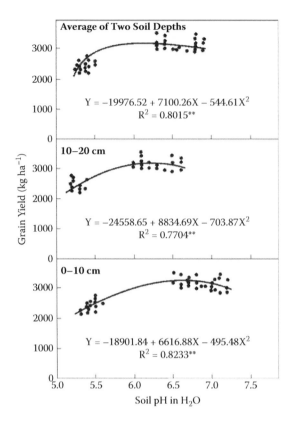

FIGURE 5.5 Relationship between soil pH and grain yield of dry bean. Values are averages of five field trails conducted for 3 years (Fageria, 2008).

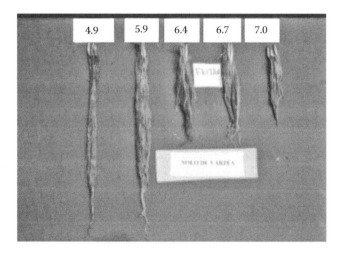

FIGURE 5.6 Influence of soil pH on root growth of dry bean.

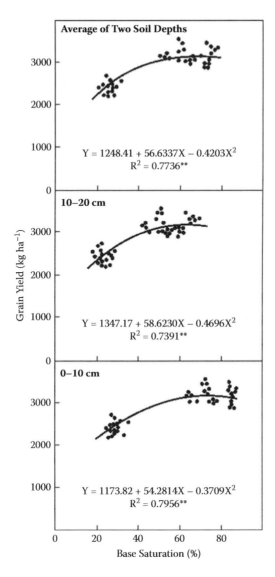

FIGURE 5.7   Relationship between base saturation and grain yield of dry bean. Values are averages of five field trails conducted for 3 years (Fageria, 2008).

required at the topsoil layer than at the lower soil layer. In the topsoil layer (0–10 cm), the lowest base saturation was 25%, and yield at this base saturation level was about 2400 kg ha⁻¹. At optimum base saturation level (73%), the yield was about 3270 kg ha⁻¹. Hence, when base saturation was raised from 25 to 73% in the topsoil layer, the yield increase was about 36%. More or less similar grain yield increase was observed when base saturation was increased from 20 to 62% in the subsoil layer. Figure 5.8 shows dry bean growth at 30%, 60%, and 70% base saturation. The increase in yield with increasing base saturation is associated with favorable cation balance in the soil (Fageria, 2002). Basic cations are usually low in the highly weathered, acid soils of

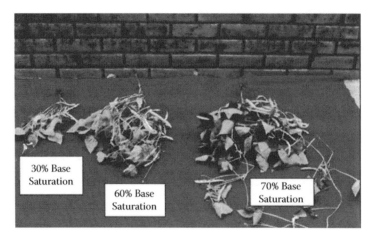

FIGURE 5.8 Dry bean plant growth at 30% base saturation (left), 60% base saturation (middle), and 70% base saturation (right) grown on a Brazilian Oxisol.

the humid and subhumid tropical bean-producing regions, and liming is the most effective practice for replenishing the soil cation pool (Dierolf et al., 1997).

Bean yield significantly increased with increasing calcium saturation in the range of 18 to 65% in the topsoil layer (0–10 cm), and 15 to 44% in the subsoil layer (10–20 cm) (Figure 5.9). Maximum yield was obtained with 53% Ca saturation in the topsoil layer, 44% Ca saturation in the 10–20 cm soil layer, and 48% Ca saturation across the two soil layers. The bean yield at the lowest Ca saturation level of 18% in the top layer was 2270 kg ha^{-1} and increased to 3200 kg ha^{-1} when Ca saturation was increased to 53% in this topsoil layer. Hence, there was a 41% increase in grain yield at 53% Ca saturation compared with 18% Ca saturation. Fageria (2001b) reported that a Ca saturation of 37% is optimum for bean production in Brazilian Oxisol in conventional cropping system. This suggests that in no-tillage systems, the optimum Ca saturation value for bean may be higher compared with conventional cropping system. However, Eckert (1987) reviewed many studies and reported that the ideal Ca saturation is around 65%. Figure 5.10 shows that soybean grain yield also increased with increasing Ca saturation in a quadratic fashion. Based on the regression equation, maximum grain yield of soybean was obtained at 57% Ca saturation. Fageria (2001b) reported that maximum grain yield of soybean was achieved with a calcium saturation of 51% in Brazilian Oxisol.

Exchangeable calcium concentration in the soil is an important indicator or parameter to determine lime requirements of a crop species. The optimum exchangeable calcium for maximum growth varied with crop species, type of soil, and other crop management practices. However, in tropical acid soils, exchangeable calcium should be more than 2 cmol$_c$ kg^{-1} for the growth of most crop plants (Fageria, 2001b; Fageria, 2002). Figure 5.11 shows the relationship between exchangeable calcium and soybean grain yield in Brazilian Oxisol. Soybean yield increased significantly and quadratically with increasing exchangeable calcium in the range of 0.5 to 3.2. Based on the regression equation, maximum grain yield was achieved at an exchangeable calcium level of 2.8.

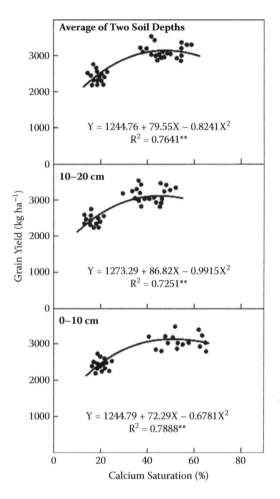

FIGURE 5.9 Relationship between calcium saturation and grain yield of dry bean. Values are averages of five field trails conducted for 3 years (Fageria, 2008).

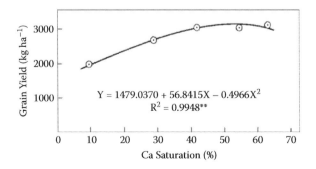

FIGURE 5.10 Relationship between calcium saturation and grain yield of soybean.

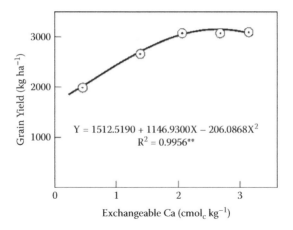

FIGURE 5.11 Relationship between exchangeable calcium content in the soil and grain yield of soybean.

Aluminum saturation significantly and linearly reduced bean yield in the top as well as in the subsoil layer (Figure 5.12). In the topsoil layer (0–10 cm), bean yield was 3130 kg ha^{-1} at zero level of aluminum saturation and decreased to 2133 kg ha^{-1} when aluminum saturation was raised to 6.5%. Hence, a decrease of 47% occurred at higher aluminum saturation in this layer. In the subsoil layer (10–20 cm), yield at zero level was 3130 kg ha^{-1} and at 10% Al saturation was 2270 kg ha^{-1}, a decrease of 38%. Across the two soil depths, the yield decrease was 42% at 8% Al saturation compared with zero level of Al saturation. Fageria and Santos (1998) and Fageria (2002) have reported the decrease in bean yield with increasing Al saturation. Thung and Rao (1999) reported that bean yield reduction could be as high as 30 to 60% in Al-toxic soils. CIAT (1992) reported that about 40% of the bean-growing areas in Latin America and 30 to 50% of bean production areas of central, eastern, and southern Africa may be affected by Al toxicity. Aluminum toxicity interferes with root development, water use, uptake, transport, and utilization of several essential nutrients (Foy, 1984). Critical aluminum saturation percentages for important crop species are given in Table 5.10.

5.7.3 Use of Appropriate Source

There are several sources of calcium (Table 5.11). Because most Ca-deficient soils are acid, a good liming program can add Ca most efficiently; both calcitic and dolomitic limestones are excellent sources (Table 5.11). Gypsum ($CaSO_4 \cdot 2H_2O$) can also supply calcium when soil pH is high enough not to need lime. Single superphosphate, which is 50% gypsum, and to a lesser extent, triple superphosphate and rock phosphates, can also add Ca to the soil. The phosphate rock (PR) has been used as a phosphorus fertilizer for decades. Because the P mineral in PR is in the form of apatite, which also contains calcium (Ca), it is possible that PR may have potential agronomic value as a source of Ca. Simmons et al. (1988) evaluated various sources of Ca in potato crop and demonstrated that applications of preplant strip $CaSO_4$ from

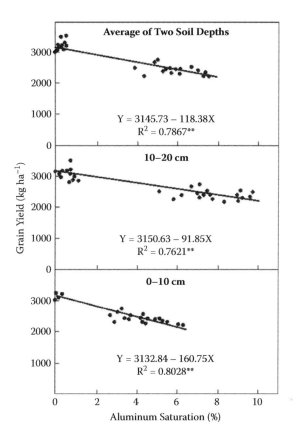

FIGURE 5.12 Relationship between aluminum saturation and grain yield of dry bean. Values are averages of five field trails conducted for 3 years (Fageria, 2008).

several sources were the most consistent means, of those evaluated, for improving potato yield and quality on sandy, low exchangeable Ca soils. Calcium sulfate was superior to finely ground dolomitic lime in all the years, although triple superphosphate performed as well as $CaSO_4$ in 1 of the 2 years it was evaluated. Calcium chloride applications resulted in reduced plant emergence at higher Ca rates (336 kg Ca ha^{-1}) used in these experiments as broadcast. It has also demonstrated that high broadcast applications of gypsum (3090 kg ha^{-1}) to peanut at early bloom improve peanut yields and quality (Hallock and Garren, 1968).

Sousa and Lobato (1996) reported that gypsum is an important amendment and can be used to improve crop production in acid soils. It can improve Ca^{2+} content of acid soils, can neutralize Al^{3+} of subsurface soils, and improves root development in the surface and subsurface soils. In cerrado soils of Brazil, gypsum is recommended on the basis of clay content for annual crop production using the following formula (Sousa and Lobato, 1996): Gypsum requirement (kg ha^{-1}) = 50 × clay content (%). In addition to Ca, gypsum can supply S to growing plants, and Ca can leach to subsoil horizons where native Ca levels may be too low to support root growth. The ionic

TABLE 5.10
Critical Al Saturation for Important Field Crops
at 90–95% of Maximum Yield

Crop	Type of Soil	Critical Al Saturation (%)
Cassava	Oxisol/Ultisol	80
Upland rice	Oxisol/Ultisol	70
Cowpea	Oxisol/Ultisol	55
Cowpea	Oxisol	42
Peanut	Oxisol/Ultisol	65
Peanut	Xanthic Halpludox	54
Soybean	Oxisol	19
Soybean	Xanthic Halpludox	27
Soybean	Oxisol/Ultisol	15
Soybean	Not given	<20
Soybean	Ultisol	20–25
Soybean	Histosol	10
Soybean	Ultisol	20
Corn	Oxisol	19
Corn	Xanthic	27
Corn	Oxisol/Ultisol	29
Corn	Oxisol/Ultisol	25
Corn	Oxisol	28
Mungbean	Oxisol/Ultisol	15
Mungbean	Oxisol/Ultisol	5
Coffee	Oxisol/Ultisol	60
Sorghum	Oxisol/Ultisol	20
Common bean	Oxisol/Ultisol	10
Common bean	Oxisol/Ultisol	8–10
Common bean	Oxisol/Ultisol	23
Cotton	Not given	<10

Source: Fageria et al. (1997), Fageria and Baligar (2001), Fageria and
Baligar (2003).

strength of the soil solution is increased by gypsum, and the increased strength lowers the activity of Al^{3+}. Further, sulfate forms ion pairs such as $AlSO_4^+$, which is less toxic to plants than Al^{3+} (Fageria and Baligar, 2003).

5.7.4 APPROPRIATE CA/MG AND CA/K RATIOS

Due to antagonism among cations in the uptake process, appropriate Ca/Mg and Ca/K ratios are important for uptake of Ca, Mg, and K by crop plants. The basic cation saturation ratio concept originated in New Jersey and was the product of the efforts of Bear and his coworkers (Bear et al., 1945; Bear and Toth, 1948; Hunter, 1949). These investigators proposed an *ideal ratio* for saturation of the cation exchange complex of 65% Ca, 10% Mg, 5% K, and 20% H. These saturations were

TABLE 5.11

Liming Materials, Their Composition, and Their Neutralizing Power

Commercial Name	Chemical Formula	Neutralizing Value (%)	Characteristics
Dolomitic lime	$CaMg(CO_3)_2$	95–109	Contains 78–120 g kg^{-1} Mg and 180–210 g kg^{-1} Ca
Calcitic lime	$CaCO_3$	100	Contains 284–320 g kg^{-1} Ca
Dolomite lime	$MgCO_3$	100–120	Contains 36–72 g kg^{-1} Mg
Burned lime	CaO	179	Fast reacting and difficult to handle
Slaked lime	$Ca(OH)_2$	136	Fast reacting and difficult to handle
Basic slag	$CaSiO_3$	86	By-product of pig-iron industry; also contains 1–7% P
Wood ash	Variable	30–70	Caustic and water soluble

Source: Fageria (1989), Bolan et al., (2003), Brady and Weil (2002), and Caudle (1991).

derived from approximately 8 years of work with alfalfa on New Jersey soils. The concept was modified by Graham (1959), who proposed saturation ranges of 65 to 85% Ca, 6 to 12% Mg, and 2 to 5% K for Missouri soils, ranges that allow for rather wide variation in actual Ca/Mg/K ratios in the soil. Further, Adams and Henderson (1962) concluded that Mg was deficient if it was present at <4% of the CEC. No significant relationship was found between the exchangeable Ca/Mg ratio and crop yields in the following ranges: 1 to 4.5 (Moser, 1953), 0.24 to 32 (Hunter, 1949), and 1 to 13.5 (Halstead et al., 1958). Later, MacLean and Finn (1967) suggested that yields were lowered by an exchangeable Ca/Mg ratio of 0.62. Lierop et al. (1979) found a critical molar soil-extracted Ca/Mg ratio of about 0.5 for onion by graphing the molar soil-extracted Ca/Mg ratio against percent of maximum yield. Onion appears to be able to absorb sufficient Ca and Mg to meet its requirement for these nutrients within at least a soil-extracted Ca/Mg range of about 0.5 to 16, as yield did not seem to be affected by unfavorable Ca/Mg ratios within that range. Figure 5.13 shows the

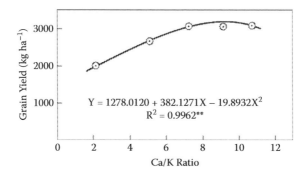

FIGURE 5.13 Relationship between Ca/K ratio and grain yield of soybean.

relationship between Ca/K ratio and grain yield of soybean grown on Brazilian Oxisol. Maximum grain yield was obtained at a Ca/K ratio of about 10.

The concept of developing a soil test interpretation on the basis of cation saturation seems reasonable in light of basic cation exchange phenomena and the effects that the degree of saturation of one cation may have on the availability of itself and other cations. Soil pH is often related to percent base saturation and is raised to acceptable levels by addition of lime, a Ca-based soil amendment (Fageria and Stone, 1999). Lime raises soil pH by neutralizing acidity but maintains the increased pH by increasing the percentage base saturation of Ca (plus Mg if dolomitic lime is used). Liebhardt (1981) noted the relationship between pH and base saturation and found that Ca+Mg saturation of 75% produced a pH 6.0 level on Delaware coastal plain soils, which was desirable from the standpoint of the corn–soybean rotation practiced on these soils. However, as the Ca + Mg saturation was increased above this level, soil pH increased and yields were reduced due to development of Mn deficiency in the crop.

5.7.5 USE OF EFFICIENT CROP SPECIES/CULTIVARS

Calcium uptake and utilization by plants depends on calcium concentration in the growth medium and is also genetically controlled (Mengel et al., 2001). Plant species have been classified as calcicoles and calcifuges in relation to calcium requirement. Calcicoles are plants that generally grow well on calcareous soils, whereas calcifuges are plants that grow well on acid, low-Ca soils. Normal leaf Ca concentration ranges from 4800 to 10,800 ug g^{-1} in numerous calcifuges and from 15,800 to 22,700 mg g^{-1} in calcicoles (Hou and Merkle, 1950). Calcifuges plants have been characterized by tolerance to low levels of Ca, low response to added Ca, low Ca^{2+} uptake rates, high K/Ca ratios in tissues, slow growth rates, and tolerance to Al. Genotype differences to Ca requirements have been reported for barley, corn hybrids, common bean, potato, pearl millet, tomato, red clover, ryegrass (Clark, 1984), and upland rice (Fageria et al., 2004). Figure 5.14, Figure 5.15, and Figure 5.16 show root and shoot growth of four lowland rice cultivars in nutrient solution. Cultivar EEA 304 was significantly better in root as well as shoot growth at low as well as high Al^{3+} levels.

Many of the plant species tolerant to acidity have their center of origin in acid soil regions, suggesting that adaptation to soil constraints is part of the evolutionary processes (Sanchez and Salinas, 1981). A typical example of this evolution is Brazilian upland rice cultivars' tolerance to soil acidity. In Brazilian Oxisols, upland rice grows very well without liming, when other essential nutrients are supplied in adequate amounts and water is not a limiting factor (Fageria et al., 2004). Fageria et al. (2004) reported that grain yield and yield components of 20 upland rice genotypes were significantly decreased at low soil acidity (limed to pH 6.4) as compared with high soil acidity (without lime, pH 4.5), demonstrating the tolerance of upland rice genotypes to soil acidity. In Table 5.12, data are presented showing grain yield and panicle number of six upland rice genotypes at two acidity levels. Fageria (1989) reported stimulation of growth of Brazilian rice cultivars at 10 mg Al^{3+} L^{-1} in nutrient solution compared with control treatment. Okada and Fischer (2001) reported

FIGURE 5.14   Root and shoot growth of four lowland rice cultivars (left to right: Suvale 1, CICA 4, EEA 304, and IRGA 408) at 10 mg Al L^{-1} in nutrient solution.

FIGURE 5.15   Root and shoot growth of four lowland rice cultivars (left to right: Suvale 1, CICA 4, EEA 304, and IRGA 408) at 20 mg Al L^{-1} in nutrient solution.

that the mechanism for the genotype difference of upland rice in the tolerance to soil acidity seemed to be related to the relationship between regulation of cell elongation and ligand-bound Ca^{2+} at the root apoplast.

A substantial number of plant species of economic importance are generally regarded as tolerant to acid soils conditions in the tropics (Sanchez and Salinas,

FIGURE 5.16   Root and shoot growth of four lowland rice cultivars (left to right: Suvale 1, CICA 4, EEA 304, and IRGA 408) at 60 mg Al L^{-1} in nutrient solution.

1981). In addition, there are cultivars tolerant to soil acidity within crop species (Garvin and Carver, 2003; Fageria et al., 2004; Yang et al., 2004). Yang et al. (2004) reported significant differences among genotypes of alfalfa, dry bean, corn, sorghum, triticale, and rice to Al toxicity. These crop species or cultivars within species can be planted on tropical acid soils in combination with reduced rate of liming in different cropping systems.

Similarly, legume cover crops of Brazilian origin grow well in acid soils (Table 5.13). Data in Table 5.13 show that dry weight of shoot of cover crops was significantly affected by acidity treatments. Maximum shoot dry weight was produced at high acidity, indicating tolerance of legume species to soil acidity. White jack bean and black mucuna bean species were most acidity tolerant, whereas Brazilian stylo and pueraria were most susceptible to soil acidity. Overall, optimal soil acidity indices were pH 5.5, H+Al 1.8 cmol$_c$ kg^{-1}, base saturation 25%, and acidity saturation

TABLE 5.12
Grain Yield and Panicle Number of 5 Upland Rice Genotypes at Two Soil Acidity Levels in Brazilian Oxisol

	Grain Yield (g pot⁻¹)		Panicle Number (pot⁻¹)	
Genotype	High Acidity (pH 4.5)	Low Acidity (pH 6.4)	High Acidity (pH 4.5)	Low Acidity (pH 6.4)
CRO97505	74.3	52.0	38.0	28.3
CNAs8983	55.2	42.9	29.0	25.7
Primavera	53.0	47.2	25.0	21.7
Canastra	51.6	38.9	32.0	26.3
Bonança	48.8	36.5	26.3	20.7
Carisma	50.8	17.5	43.3	17.7
Average	66.7	47.0	38.7	28.1

Source: Adapted from Fageria et al. (2004).

TABLE 5.13
Shoot Dry Weight (g plant⁻¹) of 14 Tropical Legume Cover Crops at Three Acidity Levels

Legume Species	High Acidity	Medium Acidity	Low Acidity	Average
Crotalaria	4.04d	1.97fg	0.91	2.31gh
Sunn hemp	12.32bc	7.76cde	4.23bc	8.10cdef
Crotalaria	4.24d	2.87efg	1.41cd	2.84gh
Crotolaria	5.80d	4.35defg	3.74bcd	4.63fgh
Crotalaria	13.05bc	6.12cdef	3.44bcd	7.53def
Calapogonio	4.88d	3.11defg	1.82cd	3.27gh
Pigeon pea	7.80cd	5.45cdefg	4.09bc	5.78efg
Lablab	14.31b	10.39bc	5.80b	10.17bcd
Mucuna bean ana	12.99bc	8.39bcd	4.32bc	8.57cde
Black mucuna bean	16.39ab	13.46b	6.48b	12.11b
Gray mucuna bean	15.13b	13.79b	4.47bc	11.13bc
White jack bean	21.43a	19.78a	13.52a	18.24ᵃ
Pueraria	3.76d	1.61fg	0.92cd	2.09h
Brazilian stylo	3.95d	0.32g	0.18d	1.48h
Average	10.00	7.09	3.95	6.69
F-test				
Soil acidity (S)		**		
Legume species (L)		**		
S X L		**		

Note: Means followed by the same letter in the same column are significantly not different by Tukey's test at the 5% probability level.

** Significant at the 1% probability level.

TABLE 5.14

Shoot Dry Weight Efficiency Index (SDWEI) of Legume Species at Medium and Low Soil Acidity Level and Their Classification to Soil Acidity Tolerance

Legume Species	SDWEI at Medium Acidity[a]	SDWEI at Low Acidity
Crotalaria breviflora	0.14S	0.11S
Crotalaria juncea L.	1.62T	1.50T
Crotalaria munocrata	0.21S	0.22S
Crotalaria spectabilis	0.43MT	0.66MT
Crotalaria ochroleuca	1.35T	1.36T
Calapogonium mucunoides	0.56MT	0.27S
Cajanus cajan L. Millspaugh	0.73MT	0.98MT
Dolichos lablab L.	2.49T	2.47T
Mucuna deeringiana L.	1.94T	1.66T
Mucuna aterrima L.	3.76T	3.22T
Mucuna cinereum L.	3.60T	2.04T
Canavalia ensiformis L. DC.	7.15T	8.80T
Pueraria phaseoloides Roxb.	0.09S	0.11S
Stylosanthes guianiensis	0.03S	0.02S

[a] T = tolerant, MT = moderately tolerant, and S = susceptible.

Source: Fageria et al. (2008).

23%. Shoot dry weight efficiency index (SDWEI) was calculated for classification of legume species to acidity tolerance by using following equation:

$$SDWEI = \left(\frac{\text{Shoot dry weight at medium or low acidity level}}{\text{Average shoot dry weight of 14 species at medium or low acidity level}} \right)$$
$$\times \left(\frac{\text{Shoot dry weight at high acidity level}}{\text{Average shoot dry weight of 14 species at high acidity level}} \right)$$

The species having SDWEI > 1 were classified as tolerant, species having SDWEI between 0.5 to 1 were classified as moderately tolerant, and species having SDWEI < 0.5 were classified as susceptible to soil acidity. These efficiency ratings were done arbitrarily, and results are presented in Table 5.14. Higher seed weight species had higher tolerance to soil acidity (Figure 5.17). Hence, seed size was also associated with acidity tolerance in tropical legume species.

5.8   CONCLUSIONS

Calcium deficiency in crop plants is most common in highly weathered acid soils. Due to weathering, Ca^{2+} is leached from the soil profile to the lower soil depths and

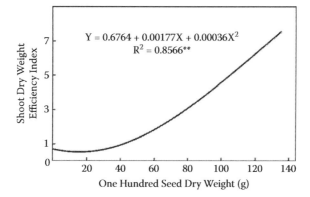

FIGURE 5.17 Relationship between seed dry weight and shoot dry weight efficiency index of legume species. Values are across 14 legume species and two acidity levels (low pH 7.0 and high pH 5.5 in water) (Fageria et al., 2008).

not available to growing roots. In addition, calcium is lost by soil erosion and plant uptake in the soil–plant systems. Calcium uptake may also be reduced by adsorption on soil colloids under some conditions. Calcium plays an important role in maintaining the integrity of the plasma membrane of plant cells, plant cell elongation, and plant cell division. Calcium is immobile in the phloem and during plant growth if this element is in short supply; thus, deficiency first appears in the young growing parts of the plant.

Plant tissue analysis and soil testing are important Ca^{2+} deficiency or sufficiency diagnostic techniques. Plant tissue analysis can also indicate soil fertility depletion by nutrients accumulated or extracted from the soil. Calcium uptake is higher in the shoot or straw compared to seed or grain. On average, about 15% of calcium is translocated to the grain of cereals, and the rest remains in the straw. In legumes, about 25% of calcium is translocated to the grain, and the rest remains in the grain. This means that incorporation of crop residues into the soil may recycle significant amounts of calcium for the succeeding crops.

Soil testing is a widely used method for making liming recommendations. Liming is the most effective and economic practice to supply Ca^{2+} to crop plants in Ca^{2+}-deficient soils. Adequate exchangeable level of Ca^{2+} in the soil varied with the crop species and genotypes within species. For example, legumes like dry bean and soybean need high levels of Ca^{2+} in the soil for optimal growth compared to cereals such as upland rice, corn, and barley. However, for tropical soils, an adequate level of exchangeable Ca^{2+} in the range of 2 to 3 $cmol_c$ kg^{-1} is sufficient for most crop plants. Liming also neutralizes Al^{3+} toxicity and improves soil pH. Optimum soil pH for the growth of most crop plants in the tropical soils is also determined by crop species. However, a pH value of about 6.5 determined in water is adequate for the growth of most crop species. Liming also improves activities of beneficial microorganisms in the soil and improves nutrient use efficiency. Liming requirements should be determined by crop response to lime for each crop under different agroecosystems.

Calcium efficiency in crop production can also be improved by adopting other practices, like the use of farmyard manures, conservation tillage, crop rotation, and control of diseases, insects, and weeds. Use of acidity-tolerant crop species and genotypes within species is another important management practice for efficient crop production on acid soils.

REFERENCES

Adams, F. 1984. Crop response to lime in the southern United States. In: *Soil acidity and liming,* 2nd edition, F. Adams, Ed., 211–265. Madison, WI: Am. Soc. Agron.

Adams, F. and J. B. Henderson. 1962. Magnesium availability as affected by deficient and adequate levels of potassium and lime. *Soil Sci. Soc. Am. Proc.* 26:65–68.

Alley, M. M. and L. W. Zelazny. 1987. Soil acidity: Soil pH and lime needs. In: *Soil testing: sampling, correlation, calibration, and interpretation,* J. R. Brown, Ed., 65–72. Madison, WI: SSSA.

Barber, S. A. 1995. *Soil nutrient bioavailability: a mechanistic approach,* 2nd edition. New York: John Wiley & Sons.

Barbosa Filho, M. P., N. K. Fageria, and F. J. P. Zimmerman. 2005. Soil fertility attributes and productivity of dry bean and soybean influenced by lime applied at soil surface and incorporated. *Ci. Agrotec.* 29:507–514.

Bear, F. E., A. L. Prince, and J. L. Malcolm. 1945. *The potassium needs of New Jersey Soils.* New Jersey. Agric. Exp. Stn. Bull 721.

Bear, F. E. and S. J. Toth. 1948. Influence of calcium on availability of other soil cations. *Soil Sci.* 65:67–74.

Bennett, W. F. 1993. Plant nutrient utilization and diagnostic plant symptoms. In: *Nutrient deficiencies & toxicities in crop plants,* W. F. Bennett, Ed., 1–7. St. Paul, MN: The American Phytopathological Society, American Phytopathological Society Press.

Bernardo, L. M., R. B. Clark, and J. W. Maranville. 1984. Nitrate/ammonium ratio effects on nutrient solution pH, dry matter yield, and nitrogen uptake of sorghum. *J. Plant Nutr.* 7:1389–1400.

Bolan, N. S., D. C. Adriano, and D. Curtin. 2003. Soil acidification and liming interactions with nutrient and heavy metal transformation and bioavailability. *Adv. Agron.* 78:215–272.

Bould, C., E. J. Hewitt, and P. Needham. 1984. *Diagnosis of mineral disorders in plants: Principles,* Vol. 1. New York: Chemical Publishing.

Brady, N. C. and R. R. Weil. 2002. *The nature and properties of soils,* 13th edition. Upper Saddle River, NJ: Prentice-Hall.

Brennan, R. F., M. D. A. Bolland, and G. H. Walton. 2007. Comparing the calcium requirements of wheat and canola. *J. Plant Nutr.* 30:1167–1184.

Bruce, R. C., L. A. Warrell, D. G. Edwards, and L. C. Bell. 1988. Effects of aluminum and calcium in the soil solution of acid soils on root elongation of Glycine max cv. Forrest. *Aust. J. Agric. Res.* 39:319–338.

Cassman, K. G. 1993. Cotton. In: *Nutrient deficiencies & toxicities in crop plants,* W. F. Bennett, Ed., 111–119. St. Paul, MN: The American Phytopathological Society, American Phytopathological Society Press.

Caudle, N. 1991. *Managing soil acidity.* Raleigh, NC: North Carolina State University, Tropsoils Publication.

Chan, K. Y., B. G. Davey, and H. R. Geering. 1979. Adsorption of magnesium and calcium by a soil with variable charge. *Soil Sci. Soc. Am. J.* 43:301–304.

CIAT (Centro International de Agriculture Tropical). 1992. *Constraints to and opportunities for improving bean production: A planning Document 1993–98 and achieving Document 1987–92*, Cali, Colombia: CIAT.

Clark, R. B. 1982. Plant response to mineral element toxicity and deficiency. In: *Breeding plants for less favorable environments*, M. N. Christian and C. F. Levis, Eds., 71–142. New York: John Wiley & Sons.

Clark, R. B. 1984. Physiological aspects of calcium, magnesium, and molybdenum deficiencies in plants. In: *Soil acidity and liming*, 2nd edition, F. Adams, Ed., 99–170. Madison, WI: ASA.

Clark, R. B. 1993. Sorghum. In: *Nutrient deficiencies & toxicities in crop plants,* W. F. Bennett, Ed., 21–26. St. Paul, MN: The American Phytopathological Society, American Phytopathological Society Press.

Dierolf, T. S., L. M. Arya, and R. S. Yost. 1997. Water and cation movement in an Indonesian Ultisol. *Agron. J.* 89:572–579.

Eckert, D. J. 1987. Soil test interpretations: basic cation saturation ratios and sufficiency levels. In: *Soil testing: sampling, correlation, calibration, and interpretation*, J. R. Brown, Ed., 53–64. Madison, WI: SSSA.

Epstein, E. and A. J. Bloom. 2005. *Mineral nutrition of plants: Principles and perspectives.* Sunderland, Massachusetts: Sinauer Associations, Inc. Publishers.

Fageria, N. K. 1983. Ionic interactions in rice plants from dilute solutions. *Plant Soil* 70:309–316.

Fageria, N. K. 1989. *Tropical soils and physiological aspects of crop yields.* EBRAPA, Brasilia-DF, Brazil.

Fageria, N. K. 2001a. Effect of liming on upland rice, common bean, corn, and soybean production in cerrado soil. *Pesq. Agropec. Bras.* 36:1419–1424.

Fageria, N. K. 2001b. Response of upland rice, dry bean, corn and soybean to base saturation in cerrado soil. *Rev. Bras. Eng. Agric. Ambien.* 5:416–424.

Fageria, N. K. 2002. Nutrient management for sustainable dry bean production in the tropics. *Commun. Soil Sci. Plant Anal.* 33:1537–1575.

Fageria, N. K. 2004. Dry matter yield and nutrient uptake by lowland rice. *J. Plant Nutr.* 6:947–958.

Fageria, N. K. 2008. Optimum soil acidity indices for dry bean production on an oxisol in no-tillage system. *Commun. Soil Sci. Plant Anal.* 39:845–857.

Fageria, N. K. and V. C. Baligar. 1997. Response of common bean, upland rice, corn, wheat, and soybean to soil fertility of an Oxisol. *J. Plant Nutr.* 20:1279–1289.

Fageria, N. K. and V. C. Baligar. 2001. Improving nutrient use efficiency of annual crops in Brazilian acid soils for sustainable crop production. *Commun. Soil Sci. Plant Anal.* 32:1303–1319.

Fageria, N. K. and V. C. Baligar. 2003. Fertility management of tropical acid soils for sustainable crop production. In: *Handbook of soil acidity,* Z. Rengel, Ed., 359–385. New York: Marcel Dekker.

Fageria, N. K., V. C. Baligar, and R. B. Clark. 2006. *Physiology of crop production.* New York: Haworth Press.

Fageria, N. K., V. C. Baligar, and C. A. Jones. 1997. *Growth and mineral nutrition of field crops*, 2nd edition. New York: Marcel Dekker.

Fageria, N. K., V. C. Baligar, and R. W. Zobel. 2007. Yield, nutrient uptake and soil chemical properties as influenced by liming and boron application in common bean in no-tillage system. *Commun. Soil Sci. Plant Anal.* 38 (in press).

Fageria, N. K., V. C. Baligar, and Y. C. Li. 2008. Differential soil acidity tolerance of tropical legume cover crops. *Commun. Soil Sci. Plant Anal.* 39:1637–1653.

Fageria, N. K. and F. Breseghello. 2004. Nutritional diagnostic in upland rice production in some municipalities of State of Mato Grosso, Brazil. *J. Plant Nutr.* 27:15–28.

Fageria, N. K., E. M. Castro, and V. C. Baligar. 2004. Response of upland rice genotypes to soil acidity. In: *The red soils of China: Their nature, management and utilization,* M. J. Wilson, Z. He, and X. Yang, Eds., 219–237. Dordrecht: Kluwer Academic Publishers.

Fageria, N. K. and H. R. Gheyi. 1999. *Efficient crop production.* Campina Grande, Brazil: Federal University of Paraiba.

Fageria, N. K. and A. B. Santos. 1998. Rice and common bean growth and nutrient concentration as influenced by aluminum on an acid lowland soil. *J. Plant Nutr.* 21:903–912.

Fageria, N. K. and A. B. Santos. 2008. Lowland rice response to thermophosphate fertilization. *Commun. Soil Sci. Plant Analy.* 39:873–889.

Fageria, N. K. and L. F. Stone. 1999. *Soil acidity management of cerrado and varzea soils of Brazil.* Santo Antônio de Goiás, Brazil: National Rice and Bean Research Center of EMBRAPA Document No. 92.

Fageria, N. K., R. J. Wright, V. C. Baligar, and J. R. P. Carvalho. 1991. Response of upland rice and common bean to liming on an Oxisol. In: *Plant-soil interactions at low pH,* R. J. Wright, V. C. Baligar, and R. P. Murrmann, Eds., 519–525. Dordrecht: Kluwer Academic Publishers.

Fageria, N. K., F. J. P. Zimmermann, and V. C. Baligar. 1995. Lime and phosphorus interactions on growth and nutrient uptake by upland rice, wheat, common bean, and corn in an Oxisol. *J. Plant Nutr.* 18:2519–2532.

Fageria, N. K. and F. J. P. Zimmermann. 1998. Influence of pH on growth and nutrient uptake by crop species in an Oxisol. *Commun. Soil Sci. Plant Anal.* 29:2675–2682.

Follett, R. H., L. S. Murphy, and R. L. Donahue. 1981. *Fertilizers and soil amendments.* Englewood Cliffs, NJ: Prentice-Hall.

Foy, C. D. 1984. Physiological effects of hydrogen, aluminum, and manganese toxicities in acid soils. In: *Soil acidity and liming,* 2nd edition, F. Adams, Ed., 57–97. Madison, WI: ASA, CSSA, and SSSA.

Foy, C. D. 1992. Soil chemical factor limiting plant root growth. *Adv Soil Sci.* 97:97–149.

Friesen, D. K., A. S. R. Juo, and H. H. Miller. 1980. Liming and lime-phosphorus-zinc interactions in two Nigerian Ultisols. I. Interactions in the soil. *Soil Sci. Soc. Am. J.* 44:1221–1226.

Garvin, D. F. and B. F. Carver. 2003. Role of the genotype in tolerance to acidity and aluminum toxicity. In: *Handbook of soil acidity,* Z. Rengen, Ed., 387–406. New York: Marcel Dekker.

Goedert, W. J. 1983. Management of the cerrado soils of Brazil: A review. *J. Soil Sci.* 34:405–428.

Goedert, W. J. 1989. Cerrado region: Agricultural potential and policy for its development. *Pesq. Agropec. Bras.* 24:1–17.

Graham, E. R. 1959. An explanation of theory and methods of soil testing. Mo. Agric. Exp. Stn. Bull 734.

Grundon, N. J., D. G. Edwards, P. N. Takkar, C. J. Asher, and R. B. Clark. 1987. Nutritional disorders of grain sorghum. Australian Center for International Agricultural Research Monograph No 2, Canberra.

Hallock, D. L. and K. H. Garren. 1968. Pod breakdown, yield and grade of Virginia type peanuts as affected by, Ca, Mg, and K sulfates. *Agron. J.* 60:253–257.

Halstead, R. L., A. J. McLean, and K. F. Nielsen. 1958. Ca: Mg ratios in soil and the yield and composition of alfalfa. *Can. J. Soil Sci.* 38:85–92.

Hanson, J. B. 1984. The functions of calcium in plant nutrition. In: *Advances in plant nutrition,* P. K. Tinker and A. Lauchli, Eds., 149–208. New York: Praeger.

Hou, H. Y. and F. G. Merkle. 1950. Chemical composition of certain calcifugous and calcicolous plants. *Soil Sci.* 69:471–486.

Hunter, S. A. 1949. Yield and composition of alfalfa as affected by variations in calcium-magnesium ratio in the soil. *Soil Sci.* 67:53–62.

Ishizuka, Y. 1978. *Nutrient deficiencies of crops.* Taipei, Taiwan: Asia and Pacific (ASPAC), Food and Fertilizer Technology Center.

James, R. O. and T. W. Healy. 1972. Adsorption of hydrolyzable metal ions at the oxide-water interface. I. Co (II) adsorption on SiO_2 and TiO_2 as model systems. *J. Colloid Interface Sci.* 40:42–52.

Kamprath, E. J. 1970. Exchangeable aluminum as a criterion for liming mineral soils. *Soil Sci. Soc. Am. Proc.* 34:252–254.

Kamprath, E. J. 1971. Potential detrimental effects from liming highly weathered soils to neutrality. *Soil and Crop Sci. Soc. Fla. Proc.* 31:200–203.

Kamprath, E. J. 1984. Crop response to lime on soils in the tropics. In: *Soil acidity and liming,* 2nd edition, F. Adams, Ed., 349–368. Madison, WI: ASA.

Lathwell, D. J. and W. S. Reid. 1984. Crop response to lime in the northwestern United States. In: *Soil acidity and liming,* 2nd edition, A. Adams, Ed., 305–332. Madison, WI: ASA.

Liebhardt, W. C. 1981. The basic cation saturation ratio concept and lime and potassium recommendations on Delaware coastal plain soils. *Soil Sci. Soc. Am. J.* 45:544–549.

Lierop, W. V., Y. A. Martel, and M. P. Cescas. 1979. Anion response to lime on acid Histosols as affected by Ca/Mg ratios. *Soil Sci. Soc. Am. J.* 43:1172–1177.

Loneragan, J. F., K. Snowball, and W. J. Simmons. 1968. Response of plants to calcium concentration in solution culture. *Aust. J. Agric. Res.* 19:845–857.

MacLean, A. D. and B. J. Finn. 1967. Amendments for oats in soil contaminated with magnesium lime dust. *Can. J. Soil Sci.* 47:253–254.

Marschner, H. 1995. *Mineral nutrition of higher plants,* 2nd edition. New York: Academic Press.

Mengel, K., E. A. Kirkby, H. Kosegarten, and T. Appel. 2001. *Principles of plant nutrition,* 5th edition. Dordrecht: Kluwer Academic Publishers.

Moore, P. A. Jr. and W. H. Patrick, Jr. 1989. Calcium and magnesium availability and uptake by rice in acid sulfate soils. *Soil Sci. Soc. Am. J.* 53:816–822.

Moser, F. 1953. The calcium-magnesium ratio in soils and its relation to crop growth. *J. Am. Soc. Agron.* 25:365–377.

Okada, K. and A. J. Fischer. 2001. Adaptation mechanisms of upland rice genotypes to highly weathered acid soils of South American savannas. In: *Plant nutrient acquisition: New perspectives,* N. Ae, J. Arihara, K. Okada, and A. Srinivasan, Eds., 185–200. Tokyo: Springer.

Pan, W. L. 2000. Bioavailability of calcium, magnesium and sulfur. In: *Handbook of soil science,* M. E. Sumner, Ed., 53–69. Boca Raton, FL: CRC Press.

Raij, B. van, and J. A. Quaggio. 1996. Method used for diagnosis and correction of soil acidity in Brazil: An overview. In: *International symposium on plant-soil interactions at low pH,* A. C. Moniz, A. M. C. Furlani, R. E. Schaffert, N. K. Fageria, C. A. Rosolem, and H. Cantalella, Eds., 205–214. Campinas/Viçosa: Brazilian Soil Science Society.

Reddy, A. S. N. and V. S. Reddy. 2002. Calcium as a messenger in stress signal transduction. In: *Handbook of plant and crop physiology,* 2nd edition, M. Pessarkli, Ed., 697–733. New York: Marcel Dekker.

Sanchez, P. A. and J. G. Salinas. 1981. Low-input technology for managing Oxisols and Ultisols in tropical America. *Adv. Agron.* 34:279–406.

Sanchez, P. A., J. H. Villachica, and D. E. Bandy. 1983. Soil fertility after clearing a tropical rainforest in Peru. *Soil Sci. Soc. Am. J.* 47:1171–1178.

Schroeder, J. I., G. J. Allen, V. Hugouvieux, J. M. Kwak, and D. Waner. 2001. Guard cell signal transduction. *Annu. Rev. Plant Physiol. Plant Molecular Biol.* 52:627–658.

Simmons, K. E., K. A. Kelling, R. P. Wolkowski, and A. Kelman. 1988. Effect of calcium source and application method on potato yield and cation composition. *Agron. J.* 80:13–21.

Snyder, F. W. and G. E. Carlson. 1984. Selecting for partitioning of photosynthetic products in crops. *Adv. Agron.* 37:47–72.

Sousa, D. M. G. and E. Lobato. 1996. *Soil amendment and fertilization for soybean growth in cerrado soils.* Brasilia, DF: EMBRAPA-CPAC. Technical Circular, 33.

Thung, M. and I. M. Rao. 1999. Integrated management of abiotic stresses. In: *Common bean improvement in the twenty-first century*, S. P. Singh, Ed., 331–370. Dordrecht: Kluwer Academic Publishers.

Thurman, L. G., K. D. Ritchy, and G. C. Naderman, Jr. 1980. Nitrogen fertilization on an Oxisol of the cerrado of Brazil. *Agron. J.* 72:261–265.

Yang, X., W. Wang, Z. Ye, Z. He, and V. C. Baligar. 2004. Physiological and genetic aspects of crop plant adaptation to elemental stresses in acid soils. In: *The red soils of China: Their nature, management and utilization*, M. J. Wilson, Z. He, and X. Yand, Eds., 171–218. Dordrecht: Kluwer Academic Publishers.

Yost, R. S., E. J. Kamprath, E. Lobato, and G. Naderman. 1979. Phosphorus response of corn on an Oxisol as influenced by rates and placement. *Soil Sci. Soc. Am. J.* 43:338–343.

6 Magnesium

6.1 INTRODUCTION

Magnesium (Mg^{2+}) is an essential macronutrient for all plant growth and development. Its adequate level in the soil is important for producing maximum economic yields. Like deficiency of calcium (Ca^{2+}), deficiency of magnesium in crop production is more common on highly weathered acid soils (Fageria and Souza, 1991). Deficiency of Mg^{2+} may also occur in coarse-textured soils of humid regions with low cation exchange capacities. Data in Table 6.1 show significant increase in root and shoot dry weight of dry bean and cowpea, except cowpea root dry weight, to increasing Mg^{2+} concentration in the range of 0.30 to 6.22 $cmol_c$ kg^{-1} of soil. The increase in root and shoot dry weight was quadratic in fashion with increasing Mg^{2+} level in the soil. More than 80% variability in root and shoot dry weight of both the crops was due to variation in Mg^{2+} level in the soil. This means that for annual crop production in Oxisols, application of magnesium is necessary for maximizing crop yields. The low Mg^{2+} content of such soils is associated with leaching of this element from the soil profile.

Due to variation in soil weathering and parent materials, Mg^{2+} contents of soils vary widely. However, Mengel et al. (2001) reported that the Mg^{2+} content of most soils generally lies in the range between 0.5 g kg^{-1} for sandy soils and 5 g kg^{-1} for clay soils (Mengel et al., 2001). The amount of Mg^{2+} adequate for plant growth varied from soil to soil, plant species to plant species, and even among cultivars within species (Fageria and Souza, 1991; Fageria et al., 1997). However, Mg^{2+} content in the range of 1.0 to 2 $cmol_c$ kg^{-1} is sufficient to produce maximum economic yield of field crops (Fageria, 2001, 2002). The availability and plant nutrition behavior of Mg^{2+} is similar to that of calcium; however, its requirements for crop growth are smaller than those of calcium. The availability of Mg^{2+} to crop plants is influenced by many soil and plant factors. These factors are the concentration of Mg^{2+} in the soil solution, degree of Mg^{2+} saturation, presence of other cations like K^+ and Ca^{2+}, soil pH, type of clay, and plant species or genotypes within species. The objective of this chapter is to discuss the role of Mg^{2+} in crop production and management strategies to improve its efficiency for maximum economic yield of crop plants.

6.2 CYCLE IN SOIL–PLANT SYSTEMS

The cycle of Mg^{2+} in soil–plant systems involves its addition to soils and depletion by several processes. The main sources of Mg^{2+} addition are liming, Mg^{2+} fertilizers, crop residues, farmyard or green manures, and liberation by weathering of parent materials. Its removal or depletion from the soil–plant system is mainly through

TABLE 6.1

Influence of Magnesium on Root and Shoot Dry Weight of Upland Rice and Dry Bean Grown on Brazilian Oxisol

Mg Level in Soil (cmol$_c$ kg^{-1})	Dry Bean Dry Weight (g plant^{-1})		Cowpea Dry Weight (g plant^{-1})	
	Root	Shoot	Root	Shoot
0.30	0.70	3.01	0.71	2.72
1.05	0.81	3.10	0.75	2.81
1.15	0.83	3.19	0.91	3.11
1.33	0.74	3.19	1.04	3.07
3.52	1.00	3.81	0.87	2.74
6.22	0.56	3.12	0.43	2.03

Regression Analysis

Mg level (X) vs. dry bean root dry wt. (Y) = $0.5898 + 0.2414X - 0.0393X^2$, $R^2 = 0.8641$*

Mg level (X) vs. dry bean shoot dry wt. (Y) = $2.7022 + 0.5490X - 0.0766X^2$, $R^2 = 0.8532$*

Mg level (X) vs. cowpea root dry wt. (Y) = $0.6811 + 0.2142X - 0.0413X^2$, $R^2 = 0.8221$[NS]

Mg level (X) vs. dry bean root dry wt. (Y) = $2.7611 + 0.2133X - 0.0538X^2$, $R^2 = 0.8938$*

*, NS Significant at the 5% probability level and nonsignificant, respectively.

uptake by crops, soil erosion, and leaching. In addition, some part of Mg^{2+} is also fixed in the soil–plant system by soil colloids and microorganisms. Sumner et al. (1978) and Grove et al. (1981) reported significant amounts of Mg^{2+} fixation in Oxisols and Ultisols when limed to pH above 7.0. Most of the Mg^{2+} is present in the soil as primary minerals, and very little exists in organic forms or in the form of organic complexes. Principal Mg^{2+}-containing soil minerals are presented in Table 6.2. Addition of Ca^{2+} increases the leaching of Mg^{2+} from the soil profile because Ca^{2+}

TABLE 6.2

Principal Magnesium-Containing Soil Minerals and Their Composition

Soil Mineral	Formula	Mg Content (g kg^{-1})
Magnesium sulfate	$MgSO_4$	200
Magnesium carbonate	$MgCO_3$	290
Serpentine	$H_4Mg_3Si_2O_9$	490
Dolomitic lime	$CaCO_3.MgCO_3$	130
Montmorillonite	$Al_5MgSi_{12}O_{30}(OH)_6$	Up to 60
Illite	$K_{0.6}Mg_{0.25}Al_{2.3}Si_{3.5}O_{10}(OH)_2$	20
Vermiculite	$Mg_3Si_4O_{10}(OH)_2.2H_2O$	120–170
Olivine	$Mg_{1.6}Fe_{0.4}SiO_4$	250
Chlorite	$Al_2Mg_5Si_3O_{10}(OH)_8$	Up to 230
Brucite	$Mg(OH)_2$	410

Source: Adapted from Barber (1995).

held more tightly on to soil colloids compared to Mg^{2+}. Under acidic conditions, Al^{3+} and H^+ ions displace Mg^{2+} from the exchange complex and leaching may occur. In addition, Mg^{2+} is not specifically bound to clay minerals, as in the case of K^+, much of which is adsorbed in the interlayer of 2:1 phyllosilicates (Mengel et al., 2001).

The affinity of cations for the exchange complex is dependent on the mineralogy of the colloids. For instance, the affinities of Ca^{2+} and Mg^{2+} for montmorillonite are similar, but the affinity of Mg^{2+} for vermiculite is much greater than the affinity of Ca^{2+} (Camberato and Pan, 2000). The fractions of Mg^{2+} in the soil are classified as nonexchangeable, exchangeable, and water-soluble or in soil solution. These three Mg forms are in equilibrium, and the nonexchangeable form is maximized compared to the other two forms. Some Mg^{2+} presents in the soil in association with organic matter, but this fraction is usually less than 1% of the total soil Mg^{2+} (Mengel et al., 2001). The transport of Mg^{2+} to the root of plants mainly occurs by mass flow and diffusion.

6.3 FUNCTIONS AND DEFICIENCY SYMPTOMS

The functions of Mg^{2+} in plants are many, and most important are its roles as enzyme activator and as component of chlorophyll molecules. Magnesium is a mineral constituent of plant chlorophyll, so it is actively involved in photosynthesis. Magnesium also aids in phosphate metabolism, plant respiration, and the activation of several enzyme systems involved in energy metabolism (Fageria and Gheyi, 1999). Magnesium aids in the formation of sugars, oils, and fats. It also activates formation of polypeptide chains from amino acids (Tisdale et al., 1985). Magnesium is also an essential element for microbial growth and was implicated in microbial ecology in early studies of soil microbiology, since magnesium carbonate applied to certain soils increases the reproduction of soil bacteria (Jones and Huber, 2007). It is also required for the preservation of ribosome structure and integrity. Up to 90% or more of cellular Mg is bound, mainly in ribosomes. It is associated with rapid growth, active mitosis, high protein levels, carbohydrate metabolism, and oxidative phosphorylation in physiological young cells (Jones and Huber, 2007).

Grass tetany is a disorder of ruminants caused by insufficient quantities of available Mg^{2+} in their feed. The chemical composition of forage is often correlated with the incidence of grass tetany in animals. Grunes and Maryland (1975) stated that the factors which give forage high potential to induce grass tetany include plant Mg^{2+} concentrations less than 2 g kg^{-1}, plant K concentrations >30 g kg^{-1}, $K/(Ca+Mg)$ equivalent ratio >2.2 in the plant, and plant N concentrations >40 g kg^{-1}.

Magnesium is a mobile element in the plant, and deficiency symptoms of this element first appear in the older leaves and tissues. Symptoms of Mg^{2+} deficiency are characterized by a light coloring between the veins or interveinal chlorosis. Leaves became brittle and margins curled. Dark necrotic spots appear, and leaves usually turn reddish purple with severe deficiency; the tips and edges may die as Mg is translocated from old to new plant tissue (Clark, 1982). Root growth is reduced in Mg-deficient plants, and Mg-deficient plants roots turn dark red (Fageria and Souza, 1991). Height and tiller number are little affected in cereals when the deficiency is moderate (Fageria and Gheyi, 1999).

Magnesium-deficient plants lack vigor, are often stunted, and usually have a delayed reproductive stage (Clark, 1993). Shoot–root ratios increase with Mg deficiency because root growth decreases more than shoot growth (Clark, 1993). Magnesium deficiency also inhibits N_2 fixation by nitrogen-fixing rhizobium bacteria. Magnesium deficiency can occur after heavy application of ammonia or K fertilization (Sinclair, 1993). Seldom is an excess of Mg directly harmful to the plant, but excess may suppress the uptake of Ca, K, and Mn and reduce plant growth (Jones and Huber, 2007).

6.4 CONCENTRATION AND UPTAKE

The concentration of an essential element in the plant tissue is an important criterion for diagnosis of its deficiency or sufficiency. Nutrient concentration in plant tissue is expressed in content per unit of plant dry weight. Several factors affect nutrient uptake or concentration in plant tissues. Among these factors, plant age, plant part analyzed, crop species, and fertility status of a soil are the most important. Crop nutrient requirements change with the advancement of plant age during the crop growth cycle. Hence, plant tissue analysis has special significance during the crop growth cycle. Figure 6.1 shows Mg^{2+} concentration in shoot of upland rice during growth cycle. The Mg^{2+} concentration significantly decreased in a quadratic fashion with the advancement of plant age. The variation in Mg^{2+} concentration in rice shoot was 87% with the variation in plant age. It was about 3 g kg^{-1} at 20 days after sowing and decreased to about 2 g kg^{-1} at harvest (130 days after sowing). This variation in Mg^{2+} concentration shows the importance of plant age in tissue analysis for nutrient deficiency or sufficiency diagnosis purposes. Table 6.3 shows Mg^{2+} sufficiency values in plant tissues of principal crop species. It can be concluded from the data of Table 6.3 that Mg^{2+} concentration was higher in legumes than in cereals. In cereals, Mg^{2+} concentration varied from 1.5 to 8 g kg^{-1} shoot dry weight; in legumes, Mg^{2+} concentration values varied from 1.7 to 15 g kg^{-1}, depending on crop species, plant part analyzed, and plant growth stage.

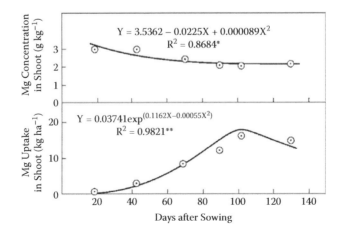

FIGURE 6.1 Magnesium concentration and uptake in upland rice as a function of plant age.

TABLE 6.3
Magnesium Sufficiency Values in the Plant Tissues of Principal Crop Species

Crop Species	Growth Stage	Plant Part	Mg Sufficiency Level (g kg⁻¹)
Wheat	Heading	Whole tops	1.5–5.0
Barley	Heading	Whole tops	1.5–5.0
Rice	100 days after sowing	Whole tops	1.7–3.0
Corn	30 to 45 days after emergence	Whole tops	3.0–8.0
Corn	Before tasseling	Leaf blade below cob	2.5–5.0
Sorghum	Vegetative	Young uppermost mature leaf blade	2.0–5.0
Soybean	Prior to pod set	Upper fully developed trifoliate	11.0–15.0
Dry bean	Early flowering	Upper fully developed trifoliate	4.0–8.0
Cowpea	Early flowering	\petiole of uppermost mature leaf blade	1.7–3.1
Peanut	Early pegging	Upper stems and leaves	3.0–8.0

Source: Adapted from Fageria et al. (1997).

Soil solution concentration of Mg^{2+} is higher than that of K^+, but uptake rate of this element is lower than that of K^+. Data in Table 6.4 show uptake rate of Mg^{2+} and K^+ by rice plants as influenced by Ca^{2+} concentration in nutrient solution. The absorption rate of Mg^{2+} was much lower compared to K^+ at a given Ca^{2+} concentration. Uptake of Mg^{2+} as well as K^+ was significantly influenced by Ca^{2+} concentration in the growth medium. For example, at the lowest Ca concentration of 6.23 μM, K^+ uptake rate was about 6.5 times higher compared to Mg^{2+} uptake rate. Similarly, at

TABLE 6.4
Influence of Calcium Concentration on Uptake of Potassium and Magnesium by Rice Plants in Nutrient Solution

Calcium Concentration (μM)	Mg²⁺ Uptake Rate (μg g⁻¹ h⁻¹ Root Dry Weight)	K⁺ Uptake Rate (μg g⁻¹ h⁻¹ Root Dry Weight)
6.23	3.80	24.49
12.47	5.06	36.09
49.90	25.70	150.67
74.79	52.59	226.56
124.75	57.41	400.00
249.50	62.49	433.42
499.0	46.46	355.25
748.0	30.76	336.00
R^2	0.66*	0.77*

* Significant at the 5% probability level.
Source: Adapted from Fageria (1973b).

the highest Ca^{2+} concentration of 748 µM, K^+ uptake rate was about 11 times higher compared to Mg^{2+} uptake rate. The higher uptake rate of monovalent cations like K^+ compared to divalent cations like Mg^{2+} and Ca^{2+} has been widely reported in the literature (Fageria, 1973a, 1976, 1983). The reasons for the high uptake rate of monovalent cations compared to divalent cations may be hydrated ion size. Berry and Ulrich (1968) reported that higher uptake rate of potassium compared to calcium and magnesium is associated with very efficient and selective potassium mechanisms of higher plants.

Nutrient uptake (concentration × dry weight) is an important criterion to know about nutrient extraction behavior of a crop from soils. Soil fertility depletion can also be estimated or measured by knowing quantity of nutrients accumulated or extracted by a crop. To estimate nutrient uptake, shoot and grain are analyzed separately for a determined nutrient, and their dry weight is also necessary to calculate uptake or accumulation. Field trial results of nutrient uptake by crops are generally expressed in kg ha^{-1}. Uptake of Mg^{2+} by shoot and grain of principal field crops grown on Brazilian Oxisols and Inceptisols are presented in Table 6.5. The Mg uptake is higher in the shoot compared to grain in the cereals as well as in legumes.

In the cereals, Mg uptake in shoot plus grain varied from 11 to 30 kg ha^{-1} depending on crop species. Crops remove from 10 to 80 kg Mg ha^{-1} depending on the yield and the particular crop (Doll and Lucas, 1973; Sanchez, 1976). Barber (1995) reported a total uptake of 45 kg Mg ha^{-1} by corn. The variation in uptake of Mg^{2+} by shoot and grain of different crops (Table 6.5) was also associated with variation in yield of shoot and grain of these crops (Fageria, 2001a, 2004; Fageria et al., 2007; Fageria and Santos, 2007). Uptake of Mg^{2+} by upland rice and dry bean during crop growth cycles is presented in Figure 6.1 and Figure 6.2. The increase in uptake of Mg^{2+} by these two crops was significantly and quadratically increased with the advancement of plant age. The quadratic uptake pattern of Mg^{2+} was associated

TABLE 6.5
Uptake of Mg^{2+} by Principal Field Crops

Crop Species	Mg Uptake in Shoot (kg ha^{-1})	Mg Uptake in Grain (kg ha^{-1})	Total Uptake (kg ha^{-1})
Upland rice[a]	15	5	20
Lowland rice[b]	15	7	22
Lowland rice[b]	15	6	21
Corn[a]	20	9	29
Dry bean[a]	7	4	11
Dry bean[a]	7	6	13
Soybean[a]	14	10	24
Soybean[a]	20	10	30

[a] Oxisols.

[b] Inceptisols.

Source: Adapted from Fageria (2001a), Fageria (2004), Fageria and Santos (2008), Fageria et al. (2007).

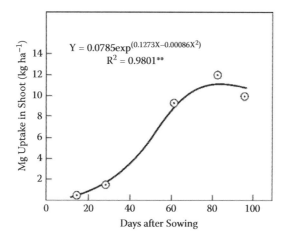

FIGURE 6.2 Magnesium uptake in dry bean plants as a function of plant age.

with dry matter yield of these two crops. The variability in uptake of Mg^{2+} by upland rice as well as by dry bean was 98% due to variation in plant age. This means that Mg uptake was highly correlated with plant age.

6.5 USE EFFICIENCY AND Mg^{2+} HARVEST INDEX

Magnesium use efficiency—defined as shoot dry matter yield or grain yield divided by Mg accumulation in the shoot or grain—of upland rice, corn, dry bean, and soybean at different growth stages is presented in Table 6.6. In upland rice, Mg use efficiency (Y) showed a quadratic association with plant age (X) (Y = 279.2875 + 3.1627X – 0.0122X^2, R^2 = 0.9231*). Based on the regression equation, maximum Mg use efficiency was achieved at 130 days' plant growth. In the case of corn, Mg use efficiency was linear with the advancement of plant age from 18 to 119 days' growth. The regression equation relating plant age (X) and Mg use efficiency in shoot (Y) was Y = 202.4203 + 3.4986X, R^2 = 0.9761**. Hence, variability in Mg use efficiency was 92% for upland rice and 98% for corn with the advancement of plant age. In the case of dry bean and soybean, Mg^{2+} use efficiency had a quadratic association with plant age, but the effect was not significant. The regression equation for dry bean was (Y = 221.8043 + 0.1319X – 0.00218X^2, R^2 = 0.1036^{NS}); for soybean, the equation was (Y = 184.8034 + 0.8373X – 0.0056X^2, R^2 = 0.2255^{NS}). In the grain, Mg^{2+} use efficiency was in the order of corn > upland rice > dry bean > soybean. Overall, Mg^{2+} use efficiency in shoot as well as grain was higher in cereals compared to legumes. The higher Mg use efficiency in cereals compared to legumes was associated with low amounts of Mg accumulated in upland rice and corn compared to dry bean and soybean. It is well known that dicotyledons generally have higher contents of bivalent cations than monocotyledons; the reverse holds true for noncovalent cations (Loneragan and Snowball, 1969).

Magnesium harvest index, defined as quantity of Mg^{2+} accumulated in grain as a percentage of total Mg accumulated in grain plus shoot, for cereals and legumes

TABLE 6.6

Magnesium Use Efficiency in Upland Rice, Corn, Dry Bean, and Soybean during Crop Growth Cycles (Values Are Averages of Two-Year Field Trials)

Upland Rice		Corn		Dry Bean		Soybean	
Age in Days after Sowing	Mg Use Efficiency (kg kg⁻¹)	Age in Days after Sowing	Mg Use Efficiency (kg kg⁻¹)	Age in Days after Sowing	Mg Use Efficiency (kg kg⁻¹)	Age in Days after Sowing	Mg Use Efficiency (kg kg⁻¹)
19	345	18	279	15	224	27	227
43	379	35	323	29	230	41	195
69	422	53	393	43	212	62	195
90	492	69	436	62	219	82	202
102	477	84	463	84	236	120	247
130	476	119	643	96	203	158	160
130 (grain)	856	119 (grain)	955	96 (grain)	434	158 (grain)	247
Regression	Quadratic		Linear		Quadratic		Quadratic
R^2	0.9231*		0.9761**		0.1036^{NS}		0.2255^{NS}

Note: Magnesium use efficiency = (Shoot or grain yield/Mg accumulated in shoot or grain).

*, **, NS Significant at the 5 and 1% probability level and not significant, respectively.

is presented in Table 6.7. Overall, Mg^{2+} harvest index was higher in legumes than in cereals. This means that the Mg^{2+} requirement of legumes is higher compared to cereals. A relation between Mg^{2+} harvest index (X) and grain yield (Y) of cereals and legumes cited in Table 6.7 was calculated. This relationship was ($Y = 15370.18 - 454.4832X + 3.6653X^2$, $R^2 = 0.3187^{NS}$). This means that Mg^{2+} accumulation in grain of cereals and legumes has no effect on grain yield.

TABLE 6.7

Grain Yield and Magnesium Harvest Index in Principal Field Crops

Crop Species	Grain Yield (kg ha⁻¹)	Mg Harvest Index (%)
Upland rice	4559	25
Lowland rice	4797	32
Lowland rice	6389	29
Corn	8501	31
Dry bean	1912	36
Dry bean	3409	46
Soybean	1441	42
Soybean	3038	33

Note: Magnesium harvest index (%) = (Mg uptake in grain/Mg uptake in grain plus straw) × 100.

6.6 INTERACTIONS WITH OTHER NUTRIENTS

Nutrient interactions at the root–soil interface are an important aspect in mineral nutrition of plants. Knowledge of the interfacial processes may lead to a better understanding of the relationship between crop yield and the nutrient level of soils. There is no easily recognizable nutrient balance within the plant for best crop production, except when an essential nutrient becomes so low as to limit growth (Ulrich et al., 1993). Ohno and Grunes (1985) reported that K^+ depresses the shoot concentration of Mg^{2+} by reducing the translocation of Mg from the root to the shoot of wheat plants. Wilkinson (1983) reported that high levels of soil K^+ depressed Mg^{2+} uptake by plants. Similarly, Ologunde and Sorensen (1982), using a sand culture system, grew sorghum with various levels of K and Mg^{2+}. They found that K^+ depressed the concentration of Mg^{2+} substantially in the shoots, but the effect of Mg^{2+} on K^+ was a slightly antagonistic effect or no effect at all. Hannaway et al. (1982), using solution culture, found that increasing levels of K^+ in solution decreased the shoot concentration of Mg^{2+} in tall fescue. Huang et al. (1990) also reported that net Mg^{2+} translocation from roots to shoots was depressed by increasing root-K^+ concentration.

Spear et al. (1978), who studied a nutrient solution culture, and Christenson et al. (1973), who worked in soils, reported interactions between K^+, Ca^{2+}, and Mg^{2+} absorption by whole plants. These studies showed that K+ and Ca^{2+} suppressed Mg^{2+} content in plant tissue, but the effect depended on the concentration of the ions and soil properties. Moore et al. (1961) and Maas and Ogata (1971) studied the influence of pH and Mg and Ca^{2+} concentrations in solutions on Mg^{2+} uptake by excised barley and corn roots, respectively. In both reports, it was concluded that enhanced solution Ca^{2+} concentrations reduced Mg^{2+} uptake rate by suppressing the Mg^{2+} transport capacity of the root rather than competing with Mg^{2+} for absorption sites. Uptake of Mg^{2+} by rice plants was also reported to be decreased by higher concentrations of Ca^{2+} and Mg^{2+} in nutrient solution by Fageria (1983). Similarly, Schwartz and Bar-Yosef (1983) reported that increasing Ca^{2+} concentration decreased Mg uptake by tomato plants in nutrient solution.

The strength with which Mg^{2+} is held on exchangeable sites is an important factor in determining interaction between cations. On 2:1 layer silicates, Mg^{2+} is held less strongly than Ca^{2+}, and both are held less strongly than K^+ (Barber, 1995). But this is not true for kaolinite and organic matter exchangeable sites. Fertilization with NO_3^- generally enhanced Mg^{2+} concentration in plants driven by the need for cation–anion balance (Wilkinson et al., 2000). When N is supplied as NO_3^-, electrical neutrality is maintained internally by its reduction in synthesizing organic acids by release from roots of anions such as OH^- or HCO_3^- or by uptake of cations. When N is supplied as NH_4^+, internal electrical neutrality is maintained by release of H^+ or by uptake of anions (Wilkinson et al., 2000). Positive interactions between P and Mg^{2+} have been reported in the literature for many crop plants (Wilkinson et al., 2000). Aluminum tolerance has been associated with greater uptake of Mg^{2+} in potato and corn cultivars (Foy, 1984). Similarly, Al^{3+} toxicity in wheat was prevented by increasing the concentrations of Mg^{2+}, Ca^{2+}, and K^+ in nutrition solution, either individually or collectively (Ali, 1973). The beneficial effects of these elements were probably due to a competitive reduction in Al–root contact or to a decrease in Al^{3+} activity rather than to the nutrient supply (Foy, 1984).

6.7 MANAGEMENT PRACTICES TO MAXIMIZE Mg^{2+} USE EFFICIENCY

The important management practices to maximize magnesium use efficiency in crop plants are liming acid soils, application of adequate rate, source, and methods of application. In addition, use of organic residues; use of efficient crops species and genotypes within species; adoption of crop rotation and conservation tillage; improved water use efficiency; and control of diseases, insects, and weeds can improve magnesium use efficiency in crop plants. Among these practices, liming acid soils is the principal method of improving Mg^{2+} content of highly weathered acid soils. Hence, liming is discussed in detail with experimental data obtained under field conditions with annual crops grown on Brazilian Oxisols.

6.7.1 LIMING ACID SOILS

Liming not only adds Mg^{2+} in acid soils but also brings many chemical and biological changes that improve magnesium use efficiency in crop plants. To improve magnesium in the acid soils, use of appropriate liming material is essential. Most liming materials produced from ground limestone also contain amounts of $MgCO_3$ ranging from only traces to as much as the equivalent of the $CaCO_3$ itself. The latter is called dolomite, and limestone with significant quantities of $MgCO_3$ is called dolomitic limestone ($CaCO_3 \cdot MgCO_3$). In dolomitic limestone, MgO content is more than 13% and CaO is more than 31%. In acid soils, dolomitic limestone is generally recommended to raise calcium levels as well as magnesium levels.

Liming increases soil pH, which is one of the most important soil chemical properties influencing plant growth. Figure 6.3 shows the beneficial effect of liming in improving grain yield of soybean grown on Brazilian Oxisol. Soybean grain yield increased significantly and quadratically with increasing soil pH in the range of 4.7

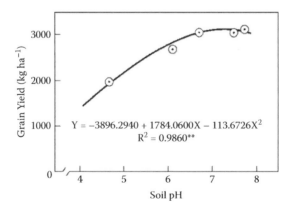

FIGURE 6.3 Relationship between soil pH and grain yield of soybean.

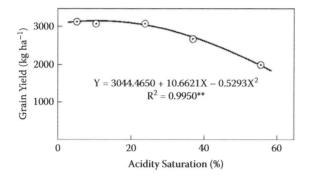

Y = 3044.4650 + 10.6621X − 0.5293X²
R² = 0.9950**

FIGURE 6.4   Grain yield of soybean as a function of acidity saturation.

to 7.7. The variability in grain yield was about 99% with the increase in soil pH, indicating the importance of this soil chemical property in improving grain yield of legume crops in Oxisols. In addition to improving soil pH, liming reduces soil acidity (H+Al), which is one of the factors in reducing yield of crops. Figure 6.4 shows the relationship between soil acidity saturation and grain yield of soybean. Soybean can tolerate a soil acidity of about 10% without reducing grain yield. However, grain yield was reduced significantly when soil acidity was higher than 10% (Figure 6.4).

Dolomitic lime improves Mg^{2+} content of acid soils significantly. Data in Table 6.8 show that there is significant increase in Mg^{2+} content of Brazilian Oxisol with the application of dolomitic lime in the 0–20 and 20–40 cm soil depth determined after harvest of four annual crops grown in rotation. Similarly, Mg saturation was also increased significantly in the two soil depths with the addition of dolomitic lime (Table 6.8). Adequate levels of Mg^{2+} and Mg^{2+} saturation determined for principal annual crops grown in rotation on Brazilian Oxisols are presented in Table 6.9. Dry bean yield increased significantly and quadratically with increasing Mg saturation

TABLE 6.8

Influence of Liming on Mg Content and Mg Saturation in Brazilian Oxisol at Two Soil Depths

Lime Rate (Mg ha⁻¹)	0–20 cm Depth		20–40 cm Depth	
	Mg (cmol_c kg⁻¹)	Mg Saturation (%)	Mg (cmol_c kg⁻¹)	Mg Saturation (%)
0	1.09	13	0.98	13
4	1.14	14	1.09	15
8	1.21	17	1.11	16
12	1.25	16	1.14	15
16	1.25	16	1.16	16
20	1.39	18	1.32	18
R²	0.23*	0.42**	0.31**	0.48**

*, ** Significant at the 5 and 1% probability level, respectively.
Source: Adapted from Fageria (2001b).

TABLE 6.9

Adequate Level of Mg and Mg Saturation for Four Annual Crops Grown in Rotation in Brazilian Oxisol (0–20 cm Soil Depth)

Crop Species	Adequate Mg^{2+} Level ($cmol_c$ kg^{-1})	Adequate Mg^{2+} Saturation (%)
Upland rice	1.2	15
Dry bean	1.4	16
Corn	1.4	18
Soybean	1.4	18

Source: Adapted from Fageria (2001a).

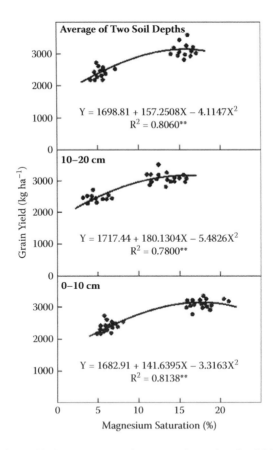

FIGURE 6.5 Relationship between magnesium saturation and grain yield of dry bean.

(Figure 6.5) in no-tillage systems in Brazilian Oxisol. Values of optimum Mg saturation for maximum yield were 21% in the top (0–10 cm) layer, 16% in the sublayer (10–20 cm), and 19% across two soil layers. Similarly, increasing Mg saturation

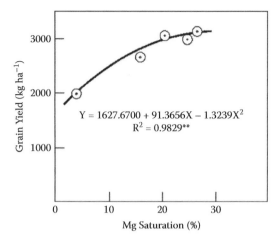

FIGURE 6.6 Relationship between magnesium saturation and grain yield of soybean.

significantly increased grain yield of soybean in Brazilian Oxisol (Figure 6.6). The variation in grain yield of soybean was almost 99% with increasing Mg saturation in the range of 4 to 27%. Fageria (2001a) reported 16% Mg saturation value for maximum grain yield of dry bean in cerrado region in conventional cropping system. Tisdale et al. (1985) reported that exchangeable Mg^{2+} normally accounts for 4 to 20% of the cation exchange capacity of soils. These authors reported that the critical Mg^{2+} saturation values of soils found necessary for optimum plant growth coincide closely with this range. These authors further reported that Mg^{2+} saturation values for maximum yield depend on type of soil, crop species, and other crop management factors. Camberato and Pan (2000) reported that exchangeable Mg^{2+} in excess of 10% of cation exchange capacity generally ensures that Mg^{2+} availability to crops will not be limiting. Eckert (1987) reviewed results of various studies and found that the optimum value of Mg saturation was reported to be in the range of 6 to 12%. Eckert (1987) also reported that Mg saturation of 10 to 15% might be satisfactory for most annual crops. Results of the present study show that the values of optimum Mg saturation in no-tillage systems are slightly higher than those reported in the conventional cropping system.

6.7.2 Appropriate Source, Rate, and Methods of Application

As discussed earlier, dolomitic lime is an inexpensive source of Mg^{2+} that is especially used in acid soils. There should be sufficient humidity in the soil for limestone reaction and liberation of Mg^{2+} for uptake of plants. Magnesium deficiencies in soil and plants can be corrected by application of magnesium sulfate, magnesium oxide, and potassium-magnesium sulfate. These sources are more effective in soils having adequate pH or higher pH soils. Soluble Mg^{2+} sources are more effective in supplying Mg^{2+} to crop plants compared with low-soluble Mg^{2+} sources (Camberato and Pan, 2000).

The best method to determine appropriate magnesium rate is relationship between exchangeable Mg^{2+} content of the soil and grain yield, known as soil test

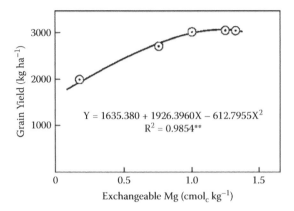

FIGURE 6.7   Relationship between soil exchangeable magnesium and grain yield of soybean.

calibration. Soil test calibration is the process of ascertaining the meaning of the soil test measurement in terms of crop response (Dahnke and Olson, 1990). Soil test calibration is used to make fertilizer recommendations. Figure 6.7 shows a soil test calibration curve of exchangeable Mg^{2+} for soybean grown on Brazilian Oxisol. According to the curve in Figure 6.7, maximum soybean yield was obtained at soil Mg^{2+} content of 1.6 $cmol_c$ kg^{-1}. To achieve this Mg level in Brazilian Oxisols, about 6 Mg ha^{-1} dolomitic lime is required (Fageria et al., 1990; Fageria, 2001b). Such soil test calibration studies should be conducted under field condition and in different agroecological zones.

The Mg/K ratio also influences crop yields. Figure 6.8 shows the relationship between Mg/K ratio and grain yield of soybean grown on Brazilian Oxisol. Based on the regression equation, maximum soybean yield was achieved at Mg/K ratio of about 5. Fageria (2001) reported Mg/K ratio of 7.0 for maximum yield of soybean and 5.0 for maximum yield of dry bean grown on Brazilian Oxisol. Excess magnesium does not affect the yield of most crops, as long as there is more exchangeable calcium than magnesium present. Hence, the ratio of calcium to magnesium can vary

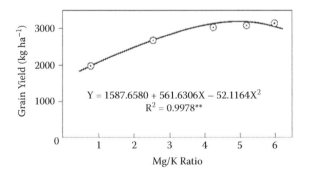

FIGURE 6.8   Relationship between Mg/K ratio and grain yield of soybean.

over the range of 1 to 20, and magnesium availability may still be sufficient as long as the total amount of magnesium is sufficiently high (Barber, 1969, 1995).

As far as methods of application of Mg^{2+} are concerned, if it is supplied by liming materials, they should be applied as a broadcast and incorporated into the soil. Liming materials will be effective in releasing Mg^{2+} for crop plants only if soil has sufficient humidity and temperature is in the adequate range. Hence, liming is always recommended in advance of planting of crops. If magnesium fertilizers are used as a source, they are more effective if applied in the band or furrow rather than broadcast.

6.7.3 OTHER MANAGEMENT PRACTICES

Other management practices that can improve Mg use efficiency in crop plants are use of crop rotation, application of farmyard manures, green manuring, improving water use efficiency, adoption of conservation tillage practices, and control of diseases, insects, and weeds. The other strategy for increasing Mg use efficiency lies in the use of efficient species or cultivars. It is widely reported that plant species differ widely in their requirement of Mg (Karlen et al., 1978; Gallagher et al., 1981; Clark, 1982; Fageria and Morais, 1987; Fageria et al., 1990). These practices are discussed in the chapters on N, P, and K for maximizing efficiencies of these nutrients; these practices are also valid for maximizing magnesium use efficiency.

6.8. CONCLUSIONS

Magnesium deficiency is common in highly weathered acid soils with low cation exchange capacity and sandy soils. Magnesium plays an important role in the activation of many enzymes and also in the process of photosynthesis. It is a mobile element in the plant; hence, deficiency first appears in the older leaves or organs of plants. Plants' magnesium requirements are less than their calcium requirements. The magnesium from soil–plant systems is depleted by removal of crop plants, soil erosion, and leaching losses. The main source of addition of magnesium is dolomitic lime in acid soils. In high-pH soils, other sources of magnesium like magnesium sulfate and magnesium oxide can be used to supply magnesium. It is also added by incorporating crop residues into the soil and using organic manures like green manures and farmyard manures. High soil K^+ and Ca^{2+} levels or high application rates of this element in soils with low Mg levels can induce Mg^{2+} deficiency. Planting magnesium-efficient crop species and genotypes within species can increase magnesium use efficiency in crop plants. Control of insects, diseases, and weeds is also important in improving magnesium use efficiency.

REFERENCES

Ali, S. M. E. 1973. *Influence of cations on aluminum toxicity in wheat* (*Triticum* aestivum). PhD diss., Oregon State University, Corvallis.

Barber, S. A. 1969. *Dolomitic or calcitic lime*. Publication AY-155. Agronomy Dept., Purdue University, West Lafayette, IN.

Barber, S. A. 1995. *Soil nutrient bioavailability: A mechanistic approach*, 2nd edition. New York: Wiley & Sons.

Berry, W. L. and A. Ulrich. 1968. Cation absorption from culture solution by sugar beets. *Soil Sci.* 106:303–308.

Camberato, J. J. and W. L. Pan. 2000. Bioavailability of calcium, magnesium, and sulfur. In: *Handbook of soil science*, M. E. Sumner, Ed., 53–69. Boca Raton, FL: CRC Press.

Clark, R. B. 1982. Plant response to mineral element toxicity and deficiency. In: *Breeding plants for less favorable environments*, M. N. Christiansen and C. F. Lewis, Eds., 71–142. New York: John Wiley & Sons.

Clark, R. B. 1993. Sorghum. In: *Nutrient deficiencies & toxicities in crop plants*. W. F. Bennett, Ed., 21–26. St. Paul, MN: The American Phytopathological Society, American Phytopathological Society Press.

Christenson, D. R., R. P. White, and E. C. Doll. 1973. Yields and magnesium uptake by plants as affected by soil pH and Ca level. *Agron. J.* 65:205–206.

Dahnke, W. C. and R. A. Olson. 1990. Soil test correction, calibration, and recommendation. In: *Soil testing and plant analysis*, 3rd edition, R. L. Westerman, Ed., 45–71. Madison, WI: Soil Science Society of America.

Doll, E. C. and R. E. Lucas. 1973. Testing soils for potassium, calcium, and magnesium. In: *Soil testing and plant analysis*, L. M. Walsh and J. D. Beaton, Eds., 133–151. Madison, WI: Soil Science Society of America.

Eckert, D. J. 1987. Soil test interpretations: basic cation saturation ratios and sufficiency levels. In: *Soil testing: sampling, correlation, calibration, and interpretation*, J. R. Brown, Ed., 53–64. Madison, WI: SSSA.

Fageria, N. K. 1973a. Absorption of magnesium and its influence on the uptake of phosphorus, potassium, and calcium by intact groundnut plants. *Plant Soil* 40:313–320.

Fageria, N. K. 1973b. *Uptake of nutrients by the rice plants from dilute solutions*. Doctoral Thesis, Louvain, Belgium: Catholic University of Louvain.

Fageria, N. K. 1976. Effect of P, Ca, and Mg concentrations in solution culture on growth and uptake of these ions by rice. *Agron. J.* 68:726–732.

Fageria, N. K. 1983. Ionic interactions in rice plants from dilute solutions. *Plant Soil* 70:309–316.

Fageria, N. K. 2001a. Response of upland rice, dry bean, corn and soybean to base saturation in cerrado soil. *Rev. Bras. Eng. Agri. Ambien.* 5:416–424.

Fageria, N. K. 2001b. Effect of liming on upland rice, common bean, corn, and soybean production in cerrado soil. *Pesq. Agropec. Bras.* 36:1419–1424.

Fageria, N. K. 2002. Nutrient management for sustainable dry bean production in the tropics. *Commun. Soil Sci. Plant Anal.* 33:1537–1575.

Fageria, N. K. 2004. Dry matter yield and nutrient uptake by lowland rice at different growth stages. *J. Plant Nutr.* 27:947–958.

Fageria, N.K., V. C. Baligar, and D. G. Edwards. 1990. Soil-plant nutrient relationships at low pH stress. In: *Crops as enhancers of nutrient use*, V. C. Baligar and R. R. Duncan, Eds., 475–507. New York: Academic Press.

Fageria, N. K., V. C. Baligar, and C. A. Jones. 1997. *Growth and mineral nutrition of field crops*, 2nd edition. New York: Marcel Dekker.

Fageria, N.K., V. C. Baligar, and R. W. Zobel. 2007. Yield, nutrient uptake and soil chemical properties as influenced by liming and boron application in common bean in no-tillage system. *Commun. Soil Sci. Plant Anal.* 38 (In Press).

Fageria, N. K. and H. R. Gheyi. 1999. *Efficient crop production*. Campina Grande, Brazil: Federal University of Paraiba.

Fageria, N. K. and O. P. Morais. 1987. Evaluation of rice cultivars for utilization of calcium and magnesium in the cerrado soil. *Pesq. Agropec. Bras.* 22:667–672.

Fageria, N. K. and A. B. Santos. 2008. Lowland rice response to thermophosphate fertilization. *Commun. Soil Sci. Plant Anal.* 39:873–889.

Fageria, N. K. and C. M. R. Souza. 1991. Upland rice, common bean, and cowpea response to magnesium application on an Oxisol. *Commun. Soil Sci. Plant Anal.* 22:1805–1816.

Foy, C. D. 1984. Physiological effects of hydrogen, aluminum, and manganese toxicities in acid soils. In: *Soil acidity and liming*, 2nd edition, F. Adams, Ed., 57–97. Madison, WI: ASA, CSSA, and SSSA.

Gallagher, R. N., M. D. Jellum, and J. B. Jones, Jr. 1981. Leaf magnesium concentration efficiency versus yield efficiency of corn hybrids. *Commun. Soil Sci. Plant Anal.* 12:345–354.

Grove, J. H., M. E. Sumner, and J. K. Syers. 1981. Effect of lime on exchangeable magnesium in variable surface charge soils. *Soil Sci. Soc. Am. J.* 45:497–500.

Grunes, D. L. and H. F. Maryland. 1975. *Controlling grass tetany.* USDA leaf let 561. U.S. Govt. Printing Office, Washington, D.C.

Hannaway, D. B., L. P. Bush, and J. E. Leggett. 1982. Mineral composition of Kenhytall fescue as affected by nutrient solution concentrations of Mg and K. *J. Plant Nutr.* 5:137–151.

Huang, J. W., D. L. Grunes, and R. M. Welch. 1990. Magnesium, nitrogen form, and root temperature effects on grass tetany potential of wheat forage. *Soil Sci. Soc. Am. J.* 82:581–587.

Jones, J. B. and D. M. Huber. 2007. Magnesium and plant disease. In: *Mineral nutrition and plant disease*, L. E. Datnoff, W. H. Elmer, and D. M. Huber, Eds., 95–100. St. Paul, MN: The American Phytopathological Society.

Karlen, D. L., R. Ellis, Jr., D. A. Whitney, and D. l. Grunes. 1978. Influence of soil moisture and plant cultivar on cation uptake by wheat with respect to grass tetany. *Agron. J.* 70:918–921.

Loneragan, J. F. and K. Snowball. 1969. Calcium requirements of plants. *Aust. J. Agric. Res.* 20:465–478.

Maas, E. V. and G. Ogata. 1971. Absorption of magnesium and chloride by excised corn roots. *Plant Physiol.* 47:357–360.

Mengel, K., E. A. Kirkby, H. Kosegarten, and T. Appel. 2001. *Principles of plant nutrition*, 5th edition. Dordrecht: Kluwer Academic Publishers.

Moore, D. P., R. Overstreet, and L. Jacobson. 1961. Uptake of magnesium and its interaction with calcium in excised barley roots. *Plant Physiol.* 36:290–295.

Ohno, T. and D. L. Grunes. 1985. Potassium magnesium interactions affecting nutrient uptake by wheat forage. *Soil Sci. Soc. Am. J.* 49:685–690.

Ologunde, O. O. and R. C. Sorensen. 1982. Influence of K and Mg in nutrient solutions on sorghum. *Agron. J.* 74:41–46.

Sanchez, P. A. 1976. *Properties and management of soils in the tropics.* New York: Wiley.

Schwartz, S. and B. Bar-Yosef. 1983. Magnesium uptake by tomato plants as affected by Mg and Ca concentration in solution culture and plant age. *Soil Sci. Soc. Am. J.* 75:267–272.

Sinclair, J. B. 1993. Soybean. In: *Nutrient deficiencies & toxicities in crop plants*, W. F. Bennett, Ed., 99–110. St. Paul, MN: The American Phytopathological Society, American Phytopathological Society Press.

Spear, S. N., D. G. Edwards, and C. J. Asher. 1978. Response of cassava, sunflower and maize to potassium concentration in solution. III. Interactions between potassium, calcium and magnesium. *Field. Crops Res.* 1:375–389.

Sumner, M. E., P. M. W. Farina, and V. J. Hurst. 1978. Magnesium fixation: A possible cause of negative yield responses to lime applications. *Commun. Soil Sci. Plant Anal.* 9:995–1008.

Tisdale, S. L., W. L. Nelson, and J. D. Beaton. 1985. *Soil fertility and fertilizers*, 4th edition. New York: MacMillan.

Ulrich, A., J. T. Morgan, and E. D. Whitney. 1993. Sugar beet. In: *Nutrient deficiencies & toxicities in crop plants,* W. F. Bennett, Ed., 91–98. St. Paul, MN: The American Phytopathological Society, American Phytopathological Society Press.

Wilkinson, S. R. 1983. Effect of soil application magnesium and other minerals on plant magnesium content. In: *Role of magnesium in animal nutrition*, J. P. Fonte, Ed., 61–79. Blacksburg, VA: Virginia Polytechnic Institute Press.

Wilkinson, S. R., D. L. Grunes, and M. E. Sumner. 2000. Nutrient interactions in soil and plant nutrition. In: *Handbook of soil science*, M. E. Sumner, Ed., 89–112. Boca Raton, FL: CRC Press.

7 Sulfur

7.1 INTRODUCTION

Sulfur (S) has long been recognized as an essential element for plant growth and development and is classified as a macronutrient. Crop responses to applied sulfur have been reported in a wide range of soils in many parts of the world (Blair et al., 1979; Follett et al., 1981; Tisdale et al., 1985; Tisdale et al., 1986; Malavolta et al., 1987; Schnug and Haneklaus, 1998; Brady and Weil, 2002). Sulfur deficiency has been reported in rice from various countries such as Indonesia, Brazil, India, Bangladesh, and Thailand (De Datta, 1981). Yasmin et al. (2007) reported that rice crops throughout the world are becoming increasingly affected by sulfur deficiency as the use of S-free fertilizers increases. Ismunadji (1985) reported higher rice grain yields with sulfur application under flooded as opposed to nonflooded conditions. Boem et al. (2007) reported yield increase of soybean with S application in the Argentinean pampas. Solberg et al. (2007) reported that in the Canadian prairie provinces, there are approximately 4 million ha of cultivated soils deficient in plant-available sulfur, and additional large areas are estimated to be potentially deficient. Wells et al. (1993) reported sulfur deficiency in rice in Arkansas, United States. In addition, in the United States, deficiency of S has been reported in the Southeast, the Northwest, California, and the Great Plains (Brady and Weil, 2002). Tisdale et al. (1986) reported that sulfur deficiency has been observed in crop plants in 22 African countries. These authors also reported that the majority of the reported S deficiencies in the African continent are in areas south of the Sahara on highly weathered soils that receive over 600 mm minimum rainfall per year. Similarly, sulfur deficiency has been reported in Asia, Australia and New Zealand, Europe, Central and South America, and the West Indies (Tisdale et al., 1986). The fungicidal effect of foliar-applied elemental S has been exploited since the end of the 19th century. In comparison, the significance of soil-applied S, independent of the form of S, for disease resistance only became evident a century later, when macroscopic S deficiency developed into a widespread nutrient disorder. The main reason was the drastic decrease of SO_2 emissions in western Europe after clean air legislation came into force (Haneklaus et al., 2007).

The main reasons for sulfur deficiency are (1) low organic matter content of the soil; (2) low mineralization rate of organic matter due to unfavorable soil environment; (3) depletion of soil reserves due to intensive cultivation, use of high yielding cultivars that remove greater amounts of S from soil, and the application of sulfur-free fertilizers; (4) declining S reserves in soil due to accelerated rate of soil erosion; (5) use of high rates of nitrogen and potassium fertilizers, which require high rate of S; (6) the decreased use of S-containing pesticides and fungicides; (7) environmental control of sulfur dioxide emissions in industrial areas; (8) depletion of S from soils by volatilization and leaching; (9) areas where soil parent materials are low in sulfur; (10) the increased use of N in the form of urea, and the decreased consumption of simple or normal superphosphate; and (11) areas of sandy soils with humid and subhumid climates.

There are three main sources of sulfur availability to plants. These sources are (1) organic matter, (2) soil minerals, and (3) sulfur gases in the atmosphere. In addition, S is also supplied by irrigation water. Work conducted at the International Rice Research Institute (1977) in the Philippines showed that flooded rice can recover 54% of the S contained in irrigation water. Wang (1979) reported that irrigation water containing >6.4 mg S L^{-1} can supply S to flooded rice in sufficient quantity to produce a yield of 4.5 Mg ha^{-1}. Blair et al. (1979) reported that in Indonesia, irrigation water used for flooded rice contains 2.8 mg S L^{-1}. If these sources are not able to supply sulfur to growing plants, sulfur-containing fertilizers must be applied to meet plants' needs.

It has been predicted that greater areas will be sulfur deficient in the future in many parts of the world because of growing high-yielding crop cultivars, use of S-free fertilizers, the high cost of chemical fertilizers, and implementation of air pollution control measures or environmental pollution concern. The worldwide trend toward the replacement of ammonium sulfate and single superphosphate with fertilizers such as urea, mono- and diammonium phosphates, and triple superphosphate, all of which are low in sulfur, seems certain to cause an upsurge in the incidence of sulfur deficiency and hence the need to apply fertilizer sulfur (Asher et al., 1983). In addition, chemical fertilizers will continue to be important for alleviating nutrient limitations in most of the cropping systems around the world. Arihara and Srinivasan (2001) reported that deficiency of secondary nutrients including sulfur is an important factor reducing nitrogen and phosphorus use efficiency and is becoming more widespread in cropping systems worldwide. Asher et al. (1983) reported that no single element is more deficient in soils for peanut production than is S; still, it has received less attention than most of the major nutrients. Under these situations, maximizing the efficiency of nutrients including sulfur by crop plants is vital for protecting and restoring environmental quality while sustaining farm yields and profitability. In modern agriculture, the challenge is to sustain soil fertility in cropping systems operating at high productivity levels. To achieve these objectives, innovative management alternatives will be essential for nutrient management. The objective of this chapter is to discuss mineral nutrition aspects of sulfur for field crops and suggest management strategies to maximize sulfur use efficiency for maximum economic yield of crops.

7.2   CYCLE IN SOIL–PLANT SYSTEMS

The behavior of sulfur in soil–plant systems is basic to understanding its deficiency and for adopting efficient and economic corrective measures. Sulfur requirements of crop plants are similar to phosphorus requirements, but sulfur has been studied much less than phosphorus. However, sulfur's role in crop plant growth and development is similar to nitrogen's. Sulfur is absorbed by plants and immobilized by microorganisms and moves in soil–plant systems like N. According to the Soil Science Society of America (1997), the sulfur cycle is defined as the sequence of transformations undergone by sulfur wherein it is used by living organisms, transformed upon death and decomposition of the organisms, and ultimately converted to its original oxidation state. Sulfur, an essential element for microorganisms and plants, is continuously being cycled between inorganic and organic forms (Castellano and Dick, 1990). The nature of the compounds formed and their transformations are strongly influenced by biologically mediated processes, which in turn are affected by environmental conditions.

The principal sources of sulfur addition to soil are chemical amendments or fertilizers, farmyard manures, or crop residues. In addition, gaseous forms of hydrogen sulfide (H_2S) and sulfur dioxide (SO_2) are released to the atmosphere by the burning of fossil fuels and are deposited to the soil by rain. Sulfur dioxide when combined with water forms sulfuric acid (H_2SO_4), and when this acid finds its way to soil, it is known as acid rain. Sulfur addition to the soil from atmospheric gases in the midwestern United States is reported to be in the range of about 10 to 20 kg ha^{-1} per year (Barber, 1995). Sulfur is also available to plants from some irrigation water sources.

The important depletion sources of sulfur from soil is uptake by crop plants, loss through erosion, surface runoff, leaching and immobilization by soil microbial activities, and adsorption by soil colloids. The adsorption is by ligand exchange with surface hydroxyl groups by which sulfate replaces OH^- coordinated to Fe^{3+} or Al^{3+} to form a covalent bond with the surface (Mengel et al., 2001). Sulfate adsorption capacity of soil colloids depends on soil pH and decreases as soil pH increases. The clay minerals adsorb sulfate in the order of kaolinite > illite > bentonite (Mengel et al., 2001). The immobilization of sulfur in the soil–plant system is controlled by the C/S ratio of the organic matter or residues. Stevenson (1986) reported that when the C/S ratio of added residues is below 200:1, there will be a net gain in inorganic SO_2^{2-}; when the ratio is over 400:1, there is a net loss. Similarly, Brady and Weil (2002) reported that a C/S ratio greater than 400:1 generally leads to sulfur immobilization by soil microorganisms. The C/N/S ratio of the soil varies widely within any given location, but the mean ratio for soils from different agroecological regions is remarkably at about 140:10:1.3 (Stevenson, 1986). The main sulfur-bearing minerals in rocks and soils are gypsum (hydrated) ($CaSO_4 \cdot H_2O$), gypsum (anhydrate) ($CaSO_4$), epsomite ($MgSO_4 \cdot 7H_2O$), mirabilite ($Na_2SO_4 \cdot 10H_2O$), pyrite and marcasite (FeS_2), sphalerite (ZnS), chalcopyrite ($CuFeS_2$), and cobaltite (CoAsS) (Tisdale et al., 1985).

Sulfur is present in soils in both organic and inorganic forms. However, the organic form is dominant in most agricultural soils. Tabatabai and Bremner (1972) reported that 95 to 98% of the sulfur in Iowa soils occurs in the organic matter. Overall, organic matter contains about 50 g S kg^{-1} or 0.5% S (Barber, 1995). Soil organic S is divided into three broad fractions: (1) ester sulfate, (2) C-bonded S (mainly amino acids), and (3) residual S (Tabatabai, 1982; Tracy et al., 1990). The ester sulfate fraction is composed of molecules, which have a C-O-S linkage. The carbon-bonded fraction is primarily made up of the amino acids (cysteine) and methionine in large molecules such as proteins. Cysteine and methionine together comprise from 11 to 31% of the total soil organic S (Scott et al., 1981).

The organic forms of sulfur must be mineralized by soil organisms if the sulfur is to be utilized by plants. The oxidation of sulfur compounds in soils occurs through chemical and biological processes. The most common species is *Thiobacillus thiooxidans*. Barber (1979) reported that 3% of the organic matter content of Indiana soils was mineralized at a rate of 2.4% per year. Assuming that the mineralized organic matter contained 0.5% sulfur, this would release about 3.5 mg S kg^{-1} of soil, or 7 kg S ha^{-1} annually. These S release figures from the mineralization of organic matter are for temperate soils and may be much higher for tropical soils. Stevenson (1986) reported that mineralization of sulfur was suppressed markedly at temperature 10°C but increased with increasing temperature from 20 to 40°C and decreased thereafter. Rates may be more rapid in alkaline soil than in acid soils. The rate of S oxidation is also lower in air-dried soils than in field-moist soils. Sulfur mineralization is considerably retarded at low moisture (<15%) and high (>80%) moisture levels. Optimum moisture content for mineralization is at 60% of the maximum water-holding capacity of a soil (Stevenson, 1986). Oxidation by bacteria to the sulfate form is favored by soil aeration and particle size of applied S, such as S ground to pass on 80–100 mesh screen (Fageria and Gheyi, 1999).

The sulfate ion (SO_4^{2-}) has a negative charge, can move easily with soil water, and is readily leached from sandy soils under conditions of high rainfall. The distribution of S in the soil profile has been related to the degree of soil development. Soils in the southeastern United States were practically devoid of SO_4^{2-} in the surface horizon but had large amounts in the subsoil (Reneau, 1983). Very little SO_4^{2-} is adsorbed in the A horizon of these soils because of their low content of hydrated oxides of Fe^{3+} and Al^{3+}, high available P, and limed status (Kamprath et al., 1956). The SO_4^{2-} buffer capacity of the soil is defined as the ability of the SO_4^{2-} in soil solution to resist change when SO_4^{2-} is added or removed from the soil (Haron and Hanson, 1988). The capacity of soils to adsorb SO_4^{2-} is dependent on a number of physical and chemical properties. Factors affecting SO_4^{2-} adsorption include pH, type of cation present, presence of competing anions, extractable Al^{3+} and Fe^{3+} fractions, extractable SO_4^{2-}, organic C, clay content, and soil horizon type. The amount of SO_4^{2-} retained or adsorbed by a soil increases with clay content. Retention of SO_4^{2-} decreases as the pH increases, and adsorption appears to be negligible in many soils when pH is above 6.5 (Tabatabai and Bremner, 1972). Hoeft et al. (1972) reported that the inclusion of pH with extractable S significantly improved the prediction of S response. Sulfate is considered to be weakly held in soils with anion adsorption strength in the following order: $OH^- > H_2PO_4^- > MoO_4^{2-} > SO_4^{2-} = CH_3CO_2 > NO_3^-$ $> Cl^-$ (Haron and Hanson, 1988). Phosphate or MoO_4^{2-} will displace or reduce the adsorption of SO_4^{2-}, but SO_4^{2-} has little effect on $H_2PO_4^-$ adsorption. Increasing electrolyte concentration also lowered SO_4^{2-} sorption by both horizons in the range of pH 3.0 to 5.2 (Courchesne, 1991). Moreover, the influence as pH on sorption was more pronounced at lower electrolyte concentration, and the presence of Ca^{2+} and K^+ favored sorption compared with Na^+. A mechanism involving the formation of a surface complex between Ca^{2+} or K^+ and SO_4^{2-} is proposed by Courchesne (1991) to explain this behavior.

In countries where large concentrations of SO_2 are present in the atmosphere or near industrial areas, the SO_2 in the atmosphere may be an important factor in S nutrition and thus in the alleviation of S deficiency. Jones et al. (1979) reported that in South Carolina, the mean estimated amount of S added to the soil from the air and precipitation varied from 11.2 to 19.8 kg ha^{-1}. Even though it is estimated that most of the S generated in industrial areas reaches the earth in a relatively short distance (Hoeft et al., 1972), there exists the potential for S additions from other distant locations as a result of global circulation (Newell, 1971). In Sweden, for example, 70% of the S is from anthropogenic sources and 77% of this S originates outside the country (Likens, 1976). The contribution of atmospheric sources of S appears to be the primary reason for variability between responses in many parts of the humid, temperate region (Tisdale et al., 1985).

Due to high organic matter content in temperate soils, S content of temperate soils is generally higher than that of tropical soils. The inorganic form of S in the soil is mainly sulfate. Inorganic sulfur is usually only 5 to 10% of total sulfur in the soil (Neptune et al., 1975; Barber, 1995). In flooded rice, inorganic S is reduced to FeS, FeS_2, and H_2S. The H_2S formation in the flooded soils occurs with microbial mineralization of organic S. The elemental S is oxidized by the bacteria of the genus *Thiobacillus*, and this reaction can be expressed as follows (Mengel et al., 2001):

$$2H_2S + O_2 \Leftrightarrow 2H_2O + 2S + 510 \text{ kJ}$$

$$2S + 3O_2 + 2H_2O \Leftrightarrow 2H_2SO_4 + 1180 \text{ kJ}$$

$$\text{Net: } 2H_2S + 4O_2 \Leftrightarrow 2H_2SO_4 + 1690 \text{ kJ}$$

Hence, sulfur oxidation produces soil acidity, and in alkaline soils, sulfur is used as an amendment for reclamation. However, elemental S is not recommended in acid soils as a source of S fertilizer due to high acidity production by oxidation. It is estimated that a maximum of approximately 3.1 kg of lime is required to neutralize H^+ ions produced from oxidation of 1 kg of elemental sulfur (Bolan and Hedley, 2003).

A survey was conducted by Fageria et al. (2003) to determine organic matter and sulfur contents of lowland rice soils in the state of Tocantins, central part of Brazil, and results are presented in Table 7.1. The organic matter content of these soils varied from 6 to 103 g kg^{-1}, with an average value of 44 g kg^{-1}. Similarly, sulfur content varied from 1.5 to 10.7 mg kg^{-1}, with an average value of 5.2 mg kg^{-1}. Wang et al. (1976) proposed 4 mg S kg^{-1} of soil as a critical S level for lowland rice, and according to this criteria, 42% of these soils were deficient in S for lowland rice production. Hence, S fertilization should be applied to these soils to maintain S at an adequate level for successful crop production. A relationship was determined between organic matter content (X) and S content (Y) of these soils. This relationship was $Y = 0.0643 + 2.4728$, $R^2 = 0.42**$. Hence, S content of these lowland soils was significantly related to organic matter content. Significant positive correlations were reported between soil organic matter content and extractable total sulfur in the soil surface layers by various workers (Pedraza and Lora, 1974; Neptune et al., 1975; Nor, 1981; Kang et al., 1981).

TABLE 7.1
Organic Matter and Sulfur Contents in Lowland Soil (Inceptisols) of Tocantins State, Central Part of Brazil

Municipality	Organic Matter (g kg⁻¹)	S (mg kg⁻¹)
Dueré	103	10.7
Dueré	79	6.0
Dueré	45	5.6
Dueré	45	3.4
Dueré	75	8.2
Dueré	45	5.6
Dueré	78	6.0
Dueré	38	6.6
Gurupi	6	2.6
Gurupi	12	3.7
Lagoa da Confusão	90	4.4
Lagoa da Confusão	23	3.0
Lagoa da Confusão	19	4.4
Lagoa da Confusão	28	3.9
Lagoa da Confusão	40	12.7
Lagoa da Confusão	37	3.7
Cristalândia	18	2.1
Figueirópolis	15	1.5
Average	44	5.2

Source: Adapted from Fageria et al. (2003).

The sulfur cycle is different in flooded or reduced soils than it is in aerobic upland soils. Flooding of a soil creates unique chemical and biological conditions that significantly affect sulfur transformation and availability to crops. The sulfur cycle in flooded soils is similar to the carbon, nitrogen, and metal cycles. Freney and Boonjawat (1983) reported that the sulfur transformations that occur in flooded soils are mainly biological and the main reactions are (1) mineralization, or decomposition of organic sulfur compounds with the release of inorganic sulfur compounds; (2) immobilization, or the conversion of inorganic sulfur to organic sulfur compounds; (3) production of sulfides, by reducing of sulfate or other inorganic sulfur compounds; (4) production of volatile sulfur compounds; and (5) oxidation of elemental sulfur and inorganic sulfur compounds.

7.3 FUNCTIONS AND DEFICIENCY SYMPTOMS

Sulfur plays many important roles in the growth and development of plants. Fageria and Gheyi (1999) summarized the important functions of the sulfur in the plant:

1. Sulfur is an important component of two amino acids, cysteine and methionine, which are essential for protein formation. Since animals cannot reduce sulfate, plants play a vital role in supplying essential S-containing amino acids to them (Streeter and Barta, 1984).
2. It plays an important role in enzyme activation.
3. It promotes nodule formation in legumes.
4. Sulfur is necessary in chlorophyll formation, although it is not a constituent of chlorophyll.
5. The maturity of seeds and fruits is delayed in the absence of adequate sulfur.
6. Sulfur is required by plants in the formation of nitrogenase.
7. It increases crude protein content of forages.
8. It improves quality of cereal crop for milling and baking.
9. It increases oil content of oilseed crops.
10. It increases winter hardiness in plants.
11. It increases drought tolerance in plants.
12. It controls certain soilborne diseases.
13. It helps in formation of glucosides that give characteristic odors and flavors to onion, garlic, and mustard.
14. Sulfur is necessary for the formation of vitamins and synthesis of some hormones and glutathione.
15. It is involved in oxidation-reduction reactions.
16. Sulfur improves tolerance to heavy metal toxicity in plants.
17. It is a component of sulfur-containing sulfolipids.
18. Organic sulfates may serve to enhance water solubility of organic compounds, which may be important in dealing with salinity stress (Clarkson and Hanson, 1980).
19. Fertilization with soil-applied S in sulfate form decreases fungal diseases in many crops (Klikocka et al., 2005; Haneklaus et al., 2007).

Sulfur deficiency symptoms are often observed in crops at early stages of growth since S can be easily leached from the surface soil (Hitsuda et al., 2005). Friesen (1991) applied ^{35}S-labeled phosphogypsum to millet (*Panicum miliaceum* L.) in a semiarid environment and determined the residual S distribution in soil layers at harvest. He found that 4% of sulfate S (SO_4-S) remained at 0 to 15 cm, 33% at 30 to 45 cm, and 31% at 45 to 60 cm depths. Ritchey and Sousa (1997) measured the amount of SO_4-S in two cerrado soils of Brazil and recorded values of 5 to 10 mg SO_4-S kg^{-1} at 0 to 15 cm and 40 to 45 mg kg^{-1} at 45 to 60 cm depths. McClung et al. (1959) examined six Brazilian red-yellow podzolic soils and showed that the organic S concentration in the A-horizon soils decreased to one-third over 20 years of cropping.

Sulfur deficiency symptoms are similar to those of nitrogen. However, N deficiency symptoms first appear in the older leaves; generally, sulfur deficiency symptoms first appear in the younger leaves because S is not easily translocated in the plant. Sulfur-deficient plants lack vigor, are stunted, are pale green to yellow in color, and have elongated thin stems. Sulfur deficiency may delay maturity in grain crops. Interveinal chlorosis may occur. Sulfur deficiency can be corrected easily by application of chemical fertilizers containing S. In rice, visual symptoms of recovery are usually noted within 5 days following fertilizer application (Wells et al., 1993). Root development is restricted, and shoot–root ratios usually decrease for plants grown under S deficiency (Clark, 1993). In peanuts, plants deficient in S are stunted but appear quite upright because of reduced branching (Smith et al., 1993). Smith et al. (1993) further reported that in peanuts sulfur deficiency causes accumulation of nitrate, amide, and carbohydrates, which in turn retards the formation of chlorophyll. Sulfur deficiency in rice produced a high percentage of unfilled grains (Yoshida and Chaudhry, 1979).

7.4 CONCENTRATION AND UPTAKE

Sulfur concentration (content per unit of dry weight) in plant tissues is an important criterion for identifying its deficiency or sufficiency in crop plants. As discussed in earlier chapters, nutrient concentration including S varied with plant age and plant part analyzed. It decreased in crop tissues over time (Yoshida and Chaudhry, 1979; Fontanive et al., 1996). Bennett (1993) reported that overall S sufficiency level in field crops tissues varied from 1.0 to 3.0 g kg^{-1} (0.1 to 0.3%). Brady and Weil (2002) reported that healthy plant foliage generally contains 1.5 to 4.5 g kg^{-1} sulfur (0.15 to 0.45%). Alcordo and Rechcigl (1993) reported that plants contain as much S as P, the usual range being from 2 to 5 g kg^{-1} (0.2 to 0.5%). Wells et al. (1993) reported that the critical concentration of S in rice varies from approximately 2.5 g S kg^{-1} at tillering to 1.0 g S kg^{-1} at heading. Yoshida (1981) reported the critical S concentration in straw of rice needed for maximum dry weight production varied from 1.6 S kg^{-1} at tillering to 0.7 g S kg^{-1} at maturity. Suzuki (1995) reported 1.0 g S kg^{-1} as a critical level in the rice shoot at tillering and 0.55 g S kg^{-1} in rough rice grains. Foth and Ellis (1988) reported that S content of plant tissues of legumes exceeding 2.6 g kg^{-1} (0.26%) may be considered optimum. Values that exceed 1.7 g kg^{-1} may be considered optimum for small grain crops (Foth and Ellis, 1988). Andrew (1977) reported that critical S concentration in the tops of tropical legume species varied from 1.4 to 2.0 g kg^{-1} (0.14 to 0.20%). Hitsuda et al. (2005) reported that critical S concentration at early stages of growth was 0.8 g kg^{-1} in corn and soybean; 1.1 to 1.3 g kg^{-1} in cotton, sorghum, and rice; and 1.4 to 1.6 g kg^{-1} in wheat, sunflower, and field bean.

Average S concentration in shoot and grain of important cereals is presented in Table 7.2. The S concentration varied from 0.84 to 2.25 in the shoot of cereals, with an average value of 1.66 g kg^{-1} (0.17%). Similarly, S concentration in the grain varied from 0.78 to 1.95 g kg^{-1}, with an average value of 1.35 g kg^{-1}. Overall, S concentration was higher in the straw of cereals compared to grain (Table 7.2). Data in Table 7.3 show adequate level of S in different crop species. It varied with crop species, plant part analyzed, and plant growth stage. Sulfur in excess may cause nutrient imbalance, but it is seldom toxic to plant tissues (Bennett, 1993).

Sulfur is absorbed as sulfate ion (SO_4^{2-}) by plants, and most part of it is reduced to the sulfhydryl (S^{2-}) or disulfide (S^{1-}) sulfur before it is incorporated into organic matter (Clarkson and Hanson, 1980; Streeter and Barta, 1984). The uptake of S depends on the rate at which it is released from organic matter, which in turn is influenced by

TABLE 7.2
Average Concentrations of S in Shoot and Grain of Principal Cereal Crops

Crop Species	S Concentration in Shoot (g kg⁻¹)	S Concentration in Grain (g kg⁻¹)
Barley	2.00	1.56
Corn	1.56	1.19
Oats	2.25	1.95
Wheat	1.67	1.25
Rice	0.84	0.78
Average	1.66	1.35

Source: Adapted from Hanson (1990); Wang et al. (1976).

TABLE 7.3
Sulfur Sufficiency Level in Plant Tissue of Principal Field Crops

Crop Species	Plant Part Analyzed and Growth Stage	S Concentration (g kg⁻¹)
Barley	Whole tops at heading	1.5–4.0
Wheat	Whole tops at heading	1.5–4.0
Rice	Uppermost mature leaves at tillering	2–6
Corn	Whole tops at 30 to 45 days after emergence	2.0–3.0
Corn	Ear leaf blade at silking	1.0–2.4
Sorghum	Whole tops at vegetative	2.5–3.0
Soybeans	Upper fully developed trifoliate at prior to pod set	2.5–4.0
Dry bean	Whole tops at early growth	1.6–6.4
Peanut	Upper stems and leaves at early pegging	2.0–3.0

Source: Adapted from Fageria et al. (1997).

the kinds of plant residues, soil moisture, and soil pH (Sinclair, 1993). Sulfur can also be absorbed by the leaves through the stomata as a gaseous sulfur dioxide (SO_2), an environmental pollutant released primarily from burning coal and fossil fuel (Smith et al., 1993). Sulfur absorbed by roots is translocated to leaves through the xylem and is reduced and assimilated into organic compounds predominantly in the leaf blades (Suzuki, 1995).

Uptake of S in the shoot and grain of principal field crops is presented in Table 7.4. The S uptake in the shoot varied from 3 to 20 kg ha⁻¹, depending on crop species and yield of each species. Similarly, S uptake in grain varied from 3 to 25 kg ha⁻¹, with an average value of 10.6 kg ha⁻¹. Total uptake of S (shoot plus grain) varied from 6 to 43 kg ha⁻¹, with an average value of 22 kg ha⁻¹. Among crop species, corn accumulated maximum S and flax minimum. The maximum and minimum values of S uptake were associated with yield of these crop species.

TABLE 7.4
Yield and Sulfur Uptake by Principal Field Crops

Crop Species	Grain Yield (kg ha⁻¹)	S Uptake in Shoot (kg ha⁻¹)	S Uptake in Grain (kg ha⁻¹)	Total (kg ha⁻¹)
Barley	5,376	11	11	22
Buckwheat	1,613	3	6	9
Corn	12,544	20	17	37
Oats	3,584	12	9	21
Rice	7,840	8	6	14
Wheat	5,376	17	6	23
Sorghum	8,960	18	25	43
Peanuts	4,480	12	11	22
Flax	1,344	3	3	6
Rapeseed	1,960	10	13	23
Soybeans	4,032	15	13	28
Sunflower	3,920	11	7	18
Average	5,086	11.6	10.6	22.2

Source: Adapted from Hanson (1990).

TABLE 7.5
Sulfur Uptake in the Straw and Grain of Lowland Rice under Different S and Two N Levels

S Rate (kg ha⁻¹)	S Uptake in Straw (kg ha⁻¹)		S Uptake in Grain (kg ha⁻¹)	
	N_{60}	N_{120}	N_{60}	N_{120}
0	0.89	0.99	1.57	1.92
25	2.93	3.82	3.30	4.37
50	3.41	4.45	3.37	3.81
100	4.12	4.72	3.50	3.80
Average	2.84	3.50	2.94	3.48

Regression Analysis

S rate (X) vs. S uptake in straw at N_{60} (Y) = 1.0108 + 0.0749X − 0.00044X², R² = 0.9690**

S rate (X) vs. S uptake in straw at N_{120} (Y) = 1.1429 + 0.1092X − 0.00074X², R² = 0.9675**

S rate (X) vs. S uptake in grain at N_{60} (Y) = 1.7055 + 5.9929X − 0.00042X², R² = 0.9103**

S rate (X) vs. S uptake in grain at N_{120} (Y) = 2.1965 + 0.0716X − 0.00056X², R² = 0.7279*

*,** Significant at the 5 and 1% probability levels, respectively.

Source: Adapted from Wang et al. (1976).

Sulfur uptake in the straw and grain of lowland rice as influenced by S rate and N rate is presented in Table 7.5. The S uptake in straw and grain was significantly increased with increasing S rate in a quadratic fashion. About 97% variability in S

TABLE 7.6

Sulfur Uptake in the Straw and Grain to Produce 1 Metric Ton Grain of Lowland Rice at Different S and Two N levels

S Rate (kg ha^{-1})	S Uptake (kg)	
	N_{60}	N_{120}
0	0.67	0.71
25	1.08	1.30
50	1.23	1.59
100	1.34	1.65
Average	1.08	1.31

Regression Analysis

S rate (X) vs. S uptake at N_{60} (Y) = 0.6860 + 0.0166X − 0.00010X^2, R^2 = 0.9878**

S rate (X) vs. S uptake at N_{120} (Y) = 0.7204 + 0.0264X − 0.00017X^2, R^2 = 0.9952**

** Significant at the 1% probability level. *Source:* Adapted from Wang et al. (1976).

uptake of straw was due to S rate at low as well as high N rates. Similarly, S uptake in the grain also increased in a quadratic fashion with increasing S rate in the range of 0 to 100 kg S ha^{-1}. Data in Table 7.5 show that S uptake in straw as well as in grain was higher at high N rate compared to low N rate, indicating improvement in S use efficiency with the application of N in lowland rice. Sulfur uptake in straw as well as grain of lowland rice significantly increased with increasing S as well as N rates (Table 7.5). Overall, to produce 1 Mg grain, lowland rice required 1.08 kg S at 60 kg N ha^{-1}, and 1.31 kg S at 120 kg N ha^{-1} (Table 7.6). Suzuki (1995) reported that to produce 1 Mg of hulled rice grains in Japan, 1.67 kg S should be absorbed. This author also reported that absorption of 10 kg S is required by rice to obtain a yield of 6000 kg ha^{-1} of hulled grains (0.07% S), 1000 kg of hulls (0.06% S), and 6500 kg of straw (0.08% S).

7.5   USE EFFICIENCY AND S HARVEST INDEX

Sulfur use efficiency (kg grain per kg S uptake in grain) and S harvest index (S translocated to grain of the total S uptake) values for important crop species are presented in Table 7.7. The S use efficiency values varied from 151 kg grain produced with the accumulation of 1 kg S in the grain of soybean to 1307 kg grain produced with the accumulation of 1 kg S in rice. Overall, S use efficiency in field crops was 480 kg grain produced with the uptake of 1 kg S in the grain. Overall, cereals had higher S use efficiency for grain production compared to legumes or oil crops, indicating higher S requirements for legumes compared to cereals. The S harvest index varied from 26% in wheat to 67% in buckwheat, with an average value

TABLE 7.7
Sulfur Uses Efficiency in Grain and Sulfur Harvest
Index of Principal Field Crops

Crop Species	S Use Efficiency (kg kg^{-1})	S Harvest Index (%)
Barley	489	50
Buckwheat	269	67
Corn	738	46
Oats	398	43
Rice	1307	43
Wheat	896	26
Sorghum	358	58
Peanuts	407	50
Flax	448	50
Rapeseed	151	57
Soybeans	310	46
Sunflower	560	39
Average	480	48

Note: S use efficiency = (Grain yield in kg/S uptake in grain in kg);
S harvest index = (S uptake in grain/S uptake in grain plus
straw) × 100.
Source: Calculated from data in Table 7.4.

of 48% in 12 crop species. Hence, it can be concluded that about 52% of S is retained in the straw of annual crops, and if it is incorporated into soil after harvest of these crops, a significant amount of S can be recycled to maintain soil fertility.

Apparent sulfur recovery efficiency (ASRE) was higher at lower S rate and decreased with increasing S rate in the range 25 to 100 kg S ha^{-1} (Table 7.8). Overall, ASRE was 13% in lowland rice grown in Brazilian Amazon Inceptisol. Immobilization, leaching of SO_4^{2-}, and reduction to sulfide forming less soluble compounds in flooded rice might be responsible for low S use efficiency by lowland rice (Wang et al., 1976).

7.6 INTERACTION WITH OTHER NUTRIENTS

Sulfur interaction with nitrogen is very common, and S requirements of crops are enhanced with the increase of N in the growth medium. The main reason for the interaction of S with N may be a significant increase in growth of plants with N addition, which may cause dilution of S in plants (Wilkinson et al., 2000). Robert and Koehler (1965) reported that insufficient S, especially when large amounts of nitrogen were applied, resulted in a decreased grain yield and a diminished S amino acid content of the grain protein. Soliman et al. (1992) reported that in calcareous soils, S reduces pH and improves uptake of micronutrients like Fe, Mn, and Zn. Uptake of P may also improve in calcareous soils with the application of S due to reduction of pH. It has been reported in Japan that zinc deficiency of lowland rice may be

TABLE 7.8

Apparent Sulfur Recovery Efficiency (ASRE) in Lowland Rice at Different S Levels

S Rate (kg ha^{-1})	ASRE (%)
0	—
25	21
50	11
100	6
Average	13

Note: ASRE (%) = (S uptake in straw and grain with S treatment − S uptake in straw and grain without S treatment or control) × 100.

induced by excess amounts of sulfides in soils (Shiratori and Miyoshi, 1971). Similarly, Suzuki (1995) reported that excess Zn induced sulfur deficiency in rice plants. Rehm and Caldwell (1968) found that sulfate uptake was higher when N was added as ammonium nitrogen, but sulfate uptake in the presence of nitrate was similar to that where no nitrogen was added.

Sulfur application was reported to decrease molybdenum content in pea plants in low-molybdenum soils, and pea plants showed molybdenum deficiency (Reisenauer, 1963). This may be attributed primarily to competition in the absorption site of plant roots between molybdate and sulfate ions, which are similar in size and charge, and, in addition, to the physiological inhibition of molybdenum utilization by sulfates within the plant body (Suzuki, 1995). Since molybdenum affects the biological nitrogen fixation, the interaction between sulfur and molybdenum has especial significance for legume crops. Tanaka et al. (1966) reported that application of gypsum to lowland rice reduced soil pH and induced iron toxicity.

7.7 MANAGEMENT PRACTICES TO MAXIMIZE S USE EFFICIENCY

Efficient fertilizers and fertilizer management techniques are essential to maximize sulfur use efficiency by crop plants and benefits to farmers. Management practices, which are discussed for maximizing use efficiencies of N, P, K, Ca, and Mg, are also applicable for sulfur. However, some of these practices are discussed in this section.

7.7.1 LIMING ACID SOILS

It is common practice to lime several weeks before planting to allow sufficient time for the lime to react with the soil. This may greatly alter the S-supplying capacity of soils, for the rise in soil pH decreases the SO_4^{2-} adsorption capacity of the soil (Korentajer et al., 1983). This effect is usually attributed to a competition between OH^- and SO_4^{2-} for adsorption sites on Al^{3+} and Fe^{3+} hydroxides (Hingston et al.,

1972); by making phosphate compounds more soluble at higher pHs, there may be more phosphate ions to compete for the sites as well (Metson and Blakemore, 1978). Glass (1989) reported that for the divalent anions like SO_4^{2-}, adsorption on soil colloids is a function of pH. Above pH 6.0 to 6.5, there is only very slight adsorption. In acid soils, however, particularly those high in Fe^{3+} and Al^{3+} oxides, adsorption may be substantial. Similarly, Bolan et al. (1988, 2003) and Marsh et al. (1987) reported that, as a general rule, many soils absorb little, if any, SO_4^{2-} once pH exceeds about 6.0.

Liming may also increase the rate of organic S mineralization by creating a more favorable environment for microbial activity (Probert, 1976; Stevenson, 1986). By increasing the mineralization of organic S into SO_4^{2-} and decreasing SO_4^{2-} adsorption, liming could accelerate the movement of SO_4^{2-} through the profile. In acid soils, this may result in decreased uptake of S by the plant, particularly if root development in the subsoil is inhibited, such as by high exchangeable Al^{3+}. Thus, rather than increasing the S availability, as observed under unleached conditions, liming a soil and subsequent leaching may decrease the S supplying capacity, particularly over a long period of time (Korentajer et al., 1983).

After liming an acid soil, Elkins and Ensminger (1971) observed an increase in the SO_4^{2-} concentration in the soil solution, and S uptake by soybean was improved. However, Bolan et al. (1988) reported that in the absence of active uptake by plants, any SO_4^{2-} released into soil solution by liming is susceptible to leaching and may be lost to subsoil horizons.

7.7.2 USE OF APPROPRIATE SOURCE, RATE, METHOD, AND TIMING OF APPLICATION

Several S carriers can be used to supply S to crop plants. Principal S carriers are listed in Table 7.9. However, dominant sources of S are single superphosphate, gypsum, and ammonium sulfate. Besides S, these sources can supply phosphorus, calcium, and nitrogen, respectively, to plants. Lysimeters studied have generally shown that the efficiency of gypsum in supplying S for plant growth depends on agroclimatic factors, particularly rainfall intensity, timing of application, SO_4^{2-} adsorption capacity, and soil liming. Jones et al. (1968) estimated that 80% of the applied gypsum S was lost by leaching during one crop season from soils with low SO_4^{2-} adsorption capacity. Leaching losses from sulfate fertilizers may be reduced by increasing the gypsum particle size (Korentajer et al., 1984) or, alternatively, by using sparingly soluble sources, i.e., anhydrite. Dehydration of gypsum at 400°C results in the formation of anhydrite, which is less susceptible to leaching than gypsum (Lobb and Rothbaum, 1969). Top-dressing of ammonium sulfate can furnish N as well as sulfur requirements of crops and avoid much leaching. Ammonium sulfate contains approximately 24% S; single superphosphate, 12% S; and triple superphosphate, approximately 1.4% S. In addition, more efficient use of S fertilizer requires better definition of the factors controlling S response to crops as well as improved plant and soil tests. Wang et al. (1976) reported that ammonium sulfate and gypsum were good sources of S for correcting S deficiency in flooded rice. Similarly, Malavolta et al. (1987) reported that ammonium sulfate and phosphogypsum were very effective sources of correcting S deficiency in annual crops in Brazil.

TABLE 7.9
Principal Sulfur Carriers

Fertilizer/Amendment	Formula	S (%)
Ammonium sulfate	$(NH_4)_2SO_4$	24.0
Ammonium polysulfide	NH_4S_x	45
Ammonium sulfate-nitrate	$(NH_4)_2SO_4 \cdot NH_4NO_3$	12.0
Ammonium thiosulfate solution	$(NH_4)_2S_2O_3 + H_2O$	26.0
Magnesium sulfate	$MgSO_4 \cdot 7H_2O$	13
Gypsum (by-product)	$CaSO_4 \cdot 2H_2O$	17.0
Single superphosphate	$Ca(H_2PO_4)_2 + CaSO_4 \cdot 2H_2O$	12.0
Triple superphosphate	$Ca(H_2PO_4)_2 \cdot H_2O$	1.4
Copper sulfate	$CuSO_4 \cdot 5H_2O$	13.0
Zinc sulfate	$ZnSO_4 \cdot H_2O$	18.0
Elemental sulfur	S	100
Sodium sulfate	Na_2SO_4	23
Potassium sulfate	K_2SO_4	18.0
Manganese sulfate	$MnSO_4 \cdot 4H_2O$	14.5
Iron sulfate	$FeSO_4 \cdot 7H_2O$	11.5
Sulfur dioxide	SO_2	50.0
Sulfuric acid	H_2SO_4	32.7
Ferrous ammonium sulfate	$Fe(NH_4)_2SO_4$	16

Source: Adapted from Follett et al. (1981); Tisdale et al. (1985); and Fageria (1989).

Yasmin et al. (2007) reported that sulfur additions, as coating, to fertilizers such as urea, diammonium phosphate, monoammonium phosphate, and triple superphosphate offer a way to introduce S back into these fertilizers. These authors tested five S sources for rice, i.e., elemental S, sulfur-coated urea, sulfur-coated diammonium phosphate, sulfur-coated triple superphosphate, and gypsum. The S rate applied was 10 kg ha^{-1}, and all sources were equally effective in increasing rice yield. Sulfur recovery efficiency by rice was 46.7% of elemental sulfur, 45.5% of sulfur-coated urea, 38% of sulfur-coated diammonium phosphate, 32.2% of sulfur-coated triple superphosphate, and 38% of gypsum. Overall, S recovery efficiency was 40%. Elemental S needs to be oxidized to SO_4 before it can be absorbed by the plants (Chien et al., 1988), and this process may take several weeks, depending on environmental conditions (Wainwright et al., 1986), the particle size of material (McCaskill and Blair, 1987), and population of S-oxidizing microorganisms (Lee et al., 1988). Another problem with elemental S is its powdered form, which is difficult to handle and apply.

Use of adequate rate of S is an important strategy to improve crop yields and maximize S use efficiency by crop plants. Wang et al. (1976) reported that 10 kg S ha^{-1} is sufficient to supply adequate S for maximum economic yield of lowland rice. These authors also reported that application of 27 kg S ha^{-1} supported two lowland rice crops. Immobilization is considered to be the major factor for the

TABLE 7.10
Lowland Rice Grain Yields as Influenced by Nitrogen and Sulfur Rates

	Grain Yield (kg ha^{-1})	
S Rate (kg ha^{-1})	N$_{60}$	N$_{120}$
0	3660	4120
25	5790	6290
50	5500	5210
100	5670	5160
Average	5160	5200

Source: Adapted from Wang et al. (1976).

reduced availability of S residual from previous crops (Wang et al., 1976). Lowland rice response to applied S at two N rates is presented in Table 7.10. Maximum grain yields of 5790 kg ha^{-1} at 60 kg N ha^{-1} and 6290 kg ha^{-1} at 120 kg N ha^{-1} were obtained with the application of 25 kg S ha^{-1}. These results also indicate that N fertilizer improves S use efficiency in lowland rice. Malavolta et al. (1987) reported that grain yield of most annual crops under Brazilian conditions can be obtained with the application of S in the range of 20–60 kg ha^{-1} either by ammonium sulfate or by phosphogypsum.

Samosir and Blair (1983), in a pot experiment, tested the S-supplying capacity of ammonium sulfate, gypsum, elemental S, and sulfur-coated urea (SCU) applied at 30 kg S ha^{-1} to flooded rice on a low-fertility Paleudult soil. At maturity, yield and S uptake were not significantly different between the gypsum, elemental S (100% < 60 mesh), and ammonium sulfate sources. These results confirm the suitability of fine (100% < 60 mesh) elemental S as a S source for rice and show that S from SCU is not available to flooded rice at least in the first crop after application. Hoeft and Walsh (1975) reported that in the first year after application, SO$_4^-$-S carriers K$_2$SO$_4$ and MgSO$_4$ were more effective than prilled elemental-S (88% S and 12% bentonite) in correcting S deficiency in alfalfa. However, in the second year after application, the various S carriers were equally effective.

As far as methods of S application are concerned, S can be applied in the band as well as broadcast. If it is applied as ammonium sulfate or superphosphate, the methods used for application of N and P should be followed. If it is applied as phosphogypsum, generally large amounts are needed and should be applied as broadcast and incorporated into the soil. Nuttall and Ukrainetz (1991) reported that S application at sowing was the optimum method for maximum yield of canola (*Brasica napus*) and that a delay in application would result in a reduction of approximately 65 kg ha^{-1} in grain yield per week. Hago and Salama (1987) also found that S applied at sowing significantly increased shoot dry weight and pod yield of groundnut (*Arachis hypogaea* L.), but S at flowering did not. Besides, rhizobium bacteria infected legume crops soon after germination (Heinrich et al., 2001), and nodulation decreases in low-S conditions (Hago and Salama, 1987) since S is one of the components of nitrogenase.

7.7.3 SOIL TEST FOR MAKING S RECOMMENDATIONS

Due to the dynamic nature of S in the soil, the soil test for S may present a problem for making fertilizer recommendations. According to Castellano and Dick (1990), the limitations of using extractable SO_4^{2-} as a soil test were shown by wide variations in SO_4^{2-} over time (2–14 mg SO_4^{2-} kg^{-1} soil) in control plots, even though there was a seed yield response in winter rape to S. Critical levels as high as 10 mg SO_4-S kg^{-1} soil have been reported for production of winter rape in the Pacific Northwest (Murphy and Auld, 1986). So, depending on the time of sampling, one would reach different conclusions about whether these soils would be responsive to S fertilization, as concluded by Castellano and Dick (1990). Further considerations that need to be addressed in S soil testing are subsoil SO_4^{2-} levels (including potential for upward transport of SO_4^{2-} due to evapotranspiration) and identification of the labile S fraction that is related to plant uptake of S.

In spite of the variation in soil test over time due to the dynamic nature of S in soil–plant systems, the soil test gives a reasonable estimate of S content of a soil and can be used as a reference point for making S fertilizers recommendations. Suzuki (1995) reported that the sulfate-sulfur extracted by $Ca(H_2PO_4)_2$ solution correlates well to the sulfur nutritional status of rice plants, and the occurrence of sulfur deficiency in plants is predicted at about 8–9 mg kg^{-1} of extractable sulfur in dry soils. Malavolta et al. (1987) reported that 90% of the maximum yield of corn under greenhouse conditions was obtained with 17 mg SO_4-S kg^{-1} soil. This is equal to about 6 mg S kg^{-1} of soil. Hue et al. (1984) reported that minimum SO_4^{2-} concentrations for maximum wheat growth was 0.25 mM in soil solution and 6.0 mg S kg^{-1} using $Ca(H_2PO_4)_2$ extracting solution. Foth and Ellis (1988) reported that the critical level of S is expected to be from 3 to 8 mg kg^{-1} of soil. Reisenauer et al. (1973) reported that cereals do not respond to sulfur application when S levels in the soil is in the range of 6–8 mg kg^{-1}. Similarly, these authors reported that legumes do not respond to applied sulfur when the level of S in the soils is in the range of 9–12 mg kg^{-1}. Hence, overall, adequate S levels for the cereals may be around 7 mg kg^{-1} and for legumes around 10 mg kg^{-1}. When S level is below critical level and still there is no response to applied S, plants may be taking S from the subsoil, rain or irrigation water might be adding S to soils, organic S might be mineralizing during crop growth, and there may be an imbalance between S and other nutrients.

A number of extracting solutions are used to extract the S from soils to define critical or adequate S levels for a determined crop species under an agroecological region. Some of these solutions are Cacl$_2$ (0.01 to 0.1 M), $Ca(H_2PO_4)_2$ (0.01 M), 0.25 M HOAc plus 0.15 M NH$_4$F, and 2N acetic acid containing 500 mg kg^{-1} of P (Foth and Ellis, 1988). According to Reisenauer et al. (1973), extractants removing the readily soluble portions of the adsorbed and organic S provide the best predictions of response in field trials. These include the strong acid (2 M HOAc), phosphorus-containing solutions employed by Cooper (1968) and Hoeft et al. (1972) and the pH 8.5 NaHCO$_3$ solution used by Bardsley and Kilmer (1963) and Cooper (1968). To be useful for routine analysis, an extractant should give a clear filtrate. In this regard, $Ca(H_2PO_4)$ appears to be one of the best extractants because P ions displace the adsorbed sulfate and Ca ions depress the extraction of soil organic matter (Reisenauer et al., 1973).

7.7.4 RECOMMENDATIONS BASED ON CROP REMOVAL, TISSUE CRITICAL S CONCENTRATION, AND CROP RESPONSES

Crops vary in their sulfur requirements. Legume crops such as alfalfa, clover, and soybeans have high sulfur requirements, as do cereals like rice, wheat, and corn. Grasses also have lower sulfur requirements compared with legumes. Spencer (1975) listed fertilizer requirements in deficient areas for certain crops, which reflect crop uptake. Cruciferous forages (40–80 kg S ha^{-1}), alfalfa (30–70 kg S ha^{-1}), and rapeseed (20–60 kg S ha^{-1}) were in the high-requirement category. The moderate-requirement category consisted of sugarcane and coffee (20–40 kg S ha^{-1}) and cotton (10–30 kg S ha^{-1}). Sugar beets, cereal forages, cereal grains, and peanuts were all in the low-requirement category—generally lower than 20 kg S ha^{-1}.

In addition to nutrients removed by a crop, the S requirement of a crop under given soil and climatic conditions can also be determined by critical level. The critical level of total plants has been reported to be 0.16% for young wheat (Hue et al., 1984) and 0.22% for the first cutting of alfalfa (Pumphrey and Moore, 1965).

Kline et al. (1989) reported that corn response to S fertilization on the coarse-textured soils of the Atlantic coastal plain occurred in only three of the 12 site–year combinations. The results of this study indicate that deficiency occurred and that subsoil sampling is needed to identify responsive sites, based on available subsoil S and physical or chemical barriers to root penetration. Average critical values for plant S for responsive sites in this study were 2.1 and 1.6 g kg^{-1} for early whole plant and ear leaves, respectively, consistent with critical plant S values (1.5 to 2.0 g kg^{-1}) used by many laboratories in the Atlantic coastal plain. Average critical N/S ratios obtained in this study were 18.7 early whole top and 20.3 ear leaves, respectively. Jones et al. (1982) reported that significant yield increase of corn and soybeans resulted from the addition of 20 kg S ha^{-1} on a Guadalupe clay soil in Philippines.

The amount of available S needed in A horizons for maximum plant growth is different for different crops. For example, Fox (1980) reported that minimum S concentration in the soil solution needed for maximum grain production for cowpea cultivars ranged from 0.06 to 0.22 mM. Fox (1980) later suggested that 0.16 mM SO$_4^{2-}$ would be adequate for most tropical crops. Minimum S concentration in plant tissue for maximum dry matter yields was 1.6 g kg^{-1}. Cotton, wheat, and corn leaves all need at least 1.5 to 1.8 g S kg^{-1} for maximum growth (Rasmussen et al., 1977; Reneau and Hawkins, 1980).

Oates and Kamprath (1985) determined response of wheat to S fertilization on sandy coastal plain soils of the southeastern United States. A rate of 20 kg S ha^{-1} as SO$_4^{2-}$ applied in late February was adequate for optimum response. Ammonium sulfate was equally effective as CaSO$_4$·2H$_2$O. Plants responded to S fertilization where nonfertilized plants had S concentration of 0.6 g kg^{-1} dry weight and N/S ratios of 21. No response to S fertilization was obtained on a Norfolk soil in which there was appreciable root growth in the 30 to 45 cm depth. That depth contained an accumulation of SO$_4^{2-}$. Hue et al. (1984) reported that there was a yield response to S for the wheat plant on an Ultisol. Minimum SO_4^{2-} concentration for maximum wheat growth was 0.25 mM in soil solution and 6.0 mg S kg^{-1} in $Ca(H_2PO_4)_2$ extracting solution. Minimum S concentration in plant tissue for maximum dry matter yields was 1.6 g kg^{-1}. These authors reported significant yield response when the concentration of S in the alfalfa tissue was 0.22% S. Kang and Osiname (1976) studied corn response to S fertilization at six locations in the forest and savanna zones of western Nigeria. Significant yield increases were observed with rates of S application from 7.5 to 30 kg S ha^{-1}. The response was more distinctly evident in the savanna than in the forest zone, where the amount of extractable S, extracted with either KH_2PO_4, $Ca(H_2PO_4)_2$, or $NH_4C_2H_3O_2$, was below 4 mg S kg^{-1} of soil. There was no response to S when extractable S levels were equal to or greater than 8.5, 10, and 12 mg kg^{-1} when soils were extracted with $Ca(H_2PO_4)_2$, $NH_4C_2H_3O$, and KH_2PO_4, respectively. The critical S level in the ear leaf was estimated at about 0.14% S.

The amount of fertilizer S needed to eliminate S deficiency in wheat is low. Rasmussen and Allmalas (1986) reported that maximum yield increase of winter wheat was obtained with 14 kg S ha^{-1}. The 14 kg S ha^{-1} application supplied adequate S to a second wheat crop, and the 28 and 42 kg S ha^{-1} applications supplied adequate S to a third wheat crop; this occurred even though increased extractable S from 14 kg S ha^{-1} was not detectable in the upper 0.6 m of soil at seeding of the second crop and increased extractable S from 28 and 42 kg S ha^{-1} was not detectable at the seeding of the third crop. Ramig et al. (1975) found 17 and 34 kg S ha^{-1} as gypsum supplied adequate S for 3 and 5 years, respectively, for a wheat green pea rotation in a 450 mm precipitation zone. Residual efficiency of S is much less where average annual precipitation exceeds 500 mm per year (Jones et al., 1968).

The need for S fertilization should be particularly related to the amount of N being applied since both nutrients are required for protein formation. Stewart and Porter (1969) found that one part S was required for every 12 to 15 parts N to ensure maximum production of both dry matter and protein in wheat, corn, and beans. Nitrogen-to-S ratios have been used as an index for S deficiency. Ratios ranging from 11:1 to 17:1 have been proposed as the critical index levels (Harward et al., 1962; Pumphrey and Moore, 1965). The protein N:S ratio, (N/S)p, is also used to establish the S to N requirement for a crop (Gaines and Phatak, 1982). The (N/S)p ratio has been reported to range from 15 to 16 for corn, 20 for soybeans, 15 for cowpea, 12 for tomatoes, and 8 to 9 for cotton and okra (Gaines and Phatak, 1982). It is more reliable than the total N–total S ratio, (N/S)t, for assessing the crops' S to N requirement because it is not influenced by the accumulation of nonprotein S and nonprotein N (Metson, 1973). When S supply is adequate, the accumulation of nonprotein S will cause the (N/S)t ratio to be lower than the (N/S)p ratio, whereas when S is deficient, nonprotein N will accumulate, resulting in a higher (N/S)t ratio (Metson, 1973). This concept assumes the near constancy of the (N/S)p ratio (Gaines and Phatak, 1982).

7.7.5 OTHER MANAGEMENT PRACTICES

In addition to the management practices discussed above, other management practices like crop rotation, use of organic manures, use of S-efficient genotypes, conservation tillage system, and improvements in water use efficiency may be important soil and crop management strategies for maximizing S use efficiency by field crops. Detailed discussions of these practices are given in the chapters on N, P, and K. Among these strategies, use of sulfur-efficient genotypes is of special importance from the environmental and economic point of view. Genotypic variation in the efficiency of utilization and/or uptake of sulfur by cultivated plants has been reported (Ferrari and Renosto, 1972; Cacco et al., 1978).

7.8 CONCLUSIONS

Sulfur deficiency is now recognized as a major constraint to crop production in various parts of the world. Sulfur deficiency not only reduces crop yields but also decreases the nutritional value or quality of food grains. Functions of sulfur in plant nutrition are similar to those of nitrogen, but the quantity of S required by field crops is much smaller compared to N. Similarly, influence of S on yield increases of field crops is not as significant as the influence of N. In addition, sulfur is added to soils through precipitation and, like N and P, through the use of fertilizers, like ammonium sulfate and phosphates. Sulfur deficiency is normally associated with low soil S content, which results from extreme weathering and leaching of sulfate-S. Another reason of S deficiency is low soil organic matter content. The SO_4^{2-} ion is negatively charged and can be leached from the topsoil layer to the lower layer easily, especially on sandy soils. However, its leaching in soil–plant systems is lower than that of NO_3^- but higher than that of potassium, calcium, and magnesium. High-analysis fertilizers being used today may contain only very small quantities of S as an impurity, and the smaller amount of S being returned to the environment as an atmospheric pollutant may also contribute to S deficiency. Sulfur is an important component of certain amino acids, coenzymes, and all proteins. It is directly involved in electron transport via sulfhydryl groups. With S deficiency, chlorophyll content in leaves decreases and the photosynthetic rate lowers.

Sulfur is an immobile nutrient in plants; hence, its deficiency symptoms first appear in the younger leaves. Many of the reactions of sulfur in soils are closely associated with organic matter and the activity of microorganisms. More than 90% of the sulfur in soil is present in the organic form, and its mineralization is essential for uptake and utilization by plants. Sulfur improves nodulation, pod yield, and seed quality in legumes. Sufficiency level of sulfur in plant tissues is in the range of 1.0 to 3.0 g kg^{-1} (0.1 to 0.3%). Overall, S content is higher in straw compared with grain of field crops. Hence, straw that is returned and incorporated to soil can add significant amounts of S. Sulfur use efficiency can be improved by liming of acid soils. Critical S level in the soils is in the range of 7 to 10 mg kg^{-1}, depending on crop species and management practices adopted. Liming increases SO_4^{2-} concentration in the soil solution by mineralization of organic matter, by release of SO_4^{2-} from organic matter by chemical hydrolysis, by release of adsorbed SO_4^{2-} from the soil surface, and by release of SO_4^{2-} from sparingly soluble Fe^{2+} and Al^{3+} hydroxy sulfates, which become more soluble at higher pH.

Use of adequate rate, source, and method of application are important management strategies to maximize S use efficiency by crop plants. Field crop S requirements varied from 10 to 60 kg S ha^{-1}, depending on crop species, type of soils, and crop management practices. Legumes require more S compared to cereals. There are several sources of S; however, the most important sulfur carriers are gypsum, single superphosphate, ammonium sulfate, and potassium sulfate. Use of S-efficient crop species and genotypes within species is another management strategy to improve S use efficiency in crop plants. Because much of the S is held in the organic fraction of the soil, maintenance of the organic matter at an adequate level in the soil is an effective practice in preventing S deficiency in field crops. In conclusion, maximizing S use efficiency not only improves crop yields but also avoids environmental contamination (air, water, and soil).

REFERENCES

Alcordo, I. S. and J. E. Rechcigl. 1993. Phosphogypsum in agriculture: A review. *Adv. Agron.* 49:55–118.

Andrew, C. S. 1977. The effect of sulfur on the growth, sulfur and nitrogen concentration of some tropical and temperate pasture legumes. *Aust. J. Agric. Res.* 28:807–820.

Arihara, J. and A. Srinivasan. 2001. Significance of nutrient uptake mechanisms in cropping systems. In: *Plant nutrient acquisition: New perspectives,* N. Ae, J. Arihara, K. Okada, and A. Srinivasan, Eds., 487–503. Tokyo: Springer.

Asher, C. J., F. P. C. Blamey, and C. P. Mamaril. 1983. Sulfur nutrition of tropical annual crops. In: *Sulfur in Southeast Asian and South Pacific agriculture,* G. J. Blair and A. R. Till, Eds., Armidale, N. S. W., Australia: University of New England.

Barber, S. A. 1979. Corn residue management and soil organic matter. *Agron. J.* 71:625–628.

Barber, S. A. 1995. *Soil nutrient bioavailability: A mechanistic approach.* New York: John Wiley & Sons.

Bardsley, C. D. and V. J. Kilmer. 1963. Sulfur supply of soils and crop yields in the southeastern United States. *Soil Sci. Soc. Am. Proc.* 27:197–199.

Bennett, W. F. 1993. Plant nutrient utilization and diagnostic plant symptoms. In: *Nutrient deficiencies & toxicities in crop plants,* W. F. Bennett, Ed., 1–7. St. Paul, MN: The American Phytopathological Society.

Blair, G. J., C. P. Mamaril, P. Umar, A. E. O. Momuat, and C. Momuat. 1979. Sulfur nutrition of rice. I. A survey of soils of South Sulawesi, Indonesia. *Agron. J.* 71:473–477.

Boem, F. H. G., P. Prystupa, and G. Ferraris. 2007. Seed number and yield determination in sulfur deficient soybean crops. *J. Plant Nutr.* 30:93–104.

Bolan, N. S. and M. J. Hedley. 2003. Role of carbon, nitrogen, and sulfur cycles in soil acidification. In: *Handbook of soil acidity,* Z. Rengel, Ed., 29–56. New York: Marcel Dekker.

Bolan, N. S., D. C. Adriano, and D. Curtin. 2003. Soil acidification and liming interactions with nutrient and heavy metal transformation and bioavailability. *Adv. Agron.* 78:215–272.

Bolan, N. S., J. K. Syers, R. W. Tillman, and D. R. Scotter. 1988. Effect of liming and phosphate additions on sulfate leaching in soils. *J. Soil Sci.* 39:493–504.

Brady, N. C. and R. R. Weil. 2002. *The nature and properties of soils,* 13th edition. Upper Saddle River, NJ: Prentice-Hall.

Cacco, G., G. Ferrari, and M. Saccomani. 1978. Variability and inheritance of sulfate uptake efficiency and ATP-sulfurylase activity in maize. *Crop Sci.* 18:503–504.

Castellano, S. D. and R. P. Dick. 1990. Cropping and sulfur fertilization influence on sulfur transformations in soil. *Soil Sci. Soc. Am. J.* 54:114–121.

Chien, S. H., D. K. Friesen, and B. W. Hamilton. 1988. Effect of application method on availability of elemental sulfur in cropping sequences. *Soil Sci. Soc. Am. J.* 52:165–169.

Clark, R. B. 1993. Sorghum. In: *Nutrient deficiencies & toxicities in crop plants*, W. F. Bennett, Ed., 21–26. St. Paul, MN: The American Phytopathological Society.

Clarkson, D. T. and J. B. Hanson. 1980. The mineral nutrition of higher plants. *Annu. Rev. Plant Physiolo.* 31:239–298.

Cooper, M. 1968. A comparison of five methods for determining the sulfur statues of New Zealand soils. *International Congress of Soil Science Trans., 9th, Adelaide, Australia* 2:263–271.

Courchesne, F. 1991. Electrolyte concentration and composition effects on sulfate sorption by two spodosols. *Soil Sci. Soc. Am. J.* 55:1576–1581.

De Datta, S. K. 1981. *Principles and practices of rice production.* New York: John Wiley & Sons.

Elkins, D. M. and L. E. Ensminger. 1971. Effect of soil pH on the availability of adsorbed sulphate. *Soil Sci. Soc. Am. Proc.* 35:931–934.

Fageria, N. K. 1989. *Tropical soils and physiological aspects of crop production.* Brasilia: EMBRAPA-DPU, Brasilia, EMBRAPA-CNPAF. Document 18. 425p.

Fageria, N. K., V. C. Baligar, and C. A. Jones. 1997. *Growth and mineral nutrition of field crops*, 2nd edition. New York: Marcel Dekker.

Fageria, N. K. and H. R. Gheyi. 1999. *Efficient crop production.* Federal University of Paraiba, Campina Grande, Brazil.

Fageria, N. K., L. F. Stone, and A. B. Santos. 2003. *Soil fertility management of lowland rice.* Santo Antônio de Goiás, Brazil: Embrapa Arroz e Feijão.

Ferrari, G. and F. Renosto. 1972. Comparative studies on the active transport by excised roots of inbred and hybrid maize. *J. Agric. Sci.* 79:105–108.

Freney, J. R. and J. Boonjawat. 1983. Sulfur transformation in wetland soils. In: *Sulfur in Southeast Asian & South Pacific agriculture*, G. J. Blair and A. R. Till, Eds., Armidale, N. S. W., Australia: University of New England.

Follett, R. H., L. S. Murphy, and R. L. Donahue. 1981. *Fertilizers and soil amendments.* Englewood Cliffs, NJ: Prentice-Hall.

Fontanive, A. V., A. M. Horra, and M. Moretti. 1996. Foliar analysis of sulphur in different soybean cultivar stages and its relation to yield. *Commun. Soil Sci. Plant Anal.* 27:179–186.

Foth, H. D. and B. G. Ellis. 1988. *Soil fertility.* New York: John Wiley & Sons.

Fox, R. L. 1980. Response to sulfur by crops growing in highly weathered soils. *Sulfur in Agriculture* 4:16–22.

Friesen, D. K. 1991. Fate and efficiency of sulfur fertilizer applied to food crops in west Africa. In: *Alleviating soil fertility constraints to increased crop production in West Africa*, A. U. Mokwaunye, Ed., 59–68. Dordrecht: Kluwer Academic Publishers.

Gaines, T. P. and S. C. Phatak. 1982. Sulfur fertilization effects on the consistency of the protein N:S ratio in low and high sulfur accumulation crops. *Agron. J.* 74:415–418.

Glass, A. D. M. 1989. *Plant nutrition: An introduction to current concepts.* Boston, MA: Jones and Bartlett Publishers.

Hago, T. M. and M. A. Salama. 1987. The effect of elemental sulfur on shoot dry-weight, nodulation and pod yield on groundnut under irrigation. *Exp. Agric.* 23:93–97.

Haneklaus, S., E. Bloem, and E. Schnug. 2007. Sulfur and plant disease. In: *Mineral nutrition and plant disease*, L. E. Datnoff, W. H. Elmer, and D. M. Huber, Eds., 101–118. St. Paul, MN: The American Phytopathological Society.

Hanson, A. A. 1990. *Practical handbook of agricultural science.* Boca Raton, Florida: CRC Press.

Haron, K. B. and R. G. Hanson. 1988. Sulfur supplying capacity of some Missouri soils. *Soil Sci. Soc. Am. J.* 52:1657–1660.

Harward, M. E., T. T. Chao, and S. C. Fang. 1962. The sulfur status and sulfur supplying power of Oregon soils. *Agron. J.* 54:101–106.

Heinrich, K., M. H. Ryder, and P. J. Murphy. 2001. Early production of rhizopine in nodules induced by *Sinorhizobium meliloti* strain L5-30. *Can. J. Microbiol.* 47:165–171.

Hingston, F. J., A. M. Posner, and J. P. Quirk. 1972. Anion adsorption by geothite and gibbsite. I. Role of the proton in determining adsorption envelopes. *J. Soil Sci.* 23:177–192.

Hitsuda, K., M. Yamada, and D. Klepker. 2005. Sulfur requirement of eight crops at early stages of growth. *Agron. J.* 97:155–159.

Hoeft, R. G., D. R. Keeney, and L. M. Walsh. 1972. Nitrogen and sulfur in precipitation and sulfur dioxide in the atmosphere in Wisconsin. *J. Environ. Qual.* 1:203–208.

Hoeft, R. G. and L. M. Walsh. 1975. Effect of carrier, rate and time of application of S on the yield, and S and N content of alfalfa. *Agron. J.* 67:427–430.

Hue, N. V., F. Adams, and C. E. Evans. 1984. Plant available sulfur as measured by soil-solution sulfate and phosphate-extractable sulfate in a Ultisol. *Agron. J.* 76:726–730.

International Rice Research Institute. 1977. Preliminary report: First international trial of nitrogen fertilizer efficiency in rice, 1975–1976. International Rice Research Institute, Los Banos, Philippines.

Ismunadji, M. 1985. Effect of sulfate application on the performance of IR36 rice variety under submerged and dry land conditions. *Indonesian J. Crop Sci.* 1:21–28.

Jones, U. S., M. G. Hamilton, and J. B. Pitner. 1979. Atmospheric sulfur as related to fertility of Ultisols and Entisols in South Carolina. *Soil Sci. Soc. Am. J.* 43:1169–1171.

Jones, M. B., W. E. Martin, and W. A. Williams. 1968. Behavior of sulfate sulfur and elemental sulfur in three California soils in Lysimeters. *Soil Sci. Soc. Am. Proc.* 32:535–540.

Jones, U. S., H. P. Samonte, and D. M. Jariel. 1982. Response of corn and inoculated legumes to urea, lime, phosphorus, and sulfur on Guadalupe clay. *Soil Sci. Soc. Am. J.* 46:328–331.

Kang, B. T., E. Okoro, D. Acquaye, and A. Osiname. 1981. Sulfur status of some Nigerian soils from the savanna and forest zones. *Soil Sci.* 132:220–227.

Kang, B. T. and O. A. Osiname. 1976. Sulfur response of maize in Western Nigeria. *Agron. J.* 68:333–336.

Kamprath, E. J., W. L. Nelson, and J. W. Fitts. 1956. The effects of pH, sulfate and phosphate concentration on the adsorption of sulfate by soils. *Soil Sci. Soc. Am. Proc.* 20:463–468.

Klikocka, H., S. Haneklaus, E. Bloem, and E. Schnug. 2005. Influence of sulfur fertilization on infection of potato tubers with *Rhizoctonia solani* and *Streptomyces scabies*. *J. Plant Nutr.* 28:819–833.

Kline, J. S., J. T. Sims, and K. L. Schilke-Gartley. 1989. Response of irrigated corn to sulfur fertilization in the Atlantic Coastal Plain. *Soil Sci. Soc. Am. J.* 53:1101–1108.

Korentajer, L., B. H. Byrnes, and D. T. Hellums. 1983. The effect of liming and leaching on the sulfur-supplying capacity of soils. *Soil Sci. Soc. Am. J.* 47:525–530.

Korentajer, L., B. H. Byrnes, and D. T. Hellums. 1984. Leaching losses and plant recovery from various sulfur fertilizers. *Soil Sci. Soc. Am. J.* 48:671–676.

Lee, A., J. H. Watkinson, and D. R. Lauren. 1988. Factors affecting oxidation rates of elemental sulfur in a soil under a ryegrass dominant sward. *Soil Biol. Biochem.* 20:809–816.

Likens, G. E. 1976. Acid precipitation. *Chem. Eng. News.* 54:29–31.

Lobb, W. R. and H. P. Rothbaum. 1969. The use of anhydrite as a fertilizer, *N. Z. J. Agric. Res.* 12:119–124.

Malavolta, E., G. C. Vitti, C. A. Rosolem, H. J. Kliemann, N. K. Fageria, P. T. G. Guimarâes, and M. L. Malavolta. 1987. Sulfur in Brazilian agriculture. *Sulfur in Agriculture* 11:2–5.

Marsh, K. B., R. W. Tillman, and J. K. Syers. 1987. Charge relationships of sulfate sorption by soils. *Soil Sci. Soc. Am. J.* 51:318–323.

McCaskill, M. R. and G. J. Blair. 1987. Particle size and soil texture effects on elemental sulfur oxidation. *Agron. J.* 79:1079–1083.

McClung, A. C., L. M. M. Freitas, and W. L. Lott. 1959. Analyses of several soils in relation to plant responses to sulfur. *Soil Sci. Soc. Am. Proc.* 23:221–224.

Mengel, K., E. A. Kirkby, H. Kosegarten, and T. Appel. 2001. *Principles of plant nutrition.* Dordrecht: Kluwer Academic Publishers.

Metson, A. J. 1973. *Sulfur in forage crops.* Sulfur Inst. Tech. Bull, 20.

Metson, A. J. and L. C. Blakemore. 1978. Sulfate retention by New Zealand soils in relation to the competitive effect of phosphate. *N.Z.J. Agric. Res.* 21:243–253.

Murphy, G. A. and D. L. Auld. 1986. Establishment and fertilizer practices for winter rape in dryland areas of northern Idaho and eastern Washington. In: *Winter rapeseed production Conf. Proc,* K. Kephart, Ed., 108–116. Pac., North West Coop Ext. Serv.

Neptune, A. M. L., M. A. Tabatabai, and J. J. Hanway. 1975. Sulfur fractions and carbon-nitrogen-phosphorus-sulfur relationships in some Brazilian and Iowa soils. *Soil Sci. Soc. Am. Proc.* 39:51–55.

Newell, R. E. 1971. The global circulation of air pollutants. *Sci. Am.* 224:32–42.

Nor, Y. 1981. Sulfur mineralization and adsorption in soils. *Plant Soil* 60:451–459.

Oates, K. M. and E. J. Kamprath. 1985. Sulfur fertilization of winter wheat grown on deep sandy soils. *Soil Sci. Soc. Am. J.* 49:925–927.

Nuttall, W. F. and H. Ukrainetz. 1991. The effect of time of s application on yield and sulphur uptake of canola. *Commun. Soil Sci. Plant Anal.* 22:269–281.

Pedraza, L. A. and S. R. Lora. 1974. Availability of sulfur for plants in two soils of the eastern plains of Colombia. *Revista ICA* 9:77–112.

Probert, M. E. 1976. Studies of available and isotopically exchangeable sulfur in some North Queensland soils. *Plant Soil* 45:461–475.

Pumphrey, F. V. and D. P. Moore. 1965. Sulfur and nitrogen content of alfalfa herbage during growth. *Agron. J.* 57:237–230.

Ramig, R. E., P. E. Rasmussen, R. R. Allmaras, and C. M. Smith. 1975. Nitrogen sulfur relations in soft white winter wheat. I. Yield response to fertilizer and residual sulfur. *Agron. J.* 67:219–224.

Rasmussen, P. E. and R. R. Allmalas. 1986. Sulfur fertilization effects on winter wheat yield and extractable sulfur in semiarid soils. *Agron. J.* 78:421–425.

Rasmussen, P. E., R. E. Ramig, L. G. Ekin, and C. R. Rhode. 1977. Tissue analysis guidelines for diagnosing sulfur deficiency in white wheat. *Plant Soil* 46:153–163.

Rehm, G. W. and A. C. Caldwell. 1968. Sulfur supplying capacity of soils and the relationship to soil type. *Soil Sci.* 105:355–361.

Reisenauer, H. M. 1963. The effect of sulfur on the absorption and utilization of molybdenum by peas. *Soil Sci. Am. Proc.* 27:553–555.

Reisenauer, H. M., L. M. Walsh, and R. G. Hoeft. 1973. Testing soils for sulphur, boron, molybdenum, and chlorine. In: *Soil testing and plant analysis,* L. M. Walsh and J. D. Beaton, Eds., 173–200. Madison, WI: Soil Science Society of America.

Reneau, R. B. 1983. Corn response to sulfur application in Coastal plain soils. *Agron. J.* 75:1036–1040.

Reneau, R. B. and G. W. Hawkins. 1980. Corn and soybeans respond to sulfur in Virginia. *Sulfur Agric.* 4:7–11.

Ritchey, K. D. and d. M. G. Sousa. 1997. Use of gypsum in management of subsoil acidity in Oxisols. In: *Plant-soil interactions at low pH; Sustainable agriculture and forestry production,* A. C. Moniz, A. M. C. Furlani, R. E. Schaffert, N. K. Fageria, C. A. Rosolem, and H. Cantarella, Eds., 165–178. Campinus, São Paulo: Brazilian Soil Science Society.

Robert, S. and F. E. Koehler. 1965. Sulphur dioxide as a source of sulphur for wheat. *Soil Sci. Soc. Am. Proc.* 29:696–698.

Samosir, S. and G. J. Blair. 1983. Sulfur nutrition of rice III. A comparison of fertilizer sources for flooded rice. *Agron. J.* 75:203–206.

Schnug, E. and S. Haneklaus. 1998. Diagnosis of sulfur nutrition. In: *Sulfur in agriculture*, E. Schnug, Ed., 1–38. Dordrecht: Kluwer Academic Publishers.

Scott, N. M., W. Bick, and H. A. Anderson. 1981. The measurement of sulphur-containing amino acids in some Scottish soils. *J. Sci. Food Agric.* 32:21–24.

Shiratori, K. and H. Miyoshi. 1971. Relationship between acidic disease and zinc deficiency of rice plant in reclaimed saline paddy fields. *J. Sci. Soil and Manure* 42:384–389.

Sinclair, J. B. 1993. Soybeans. In: *Nutrient deficiencies & toxicities in crop plants*, W. F. Bennett, Ed., 99–103. St. Paul, MN: The American Phytopathological Society.

Smith, D. H., M. A. Wells, D. M. Porter, and F. R. Fox. 1993. Peanuts. In: *Nutrient deficiencies & toxicities in crop plants*, W. F. Bennett, Ed., 105–110. St. Paul, MN: The American Phytopathological Society.

Soil Science Society of America. 1997. *Glossary of soil science terms.* Soil Science Society of America, Madison, WI.

Solberg, E. D., S. S. Malhi, M. Nyborg, B. Henriquez, and K. S. Gill. 2007. Crop response to elemental S and sulfate-S sources on S-deficient soils in the Parkland region of Alberta and Saskatchewan. *J. Plant Nutr.* 30:321–333.

Soliman, M. F., S. F. Kostandi, and M. L. van Beusichem. 1992. Influence of sulfur and nitrogen fertilizer on the uptake of iron, manganese, and zinc by corn plants in calcareous soil. *Commun. Soil Sci. Plant Anal.* 23:1289–1300.

Spencer, K. 1975. Sulphur requirements of plants In: *Sulphur in Australian agriculture*, K. D. McLachlan, Ed., 98–116. Sydney: Sydney Univ. Press, Australia.

Stevenson, F. J. 1986. *Cycles of soil: Carbon, nitrogen, phosphorus, sulfur and micronutrients.* New York: John Wiley.

Stewart, B. A. and L. K. Porter. 1969. Nitrogen-sulfur relationships in wheat, corn and beans. *Agron. J.* 61:267–271.

Streeter, J. G. and A. L. Barta. 1984. Nitrogen and minerals. In: *Physiological basis of crop growth and development*, M. B. Tesar, Ed., 175–200. Madison, WI: ASA, and CSSA.

Suzuki, A. 1995. Metabolism and physiology of sulfur. In: *Science of the rice plant: Physiology*, Vol. 2. T. Matsuo, K. Kumazawa, R. Ishii, K. Ishihara, and H. Hirata, Eds., 395–401. Tokyo: Food and Agriculture Policy Research Center.

Tabatabai, M. A. 1982. Sulfur. In: *Methods of soil analysis*, Part 2, 2nd edition, A. L. Page, Ed., 501–583. Madison, WI: ASA, and SSSA.

Tabatabai, M. A. and J. M. Bremner. 1972. Distribution of total and available sulfur in selected soils and soil profiles. *Agron. J.* 64:40–44.

Tanaka, A., R. Loe, and S. A. Navasero. 1966. Some mechanisms involved in the development of iron toxicity symptoms in the rice plant. *Soil Sci. Plant Nutr.* 12:158–162.

Tisdale, S. L., W. L. Nelson, and J. D. Beaton. 1985. *Soil fertility and fertilizers*, 4th edition. New York: Macmillan Publishing Company.

Tisdale, S. L., R. B. Reneau, Jr., and J. S. Platou. 1986. Atlas of sulfur deficiency. In: *Sulfur in agriculture*, M. A. Tabatabai, Ed., 295–322. Madison, WI: ASA, CSSA, SSSA.

Tracy, P. W., D. G. Westfall, E. T. Ellott, G. A. Paterson, and C. V. Cole. 1990. Carbon, nitrogen, phosphorus, and sulfur mineralization in plow and no-till cultivation. *Soil Sci. Soc. Am. J.* 54:457–461.

Wainwright, M., W. Navell, and S. J. Grayston. 1986. Effects of organic matter on sulfur oxidation in soil and influence of sulphur oxidation on soil nitrification. *Plant Soil* 96:369–376.

Wang, C. H. 1979. Sulfur fertilization of rice-diagnostic techniques. *Sulphur in Agriculture* 3:12–15.

Wang, C. H., T. H. Liem, and D. S. Mikkelsen. 1976. Sulfur deficiency: A limiting factor in rice production in the lower Amazon Basin. II. Sulfur requirement for rice production. Bulletin 48, New York: IRI Research Institute, Inc.

Wells, B. R., B. A. Huey, R. J. Norman, and R. S. Helms. 1993. Rice. In: *Nutrient deficiencies & toxicities in crop plants*, W. F. Bennett, Ed., 15–19. St. Paul, MN: The American Phytopathological Society.

Wilkinson, S. R., D. L. Grunes, and M. E. Sumner. 2000. Nutrient interactions in soil and plant nutrition. In: *Handbook of soil science*, M. E. Sumner, Ed., Boca Raton, FL: CRC Press.

Yasmin, N., G. Blair, and R. Till. 2007. Effect of elemental sulfur, gypsum, and elemental sulfur coated fertilizers on the availability of sulfur to rice. *J. Plant Nutr.* 30:79–91.

Yoshida, S. 1981. *Fundamentals of rice crop science*. Los Banos, Philippines: IRRI.

Yoshida, S. and M. R. Chaudhry. 1979. Sulfur nutrition of rice. *Soil Sci. Plant Nutr.* 25:121–134.

8 Zinc

8.1 INTRODUCTION

Essentiality of zinc (Zn) for higher plants was discovered by A. L. Sommer and C. B. Lipman in 1926 (Marschner, 1983; Fageria et al., 1997). Zinc is a micronutrient needed in small amounts by crop plants, but its importance in crop production has increased in recent years. It is considered to be the most yield-limiting micronutrient in crop production in various parts of the world (Sims and Johnson, 1991; Cakmak et al., 1996; Cakmak et al., 1998; Mandal et al., 2000; Fageria, 2001; Grewal, 2001; Fageria et al., 2002; Fageria et al., 2003ab; Fageria and Baligar, 2005). Duffy (2007) reported that yield losses in wheat, rice, corn, and other staple crops of up to 30% is common even in mildly zinc-deficient soils. Approximately, 50% of the soils used worldwide for cereal production contain low levels of plant-available zinc (Graham et al., 1992; Welch, 1993).

De Datta (1981) reported that zinc deficiency is the second most serious nutritional disorder limiting the yield of lowland rice in the Philippines. Similarly, zinc deficiency in rice in India has been reported by Mandal et al. (2000). Norman et al. (2003) and Slaton et al. (2005b) reported that Zn deficiency is the most common micronutrient deficiency of rice grown on neutral to alkaline soils in the United States, as well as in other rice-producing regions of the world (Fageria et al., 2003). Slaton et al. (2002) reported that 79% (>600,000 ha) of the soils used for rice production in Arkansas, USA, have soil pH > 6.0 and may require Zn fertilization for normal crop growth and yield. Clark (1982) reported that zinc deficiency is widespread throughout the United States and that zinc may potentially become deficient in nearly all intensively cropped soils. Zinc deficiency has been found throughout the United States in corn crops (Catlett et al., 2002; Shaver et al., 2007).

In Brazilian Oxisols, zinc deficiency has been reported in upland rice, corn, soybean, and wheat (Fageria and Baligar, 1997a; Fageria, 2000a). Table 8.1 shows dry matter yield of tops of five annual crops with and without zinc application grown on Brazilian Oxisol. These data show that crop responses to applied zinc was in the order of corn > soybean > dry bean > upland rice > wheat. Hence, both cereals and legumes responded to zinc fertilization. However, response varied from species to species. Figure 8.1, Figure 8.2, Figure 8.3, and Figure 8.4 show response of dry bean, soybean, wheat, and corn, respectively, to zinc fertilization in Brazilian Oxisols. Similarly, improvement in growth of root systems of upland rice (Figure 8.5), dry bean (Figure 8.6), soybean (Figure 8.7), and wheat (Figure 8.8) was also observed. Zinc deficiency in crop plants reduces not only grain yield but also the nutritional quality of the grain. Deficiency of zinc in cereal grains is also responsible for zinc

TABLE 8.1

Influence of Zinc Application on Dry Matter Yield of Shoot of Five Crop Species Grown on Brazilian Oxisol

Crop Species	Shoot Dry Weight (g plant[-1])		Increase over Control (%)
	0 mg Zn kg[-1]	5 mg Zn kg[-1]	
Upland rice[a]	1.33	0.83	60
Corn[b]	0.90	0.33	173
Wheat[c]	0.10	0.09	11
Dry bean[d]	1.67	1.00	67
Soybean[e]	0.83	0.40	108

[a] Upland rice was harvested at 63 days after sowing.
[b] Corn was harvested at 24 days after sowing.
[c] Wheat was harvested at 27 days after sowing.
[d] Dry bean was harvested at 31 days after sowing.
[e] Soybean was harvested 44 days after sowing.

Source: Adapted from Fageria and Baligar (1997a).

FIGURE 8.1 Dry bean growth (from left to right) at 0, 5, 10, and 80 mg Zn kg[-1] applied to a Brazilian Oxisol.

deficiency in humans, especially in developing countries where diets are cereal-based and poor in animal and fish products (Welch, 1993; Frossard et al., 2000). According to the World Health Organization, zinc deficiency is the fifth-leading risk factor for human disease in the developing world.

The deficiency of zinc is associated with several unfavorable environmental conditions for its uptake and utilization. In highly weathered Oxisols and Ultisols, zinc deficiency in crop plants is due to low levels of this element in the parent materials (Fageria and Gheyi, 1999; Fageria et al., 2002). In addition, liming is an essential

FIGURE 8.2   Soybean growth (from left to right) at 0, 5, and 120 mg Zn kg^{-1} of soil applied to a Brazilian Oxisol.

FIGURE 8.3   Growth of wheat (from left to right) at 0, 5, 10, and 120 mg Zn kg^{-1} soil applied to an Oxisol of Brazil.

and dominant practice to improve yield of crops on highly weathered acid soils. Liming may induce zinc deficiency due to adsorption of this element on soil colloids by increasing soil pH (Lindsay, 1979). Data in Table 8.2 show the influence of pH on Zn^{2+} uptake by upland rice grown on a Brazilian Oxisol. There was a highly significant quadratic decrease in Zn uptake with increasing soil pH in the range of 4.6 to 6.8. Zinc uptake was highly determined by soil pH, and variability in its uptake was 98% due to increase in soil pH. Zinc deficiency may be common in sandy soils due to low zinc content or low zinc-retaining capacity of these soils. High levels of phosphorus, iron, manganese, and copper may induce zinc deficiency in crop plants due to antagonistic interaction between these elements in plant absorption

FIGURE 8.4 Corn response to applied zinc in a Brazilian Oxisol (from left to right) (0, 5, and 80 mg Zn kg⁻¹ soil).

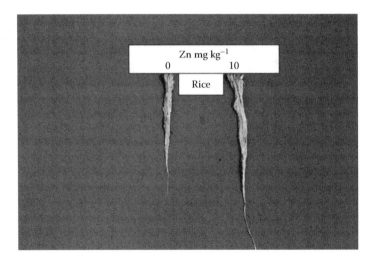

FIGURE 8.5 Upland rice root growth at two zinc levels applied to Brazilian Oxisol.

(Sumner and Farina, 1986; Wilkinson et al., 2000; Fageria et al., 2002). Soil erosion may also be responsible for loss of topsoil layers and consequently zinc deficiency in crop plants. Zinc deficiency is also reported in soils having low organic matter contents (Fageria et al., 2002). A significant positive correlation has been reported between soil extractable zinc and soil organic matter content (Alloway, 2004). Mandal et al. (1988) reported that added organic matter increased zinc bioavailability to rice plants. Clark (1982) reported that zinc deficiency is most likely to occur in plants growing on calcareous soils. Adsorption and occlusion of zinc by carbonates are the major causes of poor zinc availability and the appearance of zinc deficiency on

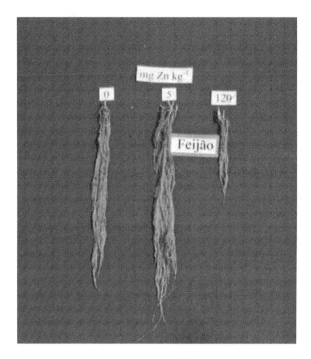

FIGURE 8.6   Dry bean root systems at 0, 5, and 120 mg Zn kg⁻¹ soil in a Brazilian Oxisol.

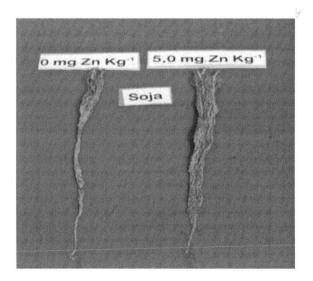

FIGURE 8.7   Soybean root growth at 0 and 5 mg Zn kg⁻¹ soil applied to a Brazilian Oxisol.

calcareous soils (Mengel et al., 2001). Zinc deficiency can be identified by analysis of soil and plant tissues or by visual symptoms.

   The soil–plant system is very dynamic, and the availability of nutrients, including zinc, changes due to physical, chemical, and biological changes. Many of the

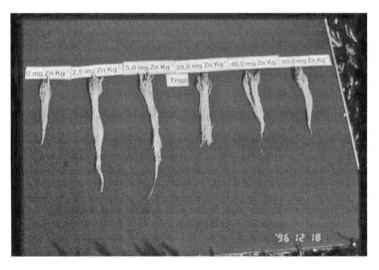

FIGURE 8.8   Root growth of wheat (from left to right) at 0, 2.5, 5, 20, 40, and 80 mg Zn kg⁻¹ levels in a Brazilian Oxisol.

TABLE 8.2
Influence of pH on Uptake of Zinc by Upland Rice Grown on a Brazilian Oxisol

Soil pH in H_2O	Zn Uptake (μg plant⁻¹)
4.6	1090
5.7	300
6.2	242
6.4	262
6.6	163
6.8	142

Regression Analysis

Soil pH (X) vs. Zn uptake (Y) = 9853.9210 − 2922.3800X + 220.5872X², R² = 0.9806**

** Significant at the 1% probability level.
Source: Adapted from Fageria (2000b).

micronutrients associated with the solid phase are not available for plant uptake (Shuman, 1991). In fact, <10% generally are in soluble and exchangeable forms (Lake et al., 1984). Due to widespread deficiency of zinc in different agroecological regions and the dynamic nature of Zn availability in soil–plant systems, an understanding of zinc behavior in soil and plants and in management practices is required to improve its bioavailability for maximum economic yields of field crops. These aspects are discussed in this chapter.

8.2 CYCLE IN SOIL–PLANT SYSTEMS

Understanding the zinc cycle in soil–plant systems is fundamental for improving zinc use efficiency in crop plants. The zinc cycle in soil–plant systems involves addition mainly by soil minerals, crop residues, and chemical fertilizers and depletion by uptake of crops and soil erosion. Some part of the zinc may also be immobilized by soil colloids and soil microbial activities. Zinc is considered a trace element in soils. The important zinc-containing minerals are sphalerite (ZnS), smithsonite ($ZnCO_3$), and hemimorphite [($Zn_4(OH)_2Si_2O_7 \cdot H_2O$)] (Tisdale et al., 1985; Barber, 1995). Zinc also substitutes for magnesium in montmorillonite-type clay minerals (Krauskopf, 1972). Relatively insoluble mineral forms account for more than 90% of the zinc in soils (Barber, 1995). Zinc is present in soil solution in the divalent form (Zn^{2+}). It does not undergo reduction in nature, due to its electropositive nature (Krauskopf, 1972). Forms of zinc in the soil that can affect its availability to plants are water-soluble zinc, exchangeable zinc, adsorbed zinc on the surface of colloids or organic matter, and zinc substituted for Mg^{2+} in the crystal lattices of clay minerals (Tisdale et al., 1985).

Zinc content of the soils varied significantly depending on type of soil, management practices adopted by the farmers, climatic conditions, crop species planted, and cropping intensity. Zinc content of the lithosphere is approximately 80 mg kg^{-1}, and the common range for soils is 10–300 mg kg^{-1} with an average content of 50 mg kg^{-1} (Lindsay, 1979). Zinc content of some Brazilian lowlands (Inceptisols) is presented in Table 8.3. Zinc values varied considerably in the soil samples analyzed. Average values of the two states were 2.0 mg kg^{-1} in the top 0–20 cm soil depth, 1.5 mg kg^{-1} in the 20–40 cm soil depth, 1.8 mg kg^{-1} in the 40–60 cm soil depth, and 1.2 mg kg^{-1} in the 60–80 cm soil depth. Overall, zinc values were lower than critical values of zinc for upland rice and corn in the range of 2 to 5 mg kg^{-1} as reported by Fageria (2000a) by Mehlich-1 extracting solution. Fageria and Breseghello (2004) conducted a survey to evaluate soil fertility and nutritional status of upland rice in 43 sites, covering 33 rural properties in the state of Mato Grosso, Brazil. These authors concluded that taking into consideration a value of 1 mg Zn kg^{-1} of soil as a critical value, 21% of the samples were deficient in zinc.

The mechanism that controls Zn solubility relationships in soils is poorly understood. Zinc is relatively immobile in the soil, and diffusion is reported to be the dominant mechanism for transporting zinc to plant roots (Tisdale et al., 1985). Chelating agents produced by microorganisms or excreted by plant roots may also play an important role in zinc transport to plant roots (Stevenson, 1986). Soil Zn may be (1) in soil solution as ionic or organically complexed species; (2) on exchange sites of reactive soil components; (3) complexed with organic matter; (4) occluded in oxides and hydroxides of Al, Fe, and Mn; and (5) entrapped in primary and secondary minerals. Zinc in the different soil fractions varies in plant availability. The Zn present in water-soluble, exchangeable fractions is readily available to plants; Zn associated with primary and secondary soil minerals is relatively unavailable to plants (Iyengar et al., 1981).

Zinc sorption in soils can be influenced by soil pH, clay minerals, organic matter content, iron and aluminum oxide content, and carbonate content. Ionic strength

TABLE 8.3

Zinc Content (mg kg⁻¹) in Brazilian Lowland (Inceptisol) Soils at Four Soil Depths

Soil Depth (cm) and Value	State of Mato Grosso	State of Mato Grosso do Sul	Average
0–20			
Minimum	0.3	1.2	0.8
Maximum	4.9	4.0	4.5
Average	1.4	2.5	2.0
20–40			
Minimum	0.1	0.5	0.3
Maximum	3.5	7.1	5.3
Average	0.8	2.1	1.5
40–60			
Minimum	0.2	0.6	0.4
Maximum	3.6	11.7	7.7
Average	0.9	2.6	1.8
60–80			
Minimum	0.2	0.6	0.4
Maximum	3.0	3.6	3.3
Average	0.8	1.5	1.2

Note: Soil samples were collected from 14 different locations covering 13 municipalities in the two states. Mehlich 1 (0.05 M HCl + 0.0125 M H_2SO_4) extracting solution was used for Zn extraction.

Source: Adapted from Fageria et al. (1997b).

and complex formation may also affect sorption of Zn with inorganic legends in soil solution. Lindsay (1979) reported that the complex $ZnSO_{40}$ can be very important in soils and can contribute significantly to total Zn in solution. On the contrary, the complexes $ZnNO_3^+$, $Zn (NO_3)_2^0$, $ZnCl^+$, $ZnCl_2^0$, $ZnCl_3^-$, and $ZnCl_4^{2-}$ usually make minor contributions to total soluble Zn. Metal chelation is another important factor that may influence the solubility and movement of metals in soils (Cleveland and Rees, 1981). Krauskopf (1972) and Lindsay (1979) reported that those forms of Zn form soluble complexes with organic ligands such as EDTA (ethylenediaminetetraacetic acid) and DTPA (diethylenetriaminepentaacetic acid). Elrashidi and O'Connor (1982a) reported that the presence of EDTA in the soil suspension significantly decreased Zn sorption by the three soils of varying physical and chemical properties. These authors also reported that Cu was more effective than Ni in decreasing Zn sorption. Kurdi and Doner (1983) reported that zinc ion sorption by different soil types conformed to a Freundlich equation but not to a Langmuir equation. Sorption differences among the soils were more related to equilibrium pH than to cation exchange capacity (CEC). These authors further reported that Zn sorption resulted in release of H and Mn. The release increased with increasing metal ion sorption, and the presence of Cu even at a concentration of 15 mg/ml completely prevented Zn sorption in one soil low in CEC.

Many studies show that Cu and Zn are strongly sorbed by soils even in the presence of high concentrations of other cations such as Ca^{2+}, Na^+, and K^+ (Cavallaro and McBride, 1984). Sorption studies in pure systems have shown that layer silicate clays bind Zn by ion exchange, fixation, and hydrolysis reactions, and the pH dependence of these reactions is believed to be due primarily to the tendency of the clays to preferentially sorb hydrolyzed species (Tiller, 1968). Organic matter also sorbs this metal, forming an inner sphere complex (Bloom and McBride, 1979). This reaction depends on pH due to both hydrolysis of the metal and the weak acid nature of the exchange sites of organic matter. Cavallaro and McBridge (1984) studied zinc sorption by acid soil clay. They concluded that microcrystalline and moncrystalline oxides in the clay fraction representing < 20% of the clay by weight provide reactive surfaces for the chemisorption of Zn. At low pH, adsorption at these surfaces may be the dominant mechanism of heavy metal immobilization, especially in the subsoil horizons. Saeed and Fox (1979) reported that unfertilized soils, which contained predominantly constant-charge colloids, absorbed more Zn than soils with variable-charge colloids. Phosphorus fertilization increased on adsorption by soils that contained colloids predominantly of the variable-charge type. These authors concluded that the results support the hypothesis that phosphate additions to soils increase zinc adsorption by increasing the negative charge on the iron and aluminum oxide system.

The role of cation exchange capacity (CEC) in controlling the amount of Zn removed from solution by soils demonstrated that the larger the CEC of a soil, the lower the saturation of the exchange sites by a given Zn concentration (Shuman, 1975). Finer-textured soils or soil horizons containing high levels of organic matter had higher Zn sorption capacities than sandy-textured, low-organic-matter soils. Soil pH is the master variable regulating the mobility of metals including Zn in soils, and high correlations exist between pH and the ease with which a metal can be extracted, a criterion commonly used to describe the lability of a metal. The solubility of all zinc minerals decreases 100-fold for each unit increase in pH (Lindsay, 1979). Adsorption is a major contributing factor to low concentrations of Zn in soil solution in Zn-deficient soils. Soil pH has shown to affect Zn adsorption, either by changing the number of sites available for adsorption or by changing the concentration of the Zn species that is preferentially adsorbed (Barrow, 1986). Iyengar et al. (1981) used a multivariable equation with six fractions and soil pH, which accounted for 94% of the Zn uptake by corn. Increasing pH decreased Zn in the water-soluble fraction (El-Kherbawy and Sanders, 1984) and in the exchangeable fraction (Iyengar et al., 1981; Sims, 1986).

Chairidchai and Ritchie (1990) studied the effect of pH on Zn adsorption in a lateritic podzolic soil and concluded that at 2.5 μmol g^{-1} of added Zn, Zn concentration decreased from 238 to 19 μmol L^{-1} when the pH value of the solution was adjusted from 3.8 to 5.6. The magnitude of the decrease with increasing pH was smaller at the lower Zn additions. The rate of increase in % Zn adsorption with increasing pH was greatest at high Zn concentration, with % Zn adsorption becoming similar (96–98%) at all initial Zn concentrations at pH 5.6 (Chairidchai and Ritchie, 1990).

The predominant Zn species in solution below pH 7.7 is Zn^{2+}, although $ZnOH^+$ is more prevalent above this pH. The neutral species $Zn(OH)_2^0$ is the major species above pH 9.1, whereas the species $Zn(OH)_3^-$ and $Zn(OH)_4^{2-}$ are never major solution species in the pH range of soils (Lindsay, 1979). The zinc species contributing significantly to total inorganic zinc in solution can be represented as follows (Lindsay, 1979):

$$\{Zn_{inorganic}\} = (Zn^{2+}) + (ZnSO_4^0) + (ZnOH^{+)} + (Zn(OH)_2^0) + (ZnHPO_4^0)$$

Friesen et al. (1980) also reported that Zn activity in two Ultisols of southeastern Nigeria declined sharply from 4 µM to about 0.5 µM when soils were limed to pH above 5.0. The observation that pHs > 6 lower free metal-ion activities including Zn in soils has been attributed to the increase in pH-dependent charge on oxides of Fe, Al, and Mn (Jenne, 1968), chelation by organic matter, or precipitation of metal hydroxides (Lindsay, 1979). Hodgson et al. (1966) reported that zinc was present in soil solution as organic as well as inorganic species. He presented the equation as follows:

$$(\text{Total soluble Zn}) = (Zn_{inorganic}) + (Zn_{organic})$$

Zinc solubility and availability to plants in flooded soils is different compared to aerobic or oxidized soils. When a soil is submerged, the concentrations of most nutrient elements in the soil increase, while the concentration of water-soluble Zn decreases (Ponnamperuma, 1977). In acid soils, the decrease in Zn concentration may be attributed to the increase in pH following soil reduction. With calcareous soils, however, the pH decreases on submergence (Ponnamperuma, 1972), and Zn solubility is supposed to increase 100-fold for each unit decrease in pH (Lindsay and Norvell, 1969). Zinc deficiency in rice usually occurs after flooding. Moore and Patrick (1988) reported that Zn deficiency in flooded rice is due to a decrease in alcohol dehydrogenase (ADH) and/or glutamate dehydrogenase (GDH) enzyme activity in the roots, since Zn is a cofactor in these enzymes. Alcohol dehydrogenase has been postulated as the most important enzyme involved in anaerobic metabolism in plant roots (Smith and Rees, 1979). Similarly, GDH is one of the main enzymes in plant roots necessary for N assimilation; hence, N contents have been found to be lower than normal in Zn-deficient rice (Sedberry et al., 1971). In reduced soil conditions, concentrations of Fe^{2+} and Mn^{2+} increase, which may also inhibit uptake of zinc by rice plants. Alloway (2004) reported that addition of zinc fertilizers to water-logged soils has been found to increase DTPA-extractable manganese but decrease the uptake and translocation of copper, iron, and phosphorus.

8.3 FUNCTIONS AND DEFICIENCY SYMPTOMS

Zinc has many biochemical functions in the plant. These functions are summarized as follows:

1. Zinc is necessary for producing chlorophyll and forming carbohydrate.
2. Zinc aids plant growth substances and enzyme systems. Zinc, along with Cu, has been shown to be a constituent of the enzyme superoxide dismutase.

3. Zinc is closely involved in the N-metabolism of the plant.
4. Zinc is required for the synthesis of tryptophane. As tryptophane is also a precursor of indoleacetic acid, the formation of this growth substance is also indirectly influenced by zinc.
5. Zinc has a possible role in plant metabolism involved in starch formation.
6. Flowering and fruiting are much reduced under conditions of severe zinc deficiency.
7. Like Ca^{2+}, Zn^{2+} also affects the integrity of biological membranes. Loss in membrane integrity of Zn-deficient plants contributes to the increase in susceptibility to fungal diseases (Sparrow and Graham, 1988).
8. Zinc improves root development in crop plants.
9. In some metalloenzymes, Zn^{2+} can interchange with Mg^{2+} and Mn^{2+} (Clarkson and Hanson, 1980).
10. A key role of zinc in gene expression and regulation has been reported (Klug and Rhodes, 1987).
11. Activity of alcohol dehydrogenase enzyme (ADH) is reduced in zinc-deficient plants (Romheld and Marschner, 1991). This enzyme plays an important role in anaerobic root respiration. Hence, growth and metabolism of flooded rice is adversely affected.
12. Zinc affects carbohydrate metabolism in plants.
13. Carbonic anhydrase (CA) is a Zn-containing enzyme involved in fixation of CO_2 in higher plants, particularly in C_4 plants (Hatch and Burnell, 1990; Badger and Price, 1994; Cakmak and Engels, 1999).
14. The reduction of photosynthesis observed in zinc-deficient plants can also be due, in part, to a major decrease in chlorophyll content and the abnormal structure of chloroplasts (Alloway, 2004).
15. Zinc-deficient wheat has been reported to develop small anthers and abnormal pollen grains (Brown et al., 1993).
16. The disease *Rhizoctonia solani* was reported to be reduced with the application of zinc in wheat plants (Thongbai et al., 2001).
17. Zinc deficiency reduces water use efficiency, which can lead to loss of turgidity and reduced growth (Duffy, 2007).

Zinc is not very mobile in the plants; hence, symptoms first appear on the younger leaves (Duffy, 2007). Zinc deficiency in corn is called "white bud" because the young bud turns white or light yellow in early growth. Plants suffering from Zn deficiency often show chlorosis in the intervenial areas of the leaf. These areas are pale green, yellow, or even white. In the monocots and particularly in maize, chlorotic bands form on either side of the midrib of the leaf. In rice plants, severe Zn deficiency resulted in brown spots or rust streaks on leaves (Fageria and Barbosa Filho, 2006). Under severe zinc deficiency, internodal length is reduced, plants are stunted, terminal growth is retarded, and new leaves develop slowly (Smith et al., 1993). Marschner and Cakmak (1989) reported that zinc deficiency intensified with high light intensity, suggesting the involvement of superoxide radicals in symptom development. Figure 8.9 and Figure 8.10 show zinc deficiency symptoms in rice leaves and corn plants, respectively.

FIGURE 8.9   Zinc deficiency symptoms in rice leaves grown on a Brazilian Oxisol.

FIGURE 8.10   Zinc deficiency in corn plants grown on a Brazilian Oxisol.

8.4   CONCENTRATION IN PLANT TISSUES AND UPTAKE

The concentration of zinc in plants can vary widely depending on crop species or genotypes within species and environmental conditions (Welch, 1995). Furthermore, plant age and plant parts analyzed are the most important factors affecting nutrient

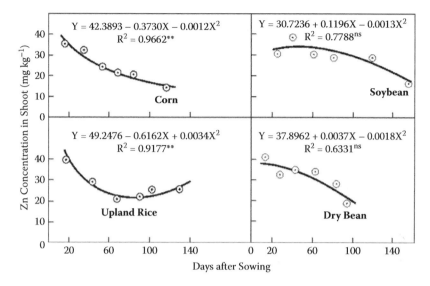

FIGURE 8.11 Relationship between plant age and zinc concentration in the shoot of four food crops.

concentration in plant tissues (Fageria et al., 1997a). Generally, nutrient concentration (content per unit dry weight) in plant tissues decreased with the advancement of plant age (Fageria, 1992). Zinc concentration (zinc content per unit of shoot dry weight) in shoot of corn and rice is significantly ($P < 0.01$) decreased with the advancement of plant age (Figure 8.11). In dry bean and soybean, Zn concentration showed a quadratic increase with the advancement of plant age; however, the increase was nonsignificant (Figure 8.11). Zinc concentration in corn shoot at 18 days after sowing was 36 mg kg^{-1} and decreased to 15 mg kg^{-1} at harvest. In rice, Zn concentration in shoot at 19 days after sowing was 40 mg kg^{-1} and decreased to 26 mg kg^{-1} at harvest. Similarly, Zn concentration in shoot of dry bean decreased from 43 mg kg^{-1} at 15 days after sowing to 19 mg kg^{-1} at harvest. In soybean, Zn concentration in shoot was 31 mg kg^{-1} at 27 days after sowing and decreased to 16 mg kg^{-1} at harvest. The decrease in Zn concentration in corn and rice with the advancement of plant age is associated with increase in shoot dry weight with increase in plant age up to certain growth stage. Such decrease in nutrient concentrations has been reported in annual crops and is known as the dilution effect (Fageria et al., 1997a).

Generally, nutrient concentrations in plant shoot are used as a criterion to identify nutrient deficiency or sufficiency during crop growth. However, nutrient concentration in grain is also important for judging crop quality for human or animal feed. Hence, quantity of grain produced by a crop species and its nutrient concentration are important for human and animal health. Grain yield and zinc concentration in the grain of four crop species are presented in Figure 8.12. Grain yield of four food crops was in the order of corn > upland rice > dry bean > soybean. Hence, cereals had higher grain yield compared with legumes. Zinc concentration in grain of corn was 18 mg kg^{-1}; in upland rice grain, 30 mg kg^{-1}; in dry bean grain, 39 mg kg^{-1}; and in soybean grain, 55 mg kg^{-1}. Fageria et al. (1997a) and Fageria (1989) reported

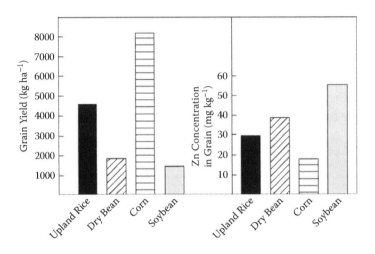

FIGURE 8.12   Grain yield and zinc concentration in grain of four food crops.

more or less similar concentration of Zn in grain of upland rice and dry bean. Zinc concentration in legume crops was higher than Zn concentration of cereals. High grain Zn concentration is considered a desirable quality factor that could increase the nutritional value of the grain for humans (Cakmak et al., 1996; Graham et al., 1992). High Zn concentration in seed is also a desirable trait for conferring seedling vigor and grain yield of the next crop when resown on Zn-deficient soil (Graham et al., 1992; Graham and Rengel, 1993; Grewal and Graham, 1999).

Graham et al. (1992) also reported that high zinc content in grain is under genetic control and is not tightly linked to agronomic zinc efficiency traits and may have to be selected for independently. Adequate zinc concentration in the tissues of important field crops is presented in Table 8.4. Similarly, adequate and toxic levels of zinc in five crop species are presented in Table 8.5. Toxic level of Zn in upland rice and

TABLE 8.4
Adequate Zinc Level in the Plant Tissues of Field Crops

Crop Species	Growth Stage and Plant Part Analyzed	Zn (mg kg^{-1})
Wheat	Heading, whole tops	15–70
Barley	Heading, whole tops	15–70
Rice	Tillering, whole tops	20–150
Corn	30 to 45 days after emergence, whole tops	20–50
Sorghum	Seedling, whole tops	30–60
Sorghum	Bloom, 3rd blade below panicle	15–30
Soybean	Prior to pod set, upper fully developed trifoliate	21–50
Dry bean	Early flowering, uppermost blade	35–100
Cowpea	42 days after sowing, leaf blade	27–32
Peanut	Early pegging, upper stems and leaves	20–50

Source: Adapted from Fageria et al. (1997a).

TABLE 8.5

Adequate and Toxic Levels of Zinc in the Plant Tissues of Five Crop Species Grown on Brazilian Oxisols

Crop Species	Adequate Level (mg kg⁻¹)	Toxic Level (mg kg⁻¹)
Upland rice	67	673
Dry bean	18	133
Corn	27	427
Soybean	20	187
Wheat	19	100

Note: Adequate level was calculated at the 90% of the maximum dry weight of shoot, and toxic level was calculated at the 10% reduction in shoot dry weight after achieving maximum weight. Plants were harvested at 6 weeks after sowing, and soil pH at harvest was about 6.0 in water in all the five experiments.

Source: Adapted from Fageria et al. (2003a).

corn tissues was much higher compared to wheat, dry bean, and soybean. Stevenson (1986) reported that excessive or toxic level of zinc in plant tissues is more than 400 mg kg⁻¹. Similarly, Jones (1991) reported excess or toxic level of zinc in plant tissues in the range of 100 to 400 mg kg⁻¹. The values of Zn concentration reported in Figure 8.11 and Table 8.4 and Table 8.5 can be used in interpretation of plant tissue analysis results for Zn at different growth stages for rice, corn, dry bean, and soybean. Fageria (2000a) reported more or similar concentration of Zn in rice, corn, dry bean, and soybean in the early stage of plant growth. The significant variation in Zn concentration suggests that plant tissue analysis should be done during different growth stages for better interpretation of such results.

Rashid and Fox (1992) studied the internal Zn requirements of several cereal and legume grain crops with special emphasis on the grain as an index for determining the Zn status of crops and soils. The order of grain yield response to Zn fertilization was wheat < sorghum < rice < millet < soybean < cowpea < corn. Fertilizer Zn requirement for near maximum grain yield was highest for cowpea (7.5 mg kg⁻¹ soil) and lowest for wheat (0.5 mg kg⁻¹ soil). Zinc concentrations (mg kg⁻¹) in recently matured leaves associated with 95% maximum yields were similar for seven species: corn and millet, 24; soybean, 22; cowpea, 21; sorghum, 20; rice, 19; and wheat, 17. Equivalent concentrations for seeds were soybean, 43; cowpea, 26; corn, 18; millet, rice, and wheat, 15; and sorghum, 10. Rashid and Cox (1992) also reported that there was little reason to prefer leaves over grain as a diagnostic tissue for Zn; therefore, Zn should be added to the list of elements for which seed analysis may be advantageous. A possible disadvantage for seeds is that foliar analysis might be useful for the current crop; seed analysis can be used only to diagnose past problems and plan for future action. With most crops, Zn values in the dry matter of leaves under 20 mg kg⁻¹ can be regarded as indicating either deficiency or approaching deficiency (Chapman, 1959). Ohki (1984) determined the Zn critical deficiency and critical toxicity level in sorghum in nutrient solution. The Zn critical deficiency and toxicity levels were 10 and 64 mg kg^{-1}, respectively, sampled at vegetative growth stage. Stevenson (1986) and Foth and Ellis (1988) reported that overall sufficiency level of Zn in annual crop tissues varied from 25 to 150 Mg kg^{-1} tissue dry weights. Welch (1995) reported that adequate zinc concentration in common food crops varied from 15 to 87 mg kg^{-1}. Zinc concentration in plant tissues in the range of 150 to 200 mg kg^{-1} may be toxic for the growth and development of higher plants (Mengel et al., 2001).

Knowledge of nutrient uptake (concentration × shoot or grain dry weight) during the growth cycle of a crop is an important aspect in managing soil fertility. Zinc uptake in shoot of corn, rice, soybean, and dry bean significantly increased with the advancement of plant age up to 89 days after sowing in corn, 108 days after sowing in soybean, 104 days after sowing in rice, and 72 days after sowing in dry bean (Figure 8.13). After these ages, Zn uptake in shoot of four crop species decreased. Zinc uptake followed more or less shoot dry matter accumulation pattern in four crops (Figure 8.14.). Fageria et al. (1997) and Karlen et al. (1988) reported similar pattern of Zn uptake in upland rice and corn, respectively.

The uptake of a nutrient by a crop is mainly determined by nutrient concentration in the growth medium and crops' yielding ability. Zinc application in soil significantly improves uptake of Zn in rice as well as in common bean plants as expected (Table 8.6). Rice plants had much higher uptake of Zn as compared to bean plants. This means that the Zn requirement for rice is much higher as compared to common bean plants grown on an Oxisol. The differences in crop species and cultivars within species in Zn uptake and utilization have been widely reported

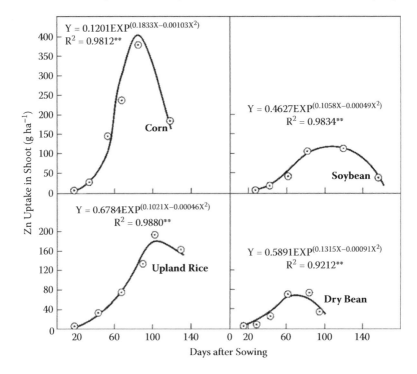

FIGURE 8.13 Relationship between plant age and zinc uptake in shoot of four food crops.

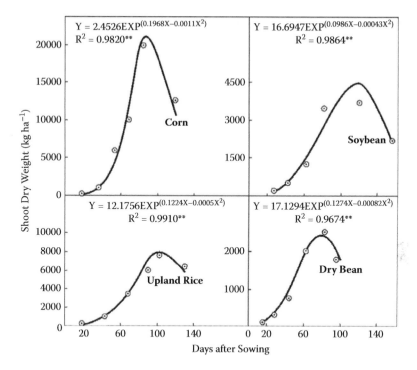

$$Y = 2.4526\text{EXP}^{(0.1968X-0.0011X^2)}$$
$$R^2 = 0.9820^{**}$$

$$Y = 16.6947\text{EXP}^{(0.0986X-0.00043X^2)}$$
$$R^2 = 0.9864^{**}$$

$$Y = 12.1756\text{EXP}^{(0.1224X-0.0005X^2)}$$
$$R^2 = 0.9910^{**}$$

$$Y = 17.1294\text{EXP}^{(0.1274X-0.00082X^2)}$$
$$R^2 = 0.9674^{**}$$

FIGURE 8.14 Relationship between plant age and shoot dry weight of four food crops.

TABLE 8.6
Zinc Uptake in the Shoot of Upland Rice and Dry Bean as Influenced by Zinc Rates Applied to a Brazilian Oxisol

Zn Rate (mg kg⁻¹)	Upland Rice (mg kg⁻¹)	Dry Bean (mg kg⁻¹)
0	28	17
5	71	56
10	87	73
20	125	92
40	363	150
80	718	215
120	863	313

Regression Analysis

Zn rate (X) vs. Zn uptake in upland rice (Y) = $-4.3774 + 10.4039X - 0.0251X^2$, $R^2 = 0.9864^{**}$

Zn rate vs. Zn uptake in dry bean (Y) = $35.5717 + 2.7936X - 0.0044X^2$, $R^2 = 0.9800^{**}$

** Significant at the 1% probability level.

Source: Adapted from Fageria (2002a).

(Cakmak et al., 1998). This may be related to better internal utilization of Zn in bean plants and to their superior ability to maintain a high growth rate at low internal Zn concentration as compared to rice plants. These results also indicate that upland rice can tolerate much higher concentration of Zn as compared to bean plants. Fageria (2000a) reported that the critical toxic level is much higher in rice plants as compared to common bean plants.

8.5   USE EFFICIENCY AND Zn HARVEST INDEX

Zinc use efficiency in shoot (kg shoot dry weight g^{-1} Zn accumulated in shoot) was significantly influenced in four crop species, except dry bean (Figure 8.15). It was linear in corn and exponential quadratic in upland rice in relation to plant age. In corn, Zn use efficiency varied from 26 to 69 kg g^{-1} at 18 to 119 days plant age. In upland rice, it varied from 25 to 40 kg g^{-1} at 19 to 130 days plant age. In soybean and dry bean, Zn use efficiency showed similar responses with the advancement of plant age. In soybean, the values of Zn use efficiency varied from 33 to 63 kg g^{-1} at 27 and 158 days plant age. Similarly, in dry bean, the values of Zn use efficiency were 24 to 51 kg g^{-1} at 15 to 96 days plant age. Hence, the Zn use efficiency in food crops can be classified in the order of corn > soybean > dry bean > upland rice. In corn, higher Zn use efficiency in shoot dry weight production was associated with high dry matter production. Zinc use efficiency data in food crops are scarce in the literature; hence, these results have special significance for scientific literature.

Nutrient use efficiency in grain production in field crops is related to quantities of nutrient absorbed and grain yield level. Zinc uptake in grain of four crops was in

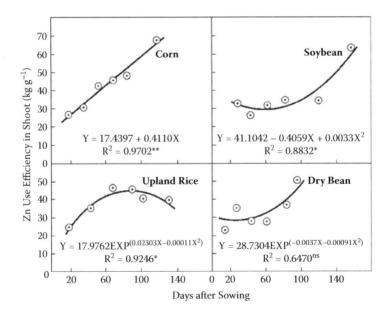

FIGURE 8.15   Relationship between plant age and zinc use efficiency in shoot of four food crops.

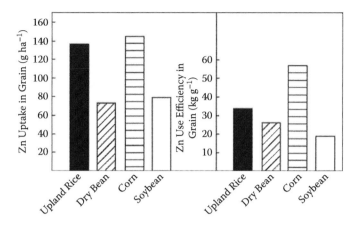

FIGURE 8.16 Zinc uptake in grain and zinc use efficiency in grain of four food crops.

the order of corn > upland rice > soybean > dry bean (Figure 8.16). The higher Zn uptake in corn and upland rice was associated with higher grain yield. In soybean, higher Zn uptake compared to dry bean was associated with high Zn concentration in the seeds of soybean. Zinc use efficiency in grain production was higher for corn and lower for dry bean and soybean (Figure 8.16). The higher Zn efficiency of cereals compared to legumes was associated with higher grain yields of corn and upland rice. Fageria (1992) reported that higher yield in crops is associated with higher nutrient use efficiency.

In addition to zinc use efficiency in shoot dry weight and grain production of annual crops as discussed earlier, zinc recovery efficiency by a crop is an important measure to know the balance of zinc in soil–plant systems. Hence, data related to zinc recovery efficiency for upland rice genotypes are presented in Table 8.7. Zinc recovery efficiency (ZnRE) varied from 8.3 to 23.1% with an average of 13% across 10 genotypes (Table 8.7). Information on Zn recovery efficiency by upland rice genotypes is scarce. However, Mortvedt (1994) reported that crop recovery of applied micronutrients is relatively low (5 to 10%) compared to macronutrients (10 to 50%). Further, he states that such low recovery of applied micronutrients is due to their uneven distribution in a soil because of low application rates, reaction with soil to form unavailable products, and low mobility in soil.

Zinc harvest index (ZnHI) [(Zn uptake in grain/Zn uptake in grain plus straw) × 100] is an important index in measuring the quantity of Zn^{2+} translocated to grain from the amount of Zn absorbed by the roots. It varied with crop species and genotypes within species. In addition, it is influenced by crop management practices. Soil Zn^{2+} levels and genotypes significantly influenced ZnHI, and interaction between soil Zn levels and genotypes was also highly significant (Table 8.8). The ZnHI varied from 0.40 to 0.76 at the low Zn^{2+} level and 0.21 to 0.37 at the high Zn^{2+} level. Average values across the genotypes were 0.60 and 0.28, respectively. The genotypes recorded higher ZnHI at low zinc level than at high soil Zn^{2+} level, and such differences are due to higher Zn accumulation in the shoot at the higher Zn^{2+} level (Fageria and Baligar, 2005). Fageria (2001b) reported that ZnHI for upland rice was 46%;

TABLE 8.7

**Zinc Recovery Efficiency as Influenced by Zinc Rate
in Lowland Rice and Upland Rice Genotypes**

Zn Rate (mg kg⁻¹)	Zn Recovery Efficiency in Lowland Rice (%)	Upland Rice Genotype	Zn Recovery Efficiency (%)
5	16.5	Bonança	15.0ab
10	6.9	Caipo	14.6ab
20	4.4	Canastra	12.4ab
40	2.4	Carajas	13.3ab
80	1.5	Carisma	11.8ab
120	3.5	CNA 8540	10.4b
Average	5.9	CNA 8557	8.5b
R^2	0.6621**	Guarani	23.1a
		Maravilha	13.1ab
		IR 42	8.3b
		Average	13.1

** Significant at the 1% probability level. Means in the same column followed by
    same letter are not different at the 5% probability level by the Tukey's test.
Source: Adapted from Fageria et al. (2003a) and Fageria and Baligar (2005).

for dry bean, 72%; for corn, 51%; and for soybean, 64%. Similarly, Fageria (1989) reported that the ZnHI for cowpea was 46%. Fageria et al. (1997a) reported that in lowland rice, ZnHI was 50% at harvest.

8.6 INTERACTION WITH OTHER NUTRIENTS

Interaction between or among nutrients can be measured in terms of crop growth and nutrient concentrations in plant tissue. Soil, plant, and climatic factors can influence interaction. In the nutrient interaction studies, all other factors should be at an optimum level except the variation in level of nutrient under investigation. The interaction varied from nutrient to nutrient and crop species to species and sometimes among cultivars of the same species. Therefore, this is a very complex issue in mineral nutrition and not well understood in annual crops grown for food production. Influence of zinc on dry matter yield and uptake of macro- and micronutrients by upland rice and common bean are presented in Table 8.9 and Table 8.10. Bowen (1969) and Kochain (1991) reported that Cu^{2+} competitively inhibits Zn^{2+} uptake and suggested that Cu^{2+} competes for the same transport site. Similarly, Alloway (2004) reported that higher concentration of copper in the soil solution, relative to zinc, can reduce the availability of zinc to a plant (and vice versa) due to competition for the same sites for absorption into the plant root. This could occur after the application of a copper fertilizer.

Zinc–copper interaction can also occur within plants, because copper nutrition affects the redistribution of zinc within plants. In common bean plants, application of zinc significantly improved uptake of N, Mg, and Cu. However, zinc showed a

---

TABLE 8.8

Zinc Harvest Index in Upland Rice Genotypes at Two Zinc Levels

Genotype	0 mg Zn kg^{-1}	10 mg Zn kg^{-1}
Bonança	0.60abc	0.25ab
Caipo	0.62ab	0.25ab
Canastra	0.64ab	0.35a
Carajas	0.76a	0.32ab
Carisma	0.60abc	0.31ab
CNA 8540	0.68a	0.37a
CNA 8557	0.46bc	0.30ab
Guarani	0.70a	0.22b
Maravilha	0.57abc	0.26ab
IR 42	0.40c	0.21b
Average	0.60	0.28
F-test		
Zn rate	**	
Genotype	**	
Zn × G	**	
CV (%)	12	

** Significant at the 1% probability level. Means in the same column followed by same letter are not different at the 5% probability level by the Tukey's test.

Source: Adapted from Fageria and Baligar (2005).

---

significant negative interaction with P uptake. Alloway (2004) reported that in rice seedlings, translocation of zinc from roots increases with manganese application. However, high manganese in combination with high iron may inhibit the absorption of zinc by rice in flooded soils and enhance zinc deficiency. Zinc-deficient plants can absorb high concentrations of boron in a similar way to zinc, enhancing phosphorus toxicity in crops; this is probably due to impaired membrane function in the root (Alloway, 2004). Graham et al. (1987) reported that low Zn treatment did not affect plant growth but enhanced B concentration to a toxic level in barley. Similarly, Zn deficiency enhanced B concentration in wheat grown on Zn-deficient soils (Singh et al., 1990). Sinha et al. (2000) noted a synergistic interaction between Zn and B in mustard (*Brassica nigra*) when both the nutrients were either in low or excess supplies.

Nitrogen fertilizer such as ammonium sulfate can have a significant acidifying effect on soils and so lead to an increase in the availability of zinc to crops in soils of relatively high pH status. Conversely, nitro-chalk (calcium nitrate) can increase the soil pH and reduce zinc availability (Alloway, 2004). Alloway (2004) also reported that application of gypsum ($CaSO_4$), which decreased the soil pH from 5.8 to 4.6, increased the zinc content of plants, but the equivalent amount of calcium applied as

TABLE 8.9

Influence of Zinc on Dry Matter and Uptake of Macronutrients by Upland Rice and Common Bean Plants

Zn Level (mg kg^{-1})	Shoot Dry Wt. (g 4 Plants^{-1})	N (g kg^{-1})	P (g kg^{-1})	K (g kg^{-1})	Ca (g kg^{-1})	Mg (g kg^{-1})
			Upland Rice			
0	1.73	42.75	2.25	28.50	7.32	4.05
5	2.05	42.25	2.20	29.00	6.75	3.77
10	2.35	40.50	2.20	30.25	6.82	3.80
20	2.15	41.50	1.87	30.75	6.95	3.75
40	2.13	39.50	2.20	25.50	6.40	3.55
80	1.97	37.50	2.27	25.50	6.47	3.77
120	1.53	40.25	2.15	25.75	6.32	3.50
F-test	**	**	NS	*	**	**
CV (%)	10	2	8	10	5	4
			Regression Analysis			
β_0	1.9770	42.7362	2.1607	29.9250	7.0708	3.8849
β_1	0.00988	−0.12598	−0.00048	−0.08676	−0.01602	−0.00591
β_2	−0.00011	0.00086	0.000006	0.00042	0.00008	0.00003
R^2	0.6693*	0.8582*	0.11^{ns}	0.6161^{ns}	0.7232*	0.5149^{ns}
			Common Bean			
0	5.37	41.00	1.80	24.25	19.20	8.22
5	5.33	40.00	1.82	22.25	19.52	8.20
10	5.27	40.25	1.45	23.25	20.65	8.37
20	5.02	41.25	1.32	27.00	19.85	8.40
40	4.67	40.50	1.10	22.50	21.45	8.45
80	4.67	40.75	1.05	24.50	18.87	8.02
120	3.97	38.50	1.02	21.25	18.70	7.72
F-test	NS	NS	**	NS	NS	NS
CV (%)	13	7	15	13	9	6
			Regression Analysis			
β_0	5.3385	40.3737	1.7625	23.5014	19.6611	8.2587
β_1	−0.01206	0.02868	−0.01990	0.04532	0.03232	0.00599
β_2	0.00001	−0.00036	0.00011	−0.00052	−0.00036	−0.00009
R^2	0.9267**	0.7272*	0.9269**	0.2810^{ns}	0.4447^{ns}	0.8822*

*, **, NS Significant at the 1 and 5% probability levels and nonsignificant, respectively.

calcium carbonate ($CaCO_3$), which increased the pH from 5.7 to 6.6, decreased the zinc content of plants.

Positive as well as negative Zn-P interactions have been reported (Sumner and Farina, 1986; Wilkinson et al., 2000). Development of zinc deficiency and its apparent accentuation following phosphorus fertilization has become known as a P-induced

TABLE 8.10
Influence of Zinc on Uptake of Micronutrients
in Rice and Common Bean Plants

Zn Level (mg kg^{-1})	Zn (mg kg^{-1})	Cu (mg kg^{-1})	Mn (mg kg^{-1})	Fe (mg kg^{-1})
		Upland Rice		
0	28	13	1075	115
5	71	14	740	125
10	87	12	663	135
20	125	12	643	113
40	363	12	785	115
80	718	11	593	128
120	863	10	440	103
F-test	**	**	**	NS
CV (%)	18	6	24	18
		Regression Analysis		
β_0	−4.3774	13.1163	846.6214	120.2178
β_1	10.40388	−0.03553	−4.48579	0.17951
β_2	−0.02510	0.00008	0.01074	−0.00248
R^2	0.9864**	0.8012*	0.5495^{ns}	0.3096^{ns}
		Common Bean		
0	17	4.75	108	125
5	56	4.50	108	115
10	73	5.00	128	140
20	92	4.75	93	95
40	150	5.50	123	150
80	215	5.00	125	133
120	313	4.25	120	133
F-test	**	NS	NS	NS
CV (%)	27	21	19	18
		Regression Analysis		
β_0	35.5717	4.6029	108.3553	119.2847
β_1	2.79364	0.02542	0.30915	0.41045
β_2	−0.00441	−0.00023	−0.00167	−0.00248
R^2	0.9864**	0.7387*	0.2001^{ns}	0.1280^{ns}

*, **, NS Significant at the 1 and 5% probability levels and nonsignificant, respectively.

Zn deficiency (Christensen and Jackson, 1981). This interaction is strong at low Zn levels, and it can be corrected or prevented by modest Zn application (Olsen, 1972). The mechanisms that have been proposed to explain the interaction include (1) P-Zn interactions within the soil, (2) dilution of Zn in plant tissues by growth response to P, (3) reduced uptake or translocation of Zn in the presence of added P, and (4) P interference with the utilization of Zn by the plant (Olsen, 1972). Loneragan et al. (1979) reported that P-enhanced bonding of Zn to oxides and hydroxides of iron (Fe) and aluminum (Al) might be important in some soils. According to Christensen and Jackson (1981), a more plausible explanation for the apparent accentuation of Zn deficiency by P fertilizer applications is that P accumulates and may become toxic in Zn-deficient plants. Zinc-deficient species such as potato—which show increased P uptake, increased mobilization and translocation of P, and increased concentration of P due to limited dry matter production—would be expected to be more subject to P toxicity than would species such as corn in which concentration of P results primarily from reduced growth. Loneragan et al. (1982) also reported that low Zn combined with high P in solution may enhance accumulation of P in older leaves to concentrations that are toxic, thereby inducing symptoms of P toxicity, which have been mistakenly identified as symptoms of Zn deficiency. Results with wheat (Webb and Loneragan, 1988), corn and potato (Christensen and Jackson, 1981), okra (Loneragan et al., 1982), and cotton (Cakmak and Marschner, 1986) support the hypothesis that Zn deficiency caused P to accumulate to toxic levels in old leaves.

An excess of P can interfere with the metabolic functions of zinc. Alloway (2004) reported that high levels of phosphorus can also lead to a reduction in vesicular arbuscular mycorrhizal infection, and this could reduce the absorbing area of the roots. In the rhizoplane, Mg, Cu, Fe, and B can decrease zinc availability (Brady and Weil, 2002). Zinc deficiency increases B uptake, and zinc application can relieve boron toxicity by reducing boron uptake (Grewal et al., 1998, 1999). Potassium increases zinc uptake through roots and influences translocation (Grunes et al., 1998). Chloride ions exaggerate the effects of zinc amendments, because the salts increase the availability of the metal (Stevens et al., 2003; Duffy, 2007).

Mandal et al. (2000) reported that high concentrations of Fe^{2+} and Mn^{2+} in the soil solution have an antagonistic effect on Zn absorption. Giordano et al. (1987) have shown the order of interference on Zn absorption and translocation by rice to be as follows: depression of Zn absorption, Fe = Cu > Mg > Mn > Ca; and depression of Zn translocation, Fe = Cu > Ca > Mn.

8.7   MANAGEMENT PRACTICES TO MAXIMIZE Zn USE EFFICIENCY

Use of appropriate method, source, and rate are important management practices to improve zinc use efficiency by crop plants. In addition, the soil test can be an important criterion for zinc fertilizer recommendations. Furthermore, use of zinc-efficient crop species or genotypes within species is another important practice to reduce the cost of crop production and environmental pollution.

8.7.1   APPROPRIATE SOURCE, METHOD, AND RATE OF APPLICATION

Application of zinc fertilizers with high water solubility is the most effective way to correct Zn deficiency (Mortvedt, 1992; Mortvedt and Gilkes, 1993; Gangloff et al., 2000; Slaton et al., 2005). Zinc fertilizer water solubility levels of about 40–50% of the total Zn are needed to meet the Zn requirements for the current crop (Mortvedt,

TABLE 8.11

Principal Zinc Carriers

Source	Formula	Zn (%)
Zinc sulfate (monohydrate)	$ZnSO_4 \cdot H_2O$	36
Zinc sulfate (heptahydrate)	$ZnSO_4 \cdot 7H_2O$	23
Zinc oxide	ZnO	78
Zinc carbonate	$ZnCO_3$	52
Zinc sulfide	ZnS	67
Zinc phosphate	$Zn_3(PO_4)_2$	51
Zinc frits	Silicates	Varies
Zinc EDTA chelates	$Na_2ZnEDTA$	14
Zinc HEDTA chelate	NaZnHEDTA	9
Raplex zinc	ZnPF	10

Source: Adapted from Tisdale et al. (1985); Foth and Ellis (1988); Fageria (1989).

1992; Amrani et al., 1999), and high correlations have been found between Zn fertilizer water solubility and plant growth and Zn uptake (Amrani et al., 1999; Shaver et al., 2007). Plant tissue Zn concentrations have been found to decrease as fertilizer Zn water solubility declines (Slaton et al., 2005a). Shaver et al. (2007) reported that Zn sources with medium (23% of total Zn, 56% water soluble) and low (37.6% total zinc, 6% water soluble) water solubility did not become more available over time, nor did they supply enough Zn for two corn crops, after which plants showed Zn deficiencies.

A good way to overcome Zn deficiencies is to apply a Zn compound broadcast or with row fertilizer. Such application often lasts several years. The most common sources of zinc are presented in Table 8.11. Zinc sulfate and zinc oxide are the most common source of Zn used for supplying Zn to alleviate zinc deficiency. In Oxisol of central Brazil, Zn deficiency is very common in upland rice and corn, and it can be alleviated by soil application of 5 kg Zn ha^{-1} through zinc sulfate (Fageria, 1984, 1989). Hergert et al. (1990) evaluated the effect of five Zn sources (ZnEDTA, Zn-NH$_3$ complex, ZnO, ZnSO$_4$, and Zn(NO$_3$)$_2$) applied at five rates (0, 0.11, 0.33, 1.12, and 3.36 kg Zn ha^{-1}) on the growth and yield of irrigated corn on sandy soils. Regression equations for the responsive sites showed that between 1 and 2 kg Zn ha^{-1} was required to attain maximum yield. Although Zn uptake indicated differences in performance of Zn sources, the differences were not reflected in grain yield. Giordano and Mortvedt (1972) reported that Zn source influenced the downward movement of fertilizer Zn in soil. The movement of fertilizer Zn 4 weeks after application was about 5 mm for ZnO, 20 mm for ZnSO$_4$, and 60 mm for ZnEDTA (ethylenediaminetetraacetic acid). The mobility of fertilizer Zn in soils also decreased as soil pH increases (Mortvedt and Giordano, 1967). This means that when Zn is surface applied to soils with no incorporation (i.e., conservation tillage), some Zn fertilizers sources may be more appropriate than others due to their solubility, mobility, or both (Slaton et al., 2005b).

TABLE 8.12

Adequate and Toxic Levels of Zinc for Annual Crops Applied to Brazilian Oxisols

Crop Species[a]	Adequate Zn Rate (mg kg⁻¹)	Toxic Zn Rate (mg kg⁻¹)
Upland rice	10	70
Dry bean	1	57
Corn	3	110
Soybean	2	59
Wheat	1	40

Note: Adequate rate was calculated at the 90% of the maximum dry weight of shoot, and toxic level was calculated at the 10% reduction in shoot dry weight after achieving maximum weight.

[a] Plants were harvested at 6 weeks after sowing, and soil pH at harvest was about 6.0 in water in all five experiments.

Source: Adapted from Fageria et al. (2003a).

Foliar application has also been used successfully, usually as a temporary or emergency measure. A concentration of about 0.5% $ZnSO_4$ can be sufficient to correct Zn deficiency, if the crop has sufficient leaf area during foliar application. Mengel (1980) suggested that foliar application of 1.4 kg Zn ha⁻¹ as $ZnSO_4$ would improve crop yields. Fenster et al. (1984) recommended a foliar application of 0.56 to 1.1 kg Zn ha⁻¹ as $ZnSO_4$ or 0.17 kg Zn ha⁻¹ as ZnEDTA in 187 L H_2O ha⁻¹ for correcting Zn deficiency. When zinc deficiency is severe, generally more than one application is necessary. Table 8.12 shows adequate and toxic rates of zinc for upland rice, dry bean, corn, soybean, and wheat applied to Brazilian Oxisols. According to these results, upland rice requires maximum zinc, and dry bean and wheat require minimum amounts. Corn tolerated a maximum rate of 110 mg Zn kg⁻¹ for 10% dry weight reduction, and wheat required a minimum rate of 40 mg Zn kg⁻¹ to show toxicity in terms of 10% yield reduction.

8.7.2 Soil Test as a Criterion for Recommendations

The soil test is a reliable criterion for making zinc fertilizer recommendations for field crops. Sims and Johnson (1991) reported that soil test for micronutrients is, in general, a well-established practice, based on fundamental principles of soil chemistry, and verified by field and greenhouse research. These authors reported critical level of Zn in soil as 0.5 to 3.0 mg kg⁻¹ by Mehlich-1 extracting solution and 0.2 to 2.0 mg kg⁻¹ by DTPA extraction method. Viets and Lindsay (1973) reviewed the methods of testing soils for zinc and recommended the use of DTPA as an extractant. The extract consists of 0.005 M DTPA (diethylenetriaminepentaacetic acid), 0.1 M triethanolamine, and 0.01 M $CaCl_2$ with a pH of 7.3 (Foth and Ellis, 1988). Lindsay and Norvell (1978) described the DTPA method in detail, and Martens and Lindsay (1990) compiled critical values of DTPA-extractable zinc for various crops, which varied from 0.5 to 1.4 mg Zn kg⁻¹ of soil. Critical level of Zn reported by

TABLE 8.13

Adequate and Toxic Level of Soil Zinc Extracted by Mehlich 1 and DTPA Extracting Solutions for Five Annual Crops Grown on Brazilian Oxisols

	Mehlich 1 Extracting Solution		DTPA Extracting Solution	
Crop Species[a]	Adequate Zn Level (mg kg^{-1})	Toxic Zn Level (mg kg^{-1})	Adequate Zn Level (mg kg^{-1})	Toxic Zn Level (mg kg^{-1})
Upland rice	5	61	4	35
Dry bean	0.7	25	0.3	25
Corn	2.0	94	1.0	60
Soybean	0.8	53	0.3	33
Wheat	0.5	27	0.3	34

Note: Adequate level was calculated at the 90% of the maximum dry weight of shoot, and toxic level was calculated at the 10% reduction in shoot dry weight after achieving maximum weight.

[a] Plants were harvested at 6 weeks after sowing, and soil pH at harvest was about 6.0 in water in all five experiments.

Source: Adapted from Fageria et al. (2003a).

tropical soil-testing laboratories are DTPA, 0.5 to 1.0 mg kg^{-1}; 0.1 M HCl, 1.0 to 5.0 mg kg^{-1}; modified Olsens, 1.5 to 10 mg kg^{-1}; and KCl, 0.85 mg kg^{-1} (Lindsay and Cox, 1985). Soil properties used to improve the interpretation of Zn soil test results include pH, organic matter, cation exchange capacity (CEC), soil test P, texture, percent clay, and percent CaCO$_3$ (Junus and Cox, 1987; Sims and Johnson, 1991).

Mehlich-1 extracting solution for Zn, Cu, Mn, and Fe analysis in the soils is a standard practice in Brazilian soil analysis laboratories for economic reasons; for P and K analysis, Mehlich-1 extracting solution is used, and micronutrients analysis can be done from the same soil solution. Fageria et al. (2003a) determined adequate and toxic levels of zinc for upland rice, dry bean, corn, soybean, and wheat grown on Brazilian Oxisols (Table 8.13). The DTPA extractable adequate level of Zn varied from 0.3 to 4 mg kg^{-1} depending on crop species. Similarly, the Mehlich-1 extractable adequate Zn level varied from 0.5 to 5 mg kg^{-1} of soil. Rice required maximum Zn for adequate growth and wheat minimum among five crop species. The toxic level of soil-extractable Zn for five crop species was much higher by Mehlich-1 extractant compared to DTPA extractant. Upland rice and corn are highly tolerant to Zn toxicity compared to wheat, dry bean, and soybean.

8.7.3 Use of Efficient Crop Species/Genotypes

Significant variation in zinc uptake and utilization has been reported among crop species and genotypes of the same species (Fageria, 1992; Mengel et al., 2001; Fageria et al., 1997a; Epstein and Bloom, 2005). Use of Zn-efficient cultivars is one of the important strategies to improve yield in deficient soils. Millikan (1961) indicated that differences in nutrient absorption and utilization were sometimes greater among varieties within species than between related species or genera. Massey and Loeffel (1966) found that concentrations in grain of 29 corn inbreeds ranged from 15.5 mg kg^{-1} to 38.4 mg kg^{-1}. Similarly, Hinesly et al. (1978) reported that concentrations of Zn in leaves of 20 corn inbreeds ranged from 61.8 mg kg^{-1} to 281.8 mg kg^{-1}. Clark (1978) reported that corn inbreeds differ in their ability to take up and use Zn, and these differences could be important for development and adaptation of plants to low-Zn conditions.

The differential responses for Zn appeared to be caused by differences in the inbreeds for translocation, requirements, and utilization of Zn and for accumulation of imbalance amounts of mineral elements that interact with Zn (Clark, 1978). Shukla and Raj (1976) evaluated eight genotypes of corn in the greenhouse using a Zn-deficient loamy sand. There was a considerable variation among the eight genotypes in the severity of Zn deficiency symptoms, growth depression, Zn concentration, P/Zn, and Fe/Zn ratios under Zn stress conditions. Zinc concentration in different genotypes under Zn stress condition varied from 7.4 to 20.5 mg kg^{-1} of dry matter. These authors reported that the differential response among the genotypes was found to be associated with their capability to exploit soil Zn and/or translocate it to the shoot.

The contribution of P/Zn and/or Fe/Zn balance in tissues to the variability in Zn responses was not significant. Safaya and Gupta (1979) reported that corn cultivars differ to Zn deficiency tested in Zn-deficient loamy sand soil of pH 8.5 and that susceptibility to Zn deficiency in corn is intimately related to the P and Fe absorption and transport mechanisms. Peaslee et al. (1981) reported that accumulation and translocation of zinc by two corn cultivars were not different, but one produced more dry matter than the other per unit of Zn concentration in shoot tissue and thereby appeared to be more efficient when Zn supply was low. Fageria and Baligar (2005) and Fageria (2001b) reported significant difference in upland and lowland rice genotypes in zinc uptake and utilization.

Alloway (2004) reported that in wheat, corn, sorghum, rice, and oats, variations in reaction to zinc deficiency might be related to different capacities of the plant species and cultivars to release zinc-mobilizing phytosiderophores (phytometallophores) from roots. Cakmak et al. (1994) demonstrated that phytosiderophores released from roots under conditions of zinc deficiency enhanced the solubility and mobility of zinc by chelating from sparingly soluble zinc compounds in calcareous soils. Kirk and Bajita (1995) reported that root-induced changes in the rhizosphere solubilize zinc and increase its uptake by rice. The solubilization involves either root-induced acidification from H$^+$ generated from the oxidation of Fe^{2+} or the intake of an excess of cations, such as NH$_4^+$, relative to anions and the concomitant release of H$^+$ from the roots. Riley and Barber (1971) reported that the rhizosphere pH could vary by as much as 2 pH units from that of the bulk soil. The ability of plant roots to modify the rhizosphere pH can vary between plant species and cultivars of the same species.

Cianzio (1999) reported that inheritance of Zn utilization suggested that there might be at least three genes involved in the expression of zinc efficiency. Breeding for greater Zn efficiency, at least in wheat, should be possible by transferring rye genes controlling the Zn efficiency trait (Rengel, 1999). Individual genes have not been deciphered yet, but transferring rye chromosomes 1R and 7R into wheat has increased its Zn efficiency (Cakmak et al., 1997b).

Land races of many grains are often zinc efficient, but this trait is lost during breeding for commercial traits, which presents an opportunity for plant gene therapy (Duffy, 2007). Indeed, genetic engineering has been successful in creating plants with the ability to hyperaccumulate zinc (Van der Zaal et al., 1999). Rye chromosomes involved in improved efficiency have been transferred into wheat and triticale to confer zinc efficiency (Cakmak et al., 1997a).

8.7.4 SYMBIOSIS WITH MYCORRHIZAE AND OTHER MICROFLORA

Micorrhizal fungi association with plant roots was first described by German scientist A. B. Frank in 1885, and he called this association "fungus roots" (Mengel et al., 2001). Mycorrhizae fungi live in symbiosis with higher plants. Plants supply carbohydrates to the fungi, and in return fungi exploit the soil volume by expanding hyphae, providing the root system with additional nutrient and water. This plant–fungal association is very old and mainly linked to plant adaptability to unfavorable environmental conditions (Mengel et al., 2001). Smith and Read (1997) and Harrier and Watson (2003) reported that over 80% of all terrestrial plant species form this type of relationship. Plant root association with mycorrhizae can improve zinc uptake by crop plants. Mycorrhizae not only improve uptake of trace elements but also protect plants from excessive uptake and avoid toxicity of these elements (Brady and Weil, 2002).

Vesicular-arbuscular mycorrhiza (VAM) has been shown to improve crop productivity of numerous crop plants in soils of low fertility (Plenchette et al., 1983). This response is usually attributed to enhanced uptake of immobile nutrients like P, Zn, and Cu (Faber et al., 1990; Fageria and Baligar, 1997b). The VAM may also improve drought resistance in crop plants (Faber et al., 1990; Fageria, 2002b). Marschner and Dell (1994) reported that Zn uptake was increased in plant roots infected with mycorrhizae compared to nonmycorrhizal roots. Mycorrhizae can increase plant capacity to take up Zn from soils low in available Zn (Rengel, 1999). The mycorrhizal roots of green gram (*Vigna radiata*) (Sharma and Srivastava, 1991), corn (Faber et al., 1990; Sharma et al., 1992), and pigeon pea (*Cjanus cajan*) (Wellings et al., 1991) have higher tissue Zn concentrations than those of nonmycorrhizal plants. Mycorrhizal associations increase water absorption and likewise increase mass flow of zinc to host plants (Al-Karaki and Hammad, 2001; Scagel, 2004). Mycorrhizae also increase zinc uptake by extending the area of soil that can be mined by the host plant, through modification of root exudates that decrease pH, and by the action of their mobilizing polyphosphates, which bind zinc and other minerals (Mehravaran et al., 2000).

In addition to mycorrhizae, other microbial population in the rhizosphere can improve Zn use efficiency in crop plants (Rengel et al., 1996, 1998; Rengel, 1997, 1999). In Zn-efficient wheat genotypes, number of efflorescent pseudomonads increased in the rhizosphere under zinc deficiency conditions (Rengel et al., 1996; Rengel, 1997). Plants and microorganisms employ metal chelators with huge affinity for zinc to improve uptake and compete with other organisms (Duffy, 2007).

8.7.5 Other Management Practices

Other management practices like using organic manures, using conservation tillage systems, improving water use efficacy, and adopting appropriate crop rotation can be useful in maximizing zinc use efficiency by field crops. These practices were discussed in the earlier chapters.

8.8 CONCLUSIONS

Among micronutrient deficiencies, zinc deficiency is the most common and widespread in crop plants. This is mainly due to intensive cropping systems; use of modern high-yielding cultivars; loss of topsoil organic matter content by erosion; use of high rates of N, P, and K fertilizers, especially in developed countries; liming of acid soils; and use of inadequate rates in most cropping systems. Zinc is an essential micronutrient for plant growth and is an important catalytic component of several enzymes. Zinc is closely involved in the N metabolism of plants, and in Zn-deficient plants, protein synthesis is significantly reduced. Zinc deficiency reduces root system development in crop plants. This may adversely affect absorption of water and nutrients and consequently growth and yield. Zinc has low mobility in plant tissues; hence, its deficiency mostly appears in the younger leaves. Zinc solubility in soils is highly pH dependent and decreases with increasing soil pH.

Overall, adequate zinc concentration in plant tissues is in the range of about 20 to 70 mg kg^{-1} dry weight and varied from crop species to crop species and among genotypes within species. Similarly, toxic level of zinc in plant tissues is higher than 100 mg kg^{-1} shoot dry weight depending on crop species or genotypes within species. For example, in upland rice tissue, toxic level of zinc is more than 600 mg kg^{-1}. Zinc concentration in shoot is generally higher than grain. Zinc concentration in grain was higher in legumes compared to cereals. However, Zn use efficiency in grain was higher in cereals compared to legumes. Higher Zn content in grain is a desirable trait for seedling vigor and human health. Zinc availability to crop plants is mainly by soil minerals or organic matter and by added chemical fertilizers. Overall, zinc recovery efficiency varied from 6 to 13% depending on rate applied and crop genotypes. Zinc has negative interactions with P, Cu, Fe, and Mn in absorption by plants.

The soil test is an important criterion to identify zinc deficiency for a given crop. Adequate soil zinc content may depend on plant species and zinc extractant used. However, 1 to 5 mg Zn kg^{-1} of soil is generally sufficient for maximum economic yields for most soils or field crops. Zinc deficiency in crop plants can be corrected by addition of 5 to 10 kg Zn ha^{-1} depending on crop species. Residual effect of zinc fertilization remains for about 3 to 5 years depending on crop intensity and soil types. Zinc deficiency can also be corrected by foliar application of 0.5% solution of soluble zinc fertilizers such as zinc sulfate. Zinc chelates are especially suitable for foliar sprays but are expensive and beyond the reach of ordinary farmers due to higher cost. However, zinc deficiency in crop plants is generally observed in the early plant growth stage, which may limit its foliar application due to low leaf area. Furthermore, in humid areas, frequent rainfall may also be a limiting factor for foliar application of this element. Hence, soil application of zinc fertilizers should be encouraged due to longer residual effects and low rate required. Zinc fertilizers should be mixed with N, P, and K fertilizers and should be applied in band to maximize its use efficiency. Seed coating with zinc solution and deepening roots in the zinc solution are other methods of zinc application. The return of plant residues to the soil is an important strategy for recycling zinc in soil–plant systems. Significant crop species or genotype differences exist in zinc uptake and utilization. Root-induced chemical changes in favor of higher zinc uptake may be the main cause of these genotypical differences. Hence, planting zinc-efficient crop species or genotypes within species is an attractive strategy for maximizing zinc use efficiency in crop plants.

REFERENCES

Al-Karaki, G. N. and R. Hammad. 2001. Mycorrhizal influence on fruit yield and mineral content of tomato grown under salt *stress*. *J. Plant Nutr.* 24:1311–1323.

Alloway, B. J. 2004. *Zinc in soils and crop nutrition*. Brussels, Belgium: International Zinc Association.

Amrani, M., D. G. Westfall, and G. A. Peterson. 1999. Influence of water solubility of granular zinc fertilizers on plant uptake and growth. *J. Plant Nutr.* 22:1815–1827.

Badger, M. and D. G. Price. 1989. The role of carbonic anhydrase in photosynthesis. *Annu. Rev. Plant Physiol. Plant Molecular Biol.* 45:369–392.

Barber, S. A. 1995. *Soil nutrient bioavailability: a mechanistic approach*. New York: John Wiley & Sons.

Barrow, N. J. 1986. Testing a mechanistic model: IV. Describing the effects of pH on zinc retention by soils. *J. Soil Sci.* 37:295–302.

Bloom, P. R. and M. B. McBride. 1979. Metal in binding and exchange with hydrogen ions in acid-washed peat. *Soil Sci. Soc. Am. J.* 43:687–692.

Bowen, J. E. 1969. Absorption of copper, zinc and manganese by sugarcane tissue. *Plant Physiol.* 44:255–261.

Brady, N. C. and R. R. Weil. 2002. *The nature and properties of soils*, 13th edition. Upper Saddle River, NJ: Prentice-Hall.

Brown, P. H., I. Cakmak, and Q. Zhang. 1993. Form and function of zinc in plants. In: *Zinc in soils and plants*, A. D. Robson, Ed., 90–106. Dordrecht: Kluwer Academic Publishers.

Cakmak, I. and C. Engels. 1999. Role of mineral nutrients in photosynthesis and yield formation. In: *Mineral nutrition of crops: Fundamental mechanisms and implications*, Z. Rengel, Ed., 141–168. New York: The Haworth Press.

Cakmak, I., K. Gulut, H. Marschner, and R. D. Graham. 1994. Effect of zinc and iron deficiency on phytosiderophore release in wheat genotypes differing in zinc efficiency. *J. Plant Nutr.* 17:1–17.

Cakmak, I., R. Derici, B. Torun, I. Tolay, H. J. Braun, and R. Schlegel. 1997a. Role of rye chromosomes in improvement of zinc efficiency in wheat and triticale. *Plant Soil* 196:249–253.

Cakmak, I., L. Ozturk, S. Eker, B. Torun, H. I. Kalfa, and A. Yilmaz. 1997b. Concentration of zinc and activity of copper/zinc superoxide dismutase in leaves of rye and wheat cultivars differing in sensitivity to zinc deficiency. *J. Plant Physiol.* 151:91–95.

Cakmak, I., B. Torun, B. Erenoglu, L. Ozturk, H. Marschner, M. Kalayci, H. Ekiz, and A. Yilmaz. 1998. Morphological and physiological differences in the response of cereals to zinc deficiency. *Euphytica* 100:349–357.

Cakmak, I., A. Yilmaz, M. Kalayci, H. Ekiz, B. Touun, B. Erenoglu, and H. J. Braaun. 1996. Zinc deficiency as a critical problem in wheat production in central Anatolia. *Plant Soil* 180:165–172.

Cakmak, I. and H. Marschner. 1986. Mechanism of phosphorus induced zinc deficiency in cotton. I Zinc deficiency enhanced uptake rate of phosphorus. *Physiol. Plant.* 68:483–490.

Catlett, K. M., D. M. Heil, W. L. Lindsay, and M. H. Ebinger. 2002. Soil chemical properties controlling zinc activity in 18 Colorado soils. *Soil Sci. Soc. Am. J.* 66:1182–1189.

Cavallaro, N. and M. B. McBride. 1984. Zinc and copper sorption and fixation by an acid soil clay: Effects of selective dissolutions. *Soil Sci. Soc. Am. J.* 48:1050–1054.

Chairidchai, P. and G. S. P. Ritchie. 1990. Zinc and sorption by a Lateritic soil in the presence of organic ligands. *Soil Sci. Soc. Am. J.* 54:1242–1248.

Chapman, H. D. 1959. The diagnosis and control of Zn deficiency and excess. *Israel J. Bot.* 8:105–130.

Christensen, N. W. and T. L. Jackson. 1981. Potential for phosphorus toxicity in zinc stressed corn and potato. *Soil Sci. Soc. Am. J.* 45:904–909.

Cianzio, S. R. 1999. Breeding crops for improved nutrient efficiency: Soybean and wheat as case studies. In: *Mineral nutrition of crops: Fundamental mechanisms and implications*, Z. Rengel, Ed., 267–287. New York: The Haworth Press.

Clark, R. B. 1978. Differential response of maize inbreeds to Zn. *Agron. J.* 70:1056–1060.

Clark, R. B. 1982. Plant response to mineral element toxicity and deficiency. In: *Breeding plants for the less favorable environments*, M. N. Christiansen and C. F. Lewis, Eds., 71–142. New York: John Wiley & Sons.

Clarkson, D. T. and J. B. Hanson. 1980. The mineral nutrition of higher plants. *Annu. Rev. Plant Physiol.* 31:239–298.

Cleveland, J. M. and T. F. Rees. 1981. Characterization of plutonium in Maxey Flats radioactive trench leachates. *Science* 212:1506–1509.

De Datta, S. K. 1981. *Principles and practices of rice production.* New York: Wiley.

Duffy, B. 2007. Zinc and plant disease. In: *Mineral nutrition and plant disease*, L. E. Datnoff, W. H. Elmer, and D. M. Huber, Eds., 155–175. St. Paul, MN: The American Phytopathological Society.

El-Kherbawy, M. I. and J. R. Sanders. 1984. Effect of pH and phosphate status of a silty clay loam on manganese, zinc, and copper concentrations in soil fractions and in clover. *J. Sci. Food Agric.* 35:733–739.

Elrashidi, M. A. and G. A. O'Connor. 1982a. Influence of solution composition on sorption of zinc by soils. *Soil Sci. Soc. Am. J.* 46:1153–1158.

Epstein, E. and A. J. Bloom. 2005. *Mineral nutrition of plants: Principles and perspectives,* 2nd edition. Sunderland, MA: Sinauer associates.

Faber, B. A., R. J. Zaboski, R. G. Burau, and K. Uriu. 1990. Zinc uptake by corn as affected by vesicular arbuscular mycorrhizae. *Plant Soil* 129:121–130.

Fageria, N. K. 1984. *Fertilization and mineral nutrition of rice.* EMBRAPA/Editora Campus, Rio de Janeiro.

Fageria, N. K. 1989. Effects of phosphorus on growth, yield and nutrient accumulation in the common bean. *Tropical Agriculture* 66:249–255.

Fageria, N. K. 1991. Response of cowpea to phosphorus on an Oxisol with special reference to dry matter production and mineral ion contents. *Tropical Agriculture* 68:384–388.

Fageria, N. K. 1992. *Maximizing crop yields.* New York: Marcel Dekker.

Fageria, N. K. 2000a. Adequate and toxic levels of zinc for rice, common bean, corn, soybean and wheat production in cerrado soil. *Rev. Bras. Eng. Agri. Ambien.* 4:390–395.

Fageria, N. K. 2000b. Upland rice response to soil acidity in cerrado soil. *Pesq. Agropec. Bras.* 35:2303–2307.

Fageria, N. K. 2001a. Nutrient management for upland rice production and sustainability. *Commun. Soil Sci. Plant Analy.* 32:2603–2629.

Fageria, N. K. 2001b. Response of upland rice, dry bean, corn, and soybean to base saturation in cerrado soil. *Rev. Bras. Eng. Agri. Ambien.* 5:416–424.

Fageria, N. K. 2002a. Influence of micronutrients on dry matter yield and interaction with other nutrients in annual crops. *Pesq. Agropec. Bras.* 37:1765–1772.

Fageria, N. K. 2002b. Nutrient management for sustainable dry bean production in the tropics. *Commun. Soil Sci. Plant Analy.* 33:1537–1575.

Fageria, N. K. and V. C. Baligar. 1997a. Response of common bean, upland rice, corn, wheat, and soybean to soil fertility of an Oxisol. *J. Plant Nutr.* 20:1279–1289.

Fageria, N. K. and V. C. Baligar. 1997. Integrated plant nutrient management for sustainable crop production. An overview. *Int. J. Tropical Agric.* 15:1–19.

Fageria, N. K. and V. C. Baligar. 2005. Growth components and zinc recovery efficiency of upland rice genotypes. *Pesq. Agropec. Bras.* 40:1211–1215.

Fageria, N. K., V. C. Baligar, and R. B. Clark. 2002. Micronutrients in crop production. *Adv. Agron.* 77:185–268.

Fageria, N. K., V. C. Baligar, and C. A. Jones. 1997a. *Growth and mineral nutrition of field crops,* 2nd edition. New York: Marcel Dekker.

Fageria, N. K. and M. P. Barbosa Filho. 2006. *Identification and correction nutrient deficiencies in rice.* Embrapa Arroz e Feijão Circular number 75, 7 pp., Santo Antonio de Goias, Brazil.

Fageria, N. K., M. P. Barbosa Filho, and L. F. Stone. 2003a. *Adequate and toxic levels of micronutrients in soil and plants for annual crops.* Paper presented at the XXIX Brazilian Soil Science Congress, July 13–18, 2003, Ribeirão Preto, São Paulo, Brazil.

Fageria, N. K. and F. Breseghello. 2004. Nutritional diagnostic in upland rice production in some municipalities of State of Mato Grosso. *J. Plant Nutr.* 27:15–28.

Fageria, N. K. and H. R. Gheyi. 1999. *Efficient crop production.* Campina Grande, Brazil: Federal University of Paraiba.

Fageria, N. K., A. B. Santos, I. D. G. Lins, and S. L. Camargo. 1997b. Characterization of fertility and particle size of varzea soils of Mato Grosso and Mato Grosso do Sul States of Brazil. *Commun. Soil Sci. Plant Analy.* 28:37–47.

Fageria, N. K., N. A. Slaton, and V. C. Baligar. 2003b. Nutrient management for improving lowland rice productivity and sustainability. *Adv. Agron.* 80:63–152.

Fenster, W. E., G. W. Rehm, and J. Grava. 1984. Zinc for Minnesota soils. Minnesota Agric. Ext. Serv. Ag-FS-O720.

Foth, H. D. and B. G. Ellis. 1988. *Soil fertility.* New York: John Wiley & Sons.

Friesen, D. K., A. S. R. Juo, and M. H. Miller. 1980. Liming and lime-phosphate-zinc interactions on two Nigerian Ultisols. I. Interactions in the soil. *Soil Sci. Soc. Am. J.* 44:1221–1226.

Frossard, E. M., M. Bucher, F. Machler, A. Mozafar, and R. Hurrell. 2000. Potential for increasing the content and bioavailability of Fe, Zn, and Ca in plants for human nutrition. *J. Sci. Food Agric.* 80:861–879.

Gangloff, W. J., D. G. Westfall, G. A. Peterson, and J. J. Mortvedt. 2000. Availability of organic and inorganic Zn fertilizers. *Colorado State University Agriculture Experiment Station Technical Bulletin* TB00-1.

Gao, X., C. Zou, F. Zhang, S. E. A. T. M. Zee, and E. Hoffland. 2005. Tolerance to zinc deficiency in rice correlates with zinc uptake and translocation. *Plant Soil* 278:253–261.

Gao, X., C. Zou, X. Fan, F. Zhang, and E. Hoffland. 2006. From flooded to aerobic conditions in rice cultivation: Consequences for zinc uptake. *Plant Soil* 280:41–47.

Giordano, P. M. and J. J. Mortvedt. 1972. Rice response to Zn in flooded and nonflooded soil. *Agron. J.* 64:521–524.

Giordano, P. M., J. C. Noggle, and J. J. Mortvedt. 1974. Zinc uptake by rice as affected by metabolic inhibitors and competing cations. *Plant Soil* 41:637–646.

Graham, R. D., J. S. Ascher, and S. C. Hynes. 1992. Selecting zinc efficient cereal genotypes for soils of low zinc status. *Plant Soil* 146:241–250.

Graham, R. D. and Z. Rengel. 1993. Genotypic variation in zinc uptake and utilization by plants. In: *Zinc in soils and plants*, A. D. Robson, Ed., 107–118. Dordrecht: Kluwer Academic Publishers.

Graham, R. D., R. M. Welch, D. L. Grunes, E. E. Cary, and W. A. Norvell. 1987. Effect of zinc deficiency on the accumulation of boron and other mineral nutrients in barley. *Soil Sci. Soc. Am. J.* 51:652–657.

Grewal, H. S. 2001. Zinc influences nodulation, disease severity, leaf drop and herbage yield of alfalfa cultivars. *Plant Soil* 234:47–59.

Grewal, H. S. and R. D. Graham. 1999. Residual effects of subsoil zinc and oilseed rape genotype on the grain yield and distribution of zinc in wheat. *Plant Soil* 207:29–36.

Grewal, H. S., R. D. Graham, and J. Stangoulis. 1998. Zinc-boron interaction effects in oilseed rape. *J. Plant Nutr.* 21:2231–2243.

Grunes, A., M. Alpaslan, and A. Inal. 1998. Critical nutrient concentrations and antagonistic and synergistic relationships among the nutrients of NFT-grown young tomato plants. *J. Plant Nutr.* 21:2035–2047.

Grunes, A., M. Alpaslan, Y. Cikili, and H. Ozcan. 1999. Effect of zinc on the alleviation of boron toxicity in tomato. *J. Plant Nutr.* 22:1061–1068.

Harrier, L. A. and C. A. Watson. 2003. The role of arbuscular mycorrhizal fungi in sustainable cropping systems. *Adv. Agron.* 20:185–225.

Hatch, M. D. and J. N. Burnell. 1990. Carbonic anhydrase activity in leaves and its role in the first step of C_4 photosynthesis. *Plant Physiol.* 93:825–828.

Hergert, G. W., G. W. Rehm, and R. A. Wiese. 1990. Field evaluations of zinc sources band applied in ammonium polyphosphate suspension. *Soil Sci. Soc. Am. J.* 48:1190–1193.

Hinesly, T. D., D. E. Alexander, E. L. Ziegler, and G. L. Barrett. 1978. Zinc and Cd accumulation by corn inbreeds grown on sludge amended soil. *Agron. J.* 70:425–428.

Hodgson, J. F., W. L. Lindsay, and J. F. Trierweiler. 1966. Micronutrient cation complexing in soil solution: II. Complexing of zinc and copper in displaced solution from calcareous soils. *Soil Sci. Soc. Am. Proc.* 30:723–725.

Iyengar, S. S., D. C. Martens, and W. P. Miller. 1981. Distribution and plant availability of soil zinc fractions. *Soil Sci. Soc. Am. J.* 45:735–739.

Jenne, E. A. 1968. Controls on Mn, Fe, Co, Ni, Cu, and Zn concentrations in soils and water: The significant role of hydrous Mn and Fe oxides. In: *Trace inorganic in water*, R. A. Baker, Ed., 337–387. Washington, DC: Adv. Chem. Ser. No. 73. ACS.

Jones, J. B., Jr. 1991. Plant tissue analysis in micronutrients. In: *Micronutrient in agriculture*, 2nd edition, J. J. Mortvedt, F. R. Cox, L. M. Shuman, and R. L. Welch, Eds., 477–521. Madison, WI: SSSA.

Junus, M. A. and F. R. Cox. 1987. A zinc soil test calibration based upon Mehlich 3 extractable zinc, pH and cation exchange capacity. *Soil Sci. Soc. Am. J.* 51:678–683.

Karlen, D. L., R. L. Flannery, and E. J. Sadler. 1988. Aerial accumulation and partitioning of nutrients by corn. *Agron. J.* 80:232–242.

Kirk, G. J. D. and J. B. Bajita. 1995. Root-induced iron oxidation, pH changes and zinc solubilisation in the rhizosphere of lowland rice. *New Phytologist* 131:129–137.

Klug, A. and D. Rhodes. 1987. Zinc fingers a novel protein motif for nucleic acid recognition. *Trends Biochem. Sci.* 12:464–469.

Kochain, L. V. 1991. Mechanisms of micronutrient uptake and translocation in plants. In: *Micronutrient in agriculture*, J. J. Mortvedt, F. R. Cox, L. M. Shuman, and R. M. Welch, Eds., 229–296. Madison, WI: Soil Science Society of America.

Krauskopf, B. 1972. Geochemistry of micronutrients. In: *Micronutrients in agriculture*, J. J. Mortvedt, Ed., 7–40. Madison, WI: SSSA.

Kurdi, F. and H. E. Doner. 1983. Zinc and copper sorption and interaction in soils. *Soil Sci. Soc. Am. J.* 47:872–876.

Lake, D. L., P. W. W. Kirk, and J. N. Lester. 1984. Fractionation, characterization, and speciation of heavy metals in sewage sludge and sludge-amended soils. A review. *J. Environ. Quality* 13:175–183.

Lindsay, W. L. 1979. *Chemical equilibrium in soils.* New York: John Wiley & Sons.

Lindsay, W. L. and F. R. Cox. 1985. Micronutrient soil test for the tropics. *Fert. Res.* 7:169–200.

Lindsay, W. L. and W. A. Norvell. 1969. Equilibrium relationships of Zn^{2+}, Fe^{3+}, Ca^{2+}, and H^+ with EDTA and DTPA in soils. *Soil Sci. Soc. Am. Proc.* 33:62–68.

Lindsay, W. L. and W. A. Norvell. 1978. Development of a DTPA soil test for zinc, iron, manganese and copper. *Soil Sci. Soc. Am. J.* 42:421–428.

Loneragan, J. F., T. S. Grive, A. D. Robson, and K. Snowball. 1979. Phosphorus toxicity as a factor in zinc-phosphorus interactions in plants. *Soil Sci. Soc. Am. J.* 43:966–972.

Loneragan, J. F., D. L. Grunes, R. M. Welch, E. A. Aduayi, A. Tengah, V. A. Lazar, and E. E. Cary. 1982. Phosphorus accumulation and toxicity in relation to zinc supply. *Soil Sci. Soc. Am. J.* 46:345–352.

Mandal, B., G. C. Hazra, and L. N. Mandal. 2000. Soil management influence on zinc desorption for rice and maize nutrition. *Soil Sci. Soc. Am. J.* 64:1699–1705.

Mandal, B., G. C. Hazra, and A. K. Pal. 1988. Transformation of zinc in soils under submerged conditions and its relation with zinc nutrition of rice. *Plant Soil* 106:121–126.

Marschner, H. 1983. General introduction to the mineral nutrition of plants. In: *Encyclopedia of plant physiology*, New Series, Vol. 15A, A. Lauchli and R. L. Bieeleski, Eds., 5–60. New York: Springer-Verlag.

Marschner, H. and I. Cakmak. 1989. High light intensity enhances chlorosis and necrosis in leaves of zinc, potassium, and magnesium deficient bean (*Phaseolus vulgaris*) plants. *J. Plant Physiol.* 134:308–315.

Marschner, H. and B. Dell. 1994. Nutrient uptake in mycorrhizal symbiosis. *Plant Soil* 159:89–102.

Martens, D. C. and W. L. Lindsay. 1990. Testing soils for copper, iron, manganese and zinc. In: *Soil testing and plant analysis*, 3rd edition, R. L. Westerman, Ed., 229–264. Madison, WI: SSSA.

Massey, H. F. and F. A. Loeffel. 1966. Variation of zinc content of grain from inbred lines of corn. *Agron. J.* 58:143–144.

Mengel, D. B. 1980. Role of micronutrients in efficient crop production. Indiana Coop. Ext. Serv. Ay-239.

Mengel, K., E. A. Kirkby, H. Kosegarten, and T. Appel. 2001. *Principles of plant nutrition*, 5th edition. Dordrecht: Kluwer Academic Publishers.

Mehravaran, H., A. Mozafar, and E. Frossard. 2000. Uptake and partitioning of P^{32} and Zn^{65} by white clover as affected by eleven isolates of mycorrhizal fungi. *J. Plant Nutr.* 23:1385–1395.

Millikan, C. R. 1961. Plant varieties and species in relation to the occurrence of deficiencies and excesses of certain nutrient elements. *Aust. Inst. Agrc. Sci.* 27:220–233.

Moore, P. A. and W. H. Patrick, Jr. 1988. Effect of zinc deficiency on alcohol dehydrogenase activity and nutrient uptake in rice. *Agron. J.* 80:882–885.

Mortvedt, J. J. 1992. Crop response to level of water soluble zinc in granular zinc fertilizers. *Fert. Res.* 33:249–255.

Mortvedt, J. J. 1994. Needs for controlled availability micronutrient fertilizers. *Fert. Res.* 38:213–221.

Mortvedt, J. J. and R. J. Gilkes. 1993. Zinc fertilizers. In: *Zinc in soils and plants*, A. D. Robson, Ed., 33–42. Dordrecht: Kluwer Academic Publishers.

Mortvedt, J. J. and P. M. Giordano. 1967. Zinc movement in soil from fertilizer granules. *Soil Sci. Soc. Am. Proc.* 31:608–613.

Norman, R. J., C. E. Wilson, Jr., and N. A. Slaton. 2003. Soil fertilization and mineral nutrition in U.S. mechanized rice culture. In: *Rice, origin, history, technology, and production*, C. W. Smith and R. H. Dilday, Eds., 331–412. Hoboken, NJ: John Wiley & Sons.

Ohki, K. 1984. Zinc nutrition related to critical deficiency and toxicity levels for sorghum. *Agron. J.* 76:253–256.

Olsen, S. R. 1972. Micronutrients interactions. In: *Micronutrients in agriculture*, J. J. Mortvedt, Ed., 243–264. Madison, WI: SSSA.

Peaslee, D. E., R. Isalangkura, and J. E. Legget. 1981. Accumulation and transformation of zinc by two corn cultivars. *Agron. J.* 73:729–732.

Plenchette, C., J. A. Fortin, and V. Furlan. 1983. Growth responses of several plant species to mycorrhizae in a soil of moderate P-fertility. I. Mycorrhizal dependency under field conditions. *Plant Soil* 70:199–209.

Ponnamperuma, F. N. 1972. The chemistry of submerged soils. *Adv. Agron.* 24:29–96.

Ponnamperuma, F. N. 1977. Physiochemical properties of submerged soils in relation to fertility. Int. Rice Res. Paper series No.8, Los Banos, Philippines.

Rashid, A. and R. L. Fox. 1992. Evaluating internal zinc requirements of grain crops by seed analysis. *Agron. J.* 84:469–474.

Rengel, Z. 1997. Root exudation and microflora populations in rhizosphere of crop genotypes differing in tolerance to micronutrient deficiency. *Plant Soil* 196:255–260.

Rengel, Z. 1999. Physiological mechanisms underlying differential nutrient efficiency of crop genotypes. In: *Mineral nutrition of crops: Fundamental mechanisms and implications*. Z. Rengel, Ed., 227–265. New York: The Haworth Press.

Rengel, Z., R. Guterridge, P. Hirsch, and D. Hornby. 1996. Plant genotype, micronutrient fertilization and take-all infection influence bacterial populations in the rhizosphere of wheat. *Plant Soil* 183:269–277.

Rengel, Z., V. Romheld, and H. Marschner. 1998. Uptake of zinc and iron by wheat genotypes differing in zinc efficiency. *J. Plant Physiol.* 152:433–438.

Riley, D. and S. A. Barber. 1971. Effect of ammonium nitrate fertilization on phosphorus uptake as related to induced pH changes at the root-soil interface. *Soil Sci. Soc. Am. Proc.* 35:301–306.

Romheld, V. and H. Marschner. 1991. Function of micronutrients in plants. In: *Micronutrient in agriculture*, J. J. Mortvedt, F. R. Cox, L. M. Shuman, and R. M. Welch, Eds., 297–328. Madison, WI: SSSA.

Saeed, M. and R. L. Fox. 1979. Influence of phosphate fertilization on zinc adsorption by tropical soils. *Soil Sci. Soc. Am. J.* 43:683–686.

Safaya, N. M. and A. P. Gupta. 1979. Differential susceptibility of corn cultivars to zinc deficiency. *Agron. J.* 71:132–136.

Scagel, C. F. 2004. Inoculation with vesicular-arbuscular mycorrhizal fungi and rhizobacteria alters nutrient allocation and flowering of harlequin flower. *HortTechnology* 14:39–48.

Sedberry, J. E., F. J. Peterson, E. Wilson, A. L. Nugent, R. M. Engler, and R. H. Brupbacher. 1971. Effects of zinc and other elements on the yield of rice and nutrient content of rice plants. Louisiana State Univ. Agric. Exp. Stn. Bull 653.

Sharma, A. K. and P. C. Srivastava. 1991. Effect of vesicular-arbuscular mycorrhizae and zinc application on dry matter and zinc uptake of green gram (*Vigna radiata* L. Wilczek). *Biology and Fertility of Soils* 11:52–56.

Sharma, A. K., P. C. Srivastava, N. Johri, and V. S. Rathore. 1992. Kinetics of zinc uptake by mycorrhizal (VAM) and non-mycorrhizal corn (*Zea mays* L.) roots. *Biol. Fert. Soils* 13:206–210.

Shaver, T. M., D. G. Westfall, and M. Ronaghi. 2007. Zinc fertilizer solubility and its effects on zinc bioavailability over time. *J. Plant Nutr.* 30:123–133.

Slaton, N. A., E. E. Gbur, C. E. Wilson, and R. J. Norman. 2005a. Rice response to granular zinc sources varying in water-soluble zinc. *Soil Sci. Soc. Am. J.* 69:443–452.

Slaton, N. A., R. J. Norman, and C. E. Wilson, Jr. 2005b. Effect of zinc source and application time on zinc uptake and grain yield of flooded-irrigated rice. *Agron. J.* 97:272–278.

Slaton, N. A., C. E. Wilson, Jr., R. J. Norman, and E. E. Gbur, Jr. 2002. Development of a critical Mehlich 3 soil zinc concentration for rice in Arkansas. *Commun. Soil Sci. Plant Anal.* 33:2759–2770.

Shukla, U. C. and H. Raj. 1976. Zinc response in corn as influenced by genetic variability. *Agron. J.* 68:20–22.

Shuman, L. M. 1975. The effect of soil properties on zinc adsorption by soils. *Soil Sci. Soc. Am. Proc.* 39:454–458.

Shuman, L. M. 1991. Chemical forms of micronutrients in soils. In: *Micronutrients in agriculture*, 2nd edition, J. J. Mortvedt, F. R. Cox, L. M. Shuman, and R. L. Welch, Eds., 113–144. Madison, WI: SSSA.

Sims, J. T. 1986. Soil pH effects on the distribution and plant availability of manganese, copper, and zinc. *Soil Sci. Soc. Am. J.* 50:367–373.

Sims, J. T. and G. V. Johnson. 1991. Micronutrient soil tests. In: *Micronutrient in agriculture*, 2nd edition, J. J. Mortvedt, F. R. Cox, L. M. Shuman, and R. L. Welch, Eds., 427–472. Madison, WI: SSSA.

Singh, J. P., D. J. Dahiya, and R. P. Narwal. 1990. Boron uptake and toxicity in wheat in relation to zinc supply. *Fert. Res.* 24:105–110.

Sinha, P., R. Jain, and C. Chatterjee. 2000. Interactive effect of boron and zinc on growth and metabolism of mustard. *Commun. Soil Sci. Plant Anal.* 31:41–49.

Smith, S. E. and D. J. Read. 1997. *Mycorrhizal symbiosis*, 2nd edition. New York: Academic Press.

Smith, A. M. and T. Rees. 1979. Pathways of carbohydrate fermentation in the roots of marsh plants. *Planta* 146:327–334.

Smith, D. H., M. A. Wells, D. M. Porter, and F. R. Coc. 1993. Peanuts. In: *Nutrient deficiencies and toxicities in crop plants*, W. F. Bennett, Ed., 105–110. St. Paul, MN: The American Phytopathological Society.

Sparrow, D. H. and R. D. Graham. 1988. Susceptibility of zinc deficient wheat plants to colonization by *Fusarium graminearum* Schw. Group 1. *Plant Soil* 112:261–266.

Stevenson, F. J. 1986. *Cycles of soil carbon, nitrogen, phosphorus, sulfur, micronutrients.* New York: John Wiley & Sons.

Stevens, D. P., M. J. McLaughlin, and T. Heinrich. 2003. Determining toxicity of lead and zinc runoff in soils; salinity effects on metal partitioning and on phytotoxicity. *Environ. Toxicol. Chem.* 22:3017–3024.

Sumner, M. E. and M. P. W. Farina. 1986. Phosphorus interactions with other nutrients and lime in field cropping systems. *Adv. Soil Sci.* 5:201–236.

Thongbai, P., R. J. Hannam, R. D. Graham, and M. J. Webb. 2001. Interaction between zinc nutritional status of cereals and Rhizoctonia root rot severity. *Plant Soil* 153:207–214.

Tiller, K. G. 1968. Stability of hectorite in weakly acidic solutions. III. Adsorption of heavy metal cations and hectorite solubility. *Clay Miner.* 7:409–419.

Tisdale, S. L., W. L. Nelson, and J. D. Beaton. 1985. *Soil fertility and fertilizers*, 4th edition. New York: Macmillan Publishing Company.

Van der Zaal, B. J., L. W. Neuteboom, J. E. Pinas, A. N. Chardonnens, H. Schat, J. A. C. Verkleij, and P. J. J. Hooykaas. 1999. Overexpression of a novel *Arabidopsis* gene related to putative zinc-transporter genes from animal can lead to enhanced zinc resistance and accumulation. *Plant Physiol.* 119:1047–1056.

Viets, F. G. and W. L. Lindsay. 1973. Testing soils for zinc, copper, manganese and iron. In: *Soil testing and plant analysis*, L. M. Walsh and J. D. Beaton, Eds., 153–172. Madison, WI: SSSA.

Webb, M. and J. F. Loneragan. 1988. Effect of zinc deficiency on growth, phosphorus concentration, and phosphorus toxicity of wheat plants. *Soil Sci. Soc. Am. J.* 52:1676–1680.

Welch, R. M. 1993. Zinc concentrations and forms in plants for human and animals. In: *Zinc in soil and plants,* A. D. Robson, Ed., 183–195. Dordrecht: Kluwer Academic Publishers.

Welch, R. M. 1995. Micronutrient nutrition of plants. *Critical Rev. Plant Sci.* 14:49–82.

Wellings, N. P., A. H. Wearing, and J. P. Thompson. 1991. Vesicular-arbuscular mycorrhizae (VAM) improve phosphorus and zinc nutrition and growth of pigeonpea in a vertisol. *Aust. J. Agric. Res.* 42:835–845.

Wilkinson, S. R., D. L. Grunes, and M. E. Sumner. 2000. Nutrient interactions in soil and plant nutrition. In: *Handbook of soil science.* Boca Raton, FL: CRC Press.

9 Copper

9.1 INTRODUCTION

Essentiality of copper (Cu) for higher plants was discovered in 1931 by C. B. Lipman and G. MacKinney (Marschner, 1995; Fageria et al., 1997a). Since then, knowledge of its role in improving crop production has improved. Copper deficiency has been reported in cereals, vegetables, grasses, and forage legumes in various parts of the world (Clark, 1982; Baligar et al., 2001; Brady and Weil, 2002; Fageria et al., 2006). The main causes of micronutrients deficiency including copper are (1) intensive cultivation; (2) higher copper requirements of modern cultivars due to higher yields; (3) loss of topsoil layers by soil erosion which contain higher organic matter and copper; (4) use of liming in acid soils, which reduces copper concentration due to increases in soil pH and adsorption of this element; (5) low rates used by farmers, especially in developing countries; (6) increased use of high-analysis fertilizers with low amounts of micronutrients; (7) low rate of application or no use of organic manures; (8) involvement of natural and anthropogenic factors that limit adequate plant availability and create element imbalances; (9) sandy soils low in organic matter content, which may become deficient in Cu because of leaching losses; and (10) no recycling of crop residues (Fageria et al., 2002). High rates of N fertilizer have greatly accentuated Cu deficiency in Europe, Australia, and India (Gartrell, 1981). The application of inorganic N and P or livestock manure to increase the fertility of certain soil types has caused major yield and quality losses in Canadian cereal crops (Evans et al., 2007). Copper deficiency can also be induced by certain foliar- and soil-applied herbicides (Evans et al., 1990).

Copper deficiency in annual crops grown on Brazilian Oxisols has been reported in recent years (Galrão, 1999; Fageria, 2001a). Data in Table 9.1 show influence of copper fertilization on shoot dry weight of upland rice and dry bean grown on Brazilian Oxisols. Dry weight of shoot of both crops was significantly increased in a quadratic fashion with increasing Cu level in the range of 0 to 96 mg kg^{-1} of soil. The variability in shoot weight of upland rice was 92%, and shoot weight of dry bean varied 84% with the application of copper fertilization. This means upland rice was more responsive to copper fertilization compared to dry bean. Figure 9.1, Figure 9.2, and Figure 9.3 show response of soybean, corn, and dry bean, respectively, to applied copper in a Brazilian Oxisol. In Brazilian Oxisols, copper deficiency is related to use of liming material to improve pH of these acid soils (Fageria, 2002ab).

Data in Table 9.2 show that uptake of copper by upland rice in Brazilian Oxisol decreased at pH higher than 6.2. Similarly, uptake of copper by dry bean and soybean also decreased in a Brazilian Inceptisol with increasing soil pH in the range

TABLE 9.1

Influence of Copper Fertilization on Dry Matter Yield of Upland Rice and Dry Bean Grown on Brazilian Oxisols

Cu Level Applied (mg kg⁻¹)	Upland Rice Shoot Dry Weight (g plant⁻¹)[a]	Dry Bean Shoot Dry Weight (g plant⁻¹)[b]
0	0.46	1.26
2	0.47	1.43
4	0.48	1.66
8	0.51	1.66
16	0.50	1.25
32	0.47	1.15
64	0.50	1.17
96	0.08	0.16

Regression Analysis

Cu level (X) vs. shoot dry wt. of rice (Y) = $0.4509 + 0.00597X - 0.000099X^2$, $R^2 =$ 0.9164**

Cu level (X) vs. shoot dry wt. of dry bean (Y) = $1.4335 + 0.00276X - 0.00015X^2$, $R^2 =$ 0.8387*

[a] Rice plants were harvested at 4 weeks after sowing, whereas,

[b] bean plants were harvested at 3 weeks after sowing.

*, ** Significant at the 5 and 1% probability levels, respectively.

Source: Adapted from Fageria (2002a).

FIGURE 9.1 Response of soybean to applied copper fertilization in a Brazilian Oxisol (Cu rates were, from left to right, 0, 2.5, 5, 10, and 60 mg kg⁻¹ of soil).

FIGURE 9.2 Response of corn to applied copper in a Brazilian Oxisol (Cu rates were, from left to right, 0, 5, 10, and 60 mg kg⁻¹ soil).

FIGURE 9.3 Dry bean response to copper fertilization applied to a Brazilian Oxisol (Cu rates used were, from left to right, 0, 2, 4, and 64 mg kg⁻¹ of soi)l

of 4.9 to 7.0. It is well known that copper bioavailability and hence copper toxicity is increased in acidic relative to calcareous soils due to increased soil solution concentration of Cu^{2+} ions, which are generally assumed to be the form of Cu^{2+} that is acquired by plants (Tyler and Olsson, 2001; Chaignon et al., 2002). El-Kherbawy and Sanders (1984) found that pH did not influence Cu^{2+} in the soil solution, but increasing the pH decreased Cu^{2+} in the exchangeable fraction. Mengel et al. (2001) reported that the concentration of Cu^{2+} in the soil solution decreases sharply with

TABLE 9.2

Influence of Soil pH on Copper Uptake by Upland Rice, Dry Bean, and Soybean Grown on Brazilian Oxisol and Inceptisol

Soil pH in H_2O[a]	Cu^{2+} Uptake in Upland Rice Shoot (μg plant^{-1})[c]	Soil pH in H_2O[b]	Cu^{2+} Conc. in Dry Bean Shoot (mg kg^{-1})[d]	Cu^{2+} Conc. in Soybean Shoot (mg kg^{-1})[d]
4.6	75	4.9	9	5
5.7	105	5.9	5	4
6.2	78	6.4	5	3
6.4	64	6.7	4	4
6.6	61	7.0	4	3
6.8	51	—		
R^2	0.89**		0.81**	0.72**

Note: Uptake = concentration × shoot dry weight and concentration = Cu^{2+} content per unit of shoot dry weight.

[a] Oxisol.

[b] Inceptisol.

[c] Upland rice plants were harvested at physiological maturity.

[d] Dry bean and soybean plants were harvested at 4 weeks after sowing.

** Significant at the 1% probability level.

Source: Adapted from Fageria (2000) and Fageria and Baligar (1999).

increasing pH, whereas the concentration of organic Cu^{2+} complexes in the soil solution is less dependent on soil pH. Higher pH may even promote the dissolution of organic Cu^{2+} complexes (McBridge, 1989).

Availability of copper is influenced by several environmental factors. These factors are (1) soil pH, (2) Cu^{2+} concentration in the soil solution, (3) crop species or genotypes within species, (4) soil temperature, (5) soil moisture, (6) soil organic matter content, (7) soil microbial activities, (8) the nature of parent material, (9) balance of macro- and micronutrients, (10) soil redox potential, (11) root morphology (length, root hairs, density, surface area), (12) soil cation exchange capacity, (13) soil texture and oxide content, and (14) crop infestation by diseases, insects, and weeds. The objective of this chapter is to discuss mineral nutrition aspects of copper to maximize its uptake and use efficiency by crop plants. This information may be useful in improving crop yields, reducing cost of crop production, and limiting environmental pollution.

9.2   CYCLE IN SOIL–PLANT SYSTEMS

The copper cycle in soil–plant systems mainly includes its addition and depletion. The main sources of addition are chemical fertilizers, crop residues, farmyard manures, and weathering soil minerals. Similarly, main copper depletion sources from soil are uptake by crop plants, removal by soil erosion, leaching by excess rainfall, and immobilization by organic matter and microbial biomass. The Cu^{2+} ion is

TABLE 9.3
Copper Content of Brazilian Inceptisol at Two Soil Depths

Cu^{2+} Value (mg kg^{-1})[a]	0–20 cm	20–40 cm
Minimum	0.3	0.2
Maximum	2.7	2.3
Average	1.3	0.9

[a] Soil samples were collected from 14 different locations covering 13 municipalities in the state of Mato Grosso and Mato Grosso do Sul, Brazil.

Source: Adapted from Fageria et al. (1997b).

strongly adsorbed by organic compounds in the soil. Hodgson et al. (1966) reported that 98% of the copper is complexed by low-molecular-weight organic compounds. The Cu^{2+} ion has a strong affinity to soil organic matter compared with other divalent cations in the order of Cu > Ni > Pb > Co > Ca > Zn > Mn > Mg (Schnitzer and Skinner, 1967; Mengel et al., 2001). Micronutrient concentrations are generally higher in the surface soil horizon and decrease with soil depth (Table 9.3). The higher copper content in the surface soil may be associated with higher organic matter content in the upper soil horizon compared to the lower soil horizon.

The average copper content of the lithosphere is reported to be 70 mg kg^{-1}, whereas soils generally range from 2 to 100 mg kg^{-1}. The average content for soils is estimated at 30 mg kg^{-1} (Lindsay, 1979). However, Havlin et al. (1999) reported that the Cu concentration in soil ranges from 1 to 40 mg kg^{-1}, with an average concentration of 9 mg kg^{-1}. An average of 50% of the Cu in soils is insoluble and unavailable, 30% is bound by organic sites, 15% is in an oxide form, and 5% is available for plant uptake (Barber, 1995). Copper-deficient soils may contain less than 2 mg Cu kg^{-1} soil, but other factors such as pH, organic matter content, and plant species also influence Cu availability (Evans et al., 2007). The predominant ion below pH 6.9 is Cu^{2+}, while $Cu(OH)_2{}^{\circ}$ is the major solution species above this pH (Lindsay, 1979). Copper solubility in soils is highly pH dependent, due to the hydrolysis of Cu^{2+} in soils at pH values above 6 and to removal of Al^{3+} and H^+ from exchangeable sites as the pH is raised in more acid soils (Cavallaro and McBridge, 1980). Copper in soil solution is found primarily as organic ligand complexes (Geering and Hodgson, 1969), with the smaller molecular species being the most effective complexes (Thornton, 1979). Therefore, Cu availability to plant roots apparently involves small organic compounds that chelate the copper ions and move them into the soil solution. These organic compounds may be exuded from plant roots (Hale et al., 1971). Extracts of soil organic matter, decayed plant material, and fulvic acid also compete well with soil clays for solution copper (Bloomfield, 1976). As Cu is strongly bound to soils, it is very immobile, and generally a higher concentration is found in the topsoil layer (Table 9.3).

The availability of Cu is strongly influenced by soil pH. However, the influence of pH on availability of Cu is less than that of Fe and Mn (Fageria, 2000).

The availability of Cu at pH 7.0 is 10 to 100 times less than at pH 6.0 (Evans et al., 2007). Precipitation of Cu was the basis for the reported decrease in Cu toxicity by liming (Alva et al., 2000). Organic and sandy soils are most likely to be Cu deficient. Soils high in organic matter (more than 10%) readily sequester Cu ions to reduce or even eliminate their availability to plants. Copper deficiency has been recognized for centuries in high-pH organic fenland-type soils in Europe, where the condition has generally been referred to as reclamation disease, when these organic soils were drained for agricultural use (Evans et al., 2007). Similarly, in North America, areas referred to as muck soils are generally recognized as chronically Cu deficient, and Cu fertilizer is generally applied to obtain good crop yields (Evans et al., 2007).

9.3 FUNCTIONS AND DEFICIENCY SYMPTOMS

Copper is a constituent of a number of important oxidase enzymes including cytochrome oxidase, ascorbic acid oxidase, and lactase. It is important in photosynthesis and protein and carbohydrate metabolism (Fageria et al., 2002). Copper is the metal component of three different forms of protein: plastocyanins, involved in electron transfer; peroxidases, which oxidize monophenols to diphenols; and multi-Cu proteins, which act as oxidases (Sandermann and Boger, 1983; Evans et al., 2007). The copper atom is involved in the mechanism of detoxification of O_2^- generated in photorespiration (Halliwell, 1978). Copper is involved in the process of photosynthesis, and in copper-deficient plants, rate of photosynthesis is reduced. In copper-deficient plants, soluble carbohydrates decreased during the vegetative growth stage, which may reduce pollen formation and fertilization; consequently, yields may be reduced (Marschner, 1995). In legumes, copper deficiency reduces nodulation and N_2 fixation (Hallsworth et al., 1964; Cartwright and Hallsworth, 1970).

Copper is involved in cell wall formation and, like other nutrients, in electron transport and oxidation reactions (Tisdale et al., 1985). Cell wall lignification is reduced in copper-deficient plants, which may result in crop lodging in cereals (Marschner, 1995). In addition, in copper-deficient plants, lignification of the anther cell walls is reduced or absent (Dell, 1981). Copper is important in grain or seed formation in cereals and legumes. Dell (1981) and Marschner (1995) reported that reduced seed set in copper-deficient plants might be the result of the inhibition of pollen release from the stamina, since lignification of the anther cell walls is required for rupture of the stamina and subsequent release of the pollen. In wheat, copper deficiency delayed maturity, shriveled grain, and produced distorted, twisted, incompletely emerged heads; awns may also result (Wiese, 1993).

Copper improves root development in dry bean and soybean grown on Brazilian Oxisol (Table 9.4; Figure 9.4) and hence has special importance for improving crop yields (Fageria, 1992, 2004). The plant root plays a major role in controlling plant growth and development due to its importance in absorbing water and nutrients. Further, roots provide mechanical support to plants from the seedling stage through the maturity stage of growth and development. In addition, rooting plays an important role in the reduction of N, which can then be supplied to the shoot through xylem, either as inorganic N or as amino acids and amides. Roots are also known to summarize growth substances such as cytokinins, which may be important in leaf

TABLE 9.4
Influence of Copper Fertilization on Root Length of Dry Bean Grown on Brazilian Oxisols

Cu Level Applied (mg kg⁻¹)	Maximum Root Length of Dry Bean (cm)
0	25
2	30
4	27
8	28
16	28
32	24
64	30
96	14
Regression	Quadratic
R^2	0.4222**

Note: Dry bean plants were harvested at 3 weeks after sowing.
** Significant at the 1% probability level.
Source: Adapted from Fageria (2002b).

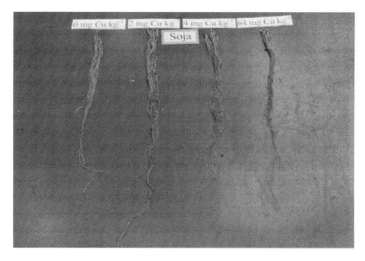

FIGURE 9.4 Soybean root development under different copper levels applied to Brazilian Oxisol. Left to right: 0, 2, 4, and 64 mg Cu kg⁻¹ soil.

function and possibly in grain development (Evans and Wardlaw, 1976; Fageria et al., 1997a).

Copper is not mobile in the plants; hence, its deficiency first appears in the younger leaves (Fageria and Barbosa Filho, 2006). Copper-deficient leaves appear bluish green and then become chlorostic near the tips. The chlorosis develops downward along both sides of the midrip; it is followed by dark-brown necrosis of the

FIGURE 9.5 Copper deficiency symptoms in rice leaves.

tips. The new emerging leaves fail to unroll and appear needlelike for the entire leaf or, occasionally, for half the leaf, with the basal end developing normally. In severe cases, the leaf tips become white and the leaves are narrow and twisted. A typical feature of Cu^{2+} deficiency in cereals is the bushy habit of the plants with white twisted tips and a reduction in panicle formation (Fageria and Gheyi, 1999). In peanuts, copper deficiency deforms young leaves, and their color is greenish yellow or chlorotic. Terminal leaflets are small, and margins curl inwards, giving a cupped appearance (Smith et al., 1993). Figure 9.5 shows copper deficiency in rice plant leaves.

9.4 CONCENTRATION IN PLANT TISSUES AND UPTAKE

Copper is taken up by the plants in only very small quantities. The copper content of most plants is generally between 2 and 20 mg kg^{-1} in the dry plant material (Mengel et al., 2001). About 5–8 mg Cu kg^{-1} dry plant tissue may be considered as the critical level for most field crops (Fageria et al., 1997a). Plant age is one of the important factors that determine copper concentration in plant tissues. Data in Figure 9.6 show that copper concentration in upland rice shoot decreases significantly with the advancement of plant age. At 19 days of plant growth, Cu^{2+} concentration in rice plant tissue was about 20 mg kg^{-1} and decreased to about 5 mg kg^{-1} at harvest (about 130 days after sowing). Similarly, in the shoot of dry bean, Cu^{2+} concentration decreased significantly with increasing plant age (Figure 9.7). It was about 12 mg kg^{-1} at 15 days after sowing and decreased to 5 mg kg^{-1} at harvest (96 days after sowing). These results indicate that Cu^{2+} concentration in crop plants should be determined at different growth stages for use in nutritional deficiency or sufficiency diagnostic techniques. In addition, these results also indicate that copper concentration was higher in the shoot of cereal compared to legume.

Table 9.5 presents data about the adequate levels of copper in plant tissues of principal crop species. Similarly, in Table 9.6, data are presented about adequate and toxic level of copper in principal crop species grown on Brazilian Oxisols. Overall,

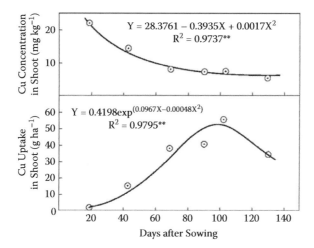

FIGURE 9.6 Copper concentration and uptake in upland rice grown on Brazilian Oxisol.

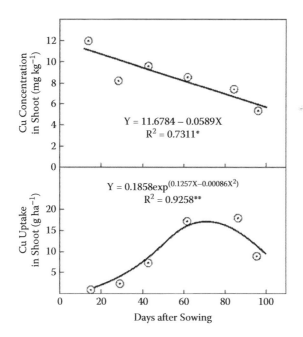

FIGURE 9.7 Copper concentration and uptake in shoot of dry bean grown on Brazilian Oxisol.

the range between adequate and toxic level in five crop species was 10 to 15 mg Cu kg^{-1} shoot dry weights. This means that copper range between adequate and toxic level is very narrow, and care should be taken when applying copper fertilizers for correcting copper deficiency in crop plants. Pettersson (1976) found that 10 μmol L^{-1} copper in solution culture depressed plant growth. The copper concentration used in many nutrient solutions is 0.5 μmol L^{-1} (Barber, 1995). Hence, there appears to be

TABLE 9.5

Adequate Level of Copper in Plant Tissues of Crop Plants

Crop Species	Growth Stage and Plant Part Analyzed	Adequate Cu Level (mg kg⁻¹)
Wheat	Heading, leaf blade	>2.2
Barley	Stem extension, whole tops	4.8–6.8
Rice	Tillering, uppermost mature leaves	5–20
Corn	30 to 45 days after emergence, whole tops	7–20
Sorghum	Seedling, whole tops	8–15
Sorghum	Bloom, 3rd blade below panicle	2–7
Soybean	Prior to pod set, upper fully developed trifoliate	10–30
Dry bean	Early flowering, uppermost blade	5–15
Peanut	Early pegging, upper stems and leaves	10–50

Source: Adapted from Fageria et al. (1997a).

TABLE 9.6

Adequate and Toxic Levels of Copper in the Plant Tissues of Five Crop Species Grown on Brazilian Oxisols

Crop Species	Adequate Level (mg kg⁻¹)	Toxic Level (mg kg⁻¹)
Upland rice	15	26
Dry bean	6	10
Corn	7	11
Soybean	7	10
Wheat	14	17
Average	10	15

Note: Adequate level was calculated at the 90% of the maximum dry weight of shoot and toxic level was calculated at the 10% reduction in shoot dry weight after achieving maximum weight. Plants were harvested at 4 weeks after sowing and soil pH at harvest was about 6.0 in water in all five experiments.

Source: Adapted from Fageria et al. (2003).

a relatively narrow range over which copper concentration is satisfactory for plant growth. Jones (1991) reviewed the literature on plant tissue test and reported that Cu^{2+} sufficiency or normal level in various crop plants was in the range of 5–30 mg kg⁻¹ and toxic level was in the range of 20–100 mg kg⁻¹. This author also reported that average copper in crop plants was about 6 mg kg⁻¹.

Copper uptake rates are lower than for most other micronutrients (Barber, 1995). The uptake of copper depends on grain and shoot dry matter yield. Fageria (2001) reported that uptake of copper in crop plants varied with plant age and also depended on dry matter accumulation. Figure 9.6 and Figure 9.7 show that copper uptake in shoot of upland rice and dry bean increased in a quadratic fashion with the advancement of plant age. At harvest, uptake of copper decreased in both the crops due to translocation to grains. In upland rice, copper uptake was maximum (about 50 g ha^{-1}) at about 100 days after sowing, and in dry bean, copper uptake was maximum (about 15 g ha^{-1}) at about 70 days after sowing. This means that copper uptake was higher in upland rice compared to dry bean, and this was associated with higher dry matter yield of upland rice (Fageria, 2001b).

9.5   USE EFFICIENCY AND Cu HARVEST INDEX

Crop recovery values for micronutrients generally range from only 5 to 10 % (Mortvedt, 1994). Some of the reasons for the low efficiency of micronutrient fertilizers are poor distribution of the low rates applied to soils, fertilizer reactions with soil to form unavailable reaction products, and low mobility in the soil to reach plant roots (Martens and Lindsay, 1990). Copper use efficiency (dry weight of shoot or grain/copper uptake in shoot or grain) of upland rice, dry bean, corn, and soybean during the crop growth cycle is presented in Table 9.7 and Table 9.8. In upland rice and dry bean, copper use efficiency was significantly increased in a quadratic fashion (Table 9.7). The variation in copper use efficiency was about 98% in rice due to plant age and about 93% in dry bean due to plant age. At harvest, copper use efficiency in grain of rice was 84.4 kg rice produced with the accumulation of 1 g of copper in the grain. In the dry bean, 85.3 kg grain was produced with the accumulation of 1 g of copper in the grain. This means that copper use efficiencies in grain production in upland rice and dry bean were more or less similar.

Copper use efficiency in corn also experienced a significant quadratic increase in the shoot with the advancement of plant age (Table 9.8). There was 98% variability in copper use efficiency in shoot of corn with the advancement of plant age.

TABLE 9.7
Copper Use Efficiency (CuUE) in Shoot and Grain of Upland Rice and Dry Bean during Crop Growth Cycle

Plant Age in Days	Upland Rice (kg g^{-1})	Plant Age in Days	Dry Bean (kg g^{-1})
19	45.3	15	84.5
43	70.9	29	125.8
68	120.7	43	103.5
90	140.3	62	120.9
102	141.7	84	135.7
130	187.7	96	216.2
130 (grain)	84.4	96 (grain)	85.3

Regression Analysis

Plant age (X) vs. CuUE in rice shoot (Y) = 16.6851 + 1.4727X − 0.0014X^2, R^2 = 0.9791**
Plant age (X) vs. CuUE in dry bean shoot (Y) = 7.0336 + 2.7758X − 0.0108X^2, R^2 = 0.9289*

Note: Copper use efficiency (CuUE kg g^{-1}) = (Shoot or grain dry weight in kg/Uptake of Cu in shoot or grain in g).

*, ** Significant at the 5 and 1% probability levels, respectively.

TABLE 9.8
Copper Use Efficiency (CuUE) in Shoot and Grain of Corn and Soybean during Crop Growth Cycle

Plant Age in Days	Corn (kg g^{-1})	Plant Age in Days	Soybean (kg g^{-1})
18	76.6	27	73.3
35	74.3	41	58.8
53	105.9	62	94.8
69	156.4	82	79.2
84	194.9	120	45.7
119	250.0	158	98.2
119 (grain)	720.9	158 (grain)	46.9

Regression Analysis

Plant age (X) vs. CuUE in corn shoot (Y) = 38.9920 + 1.2870X − 0.0045X^2, R^2 = 0.9622**

Plant age (X) vs. CuUE in soybean shoot (Y) = 85.6260 − 0.4212X + 0.0027X^2, R^2 = 0.1011^{NS}

Note: Copper use efficiency (CuUE kg g^{-1}) = (Shoot or grain weight in kg/Uptake of Cu in shoot or grain in g)

**, NS Significant at the 1% probability level and nonsignificant, respectively.

However, copper use efficiency was nonsignificant in soybean with the advancement of plant age. Copper use efficiency for grain production was about 721 kg grain per g of copper uptake in corn. In soybean, copper use efficiency for grain production was about 47 kg grain per g of copper uptake. Hence, copper use efficiency for grain production was in the order of corn > dry bean > upland rice > soybean. Copper harvest index (CuHI) (Cu translocation to grain/copper uptake in grain plus shoot) was maximum for dry bean and minimum for corn (Table 9.9). The lowest CuHI in corn was due to highest grain and straw yield compared to other crops. From a human nutritional point of view, the crops can be classified on the basis of CuHI as dry bean > rice > soybean > corn.

TABLE 9.9
Copper Uptake in Shoot and Grain and Copper Harvest Index (CuHI) in Upland Rice, Dry Bean, Corn, and Soybean Grown on Brazilian Oxisol

Crop Species	Uptake in Shoot (g ha^{-1})	Uptake in Grain (g ha^{-1})	Total (g ha^{-1})	CuHI
Upland rice	34.78	56.64	91.41	0.62
Dry bean	8.01	22.41	30.42	0.74
Corn	53.32	13.75	67.07	0.21
Soybean	53.15	30.79	83.94	0.37

Note: CuHI = (Cu uptake in grain/Cu uptake in grain plus shoot).

Source: Adapted from Fageria (2001b).

9.6 INTERACTION WITH OTHER NUTRIENTS

Understanding copper's interactions with other nutrients is an important aspect in improving crop efficiency in crop plants and consequently crop yields. Nitrogen application accentuates copper deficiency in crop plants; under such conditions, application of copper fertilization is required for maximum economic yields (Marschner, 1995). Critical deficiency level of copper in plant tissues increases with increasing nitrogen supply (Marschner, 1995).

Fageria (2002a) studied uptake of macro- and micronutrients in upland rice and dry bean plants grown on Brazilian Oxisols. Copper had a significant synergistic effect on the uptake of P, K, and Mn in the upland rice plants (Fageria, 2002a). However, concentrations of Ca, Mg, and Fe were significantly decreased in rice plants with the application of copper. Copper did not significantly influence uptake of Zn in rice plants. In bean plants, copper application significantly increased uptake of Zn and had no significant effect on uptake of P, K, Ca, Mg, Mn, and Fe. Bowen (1969), however, reported that copper uptake appears to be a metabolically mediated process, and there is evidence that copper strongly inhibits the uptake of zinc and vice versa. Copper in excess interferes with plants' capacity to absorb and/or translocate other nutrients (Fageria, 2002a), inhibiting root elongation and adversely affecting the permeability of the root cell membrane (Woolhouse and Walker, 1981). Copper in excess also has a destructive effect on the integrity of the chloroplast membrane, leading to a decrease in photosynthetic activity (Eleftheriou and Karataglis, 1989), which may affect uptake of nutrients.

Ouzounidou et al. (1995) reported no change in Ca^{2+} and K^+ and a decrease in iron uptake by corn plants with the application of copper. Similarly, according to Lexmond and van der Vorm (1981), P and Fe contents in both corn root and shoot were decreased when copper was raised from 0.16 to 3.1 uM. Mocquot et al. (1996) observed no significant change in the uptake of Ca, K, P, and Fe in corn plants. In soybean plants, copper rates significantly increased uptake of P and Mg and had no significant effects on other elements analyzed. In wheat plants, copper had a significant positive effect on uptake of P and significant negative effects on uptake of Ca and Mg.

9.7 MANAGEMENT PRACTICES TO MAXIMIZE Cu USE EFFICIENCY

Adopting appropriate crop management practices is important to maximize crop yields, reduce cost of fertilizer application, and maximize nutrient use efficiency. Some of the important management practices to maximize copper use efficiency in crop plants are use of adequate method, rate, and source of application. Use of copper-efficient species or genotypes within species is another practice to maximize copper use efficiency and reduce cost of crop production. Some of the practices discussed in earlier chapters—such as supplying sufficient organic matter, maintaining adequate soil moisture, and controlling diseases, insects, and weeds—are also applicable for copper.

TABLE 9.10
Principal Copper Carriers

Common Name	Formula	Cu Content (%)
Copper sulfate pentahydrate	$CuSO_4.5H_2O$	25
Copper sulfate monohydrate	$CuSO_4.H_2O$	35
Cupric oxide	CuO	75
Copper chloride	$CuCl_2$	17
Copper frits	Frits	40–50
Copper EDTA chelate	$Na_2CuEDTA$	13
Copper HEDTA chelate	NaCuHEDTA	9
Cuprous oxide	Cu_2O	89

Source: Fageria (1989), Tisdale et al. (1985), Foth and Ellis (1988).

9.7.1 APPROPRIATE METHOD AND SOURCE

Copper deficiency in crop plants can be corrected by applying Cu in the soil as broadcast or band or by foliar application. Soil application in Cu-deficient soil through an appropriate source is better than foliar application. Foliar application of Cu should be used in emergency situations. Lower rates of Cu application are required for correcting Cu deficiency with banded $CuSO_4$ than with broadcast $CuSO_4$. Rates of band-placed $CuSO_4$ required for correcting Cu deficiency are as low as 1.1 kg Cu ha^{-1} for vegetables and as high as 6.6 kg Cu ha^{-1} for highly sensitive crops (Martens and Westermann, 1991). If copper deficiency occurs during crop growth, foliar application of $CuSO_4$ can alleviate this problem. Mengel (1980) recommended foliar application of 2.2 kg Cu as $CuSO_4$ in 280 L water ha^{-1} for correcting Cu deficiencies in corn, soybean, and wheat. Principal copper carriers are given in Table 9.10. Among these sources, all are water soluble, except cupric oxide (CuO).

9.7.2 ADEQUATE RATE

Appropriate rate of copper application is essential for economic and environmental reasons. The best method of defining appropriate rate is to refer to crop response curves to applied copper fertilizers. Data on crop responses to applied copper fertilizers are limited under field as well as greenhouse conditions. However, Fageria (2001a) developed crop response curves to applied copper fertilization for principal food crops (Figure 9.8). There was a quadratic increase in dry weight of upland rice, common bean, corn, and wheat with increasing copper rate, but at higher copper rates, dry weight was decreased in all four crop species. In the case of soybean, there was a quadratic decrease in shoot dry weight with increasing copper rate in the growth medium. Adequate copper rate, calculated on the basis of 90% maximum yield, was 3 mg kg^{-1} for upland rice, 2 mg kg^{-1} for common bean, 3 mg kg^{-1} for corn, and 12 mg kg^{-1} for wheat. For soybean, copper application was not required, and the original soil level of 1 Cu mg kg^{-1} extracted by Mehlich-1 extracting solution was sufficient to produce maximum plant growth. Martens and Westermann (1991) reported

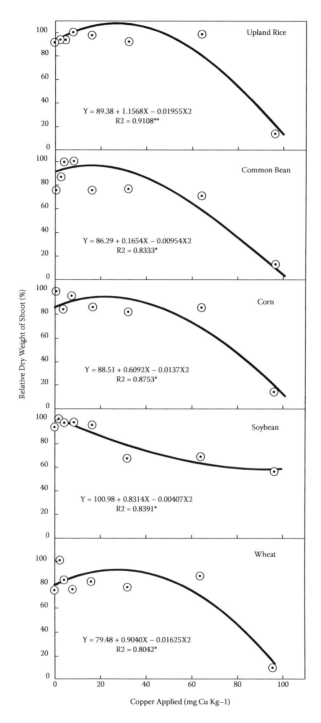

FIGURE 9.8 Relationship between copper applied in the soil and relative dry weight of shoot of five crop species (Fageria, 2001a).

that application of 1.7 to 7.25 mg Cu kg^{-1} of soil as broadcast $CuSO_4$ could correct copper deficiency in most annual crops. Similarly, Follett et al. (1981) reported that 1 to 7 kg ha^{-1} copper as copper sulfate is sufficient for most crop plants. Differences in broadcast rates of Cu required for correcting Cu deficiency reflect variations in soil properties, plant requirements, and concentrations of extractable soil Cu. Copper applied to soil has a prolonged residual effect (3 to 5 years depending on cropping intensity). Hence, soil application of copper is generally recommended.

Critical toxic levels of applied Cu rate were 51 mg kg^{-1} for upland rice, 37 mg kg^{-1} for common bean, 48 mg kg^{-1} for corn, 15 mg kg^{-1} for soybean, and 51 mg kg^{-1} for wheat. This means that upland rice and wheat crops were more tolerant to copper toxicity and that soybean was most susceptible to toxicity. In the literature, practically no data are available regarding toxic rate of application for annual crops grown on Oxisols of central Brazil.

Soil test is an important criterion in determining copper fertilizer rate for crop plants (Fageria et al., 1997a). The DTPA extracting solution is the most commonly used solution for soil test of copper. However, in Brazil, Mehlich-1 extracting solution is commonly used in routine micronutrient analysis for economic reasons. Adequate soil test level of copper was established as 2 mg kg^{-1} for upland rice, 1.5 mg kg^{-1} for common bean, 2.5 mg kg^{-1} for corn, 1 mg kg^{-1} for soybean, and 10 mg kg^{-1} for wheat, using Mehlich-1 extracting solution (Fageria, 2001a). Adequate level of copper was 1 mg kg^{-1} for upland rice, 0.5 mg kg^{-1} for common bean, 1.5 mg kg^{-1} for corn, 0.5 mg kg^{-1} for soybean, and 8.5 mg kg^{-1} for wheat, when DTPA extracting solution was used (Figure 9.9). Adequate soil test levels of 0.1 to 10 mg kg^{-1} for Mehlich-1 extracting solution have been reported for annual crops (Sims and Johnson, 1991). Galrão (1999) reported 0.6 mg Cu kg^{-1} as an adequate level in the soil using DTPA extracting solution for soybean crop grown on a Brazilian Oxisol.

9.7.3 USE OF EFFICIENT CROP SPECIES/GENOTYPES

Crop plants also differ in their sensitivity to Cu deficiency. In general, the most responsive crops to Cu fertilizers are oats, spinach, wheat, alfalfa, flax, and barley. In the medium range are cabbage, cauliflower, sugar beet, and corn, while beans, grass, potatoes, and soybeans show a low response (Mengel et al., 2001). Penning et al. (1989) also reported that pea, canola (rapeseed), and rye are also Cu-efficient species and show little response to applied Cu. Similarly, many varieties of triticale, an interspecific hybrid cereal containing a combination of wheat and rye genes, retain rye's superior ability to extract Cu from soil (Evans et al., 2007). Genotype differences have been reported for copper utilization in crop plants (Baligar et al., 2001; Fageria et al., 2002). Data in Table 9.11 show that copper use efficiency by upland rice genotypes varied significantly among genotypes and soil acidity levels.

Knowledge of molecular genetics in copper use efficiency in plants has been improved in the last two decades. Cianzio (1999) reported that efficiency of rye to utilize Cu^{2+} was inherited by the amphiploid hybrid triticale and, therefore, immediately available to wheat in a genetically related background. Copper efficiency in rye and triticale is determined by a single gene carried on the 5RL chromosome arm (Graham, 1978). The 5RL-translocation lines of rye were the donor parents in backcrosses to transfer the trait to locally adapted wheat genotype. The highly

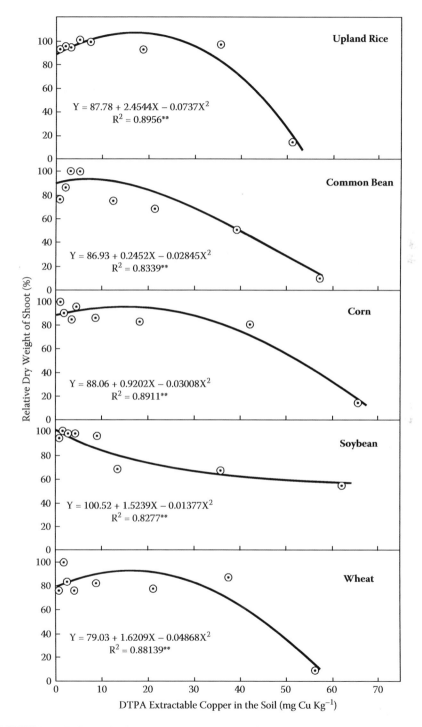

FIGURE 9.9 Relationship between DTPA extractable Cu and relative dry weight of shoot of five crop species (Fageria, 2001a).

TABLE 9.11

Copper Use Efficiency (mg Grain per µg Nutrient Accumulated in the Plant) by 10 Upland Rice Genotypes at Two Acidity Levels in Brazilian Oxisol

| Upland Rice Genotype | Cu-Use Efficiency (mg Grain µg⁻¹ Cu Uptake) | |
	High Acidity (Soil pH 4.5)	Low Acidity (Soil pH 6.4)
CRO 97505	150bc	197
CNAs 8983	207a	182
CNAs 8938	191ab	184
CNAs 8989	138bcd	176
CNAs 8824	86de	122
CNAs 8952	96cde	152
CNAs 8950	79e	141
CNAs 8540	76e	144
Primavera	117cde	192
Carisma	127cde	114
Average	127	160
F-test		
Acidity (A)	*	
Genotype (G)	**	
A X G	*	
CV (%)	18	

*, ** Significant at the 5 and 1% probability levels, respectively. Means in the same column followed by the same letter are not significantly different at the 5% probability level by Tukey's test.

Source: Adapted from Fageria et al. (2004).

copper-efficient wheat lines developed have been top-yielding lines on copper-deficient soils (Graham, 1988). At nondeficient sites, the lines have yielded similarly to their respective recurrent parent, indicating no suppressive effects on yield or in baking quality by the presence of the 5RL chromosome segment (Cianzio, 1999).

9.8 CONCLUSIONS

Crop plants require less copper than any nutrient except molybdenum. In higher plants, most of the functions of copper are associated with its participation in enzymatic reactions. It is also important for protein synthesis, chlorophyll formation, and N metabolism. The role of copper in higher plants is also associated with photosynthesis, nodulation formation, and N_2 fixation in legumes. It is strongly adsorbed by organic compounds in the soil. Copper solubility in soil is strongly pH dependent and decreases with increasing pH. Overall, adequate copper concentration in plant tissue varied from 6 to 26 mg kg⁻¹, and copper uptake is lower than 50 g ha⁻¹. The range between adequate and toxic level of copper depends on crop species. However, overall range for annual crop is in the range of 10 to 15 mg kg⁻¹. Copper deficiency in crop plants can be identified by visual copper deficiency symptoms and soil and plant tissue tests. Soil test is the most common practice to identify copper deficiency in a given soil for a determined crop species. Soil as well as foliar application of soluble copper products can correct copper deficiency. Copper sulfate is one of the most common and economic sources of copper fertilization of crop plants. Significant progress has been made in breeding wheat genotypes for copper use efficiency in Australia. Use of copper-efficient crop species or genotypes within species is an important strategy in correcting copper deficiency. This strategy not only reduces fertilizer cost but also reduces risk of environmental pollution.

REFERENCES

Alva, A. K., B. Huang, and S. Paramasivam. 2000. Soil pH affects copper fractionation and phytotoxicity. *Soil Sci. Soc. Am. J.* 64:955–962.

Baligar, V. C., N. K. Fageria, and Z. L. He. 2001. Nutrient use efficiency in plants. *Commun. Soil Sci. Plant Anal.* 32:921–950.

Barber, S. A. 1995. *Soil nutrient bioavailability: A mechanistic approach,* 2nd edition. New York: John Wiley & Sons.

Bloomfield, C., W. I. Kelso, and G. Pruden. 1976. Reactions between metals and humified organic matter. *J. Soil Sci.* 27:16–31.

Bowen, J. E. 1969. Adsorption of copper, zinc and manganese by sugar cane tissue. *Plant Physiol.* 44:255–261.

Brady, N. C. and R. R. Weil. 2002. *The nature and properties of soils*, 13th edition. Upper Saddle River, NJ: Prentice-Hall.

Cartwright, B. and E. G. Hallsworth. 1970. Effects of copper deficiency on root nodules of subterranean clover. *Plant Soil* 33:685–698.

Cavallaro, N. and M. B. McBridge. 1980. Activities of Cu^{2+} and Cd^{2+} in soil solutions as affected by pH. *Soil Sci. Soc. Am. J.* 44:729–732.

Chaignon, V., F. Bedin, and P. Hinsinger. 2002. Copper bioavailability and rhizosphere pH changes as affected by nitrogen supply for tomato and oilseed rape cropped on an acidic and a calcareous soil. *Plant Soil* 243:219–228.

Cianzio, S. R. 1999. Breeding crops for improved nutrient efficiency: Soybean and wheat as case studies. In: *Mineral nutrition of crops: Fundamental mechanisms and implications*, Z. Rengel, Ed., 267–287. New York: The Haworth Press.

Clark, R. B. 1982. Plant response to mineral element toxicity and deficiency. In: *Breeding plants for the less favorable environments*, M. N. Christiansen and C. F. Lewis, Eds., 71–142. New York: John Wiley & Sons.

Dell, B. 1981. Male sterility and outer wall structure in copper deficient plants. *Ann. Bot.* 48:599–608.

Eleftheriou, E. P. and S. Karataglis. 1989. Ultrastructural and morphological characteristics of cultivated wheat growing on copper polluted fields. *Bot. Acta.* 102:134–140.

El-Kherbawy, M. I. and J. R. Sanders. 1984. Effect of pH and phosphate status of a silty clay loam on manganese, zinc, and copper concentrations in soil fractions and in clover. *J. Sci. Food Agric.* 35:733–739.

Evans, I., J. Demulder, D. Maurice, D. Penney, and E. Solberg. 1990. Take all and other diseases of wheat affected by copper deficiency and herbicide application. *Can. J. Plant Pathol.* 14:241–242.

Evans, I., E. Solberg, and D. M. Huber. 2007. Copper and plant disease. In: *Mineral nutrition and plant disease*, L. E. Datnoff, W. H. Elmer, and D. M. Huber, Eds., 177–188. St. Paul, MN: The American Phytopathological Society.

Evans, L. T. and I. F. Wardlaw. 1976. Aspects of the comparative physiology of grain yield in cereals. *Adv. Agron.* 28:301–359.

Fageria, N. K. 1989. *Tropical soils and physiological aspects of crops*. Brasilia: EMBRAPA Publication Department.

Fageria, N. K. 1992. *Maximizing crop yields*. New York: Marcel Dekker.

Fageria, N. K. 2000. Upland rice response to soil acidity in cerrado soil. *Pesq. Agropec. Bras.* 35:2303–2307.

Fageria, N. K. 2001a. Adequate and toxic levels of copper and manganese in upland rice, common bean, corn, soybean, and wheat grown on an Oxisol. *Commun. Soil Sci. Plant Anal.* 32:1659–1676.

Fageria, N. K. 2001b. Response of upland rice, dry bean, corn and soybean to base saturation in cerrado soil. *Rev. Bras. Eng. Agri. Ambien.* 5:416–424.

Fageria, N. K. 2002a. Influence of micronutrients on dry matte yield and interaction with other nutrients in annual crops. *Pesq. Agropec. Bras.* 37:1765–1772.

Fageria, N. K. 2002b. Micronutrients influence on root growth of upland rice, common bean, corn, wheat and soybean. *J. Plant Nutr.* 25:613–622.

Fageria, N. K. 2004. Influence of dry matter and length of roots on growth of five field crops at varying soil zinc and copper levels. *J. Plant Nutr.* 27:1517–1523.

Fageria, N. K. and V. C. Baligar. 1999. Growth and nutrient concentrations of common bean, lowland rice, corn, soybean, and wheat at different soil pH on an Inceptisol. *J. Plant Nutr.* 22:1495–1507.

Fageria, N. K., V. C. Baligar, and R. B. Clark. 2002. Micronutrients in crop production. *Adv. Agron.* 77:185–268.

Fageria, N. K., V. C. Baligar and R. B. Clark. 2006. *Physiology of crop production*. New York: The Haworth Press.

Fageria, N. K., V. C. Baligar, and C. A. Jones. 1997a. *Growth and mineral nutrition of field crops*, 2nd edition. New York: Marcel Dekker.

Fageria, N. K. and M. P. Barbosa Filho. 2006. *Identification and correction nutrient deficiencies in rice*. Embrapa Arroz e Feijão Circular number 75, 7p, Santo Antonio de Goias, Brazil.

Fageria, N. K., M. P. Barbosa Filho, and L. F. Stone. 2003. Adequate and toxic levels of micronutrients in soil and plants for annual crops. Paper presented at the XXIX Brazilian Soil Science Congress, July 13–18, 2003, Ribeirão Preto, São Paulo, Brazil.

Fageria, N. K., E. M. Castro, and V. C. Baligar. 2004. Response of upland rice genotypes to soil acidity. In: *The red soils of China: Their nature, management and utilization*, M. J. Wilson, Z. He, and X. Yang, Eds., 219–237. Dordrecht: Kluwer Academic Press.

Fageria, N. K. and H. R. Gheyi. 1999. *Efficient crop production*. Campina Grande, Brazil: Federal University of Paraiba.

Fageria, N. K., A. B. Santos, I. D. Lins, and S. L. Camargo. 1997b. Characterization of fertility and particle size of varzea soils of Mato Grosso and Mato Grosso do Sul States of Brazil. *Commun. Soil Sci. Plant Anal.* 28:37–47.

Follett, R. H., L. S. Murphy, and R. L. Donahue. 1981. *Fertilizers and soil amendments*. Englewood Cliffs, New Jersey: Prentice-Hall.

Foth, H. D. and B. G. Ellis. 1988. *Soil fertility*. New York: John Wily & Sons.

Galrão, E. Z. 1999. Methods of copper application and evaluation of its availability for soybean grown on a cerrado Red Yellow Latosol. *R. Bras. Ci. Solo.* 23:265–272.

Gartrell, J. W. 1981. Distribution and correction of copper deficiency in crops and pastures. In: *Copper in soils and plants*, J. F. Loneragan, A. D. Robson, and R. D. Graham, Eds., 313–349. New York: Academic Press.

Geering, H. R. and J. F. Hodgson. 1969. Micronutrient cation complexes in soil solution III. Characterization of soil solution ligands and their complexes with Zn^{2+} and Cu^{2+}. *Soil Sci. Soc. Am. Proc.* 33:54–59.

Graham, R. D. 1978. Nutrient efficiency objectives in cereal breeding. In: *Plant nutrition,* A. R. Ferguson, R. L. Bieleski, and I. B. Ferguson, Eds., 165–170. Auckland, New Zealand Division of Scientific and Industrial Research.

Graham, R. D. 1988. Development of wheats with enhanced nutrient efficiency: Progress and potential. In: *Wheat production constraints in tropical environments: Proceedings of the International Conference, January 19–23, 1987,* A. R. Klatt, Ed., 305–320. Mexico City: International Maize and Wheat Improvement Center.

Hale, M. G., C. D. Foy, and F. J. Shay. 1971. Factors affecting root exudation. *Adv. Agron.* 23:89–109.

Halliwell, B. 1978. Biochemical mechanisms accounting for the toxic action of oxygen on living organisms: The key role of superoxide dismutase. *Cell Biol. Int. Rep.* 2:113–128.

Hallsworth, E. G., E. A. N. Greenwood, and M. G. Yates. 1964. Studies on the nutrition of forage legumes. III. The effect of copper on nodulation of *Trifolium subterraneum* L. and *Trifolium repens. Plant Soil* 20:17–33.

Havlin, J. L., J. D. Beaton, S. L. Tisdale, and W. L. Nelson. 1999. *Soil fertility and fertilizers,* 6th edition. Upper Saddle River, NJ: Prentice Hall.

Hodgson, J. F., W. L. Lindsay, and J. T. Trierweiler. 1966. Micronutrient cation complexing in soil solution. II. Complexing of zinc and copper in displaced solution from calcareous soils. *Soil Sci. Soc. Am. Proc.* 30:723–726.

Jones, J. B., Jr. 1991. Plant tissue analysis in micronutrients. In: *Micronutrient in agriculture,* 2nd edition, J. J. Mortvedt, F. R. Cox, L. M. Shuman, and R. M. Welch, Eds., 477–521. Madison, WI: SSSA.

Lexmond, T. M. and P. D. J. van der Vorm. 1981. The effect of pH on copper toxicity to hydroponically grown maize. *Neth. J. Agric. Sci.* 29:217–238.

Lindsay, W. L. 1979. *Chemical equilibria in soils.* New York: Wiley-Interscience.

McBridge, M. B. 1989. Reactions controlling heavy metal solubility in soils. *Adv. Soil Sci.* 10:1–56.

Marschner, H. 1995. *Mineral nutrition of higher plants,* 2nd edition. New York: Academic Press.

Martens, D. C. and W. L. Lindsay. 1990. Testing soils for copper, iron, manganese and zinc. In: *Soil testing and plant analysis,* 3rd edition, R. L. Westerman, Ed., 229–264. Madison, WI: SSSA.

Martens, D. C. and D. T. Westermann. 1991. Fertilizer application for correcting micronutrient deficiencies. In: *Micronutrients in agriculture,* 2nd edition, J. J. Mortvedt, Ed., 549–592. Madison, WI: SSSA.

Mengel, D. B. 1980. *Role of micronutrients in efficient crop production.* Indiana Coop. Ext. Serv. Ay-239. West Lafayette, IN: Purdue University.

Mengel, K., E. A. Kirkby, H. Kosegarten, and T. Appel. 2001. *Principles of plant nutrition,* 5th edition. Dordrecht: Kluwer Academic Publishers.

Mocquot, B., J. Vangronsveld, H. Clijsters, and M. Mench. 1996. Copper toxicity in young maize (*Zea mays L.*) plants: effects on growth, mineral and chlorophyll contents, and enzyme activities. *Plant Soil* 182: 287–300.

Mortvedt, J. J. 1994. Needs for controlled availability micronutrient fertilizers. *Fert. Res.* 38:213–221.

Ouzounidou, G., M. Ciamporova, M. Moustakas, and S. Karrtaglis. 1995. Response of maize plants to copper stress. I. Growth, mineral content and ultrastructure of roots. *Environ. Exp. Bot.* 35:167–176.

Penning, L. J., D. J. MacPherson, and S. S. Mahli. 1989. Stem melanosis of some wheat, barley, and oat cultivars on a copper deficient soil. *Can. J. Plant Pathol.* 11:65–67.

Pettersson, O. 1976. Heavy metal ion uptake by plants from nutrient solutions with metal ion, plant species, and growth period variations. *Plant Soil* 45:445–459.

Sandermann, G. and P. Boger. 1983. The enzymatological function of heavy metals and their role in the electron transfer processes of plants. In: *Encyclopedia of plant physiology*, New Ser. Vol. 15A. A. Lauchli and R. L. Bicleski, Eds., 563–596. Berlin: Springer-Verlag.

Schnitzer, M. and S. I. M. Skinner. 1967. Organic metallic interactions in soils. Stability constants of Pb-2, Ni-2, Mn-2, CO-2, Ca-2 and Md-2 fulvic acid complex. *Soil Sci.* 103:247–252.

Sims, J. T. and G. V. Johnson. 1991. Micronutrient soil tests. In: *Micronutrient in agriculture,* 2nd edition, J. J. Mortvedt, Ed., 427–476. Madison, WI: SSSA.

Smith, D. H., M. A. Wells, D. M. Porter, and F. R. Cox. 1993. Peanuts. In: *Nutrient deficiencies and toxicities in crop plants,* W. F. Bennett, Ed., 105–110. St. Paul, MN: APS Press, The American Phytopathological Society.

Thornton, I. 1979. Copper in soils and sediments. In: *Copper in the environment, Part 1, Ecological cycling,* J. O. Nriaju, Ed., 172–216. New York: John Wiley and Sons.

Tisdale, S. L., W. L. Nelson, and J. D. Beaton. 1985. *Soil fertility and fertilizers,* 4th edition. New York: Macmillan Publishing Company.

Tyler, G. and T. Olsson. 2001. Concentrations of 60 elements in the soil solution as related to the soil acidity. *Eur. Soil Sci.* 52:151–165.

Wiese, M. V. 1993. Wheat and other small grains. In: *Nutrient deficiencies and toxicities in crop plants,* W. F. Bennett, Ed., 27–33. St. Paul, MN: APS Press, The American Phytopathological Society.

Woolhouse, H. M. and S. Walker. 1981. The physiological basis of copper toxicity and copper tolerance in higher plants. In: *Copper in soils and plants,* J. F. Loneragan, A. D. Robson, and R. D. Graham, Eds., 265–285. Sydney, Australia: Academic Press.

10 Iron

10.1 INTRODUCTION

Iron (Fe) plays many important roles in the growth and development of higher plants. Iron has many functions in plants; however, its main role is its participation in many plant metabolic functions, and it is also a component of many enzymes. Uptake of iron by crop plants exceeds uptake of all other essential micronutrients except chlorine. Iron deficiency and toxicity (flooded rice) are important yield-limiting factors in crop production around the world (Dudal, 1976; Follett et al., 1981; Clark, 1982; Baligar et al., 2001; Fageria et al., 2003; Naeve, 2006). Iron deficiency chlorosis reduces total soybean production in the United States by several million metric tons (Naeve and Rehm, 2006). Corn grown on calcareous, high-pH soils is susceptible to Fe deficiency, which can reduce grain yield by as much as 20% (Godsey et al., 2003). Any factor that decreases the availability of Fe in a soil or competes in the plant absorption process contributes to Fe deficiency (Fageria, 1992; Fageria et al., 1997, 2006; Mengel et al., 2001). Iron deficiency occurs in a variety of soils. Affected soils usually have a pH higher than 6.0 (Fageria et al., 1994). Iron-deficient soils are often sandy, although deficiencies have been found on fine-textured soils, mucks, and peats (Brown, 1961). Iron deficiency is potentially a problem on most calcareous soils (Marschner and Romheld, 1995; Lucena and Chaney, 2007). Calcareous soils are widespread throughout the world (Lombi et al., 2004). The United Nations Food and Agriculture Organization (FAO) estimated the extent of calcareous soils at 800 million ha worldwide, mainly concentrated in areas with arid or Mediterranean climates (Land, FAO, and Plant Nutrition Management, 2000). These soils are important in terms of agricultural production in many areas of the world. For instance in South Australia, about 40% of the wheat is produced on the Eyre Peninsula, which contains over a million hectares of calcareous soils (Holloway et al., 2001). In addition, in Europe alone, more than 60 million US$ are spent every year on Fe chelates to treat Fe chlorosis (Alvarez-Fernandez et al., 2005).

Soybean is susceptible to iron deficiency chlorosis when grown on calcareous soils in the North Central region of the United States (Franzen and Richardson, 2000). In northern China, Fe deficiency chlorosis is a severe and common problem in peanut-producing areas on calcareous soils, which greatly suppresses the overall growth and yield of peanut (Zuo et al., 2000; Gao and Shi, 2007). It is estimated that as much as one-third of the world's land surface is calcareous and potentially associated with iron nutrition deficiency of crops (Follett et al., 1981). Several soil factors have been reported to be associated with iron deficiency in crop plants (Naeve and Rehm, 2006). These include carbonate, specifically HCO_3^-, content (Inskeep and Bloom, 1987; Coulombe et al., 1984a; Naeve and Rehm, 2006); ionic strength of soil solution, as measured by electrical conductivity (Franzen and Richardson, 2000; Naeve and Rehm, 2006); iron (Fe) oxide concentration (Morris et al., 1990); chromium (Cr)-soluble salts; and soil water content (Hansen et al., 2003). Hansen et al. (2003) found that soil moisture content was positively correlated with iron deficiency chlorosis symptoms in producers' fields, indicating that yearly variations in rainfall patterns could affect iron deficiency chlorosis symptoms.

Iron deficiency is a complex disorder and occurs in response to multiple soil, environmental, and genetic factors (Wiersma, 2005). Factors that can contribute to Fe deficiency in plants include low Fe supply from soil; high lime and P application; high levels of heavy metals such as Zn, Cu, and Mn; low and high temperature; high levels of nitrate nitrogen; poor aeration; unbalanced cation ratios; and root infection by nematodes (Fageria et al., 1990). Iron deficiency has been observed in soybean at early growth stages with increasing salinity levels in the rhizosphere with high soil moisture in Brazilian cerrado regions (author's personal observation). However, this deficiency was automatically corrected with decreasing soil moisture and sufficient sunshine for 5 to 6 days. Naeve and Rehm (2006) also reported that iron deficiency in soybean is often most severe early in the growing season and gradually disappears in the mid-growing season. This means that soil, climate, and management practices are responsible for Fe deficiencies in crop plants. Iron toxicity is a serious problem in flooded rice (Fageria, 1984; Fageria et al., 1990). In many parts of the world (Africa, South America, and Asia) where rice is grown on acid soils having great potential for rice production, Fe toxicity is or will be a serious problem (Fageria, 1984, 1989; Fageria et al., 1990).

Iron deficiency is frequently observed in upland rice grown after dry bean and soybean in Brazilian Oxisols with pH higher than 6.5 (Fageria et al., 1994, 2002; Fageria, 2000). This deficiency is not due to low level of iron in the soil but it is not readily available to plants due to precipitation. It is also associated with some genotypes that are more susceptible to iron deficiency. Table 10.1 shows that iron uptake by upland rice significantly decreases with increasing soil pH in a Brazilian Oxisol. Growth of upland rice with increasing soil pH (in the range of 4.6 to 6.9) decreases, and iron deficiency symptoms are very clear in the higher pH range (Figure 10.1). Increasing soil pH in the range of 4.9 to 7.0 decreased Fe content of soil significantly in a linear fashion in a Brazilian Inceptisol (Table 10.2). Similarly, uptake of Fe by dry bean and wheat also decreased significantly in a quadratic fashion with increasing soil pH in the Inceptisol (Table 10.2). Hence, these results indicate that soil pH is an important chemical property in determining uptake of iron in the Oxisols and Inceptisols. Figure 10.2 and Figure 10.3 show iron deficiency symptoms in upland rice grown on Brazilian Oxisols.

TABLE 10.1
Iron Uptake by Upland Rice under Different Soil pH

Soil pH in H_2O	Fe Uptake μg Plant^{-1}
4.6	4541
5.7	1856
6.2	1978
6.4	1634
6.6	1663
6.8	1569
R^2	0.9735**

** Significant at the 1% probability level.
Source: Adapted from Fageria (2000).

FIGURE 10.1 Iron deficiency in upland rice grown in Brazilian Oxisol. Soil pH in water from left to right: 4.6, 5.7, 6.2, 6.6, and 6.8.

TABLE 10.2

Iron Concentration in Soil and Uptake by Dry Bean and Wheat Crops under Different Soil pH in Brazilian Inceptisol

Soil pH in H$_2$O	Fe Content in Soil (mg kg^{-1})	Fe Uptake in Shoot of Dry Bean (mg kg^{-1})	Fe Uptake in Wheat Shoot (mg kg^{-1})
4.9	297	327	153
5.9	231	107	100
6.4	187	93	97
6.7	158	107	97
7.0	148	83	100

Regression Analysis

Soil pH (X) vs. Fe content of soil (Y) = 673.50 − 75.69X, R^2 = 0.55**

Soil pH (X) vs. Fe uptake in dry bean shoot (Y) = 3648.99 −1074.90X + 81.05X^2, R^2 = 0.67**

Soil pH (X) vs. Fe uptake in wheat shoot (Y) = 1117.74 −315.77X + 24.36X^2, R^2 = 0.78**

** Significant at the 1% probability level.

Source: Adapted from Fageria and Baligar (1999).

Bolan et al. (2003) reported that with the exception of Mo, plant availability of most other trace elements decreases with liming due to decrease in the concentration of these elements in soil solution. The decrease in Fe uptake with increasing pH is referred to as lime-induced chlorosis. The effect of pH >6.0 in lowering free metal ion activities in soils has been attributed to the increase in pH-dependent surface charge on oxides of Fe, Al, and Mn (Stahl and James, 1991) or precipitation of metal hydroxides (Lindsay, 1979). The solubility of Fe decreases by ~1000-fold for each unit increase of soil pH in the range of 4 to 9 compared to ~100-fold decreases in the

FIGURE 10.2 Iron deficiency symptoms in upland rice grown on Brazilian Oxisol with pH higher than 6.5.

FIGURE 10.3 Iron deficiency symptoms in upland rice in the early growth stage in Brazilian Oxisol with pH higher than 6.5.

activity of Mn, Cu, and Zn (Lindsay, 1979). With increasing pH, iron is precipitated and availability is reduced according to following equation (Fageria et al., 1994):

$$Fe^{3+} + 3OH^- \Leftrightarrow Fe(OH)_3 \text{ (precipitated)}$$

Iron deficiency and toxicity are important yield-limiting factors in crop production around the world. In addition, iron has special importance in human health. Iron deficiency anemia can be found worldwide, especially in Asia, where rice is the staple food for most of the population. The grain iron concentration varied with genotypes, from 12 to 25 mg Fe kg^{-1} in unhusked rice and 7 to 19 mg Fe kg^{-1} in brown rice (Promuthai and Rerkasem, 2001). In regard to the importance of iron for plants and human being, in this chapter, an attempt is made to summarize the information on the diagnosis of Fe deficiency and toxicity, chemistry of Fe in soil–plant systems, concentration and uptake of Fe, and management practices to correct iron deficiency in crop plants. Information presented in this chapter should provide a basis for correcting this nutritional disorder in crop plants and indicates the new research efforts needed to solve Fe stress problems.

10.2 CYCLE IN SOIL–PLANT SYSTEMS

The iron cycle, like the cycles of other nutrients, involves addition and depletion of this nutrient from soil–plant systems. The main addition sources are inorganic fertilizers, crop residues, organic manures, and liberation from parent materials. The main depletion sources from soil–plant systems are uptake by crops, soil erosion, and immobilization by organic and microbial biomass. The most important aspect of iron chemistry in the soil is not the amount of the metal that is present but the solubility and availability to plants of that which is present (Follett et al., 1981).

Iron, an element essential for all organisms, often occurs in highly insoluble forms within the soil (Emery, 1982). The solubility of Fe in soils is of considerable interest to soil and plant scientists, but a thorough understanding of its chemistry has remained elusive (Schwab and Lindsay, 1983). Iron is a major constituent of the lithosphere, comprising approximately 5.1%; the average content of soil is estimated at 3.8% (Lindsay, 1979). Iron minerals commonly found in soils include goethite ($FeOOH$), hematite (Fe_2O_3), lepidocrocite ($FeOOH$), maghemite (Fe_2O_3), and magnetite (Fe_3O_4). Hematite gives a red color to soils, while geothite imparts a yellow color (Lindsay, 1979; Barber, 1995). Amorphous iron as $Fe(OH)_3$ is probably the most significant supplier of iron for uptake by plants (Barber, 1995). During weathering, these minerals decompose, and the Fe released precipitates as ferric oxides and hydroxides.

Iron is included in a group of heavy metal cations that are held in soils principally on organic or inorganic surfaces or substituted as accessory constituents in common soil minerals (Hodgson, 1963). The divalent form of Fe^{2+} is less strongly held by soil surfaces than Co, Cu, and Zn. When these elements are oxidized to higher valence states, they can form very insoluble oxides and phosphates, which renders these elements much less available to processes of leaching as well as to plants. The solubility of Fe in soils is controlled by $Fe(OH)_3$ in well-oxidized soils, by $Fe_3(OH)_8$ (ferrosic hydroxide) in moderately oxidized soils, and by $FeCO_3$ (siderite) in highly reduced soils (Lindsay and Schwab, 1982). The Fe^{3+} hydrolysis species $Fe(OH)_2{}^+$ and $Fe(OH)_3{}^o$ are the major solution species of inorganic Fe, but they are maintained at too low levels to supply available Fe to plants.

The dissolution and precipitation of ferric oxides is the major factor controlling the solubility of Fe in well-aerated soils. The activity of Fe^{3+} maintained by these oxides is highly pH dependent. The solubility of Fe^{3+} decreases 1000-fold for each unit increase in pH and is decreased to levels below 10^{-20} M as pH rises above 7.5 (Lindsay and Schwab, 1982). Calcium carbonate buffers soils in the general pH range of 7.4 to 8.5. In this pH range, iron oxides attain their minimum solubility, and Fe deficiency in plants is most severe. The Fe^{3+} oxides have different solubilities, which decrease in the order $Fe(OH)_3$ (amorph) > $Fe(OH)_3$(soil) > Fe_2O_3 (maghemite) > $FeOOH$ (lepidocrocite) > Fe_2O_3 (hematite) > $FeOOH$ (goethite) (Lindsay, 1979).

Lindsay (1979) showed that Fe solubility in well-aerated soils tends to approach that of $Fe(OH)_3$ (soil) or "soil-Fe." Its solubility is represented by the following reaction (Lindsay, 1979):

$$Fe(OH)_3 \text{ (soil)} + 3H^+ \Leftrightarrow Fe^{3+} + 3H_2O$$

$$3H_2O \Leftrightarrow 3H^+ + 3OH^-$$

$$Fe(OH)_3 \text{ (soil)} \Leftrightarrow Fe^{3+} + 3OH^-$$

If soluble Fe salts are added to well-aerated soils, they quickly dissolve to precipitate $Fe(OH)_3$ (amorph). Over a period of several weeks, the solubility of Fe decreases slowly and approaches that of soil-Fe.

Ferric ion (Fe^{3+}) hydrolyzes readily in aqueous media to yield a series of hydrolysis products, mainly $FeOH^{2+}$, $Fe(OH)_2^+$, $Fe(OH)_3^0$, $Fe(OH)_4^-$, and $Fe_2(OH)_2^{4+}$ (Lindsay, 1979). The sum of these various hydrolysis species and of Fe^{3+} gives the total soluble inorganic Fe. Measured Fe concentrations in soil solutions are usually in the range 10^{-8} to 10^{-6} M (Bradford et al., 1971; O'Connor et al., 1971; Uren, 1984) and therefore are higher than those predicted from thermodynamic equilibrium calculations. This higher Fe solubility is mainly due to soluble organic complexes (Olomu et al., 1973). Besides concentration, nutrient mobility is also important. The mobility of an element in soil is a reflection of its solution concentration. As such, any factor that affects the solubility of an element in the same way affects the movement.

The presence of soluble substances leached from organic residues or produced through microbial action influences the movement of Fe in two ways: (1) by stabilizing hydrosols of Fe in the soil solution or (2) by forming strictly soluble organic complexes (Hodgson, 1963). From the standpoint of soil fertilization, micronutrient amendments do not generally move far in the soil profile, but organic chelates can promote the movement of Fe along with other heavy metals.

One of the most important factors affecting the mobilization and immobilization of Fe in soils is drainage. The manner and degree to which soil aeration alters the chemistry of Fe is well known. Oxidation of Fe in soils is, of course, a reversible process. As drainage becomes impeded and the oxidation potential (Eh) approaches 0.2 volts, oxides of Fe^{3+} can be reduced. When a soil is submerged, the gaseous exchange between soil and air is interrupted and microorganisms consume trapped oxygen within a few hours. After oxygen depletion, the soil undergoes microbe-mediated sequential reduction.

The redox chemistry of Fe in flooded soils has received a considerable amount of attention. One reason for this interest is the dominating effect that Fe has on the chemistry of flooded soils. Patrick and Reddy (1978) reported that the amount of Fe that can undergo reduction usually exceeds the total amount of other redox elements by a factor of 10 or more. The soluble Fe concentration in reduced soils increases due to dissolution of free Fe oxides, the so-called reductant-soluble Fe (Patrick and Mahapatra, 1968). Sah et al. (1989) demonstrated that it is difficult to predict the nature, rate, and magnitude of Fe transformation in a flooded soil without considering temperature and energy sources (organic matter), since these factors determine microbial activity and, ultimately, the Eh (redox potential) of soil during flooding. Moore and Patrick (1989) studied iron availability and uptake by rice in 134 flooded acid sulfate soils in the central plains region of Thailand and in a growth chamber experiment utilizing 50 of the same soils. The results indicated that Fe^{2+} activities in flooded acid sulfate soils are seldom in equilibrium with pure Fe solid phases under natural conditions. This is believed to be due to (1) transient redox condition, (2) the presence of ill-defined ferric oxides or hydroxides, and/or (3) cation exchange reactions.

Increases in water-soluble and exchangeable iron are favored by a decrease in both redox potential and pH. Gotoh and Patrick (1974) reported that the critical redox potentials for iron reduction and consequent dissolution was between +300 mV and +100 mV at pH 6 and 7 and –100 mV at pH 8, while at pH 5, appreciable reduction occurred at +300 mV. The distribution between water-soluble and exchangeable iron fractions was highly pH dependent. With a decrease in pH at a given redox potential, increasing the relative amount of ferrous iron in the soil solution was at the expense of that on the exchange complex. Gotoh and Patrick (1974) also reported that a thermodynamic approach to the equilibria between solid-phase ferric oxyhydroxide and a water-soluble species of iron (Fe^{2+}) indicated that it was largely governed by the Fe^{2+}-$Fe(OH)_3$ system in which ferric oxyhydroxide was a mixture of goethite and amorphous material.

10.3   FUNCTIONS AND DEFICIENCY SYMPTOMS

Iron is essential for the synthesis of chlorophyll. It is involved in nitrogen fixation, photosynthesis, and electron transfer (Bennett, 1993). As an electron carrier, it is involved in oxidation-reduction reactions (Follett et al., 1981). It is required in protein synthesis and is a constituent of hemoprotein. It is also a component of many enzymes and involved in respiratory enzyme systems as a part of cytochrome and hemoglobin (Bennett, 1993). The catalytic function of iron depends on its electronic structure, which can undergo reversible changes through several oxidation states differing by one electron. Depending on the ligand environment, the redox potential of iron-containing enzymes covers a range of nearly 1000 mV (Bullen and Griffith, 1999; Expert, 2007). Chlorophyll synthesis, thylakoid synthesis, and photosynthesis are dependent on the integrity of many iron-containing proteins, including heme and iron sulfur proteins (Imsande, 1998). The activity of ribulose-1,5-biphosphate carboxylase/oxidase (rubisco) is also reduced under iron deficiency (Expert, 2007). Iron improved the root system of upland rice grown on an Oxisol (Figure 10.4). Fageria (1992) also reported that iron supply at an adequate level improved the root system of rice in nutrient solution.

Iron is immobile in the plant and not translocated from older to newer plant tissues. Hence, deficiency symptoms first appear on the younger tissues or leaves. The deficiency is exhibited as a chlorosis developing intervenally in the new leaves. The area between the veins becomes light green and turns yellow as the deficiency advances. The veins usually remain green except with extremely severe deficiency, where the entire leaf becomes white and translucent. In the beginning, Fe deficiency is similar to Mn deficiency, but at an advanced stage, Fe-deficient leaves are bleached while Mn-deficient leaves form intervenal necrosis, resulting in dead brown tissue

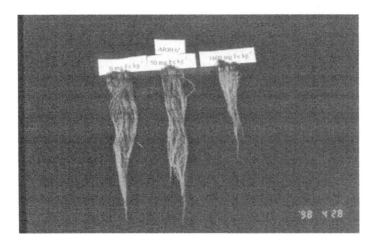

FIGURE 10.4 Root growth of upland rice at 0 mg Fe kg^{-1} soil at left, 50 mg Fe kg^{-1} in the middle, and 1600 mg Fe kg^{-1} of soil at the right.

(Fageria, 1984). Most crop plants are more susceptible to Fe deficiency in the early stage of growth. Plants become stunted in the early seedling stages. If the deficiency is severe and prolonged, plants die. Iron deficiency under field conditions commonly appears as irregularly shaped yellow areas sporadically fading in and out of areas with normally colored plants. The growth of an iron-deficient plant is substantially reduced compared with that of a normal plant. In identifying Fe deficiency in field crops, visual symptoms may be more helpful than plant analysis, because it has been observed that plant Fe content may be greater in chlorotic leaves than in nonchlorotic leaves and that plant Fe may be held in an inactive form as ferritin (Seckback, 1982) or Fe-phosphates (Loeppert and Hallmark, 1985). Iron deficiency delayed flowering and maturity dates in crop plants. Iron-susceptible crops are sorghum, soybean, common bean, proso millet, and rice, while moderately susceptible crops are corn, oats, barley, peanut, alfalfa, and potato. Relatively Fe-efficient crops are wheat, sunflower, and amaranth.

10.4 IRON TOXICITY

Iron toxicity is most common in flooded rice grown on acid soils. Due to flooding, reducing conditions develop and Fe^{3+} is reduced to Fe^{2+}. The reduction of hydrous Fe oxides to Fe^{2+} occurs according to following equation (Ponnamperuma, 1972):

$$Fe(OH)_3 + e^- \Leftrightarrow Fe^{2+} + 3H_2O$$

In this reduction process, H^+ is consumed and soil pH increases. Due to Fe^{3+} reduction to Fe^{2+}, concentration and uptake are increased. Iron toxicity in lowland rice has been reported in South America, Asia, and Africa (Sahu, 1968; Barbosa Filho et al., 1983; Fageria, 1984; Fageria and Rabelo, 1987; Fageria et al., 2003; Sahrawat, 2004). Metal toxicity can be expressed in two ways. Direct toxicity occurs when an excess of the element is absorbed and becomes lethal to the plant cell. On the other hand, toxicity can be related to nutritional imbalance. When excess Fe is present in the growth medium, it may inhibit uptake, transport, and utilization of many other nutrients and induce nutritional deficiency. In flooded rice grown on acid soils, the second type of toxicity is most common. The most important nutrient deficiencies observed in irrigated or flooded rice in Brazil are P, K, and Zn (Fageria et al., 1990). Ottow et al. (1982) also reported that Fe toxicity in rice is the result of multiple nutritional stresses that occur in plants with low levels of K, P, Ca, and Mg.

Iron toxicity causes some physiological disorders in rice plants. These physiological disorders are known as *bronzing, akagare type I*, and *akicochi* (Yoshida, 1981; Tadano, 1995). Bronzing is discoloration of rice leaves. Many small brown spots appear on the leaves. These symptoms start on the tips of lower leaves and spread to the basal parts. In severe cases, the brown discoloration appears even on the upper leaves. The bronzing symptoms vary with the cultivars and may be purplish orange, yellowish brown, reddish brown, brown, or purplish brown (Tadano, 1995).

Akagare type I disorder caused by iron toxicity in rice has been reported in Japan, South Korea, and Sri Lanka (Tadano, 1995). The symptoms of this disorder are dark green leaves and the appearance of small reddish brown spots around the tips of older leaves. Under severe iron toxicity, these symptoms spread all over the plant leaves, and leaves die from the tips. The roots of affected plants turn light brown, dark reddish brown, or black depending on soil types (Tadano, 1995). Yoshida (1981) reported that the main cause of akagare disorder is K deficiency in rice-growing soils.

Akicochi disorder in rice caused by iron toxicity produced small brown spots on the older leaves and may spread to the whole plant foliage under severe iron toxicity. Akicochi disorder symptoms are similar to those of helminthosporium, a fungus disease of rice. Akicochi disorder mainly occurs when soils are deficient in silicon and K. The presence of adequate amounts of silicon and K in the soil solution increases the oxidizing power of rice roots and decreases the excessive uptake of iron (Tadano, 1976; Yoshida, 1981). Application of silicates is normally considered beneficial when the silica content in rice straw is lower than 11% (Yoshida, 1981). Similarly, application of K is desirable when K content of rice soils is lower than 60 mg kg^{-1} (Fageria, 1984). Unlike deficiency symptoms, Fe toxicity symptoms first appear in older leaves. Both top growth and root growth are reduced under higher iron levels (Figure 10.5 and Figure 10.6). Iron toxicity symptoms resemble zinc and P deficiency symptoms (Figure 10.7).

10.4.1 MANAGEMENT PRACTICES TO AMELIORATE FE TOXICITY

Effective measures to ameliorate Fe^{2+} toxicity include periodic surface drainage to oxidize reduced Fe^{2+}, liming of acid soils, use of adequate amounts of essential nutrients, and planting of iron toxic cultivars/genotypes. The Fe-excluding ability of rice plants is lowered by deficiencies of P, K, Ca, and Mg (Fageria et al., 1984; Obata, 1995). In particular, K deficiency readily induces Fe toxicity (Fageria et al., 2003). Among these management practices, use of tolerant cultivars is the most economical and environmentally sound practice. Breeding programs have identified lines

FIGURE 10.5 Iron toxicity in lowland rice in nutrient solution. Left to right, iron concentration was 2.5, 20, 40, and 100 mg Fe L^{-1} of nutrient solution.

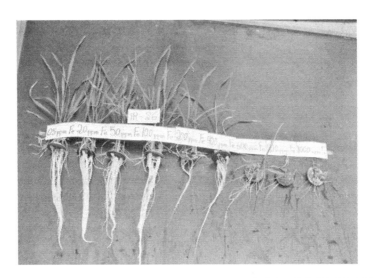

FIGURE 10.6 Roots and tops growth of lowland rice (cv. IR 26) at 0.5, 2.0, 5.0, 10.0, 20.0, 40.0, 60.0, 80.0, and 100 mg Fe L^{-1} nutrient solution.

that show fewer symptoms of toxicity and more vigorous growth on Fe-toxic soils (Masajo et al., 1986; Winslow et al., 1989). Rice roots exude oxygen, and ferrous iron is oxidized at the root surface and deposited there as ferric compounds, allowing little iron to enter the plant. Thus, the oxidizing power of the roots is correlated with the susceptibility to iron toxicity. Figure 10.8 shows that root and top growth of two lowland rice lines are different at 100 mg Fe L^{-1} iron in the nutrient solution. Line CNA 808951, at right, is much more tolerant to iron toxicity than line CNA 810208, at left.

FIGURE 10.7   Iron toxicity symptoms in flooded rice under field conditions.

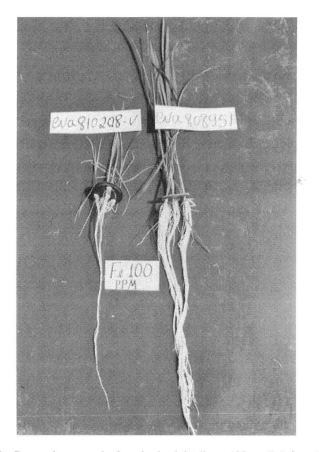

FIGURE 10.8   Root and top growth of two lowland rice line at 100 mg Fe L^{-1} nutrient solution.

Genetic variability in rice cultivars to Fe^{2+} toxicity has been reported (Gunawarkena et al., 1982; Fageria et al., 1984, 1990; Fageria and Rabelo, 1987). Data in Table 10.3 show differences in the shoot dry weight of eight lowland rice cultivars cultivated in solution. Based on the shoot dry weight in Table 10.3, the iron

TABLE 10.3

Shoot Dry Weight (g/4 Plants) of Lowland Rice Cultivars as Influenced by Iron Concentration in Nutrient Solution

Cultivar	Fe Concentration (mg L⁻¹)			
	2	40	60	80
BG 90-2	4.50	2.47	1.50	1.36
Suvale 1	5.88	2.12	1.78	0.63
Paga Divida	4.66	2.35	1.68	1.48
CICA 8	3.17	1.21	0.39	0.18
Bluebelle	2.00	1.44	1.19	0.76
IR 36	1.85	0.40	0.58	0.26
IR 22	1.46	0.67	0.25	0.19
IR 26	1.26	0.18	0.15	0.05
Average	3.10	1.36	0.94	0.61

Source: Adapted from Fageria et al. (1984).

tolerance index (ITE) was calculated and cultivars were classified as tolerant (T), moderately tolerant (MT), or susceptible (S) to iron toxicity (Table 10.4). Cultivars BG 90-2, Suvale 1, and Paga Divida were tolerant at three Fe concentrations. On the other hand, cultivars IR 36, IR 22, and IR 26 were susceptible to three Fe concentrations. Tadano (1976) reported that three mechanisms may be responsible for iron toxicity variability in rice cultivars. These mechanisms are (1) oxidation of Fe^{2+} in the rhizosphere, (2) exclusion of Fe^{2+} at the root surface, and (3) retention of Fe in the root tissues, which prevents translocation of Fe from the root to the shoot.

10.5 CONCENTRATION AND UPTAKE

Knowledge of iron concentration (content per unit dry matter) in plant tissue is a valuable index to identify Fe deficiency or sufficiency in crop plants. According to Jarrell and Beverly (1981), the basic principle of the use of plant analysis is that the chemical composition of the plant reflects the nutrient supply in relation to growth. Iron concentration in plant tissues varied with plant age, crop species, and plant part analyzed. In addition, crop management practices also change iron concentration and uptake by crop plants (Fageria et al., 1997). Data in Figure 10.9 and Figure 10.10 show concentration of Fe in the shoot of upland rice and dry bean, respectively, during crop growth cycles. In upland rice, Fe concentration in the early stage of growth (19 days after sowing) was about 1200 g kg⁻¹ and decreased to about 100 g kg⁻¹ at maturity (130 days after sowing). The variation in Fe concentration was about 91% due to plant age. Similarly, in dry bean, the decrease was not as extensive as in upland rice (Figure 10.10). At 15 days after sowing, the Fe concentration in bean shoot was about 750 mg kg⁻¹ and decreased to about 150 mg kg⁻¹ at harvest (96 days after sowing). The variability in Fe concentration was 98% with the advancement of

TABLE 10.4
Iron Tolerance Index (ITE) and Classification of Lowland Rice Cultivars to Iron Toxicity in Nutrient Solution

Cultivar	Fe Concentration (mg L^{-1})		
	40	60	80
BG 90-2	2.64(T)	2.32(T)	3.24(T)
Suvale 1	2.96(T)	3.59(T)	1.96(T)
Paga Divida	2.60(T)	2.69(T)	3.65(T)
CICA 8	0.91(MT)	0.42(S)	0.30(S)
Bluebell	0.68(MT)	0.82(MT)	0.80(MT)
IR 36	0.18(S)	0.37(S)	0.25(S)
IR 22	0.23(S)	0.13(S)	0.15(S)
IR 26	0.05(S)	0.06(S)	0.03(S)

Note: ITE = (Shoot dry weight at 2 mg Fe L^{-1} level/average shoot dry weight of 8 cultivars at 2 mg Fe L^{-1}) × (Shoot dry weight at 40, 60, or 80 mg Fe L^{-1} levels/average shoot dry weight of 8 cultivars at 40, 60, or 80 mg Fe L^{-1} levels). Shoot dry weight values were taken from Table 10.3. The species having ITE > 1 were classified as tolerant, species having ITE between 0.5 and 1 were classified as moderately tolerant, and species having ITE < 0.5 were classified as susceptible to soil acidity. These efficiency indices were done arbitrarily.

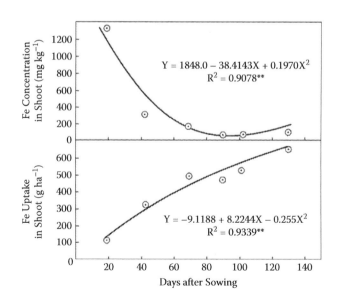

FIGURE 10.9 Iron concentration and uptake in the shoot of upland rice grown on Brazilian Oxisol.

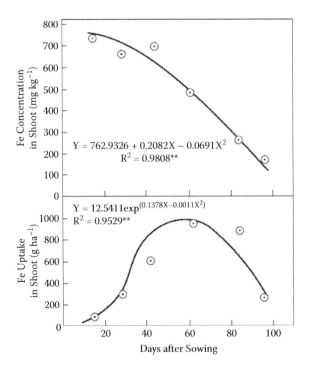

$Y = 762.9326 + 0.2082X - 0.0691X^2$

$R^2 = 0.9808^{**}$

$Y = 12.5411\exp^{(0.1378X - 0.0011X^2)}$

$R^2 = 0.9529^{**}$

Days after Sowing

FIGURE 10.10   Iron concentration and uptake in dry bean shoot at different growth stages.

plant age in dry bean (Figure 10.10). These results clearly show the importance of plant age in determining Fe concentration in plant tissues. Critical levels of mineral nutrients in vegetative or seed tissue often are calculated to determine nutrient concentrations associated with maximum or improved seed quality (Jones et al., 1980; Hitsuda et al., 2004; Wiersma, 2005). Marschner (1995) reported that the critical deficiency level of iron in plant tissues is in the range of 50–150 mg kg⁻¹ of dry weight. Adequate levels of Fe in plant tissues of principal crop species are presented in Table 10.5.

Plants require a continuous supply of Fe to maintain proper growth since very little accumulated Fe is mobilized from older to younger tissues (Karlen et al., 1982; Burton et al., 1998). Uptake of Fe in both crop species increased significantly in a quadratic fashion with the advancement of plant age (Figure 10.9 and Figure 10.10). In upland rice, Fe uptake was about 100 g ha⁻¹ at 19 days after sowing and increased to about 650 g ha⁻¹ at physiological maturity (130 days after sowing). Similarly, in dry bean, Fe uptake was about 100 g ha⁻¹ at 15 days after sowing, increased to 1000 g ha⁻¹ at 62 days after sowing, and then decreased to about 250 g ha⁻¹ at harvest. The decrease in Fe uptake at harvest was associated with translocation to grain and also with loss or fall-down of leaves.

The primary, if not the sole, form in which iron is absorbed by plants (except gramineous monocots) is Fe^{2+} (Chaney et al., 1972). However, in normal aerated soils, Fe^{3+} (ferric ion) is the form present. Iron (Fe^{3+}), therefore, has to be reduced to Fe^{2+} (ferrous ion) before plant roots can absorb it. Reduction of Fe^{3+} to Fe^{2+} is

TABLE 10.5

Adequate Level of Iron in Plant Tissues of Crop Plants

Crop Species	Growth Stage and Plant Part Analyzed	Adequate Fe Level (mg kg^{-1})
Wheat	Preheading, upper leaf blade	25–100
Barley	Heading, whole tops	25–100
Rice	Tillering, whole tops	70–300
Corn	30 to 45 days after emergence, whole tops	50–300
Sorghum	Seedling, whole tops	160–250
Sorghum	Bloom, 3rd blade below panicle	65–100
Soybean	Prior to pod set, upper fully developed trifoliate	51–350
Dry bean	Early flowering, uppermost blade	100–300
Peanut	Early pegging, upper stems and leaves	100–250

Source: Adapted from Fageria et al. (1997).

known to be an obligatory step in the Fe uptake by Fe-efficient species (Chaney et al., 1972). Although research has demonstrated that reduction does take place and that for certain species the release of reductants is enhanced under iron deficiency stress, the exact area in, on, or near the root where this reduction takes place is unknown (Romheld and Marschner, 1985). Chaney et al. (1972) showed that it was exocellular. Bienfait et al. (1983) proposed that an enzyme in the plasmalemma of the cortex or epidermis cell manifests the ferric reeducating activity in iron-deficient bean plants. Romheld and Marschner (1983), working with plants, indicated enzymatic reduction in the plasmalemma of cortical cells of the roots. Plant roots under iron stress absorbed seven to ten times more iron than unstressed roots (Korcak, 1987). Romheld and Marschner (1979), however, demonstrated that most of the proton release and reductant exudation from sunflower roots accrued in the 0–1.5 cm region of the root tip.

Marschner et al. (1986) reported that in the plant kingdom at least two different strategies exist in the Fe deficiency-induced root response, which leads to enhancement of both iron mobilization in the rhizosphere and uptake rate of iron. *Strategy I* is found in all dicots and monocots, with the exception of the grasses (graminaceous species, e.g., barley, corn, sorghum). Strategy I is characterized, in all instances, by an increase in the activity of a plasma membrane–bound reductase, leading to enhanced rates of Fe^{3+} reduction and corresponding splitting of Fe^{3+}-chelates at the plasma membrane (Figure 10.11). Often, the net rate of H^+ extrusion (i.e., acidification of the rhizosphere) is also increased. This acidification facilitates iron uptake by both enhancement of the reductase activity and solubilization of iron in the rhizosphere. This enhanced proton efflux is driven by an increase in plasma membrane H^+-ATPase activity (Romheld, 1987; Schmidt, 1999). An additional mobilization of sparingly soluble iron in the rhizosphere may occur by release of reducing and/or chelating substances (e.g., phenolics) from the Fe-deficient root in response to the acidification. The efficiency of strategy I relies mainly on the supply of Fe chelates to the reductase and its activity respectively. Thus, high HCO_3^- concentrations in the rhizosphere strongly depress the efficiency of strategy I.

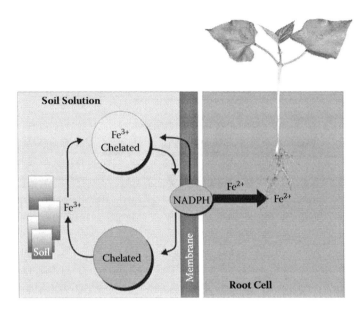

FIGURE 10.11 Mechanism of iron uptake in chelated form by dicotyledonous plants (Strategy I). Adapted from Brady and Weil (2002).

The graminaceous plants—including rice, wheat, barley, rye, and corn—adopt a different strategy, called *strategy II*, which is similar to the strategy of bacterial systems for ion acquisition (Mori, 1999; Expert, 2007). These plants summarize and excrete phytosiderophores into the rhizosphere. Chemically, phytosiderophores belong to the class of mugineic acids. They mobilize iron by complexation with Fe^{3+}; their ferric chelates are then taken up by the root through specific transporters and reduced in the cell (Expert, 2007). This strategy is depicted in Figure 10.12. The release of phytosiderophores is only slightly depressed by high substrate pH and is positively correlated to genotypical differences between species in their resistance to "lime chlorosis." The uptake of Fe phytosiderophores by grasses is mediated by a highly specific system that is absent in species with strategy I. The affinity of this specific system for synthetic or microbial Fe chelates (Fe-siderophores) is very low; it may even be absent. The utilization of iron from these chelates depends on their rate of decay (chelate stability) and thus the supply of inorganic Fe^{3+} as substrate for the phytosiderophores. High substrate pH only slightly depresses the uptake rate of Fe phytosiderophores by grasses. The principal differences between strategy I and strategy II have important ecological implications and also require systematic consideration in the development of screening methods for higher resistance to lime chlorosis. Marschner and Romheld (1995) have provided a comprehensive review of strategies I and II used by plants for Fe acquisition.

Various biologically produced chelating agents enhance iron availability to plants and soil microorganisms. One important group of chelating agents is the siderophores (from the Greek *sideros*, for "iron," and *phoros*, for "carrier"), low-molecular-weight peptide derivatives with extremely high stability constants for Fe^{3+}. Emery (1982) reported that ferrichrome, a corn iron siderophore, will pull Fe^{3+} from Pyrex

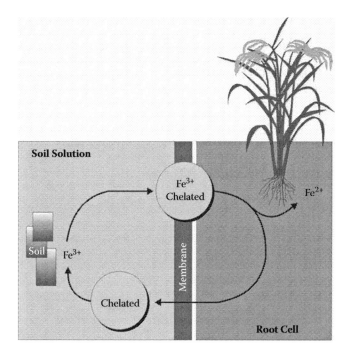

FIGURE 10.12   Mechanism of iron uptake in chelated form by monocotyledonous plants (Strategy II). Adapted from Fageria et al. (2003) and Brady and Weil (2002).

glassware or stainless steel. Siderophores are produced and excreted by almost all aerobic and facultative aerobic microbial species that have been studied (Neilands, 1977). Two classes of siderophores are distinguished by the chemical group that bends the iron: catechols, produced by bacteria; and hydroxymates, also produced by bacteria but primarily by fungi (Powell et al., 1982). Perry et al. (1984) also reported that hydroxymate siderophores are high-affinity Fe^{3+} chelators, which are produced and secreted into soil by various soil microfloras, including mycorrhizal fungi. Microorganisms generally produce siderophores only when they are iron limited, and there is little question that these chelating agents are crucial to the iron nutrition of microbes (Neilands, 1977; Emery, 1982). In addition to iron supply, siderophores may also protect higher plants from soilborne pathogens (Perry et al., 1984).

Two of the most important factors that influence the mobility of micronutrients and heavy metals in soils are pH and organic matter content. Modification in pH level is usually achieved by the addition of lime or S. The addition of lime decreases the availability of most micronutrients except Mo, while S has the opposite effect. The uptake of Fe and Mn is more markedly dependent on pH than that of Zn and Cu (Albasel and Cottonie, 1985).

10.6   USE EFFICIENCY AND Fe HARVEST INDEX

Iron use efficiency values, defined as the shoot dry matter yield or grain yield per unit of Fe uptake in shoot or grain, for upland rice and dry bean are presented in

TABLE 10.6

Iron Use Efficiency (FeUE) in Shoot and Grain of Upland Rice and Dry Bean during Crop Growth Cycle

Plant Age in Days	Upland Rice (kg g⁻¹)	Plant Age in Days	Dry Bean (kg g⁻¹)
19	0.8	15	1.4
43	3.5	29	1.4
68	7.0	43	1.6
90	13.9	62	5.4
102	14.8	84	4.6
130	10.6	96	6.6
130 (grain)	46.8	96 (grain)	13.5

Regression Analysis

Plant age (X) vs. FeUE in rice shoot (Y) $= -6.3432 + 0.3384X - 0.00149X^2$, $R^2 = 0.8482*$

Plant age (X) vs. FeUE in dry bean shoot (Y) $= -0.1662 + 0.0668X$, $R^2 = 0.8299**$

Note: Iron use efficiency (FeUE kg g⁻¹) = (Shoot or grain dry weight in kg/Uptake of Fe in shoot or grain in g).

*, ** Significant at the 5 and 1% probability levels, respectively.

Table 10.6. The Fe use efficiency in upland rice increased quadratically and was higher in grain compared to shoot. Similarly, in dry bean, the relationship between Fe use efficiency and plant age was linear. The Fe use efficiency was higher in grain compared to shoot. The Fe use efficiency in corn and soybean displayed a significant quadratic response with the advancement of plant age (Table 10.7). This type of response may be associated with the increase in dry matter of shoot quadratically with the advancement of plant age. Like rice and dry bean, the Fe use efficiency (kg grain or shoot dry weight per g Fe uptake) was higher for grain compared to shoot dry matter. The Fe use efficiency of grain was higher for the cereals (rice and corn) compared with legumes (dry bean and soybean). The higher Fe use efficiency in cereals may be associated with lower Fe uptake compared to legumes. Overall, iron harvest index (Fe uptake in grain/Fe uptake in grain plus shoot) was higher in cereals compared to legumes (Table 10.8). This means that from the human nutrition point of view, cereals like corn and lowland rice are superior in furnishing iron compared to legumes like dry bean and soybean.

10.7 INTERACTION WITH OTHER NUTRIENTS

Iron's interaction with other nutrients influences its availability to plants. Iron availability is reported to be decreased with the presence P, Zn, Cu, and Mn in the growth medium (Follett et al., 1981). The specific absorption rate of Fe decreased with increasing P supply due to physiological interaction of P and Fe. The inhibition of Fe uptake by P may be related to its competing with the roots for Fe^{2+} and interfering with reduction of Fe^{3+} in solution (Chaney and Coulombe, 1982). According to Elliott and Lauchli (1985), phosphate-induced Fe deficiency in maize evidently involved

TABLE 10.7
Iron Use Efficiency (FeUE) in Shoot and Grain of Corn and Soybean during Crop Growth Cycle

Plant Age in Days	Corn (kg g⁻¹)	Plant Age in Days	Soybean (kg g⁻¹)
18	1.2	27	1.9
35	1.1	41	2.2
53	8.9	62	2.8
69	7.9	82	4.8
84	9.5	120	4.9
119	22.9	158	4.2
119 (grain)	41.7	158 (grain)	20.2

Regression Analysis

Plant age (X) vs. FeUE in corn shoot (Y) = $0.5266 + 0.0146X + 0.0014X^2$, $R^2 = 0.9323*$

Plant age (X) vs. FeUE in soybean shoot (Y) = $-0.5985 + 0.0889X - 0.00037X^2$, $R^2 = .8836*$

Note: Iron use efficiency (FeUE kg g⁻¹) = (Shoot or grain weight in kg/Uptake of Fe in shoot or grain in g).

*,** Significant at the 5 and 1% probability level, respectively.

TABLE 10.8
Iron Uptake and Harvest Index in Principal Field Crops

Crop Species	Fe Uptake in Shoot (g ha⁻¹)	Fe Uptake in Grain (g ha⁻¹)	Total	Fe Harvest Index
Upland rice	654	117	771	0.15
Lowland rice	980	449	1429	0.31
Corn	206	206	412	0.50
Dry bean	268	144	412	0.35
Dry bean	1010	275	1285	0.21
Soybean	778	190	968	0.20

Note: Fe harvest index = (Fe uptake in grain/Fe uptake in grain plus shoot).

both inhibition of Fe absorption by roots and inhibition of Fe transport from roots to shoots.

Iron–molybdenum interactions have been also observed in several plant species but are not well understood. Higher amounts of Mo in the growth medium produced iron chlorosis (Follett et al., 1981). Higher levels of Ca^{2+} inhibit iron uptake, which may be one reason why iron deficiencies for some plants occur on high-calcium soils (Barber, 1995). Nitrogen uptake in the form of NO_3^- can increase soil pH due to liberation of OH^- ions, thus decreasing iron solubility and uptake. If NH_4^+ N is absorbed, the reverse will occur: decrease in soil pH may solubilize iron in the soil, and its uptake may increase. Deficiency of K may disrupt the movement of iron within plants. Lack of K causes iron to accumulate in the stem of corn (Tisdale et al., 1985).

10.8 MANAGEMENT PRACTICES TO MAXIMIZE Fe USE EFFICIENCY

Many management practices have been adopted to overcome iron deficiency problems in crop plants: foliar spray (Goos and Johnson, 2000), Fe seed treatment (Karkosh et al., 1988; Goos and Johnson, 2001; Wiersma, 2005), increased seeding rate (Goos and Johnson, 2000), and companion crops (Naeve, 2006). Planting seed with higher Fe concentration may be especially important in preventing or delaying the onset of iron deficiency chlorosis during seedling establishment and early plant growth (Tiffin and Chaney, 1973). All these practices have had limited success in correcting iron deficiency. Use of foliar spray and Fe-efficient cultivars have been proven to be effective practices in correcting iron deficiency in crop plants (Goos and Johnson, 2000, 2001; Naeve and Rehm, 2006; Fageria et al., 2006). For correcting nutritional disorders in crop plants, including Fe, knowledge of inorganic fertilizer sources and methods of application, as well as knowledge of the soil test to determine the level of nutrient in the soil, is essential. Hence, these practices are discussed in this section.

10.8.1 SOURCE, METHOD, AND RATE OF APPLICATION

Principal iron carriers are presented in Table 10.9. Soil application of inorganic Fe sources is ineffective for controlling Fe chlorosis except at very high rates. These sources either are too insoluble to be effective at low to moderate application rates or become unavailable because of insoluble reaction products formed in soils (Martens and Westerman, 1991). In the 1950s, an effective remedy for lime-induced chlorosis was developed, but the high cost of FeEDDHA (ethylenediamine di-o-hydroxyphenylacetic acid), marketed by Ciba-Geigy Agricultural Chemicals as sequestrate 138, has both hindered its extensive use for field crops and spawned intensive efforts by the chemical and fertilizer industries to develop an inexpensive substitute (Barak

TABLE 10.9
Principal Iron Carriers

Common Name	Formula	Fe Content (%)
Ferrous sulfate	$FeSO_4 \cdot 7H_2O$	19
Ferric sulfate	$Fe_2(SO_4)_3 \cdot 4H_2O$	23
Ferrous oxide	FeO	77
Ferric oxide	Fe_2O_3	69
Ferrous ammonium sulfate	$(NH_4)_2SO_4 \cdot FeSO_4 \cdot 6H_2O$	14
Iron frits	Varies	10–40
Iron ammonium polyphosphate	$Fe(NH_4)HP_2O_7$	22
Ferrous ammonium phosphate	$Fe(NH_4)PO_4 \cdot H_2O$	29
Iron-EDTA chelate	NaFeEDTA	9–12
Iron-HEDTA chelate	NaFeHEDTA	5–9
Iron-EDDHA chelate	NaFeEDDHA	6

Source: Adapted from Fageria (1989), Tisdale et al. (1985), Foth and Ellis (1988).

and Chen, 1982). FeEDDHA has become the standard to which all alternative fertilizers are compared.

The most economical and attractive strategies to correct iron deficiency in crop plants at present are foliar applications and use of efficient cultivars. Application of 1 to 2% ferrous sulfate ($FeSO_4 \cdot 7H_2O$ or $FeSO_4 \cdot H_2O$) solution can correct Fe deficiency in most crop plants. If the deficiency is severe, more than one application is sometimes necessary. One practical difficulty in foliar Fe application is that Fe deficiency normally occurs in the early growth stage of crops, and plants may not have sufficient leaf area to absorb the applied element. Selecting Fe-efficient plant species or cultivars is another effective management practice.

Use of iron sulfate to calcareous soils quickly reacts with $CaCO_3$ to form Fe oxides that are less available for plant uptake. However, some researchers have reported that application of ferrous sulfate in band can correct iron chlorosis in annual crops. For example, Godsey et al. (2003) reported that application of 81 kg ha^{-1} $FeSO_4 \cdot H_2O$ applied in seed row was the most consistent treatment for correcting Fe deficiency in corn. Similarly, Hergert et al. (1996) reported that corn yield was increased of nearly 3.6 Mg ha^{-1} on two soils with pH greater than 8.2 by placing 85 kg ha^{-1} $FeSO_4 \cdot 7H_2O$ directly in the seed row. Mathers (1970) indicated that 112 kg ha^{-1} $FeSO_4 \cdot 7H_2O$ banded 20 cm beneath the soil surface and directly under the seed increased grain yield of sorghum by 1.0 Mg ha^{-1} compared with the control treatment. However, some researchers have experienced mixed results with the addition of $FeSO_4 \cdot 7H_2O$. Yield response varies greatly from year to year, and some research has indicated no significant response to the addition of $FeSO_4 \cdot 7H_2O$ (Olson, 1950; Mortvedt and Giordano, 1970).

10.8.2 Soil Test to Identify Critical Fe Level

Generally, soil test is used to identify nutrient deficiency and correct nutritional disorders in crop plants. As far as Fe is concerned, several extractants have been compared to establish critical Fe level in soils and plant tissues (Fageria et al., 1990). Norvell (1984) compared five chelating agents, DTPA (diethylenetriaminepenta-acetic acid), EDTA (ethylenediaminetetraacetic acid), HEDTA (hydroxyethylethyl-enediaminetriacetic acid), NTA (nitrilotriacetic acid), and EGTA [ethyleneglycol-bis (2-aminoethylether) tetraacetic acid]. The DTPA soil test extractant was prepared according to the method of Lindsay and Norvell (1978) to contain 0.005 M DTPA and 0.01 M $CaCl_2$. The pH was buffered by 0.1 M triethanolamine (TEA) adjusted to pH 7.3 with HCl. The other chelating extractants were prepared to contain 0.005 M chelating agent and 0.01 M $CaCl_2$ buffered at pH 5.3 by a mixture of acetic acid and ammonium acetate, which was 0.1 M in total acetate. These authors concluded that DTPA and HEDTA were among the most effective ligands for extracting Fe, Mn, Zn, Cu, Al, Ni, and Cd. The DTPA extractant can be used for neutral and alkaline as well as acid soils (Norvell, 1984). Critical levels for DTPA in the United States and tropics ranged from 2.5 to 5.0 mg kg^{-1} extractable Fe (Sims and Johnson, 1991). For irrigated corn, DTPA extractable Fe critical level is reported to be 4.5 mg kg^{-1} (Godsey et al., 2003).

10.8.3 Use of Efficient Crop Species/Genotypes

Plants differ in their uptake of Fe, and it has been suggested that monocotyledonous species are less Fe efficient than dicotyledonous species (Brown and Jones, 1976). The results of detailed experiments by Chaney et al. (1972) provided evidence regarding the chemical nature of enhanced Fe uptake by efficient soybean in nutrient solutions. Chaney et al. (1972) concluded the following from these results:

1. Fe^{2+} is the primary form of Fe available to plants.
2. Fe^{3+} is reduced to Fe^{2+} by soybean-chlorotic, efficient soybean prior to uptake. This may occur at specific sites of reduction on the root, by enzymatic reduction, and in solution by chemical reducing agents excreted by the plant into the rhizosphere. Phenolic, malic, and caffeic acid may be such compounds released by plant roots.
3. Reduction makes Fe^{3+} complexes such as Fe^{3+} EDDHA unstable and increases the availability of Fe to plants.
4. Strong Fe^{2+} chelators can restrict Fe uptake by plant roots under the reducing conditions imposed by the Fe-stressed plants.

Experiments using a similar approach (Romheld and Marschner, 1983, 1985) supported many of these conclusions. Chlorotic soybean roots also lower the pH of their environments, and this increased acidity could make Fe more available from Fe^{3+} EDDHA without imposing reduction conditions.

At present, studies on physiological responses to Fe deficiency and screening for Fe-efficient genotypes have already made significant progress in some crops, such as soybean, dry bean, sorghum, and tomato (Chaney et al., 1989; Jolley et al., 1996; Fairbanks, 2000). Peanut cultivars of different Fe absorption efficiencies have been identified in Israel by Hartzook (1982) and Hartzook et al. (1974). Similarly, Gao and Shi (2007) reported significant differences in resistance to Fe deficiency chlorosis in peanut cultivars grown on calcareous soils in China.

The ability of plant species to extract Fe from soils varies widely. Kashirad and Marschner (1974) showed that sunflower plants reduced the pH of solution and thus overcame Fe chlorosis. However, corn plants did not reduce the pH and failed to utilize precipitated ferric compounds to overcome Fe chlorosis. Mortvedt et al. (1992) suggested that use of hydrated polymers as a matrix for Fe fertilizers may provide a novel method for improving the efficiency of Fe uptake by plants in calcareous soils. Root Fe^{3+}-reducing activity of soybean genotypes was correlated with genotypic resistance to Fe chlorosis measured in field nurseries and was used as a method for identifying chlorosis-resistant (Fe-efficient) genotypes (Rengel, 1999). Soybean genotypes vary in the magnitude and timing of Fe stress responses; resistant genotypes reduce Fe^{3+} sooner and in larger quantities than susceptible genotypes (Jolley et al., 1992; Ellsworth et al., 1998). Rye, a very tolerant species to Fe deficiency in calcareous soils, exudes mugineic acid, hydroxymugineic acid (HMA), and DMA into the rhizosphere. It has been suggested that the capacity of rye to exude HMA may be related to superior tolerance of rye to Fe, as well as Zn deficiency, when compared to wheat, which exudes only DMA (Cakmak et al., 1996; Rengel, 1999).

TABLE 10.10

Crop Species Classification to Iron Use Efficiency

Efficient	Moderately Efficient	Inefficient
Wheat	Corn	Sorghum
Sunflower	Oats	Soybean
Dry bean	Barley	Upland rice
Amaranth	Alfalfa	Peanut
	Potato	Millet

Source: Adapted from Clark (1991) and Fageria et al. (1994).

TABLE 10.11

Classification of Upland Rice Genotypes for Iron Use Efficiency

Genotype	Grain Yield (kg ha^{-1})	Rating	Classification
CNA 7460	2679	3	Highly efficient
CNA 6895	2890	3	Highly efficient
CNA 6843-1	2570	5	Moderately efficient
Aaguaia	2539	7	Moderately efficient
CNA 7286	2398	7	Moderately efficient
CNA 7475	2375	7	Moderately efficient
CNA 7449	2218	9	Inefficient

Source: Fageria et al. (1994).

Aqueous solutions of $FeSO_4$, $Fe_2(SO_4)_3$, or other inorganic Fe salts could be absorbed by polymers to form gels that are band applied to soil. Because some polymers are only slowly biodegraded in soil, it is possible that these hydrogels could provide available Fe for crops for a long period of time. Technologies to correct Fe deficiencies in field, vegetable, and tree crops were reviewed by Mortvedt (1991). Similarly, controlled availability of micronutrients including iron was discussed by Mortvedt (1994). Readers may refer to this article for detailed information about the subject. Crop species that are efficient or inefficient in iron use are cited in Table 10.10. Similarly, upland rice genotypes classified as efficient or inefficient in iron uptake are listed in Table 10.11.

10.9   BREEDING FOR Fe EFFICIENCY

Breeding for mineral efficiency or stresses requires genetic information, especially the number of genes involved. Genetic studies on the inheritance of Fe deficiency indicated that the trait was conditioned by a single gene, with efficiency being dominant over inefficiency (Cianzio, 1999). Cianzio and Fehr (1980) reported that a single major gene and several modifying genes could explain segregation from a cross of Fe-efficient and Fe-inefficient genotypes. Recurrent selection has been used for the development of improved Fe-efficient soybean germplasm (Cianzio, 1999). Cianzio and Fehr (1980) reported that for breeding purposes, the trait is therefore considered a quantitative character. Cianzio (1999) further reported that selection for improved Fe efficiency must always be considered simultaneously with selection for yield to obtain genotypes that can be released for use by farmers.

Iron deficiency in soybean and sorghum has been successfully corrected by use of iron-efficient genotypes of these crops in the United States. In a survey by Hansen et al. (2003) of Minnesota (USA) growers, respondents identified their current management practices to prevent iron deficiency chlorosis. Producers used variety selection (70%), seeding management (42%, including seeding rate, row spacing, and planting date), and fertility management (11%, including foliar, seed, and soil treatments; Naeve, 2006). This means that use of iron-efficient genotypes is the most dominant practice to correct iron deficiency in the region.

For breeding iron-efficient cultivars, appropriate selection methodology is the first step to achieve meaningful results. The results of genotypes versus environment (G × E) studies with iron chlorosis are not consistent. For example, Niebur and Fehr (1981) evaluated soybean genotype responses to calcareous and noncalcareous environments with the addition of foliar-applied chelated Fe to remove the confounding effect of iron deficiency chlorosis. They concluded that selection for yield quality and harvest traits can be made among iron deficiency chlorosis–tolerant lines in noncalcareous environments. Jessen et al. (1988) and Coulombe et al. (1984b) found field iron deficiency scores for soybean genotypes to be highly correlated with analogous scores from nutrient solution systems. Both systems used increased HCO_3^- concentration in nutrient solutions to increase iron deficiency severity. Charlson et al. (2003), however, found visual iron deficiency chlorosis scores among F2:4 derived lines of soybean to vary substantially when grown at two locations with calcareous soils. Similarly, Morris et al. (1990) found a strong interaction among soybean genotypes and soil factors, indicating a better understanding of G × E effects among Fe-efficient genotypes. Naeve and Rehm (2006) found soybean genotype × environment interactions for visual iron deficiency scores even when significant effects on yield were not found, indicating that variety screening based on iron deficiency chlorosis scores requires multiple locations to be predictive.

10.10 CONCLUSIONS

Iron requirements of plants exceed their requirements for all other micronutrients except chlorine. Iron deficiency is most common on neutral to alkaline and sandy soils and toxicity under reduced conditions (flooded rice). Chlorosis caused by iron deficiency is referred to as lime-induced chlorosis. In flooded rice, Fe^{3+} reduces to Fe^{2+} in anaerobic soils, and the concentration of Fe^{2+} in the soil solution significantly increases and leads to toxicity. Iron is involved in nitrogen fixation, chlorophyll formation, protein synthesis, enzyme systems, plant respiration, and photosynthesis and energy transfer. It is an immobile nutrient in plants; hence, its deficiency first starts in the younger tissues or leaves. Solubility and uptake of iron is highly pH dependent, decreasing with increasing soil pH. In addition, the concentration and reactivity of soil carbonates and the concentration and quality of soil salts are also associated with iron chlorosis.

Soil and plant tissue tests are generally performed to identify Fe deficiency or sufficiency. However, soil and plant tests for Fe availability should be used with caution because many factors aside from an extractable amount of Fe affect its availability. For example, in Oxisols of Brazil, when soil pH is raised to around 6.0, iron deficiency is frequently observed in upland rice. Soil analysis always reveals more than 50 mg Fe kg^{-1} of soil. In this case, Fe availability is the main problem rather than iron content of the soil. Similarly, inactivation of iron in the plant is very common, and iron-deficient leaves may have higher concentration of Fe as compared to normal leaves.

Generally, plants employ two strategies to overcome Fe deficiency under Fe stress conditions. These strategies are designated as strategies I and II. Strategy I, which is characteristic for dicots and nongraminaceous monocots, involves increased reduction of Fe^{3+} to Fe^{2+} at the root–cell plasma membrane, acidification of the rhizosphere, enhanced exudation of reducing and/or chelating compounds into the rhizosphere, and changes in root morphology to improve Fe uptake. Strategy II is employed by grasses or graminaceous species and involves increased exudation of phytosiderophores into the rhizosphere. Both strategies are activated under iron stress conditions.

Iron deficiency is very difficult to correct by inorganic fertilization, because Fe fertilizers applied to the soil are rapidly converted to insoluble forms and have very limited effectiveness. Iron chelates are very effective in correcting iron deficiency, but due to high cost, their use is very limited in annual crops. Foliar application is one of the most effective measures to correct iron deficiency. In addition, planting iron-efficient species or genotypes within species is another important management strategy to correct iron deficiency. This strategy is most commonly used in the United States to correct iron deficiency in sorghum and soybean.

REFERENCES

Albasel, N. and A. Cottonie, 1985. Heavy metals uptake from contaminated soils as affected by peat, lime, and chelates. *Soil Sci. Soc. Am. J.* 49:386–390.

Alvarez-Fernandez, A., S. Garcia-Marco, and J. J. Lucena. 2005. Evaluation of synthetic iron (III)-chelates (EDDHA/Fe^{3+}), EDDHMA/Fe^{3+} and EDDHSA/Fe^{3+}) to correct iron chlorosis. *European J. Agron.* 22:119–130.

Baligar, V. C., N. K. Fageria, and Z. L. He. 2001. Nutrient use efficiency in plants. *Commun. Soil Sci. Plant Anal.* 32:921–950.

Barber, S. A. 1995. *Soil nutrient bioavailability: A mechanistic approach.* New York: John Wiley & Sons.

Barak, P. and Y. Chen. 1982. The evaluation of iron deficiency using a bioassay-type test. *Soil Sci. Soc. Am. J.* 46:1019–1022.

Barbosa Filho, M. P., N. K. Fageria, and L. F. Stone. 1983. Water management and liming in relation to grain yield and iron toxicity. *Pesq. Agropec. Bras.* 18:903–910.

Bennett, W. F. 1993. Plant nutrient utilization and diagnostic plant symptoms. In: *Nutrient deficiencies and toxicities in crop plants*, W. F. Bennett, Ed., 1–7. St. Paul, MN: The APS Press, The American Phytopathological Society.

Bienfait, H. F., R. J. Bino, A. M. Blick, J. F. Duivenvoorden, and J. M. Fontain. 1983. Characterization of ferric reducing activity in roots of Fe-deficient *Phaseolus vulgaris*. *Physiol. Plant.* 59:196–202.

Bolan, N. S., D. C. Adriano, and D. Curtin. 2003. Soil acidification and liming interactions with nutrient and heavy metal transformation and bioavailability. *Adv. Agron.* 78:215–272.

Brady, N. C. and R. R. Weil. 2002. *The nature and properties of soils*, 13th edition. Upper Saddle River, NJ: Prentice-Hall.

Bradford, G. R., F. L. Blair, and V. Hunsaker. 1971. Trace and major element contents of soil saturation extracts. *Soil Sci.* 112:225–230.

Brown, J. C. 1961. Iron chlorosis in plants. *Adv. Agron.* 13:329–369.

Brown, J. C. and W. E. Jones. 1976. A technique to determine iron efficiency in plants. *Soil Sci. Soc. Am. J.* 40:398–405.

Bullen, J. J. and E. Griffith. 1999. *Iron and infection: Molecular, physiological and clinical aspects.* New York: John Wiley & Sons.

Burton, J. W., C. Harlow, and E. C. Theil. 1998. Evidence for reutilization of nodule iron in soybean seed development. *J. Plant Nutr.* 21:913–927.

Cakmak, I., L. Ozturk, S. Karanlik, H. Marschner, and H. Ekiz. 1996. Zinc-efficient wild grasses enhance release of phytosiderophores under zinc deficiency. *J. Plant Nutr.* 19:551–563.

Chaney, R. L., P. F. Bell, and B. A. Coulombe. 1989. Screening strategies for improved nutrient uptake and use by plants. *Hort. Science* 24:565–572.

Chaney, R. L., J. C. Brown, and L. O. Tiffin. 1972. Obligatory reduction of ferric chelates in iron uptake by soybeans. *Plant Physiol.* 50:208–213.

Chaney, R. L. and B. A. Coulombe. 1982. Effect of phosphate on regulation of Fe-stress response in soybean and peanut. *J. Plant Nutr.* 5:469–487.

Charlson, D. V., S. R. Cianzio, and R. C. Shoemaker. 2003. Association of SSR markers with soybean resistance to iron deficiency chlorosis. *J. Plant Nutr.* 26:2267–2276.

Cianzio, S. R. 1999. Breeding crops for improved nutrient efficiency: Soybean and wheat as case studies. In: *Mineral nutrition of crops: Fundamental mechanisms and implications*, Z. Rengel, Ed., 267–287. New York: The Haworth Press.

Cianzio, S. R. and W. R. Fehr. 1980. Genetic control of iron deficiency chlorosis in soybeans. *Iowa State J. Res.* 54:367–375.

Clark, R. B. 1982. Plant response to mineral element toxicity and deficiency. In: *Breeding plants for the less favorable environments*, M. N. Christiansen and C. F. Lewis, Eds., 71–142. New York: John Wiley & Sons.

Coulombe, B. A., R. L. Chaney, and W. J. Wiebold. 1984a. Bicarbonate directly induces iron chlorosis in susceptible soybean cultivars. *Soil Sci. Soc. Am. J.* 48:1297–1301.

Coulombe, B. A., R. L. Chaney, and W. J. Wiebold. 1984b. Use of bicarbonate in screening soybeans for resistance to iron chlorosis. *J. Plant Nutr.* 7:411–425.

Dudal, R. 1976. Inventory of the major soils of the world with special reference to mineral stress hazards. In: *Plant adaptation to mineral stress in problem soils*, M. J. Wright, Ed., 3–13. Ithaca, NY: Cornell University Press.

Elliott, G. C. and A. Lauchli. 1985. Phosphorus efficiency and phosphate-iron interaction in maize. *Agron. J.* 77:339–403.

Ellsworth, J. W., V. D. Jolley, D. S. Nuland, and A. D. Blaylock. 1998. Use of hydrogen release or a combination of hydrogen release and iron reduction for selecting iron-efficient dry bean and soybean cultivars. *J. Plant Nutr.* 21:2639–2651.

Emery, T. 1982. Iron metabolism in humans and plants. *Am. Sci.* 27:626–632.

Expert, D. 2007. Iron and plant disease. In: *Mineral nutrition and plant disease*, L. E. Datnoff, W. H. Elmer, and D. M. Huber, Eds., 119–137. St. Paul, MN: The American Phytopathological Society.

Fageria, N. K. 1984. *Fertilization and mineral nutrition of rice*. EMBRAPA/CNPAF, Goiania, Editora Campus, Rio de Janeiro, Brazil.

Fageria, N. K. 1989. *Tropical soils and physiological aspects of crops*. Brasilia: EMBRAPA Publication Department.

Fageria N. K. 1992. *Maximizing crop yields*. New York: Marcel Dekker.

Fageria, N. K. 2000. Upland rice response to soil acidity in cerrado soil. *Pesq. Agropec. Bras.* 35:2303–2307.

Fageria, N. K. 2002a. Influence of micronutrients on dry matte yield and interaction with other nutrients in annual crops. *Pesq. Agropec. Bras.* 37:1765–1772.

Fageria, N. K. 2002b. Micronutrients influence on root growth of upland rice, common bean, corn, wheat and soybean. *J. Plant Nutr.* 25:613–622.

Fageria, N. K. and V. C. Baligar. 1999. Growth and nutrient concentrations of common bean, lowland rice, corn, soybean, and wheat at different soil pH on an Inceptisol. *J. Plant Nutr.* 22:1495–1507.

Fageria, N. K., V. C. Baligar, and R. B. Clark. 2002. Micronutrients in crop production. *Adv. Agron.* 77:185–268.

Fageria, N. K., V. C. Baligar, and R. B. Clark. 2006. *Physiology of crop production*. New York: The Haworth Press.

Fageria, N. K., V. C. Baligar, and C. A. Jones. 1997. *Growth and mineral nutrition of field crops*, 2nd edition. New York: Marcel Dekker.

Fageria, N. K., V. C. Baligar, and R. J. Wright. 1990. Iron nutrition of plants: An overview on the chemistry and physiology of its deficiency and toxicity. *Pesq. Agropec. Bras.* 25:553–570.

Fageria, N. K., M. P. Barbosa Filho, and J. R. P. Carvalho. 1981. Influence of iron on growth and absorption of P, K, Ca and Mg by rice plant in nutrient solution. *Pesq. Agropec. Bras.* 16:483–488.

Fageria, N. K., M. P. Barbosa Filho, J. R. P. Carvalho, P. H. N. Rangel, and V. A. Cutrim. 1984. Preliminary screening of rice cultivars for tolerance to iron toxicity. *Pesq. Agropec. Bras.* 19:1271–1278.

Fageria, N. K., C. M. Guimarães, and T. A. Portes. 1994. Iron deficiency in upland rice. *Lav. Arrozeira* 47:3–5.

Fageria, N. K. and N. A. Rabelo. 1987. Tolerance of rice cultivars to iron toxicity. *Journal of Plant Nutrition* 10:653–661.

Fageria, N. K., L. F. Stone, and A. B. Santos. 2003. Soil fertility management for irrigated rice. Embrapa Arroz e Feijão, Santo Antônio de Goiás, Brazil.

Fairbanks, D. J. 2000. Development of genetic resistance to iron-deficiency chlorosis in soybean. *J. Plant Nutr.* 23:1903–1913.

Follett, R. H., L. S. Murphy, and R. L. Donahue. 1981. *Fertilizers and soil amendments*. Englewood Cliffs, New Jersey: Prentice-Hall.

Foth, H. D. and B. G. Ellis. 1988. *Soil fertility*. New York: John Wiley & Sons.

Franzen, D. W. and J. L. Richardson. 2000. Soil factors affecting iron chlorosis of soybean in the Red River Valley of North Dakota and Minnesota. *J. Plant Nutr.* 23:67–78.

Gao, Li and Y. Shi. 2007. Genetic differences in resistance to iron deficiency chlorosis in peanut. *J. Plant Nutr.* 30:37–52.

Godsey, C. B., J. P. Schmidt, A. J. Schlegel, R. K. Taylor, C. R. Thompson, and R. J. Gehl. 2003. Correcting iron deficiency in corn with seed row applied iron sulfate. *Agron. J.* 95:160–166.

Goos, R. J. and B. E. Johnson. 2000. A comparison of three methods for reducing iron deficiency chlorosis in soybean. *Agron. J.* 92:1135–1139.

Goos, R. J. and B. E. Johnson. 2001. Seed treatment, seeding rate, and cultivar effects on iron deficiency chlorosis of soybean. *J. Plant Nutr.* 24:1255–1268.

Gotoh, S. and W. H. Patrick, Jr. 1974. Transformation of iron in a waterlogged soil as influenced by redox potential and pH. *Soil Sci. Soc. Am. Proc.* 38:66–71.

Hansen, N. C., M. A. Schmitt, J. E. Anderson, and J. S. Strock. 2003. Iron deficiency of soybean in the upper Midwest and associated soil properties. *Agron. J.* 95:1595–1601.

Hartzook, A. 1982. The problem of iron deficiency in peanut (*Arachis hypogaea* L.) on basic and calcareous soils in Israel. *J. Plant Nutr.* 5:923–926.

Hartzook, A. 1984. The performance of iron absorption efficient peanut cultivars on calcareous soils in the Lakhish and Beisan valley region in Israel. *J. Plant Nutr.* 7:407–409.

Hergert, G. W., P. T. Nordquist, J. L. Petersen, and B. A. Skates. 1996. Fertilizer and crop management practices for improving maize yields on high pH soils. *J. Plant Nutr.* 19:1223–1233.

Hitsuda, K., G. J. Sfredo, and D. Klepker. 2004. Diagnosis of sulfur deficiency in soybean using seeds. *Soil Sci. Soc. Am. J.* 68:1445–1451.

Hodgson, J. F. 1963. Chemistry of the micronutrient elements in soils. *Adv. Agron.* 15:119–159.

Holloway, R. E., I. Bertrand, A. J. Frischke, D. M. Brace, M. J. McLaughlin, and W. Shepperd. 2001. Improving fertilizer efficiency on calcareous and alkaline soils with fluid sources of P, N, and Zn. *Plant Soil* 236:209–219.

Imsande, J. 1998. Iron, sulfur, and chlorophyll deficiencies: A need for an integrative approach in plant physiology. *Physiol. Plant.* 103:139–144.

Inskeep, W. P. and P. R. Bloom. 1987. Soil chemical factors associated with soybean chlorosis in calciaquolls of western Minnesota. *Agron. J.* 79:779–786.

Jarrell, W. M. and R. B. Beverly. 1981. The dilution effect in plant nutrition studies. *Adv. Agron.* 34:197–224.

Jessen, H. J., M. B. Dragonuk, R. W. Hintz, and W. R. Fehr. 1988. Alternative breeding strategies for the improvement of iron efficiency in soybean. *J. Plant Nutr.* 11:717–726.

Jolley, V. D., K. A. Cook, N. C. Hansen, and W. B. Stevens. 1996. Plant physiological responses for genotypic evaluation of iron efficiency in Strategy I and Strategy II plants: A review. *J. Plant Nutr.* 19:1241–1255.

Jolley, V. D., D. J. Fairbanks, W. B. Stevens, R. E. Terry, and J. H. Orf. 1992. Using root-reduction capacity for genotype evaluation of iron efficiency in soybean. *J. Plant Nutr.* 15:1679–1690.

Jones, J. B., Jr. 1991. Plant tissue analysis in micronutrients. In: *Micronutrient in agriculture*, 2nd edition, J. J. Mortvedt, F. R. Cox, L. M. Shuman, and R. M. Welch, Eds., Madison, WI: SSSA.

Jones, M. B., J. E. Ruckman, W. A. Williams, and R. L. Koenigs. 1980. Sulfur diagnostic criteria as affected by age and defoliation of subclover. *Agron. J.* 72:1043–1046.

Karkosh, A. E., A. K. Walker, and J. J. Simons. 1988. Seed treatment for control of iron deficiency chlorosis of soybean. *Crop Sci.* 28:369–370.

Karlen, D. L., P. G. Hunt, and T. A. Matheny. 1982. Accumulation and distribution of P, Fe, Mn and Zn by selected determinate soybean cultivars grown with and without irrigation. *Agron. J.* 74:297–303.

Kashirad, A. and H. Marschner. 1974. Effect of pH and phosphate on iron nutrition of sunflower and corn plants. *Agrochimica* 18:497–508.

Korcak, R. F. 1987. Iron deficiency chlorosis. *Hort. Rev.* 9:133–186.

Land, FAO and Plant Nutrition Management. 2000. Prosoil-problem soil database. htp:www.fao.org/ag/AGL/agll/prosoil/default.htm (verified Dec. 16, 2003). FAO, Rome, Italy.

Lindsay, W. L. 1979. *Chemical equilibria in soils.* New York: Wiley-Interscience.

Lindsay, W. L. and W. A. Norvell. 1978. Development of a DTPA soil test for zinc, iron, manganese and copper. *Soil Sci. Soc. Am. J.* 42:421–428.

Lindsay, W. L. and A. P. Schwab. 1982. The chemistry of iron in soils and its availability to plants. *J. Plant Nutr.* 5:321–340.

Loeppert, R. H. and C. T. Hallmark. 1985. Indigenous soil properties influencing the availability of iron in calcareous soils. *Soil Sci. Soc. Am. J.* 49:597–603.

Lombi, E., M. J. McLaughlin, C. Johnston, R. D. Armstrong, and R. E. Holloway. 2004. Mobility and lability of phosphorus from granular and fluid monoammonium phosphate differs in a calcareous soil. *Soil Sci. Soc. Am. J.* 68:682–689.

Lucena, J. J. and R. L. Chaney. 2007. Response of cucumber plants to low doses of different synthetic iron chelates in hydroponics. *J. Plant Nutr.* 30:795–809.

Marschner, H. 1995. *Mineral nutrition of higher plants*, 2nd edition. New York: Academic Press.

Marschner, H. and V. Romheld. 1995. Strategies of plants for acquisition of iron. In: *Iron nutrition in soils and plants*, J. Abadia, Ed., 375–378. Dordrecht: Kluwer Academic Publishers.

Martens, D. C. and W. L. Lindsay. 1990. Testing soils for copper, iron, manganese and zinc. In: *Soil testing and plant analysis*, 3rd edition, R. L. Westerman, Ed., 229–264. Madison, WI: SSSA.

Martens, D. C. and D. T. Westermann. 1991. Fertilizer application for correcting micronutrient deficiencies. In: *Micronutrients in agriculture*, 2nd edition, J. J. Mortvedt, Ed., 549–592. Madison, WI: SSSA.

Masajo, T. M., K. Alluri, A. O. Abifarin, and D. Janakiram. 1986. Breeding for high stable yields in Africa. In: *The wetland and rice in subsaharan Africa,* A. S. R. Juo and J. A. Lowe, Eds., 111–112. Ibadan, Nigeria: Int. Inst. of Tropical Agric.

Marschner, H. 1995. *Mineral nutrition of higher plants*, 2nd edition. New York: Academic Press.

Marschner, H., V. Romheld, and M. Kissel. 1986. Different strategies in higher plants in mobilization and uptake of iron. *J. Plant Nutr.* 9:695–713.

Marschner, H. and V. Romheld. 1995. Strategies of plants for acquisition of iron. In: *Iron nutrition in soils and plants*, J. Abadia, Ed., 375–388. Dordrecht: Kluwer Academic Publishers.

Mathers, A. C. 1970. Effect of ferrous sulfate and sulfuric acid on grain sorghum yields. *Agron. J.* 62:555–556.

Mengel, K., E. A. Kirkby, H. Kosegarten, and T. Appel. 2001. *Principles of plant nutrition*, 5th edition. Dordrecht: Kluwer Academic Publishers.

Moore, P. A. Jr. and W. H. Patrick, Jr. 1989. Iron availability and uptake by rice in acid sulfate soils. *Soil Sci. Soc. Am. J.* 53:471–476.

Mori, S. 1999. Iron acquisition by plants. *Curr. Opin. Plant Biol.* 2:250–253.

Morris, D. R., R. H. Loeppert, and T. J. Moore. 1990. Indigenous soil factors influencing iron chlorosis of soybean in calcareous soils. *Soil Sci. Soc. Am. J.* 54:1329–1336.

Mortvedt, J. J. 1991. Correcting iron deficiencies in annual and perennial plants. Present technologies and future prospects. *Plant Soil* 130:273–279.

Mortvedt, J. J. 1994. Needs for controlled availability micronutrient fertilizers. *Fert. Res.* 38:213–221.

Mortvedt, J. J. and P. M. Giordano. 1970. Crop response to iron sulfate applied with fluid polyphosphate fertilizers. *Fert. Solutions* 14:22–27.

Mortvedt, J. J., R. L. Mikkelsen, and J. J. Kelsoe. 1992. Crop response to ferrous sulfate in banded gels of hydrophilic polymers. *Soil Sci. Soc. Am. J.* 56:1319–1324.

Naeve, S. L. 2006. Iron deficiency chlorosis in soybean: Soybean seeding rate and companion crop effects. *Agron. J.* 98:1575–1581.

Naeve, S. L. and G. W. Rehm. 2006. Genotypes by environment interactions within iron deficiency chlorosis tolerant soybean genotypes. *Agron. J.* 52:84–86.

Neilands, J. B. 1977. Siderophores: Biochemical ecology and mechanism of iron transport in enterobacteria. *Adv. Chem.* 162:3–32.

Niebur, W. S. and W. R. Fehr. 1981. Agronomic evaluation of soybean genotypes resistance to iron deficiency chlorosis. *Crop Sci.* 21:551–554.

Norvell, W. A. 1984. Comparison of chelating agents as extractants for metals in diverse soil materials. *Soil Sci. Soc. Am. J.* 48:1285–1292.

Obata, H. 1995. Physiological functions of micro essential elements. In: *Science of rice plant: Physiology*, Vol. 2, T. Matsu, K. Kumazawa, R. Ishii, K. Ishihara, and H. Hirata, Eds., 402–419. Tokyo: Food and Agricultural Policy Research Center.

O'Connor, G. A., W. L. Lindsay, and S. R. Olsen. 1971. Diffusion of iron chelated in soil. *Soil Sci. Soc. Am. Proc.* 35:407–410.

Olomu, M. O., C. J. Racz, and C. M. Cho. 1973. Effect of flooding on the Eh, pH and concentrations of Fe and Mn in several Manitoba soils. *Soil Sci. Soc. Am. Proc.* 37:220–224.

Olson, R. V. 1950. Effects of acidification, iron oxide addition, and other soil treatments on sorghum chlorosis and iron absorption. *Soil Sci. Soc. Am. Proc.* 15:97–101.

Ottow, J. C. G., G. H. Benckiser, I. J. Watanabe, and S. Santiago. 1982. A multiple nutritional stress as the prerequisite for iron toxicity of wetland rice. *Trop. Agric.* 60:102–106.

Patrick, W. H., Jr. and I. C. Mahapatra. 1968. Transformation and availability of nitrogen and phosphorus in waterlogged soils. *Adv. Agron.* 20:323–359.

Patrick, W. H., Jr., and C. N. Reddy. 1978. Chemical changes in rice soils. In: *Soils and rice*, IRRI, Ed., 361–379. Los Banos, Philippines: IRRI.

Perry, D. A., S. L. Rose, D. Pitz, and M. M. Schoenberger. 1984. Reduction of natural ferric iron chelators in disturbed forest soils. *Soil Sci. Soc. Am. J.* 48:379–382.

Ponnamperuma, F. N. 1972. The chemistry of submerged soils. *Adv. Agron.* 24:29–96.

Powell, P. E., P. J. Szaniszlo, G. R. Cline, and C. P. P. Reid. 1982. Hydroxymate siderophores in the iron nutrition of plants. *J. Plant Nutr.* 5:653–673.

Promuthai, C. and B. Rerkasem. 2001. Grain iron concentration in Thai rice germplasm. In: *Plant nutrition: Food security and sustainability of agro-ecosystems,* W. J. Horst et al., Eds., 350–351. Dordrecht: Kluwer Academic Publishers.

Rengel, Z. 1999. Physiological mechanisms underlying differential nutrient efficiency of crop genotypes. In: *Mineral nutrition of crops: Fundamental mechanisms and implications*, Z. Rengel, Ed., 227–265. New York: The Haworth Press.

Romheld, V. 1987. Different strategies for iron acquisition in higher plants. *Physiol. Plant.* 70:231–234.

Romheld, V. and H. Marschner. 1979. Fine regulation of iron uptake by the Fe-efficient plant *Helianthus annus.* In: *The soil-root interface,* J. L. Harley and R. S. Russell, Eds., 405–417. New York: Academic Press.

Romheld, V. and H. Marschner. 1983. Mechanism of iron uptake by peanut. I. Fe^{3+} reduction, chelates splitting, and release of phenolics. *Plant Physiol.* 71:949–954.

Romheld, V. and H. Marschner. 1985. Mobilization of iron in the rhizosphere of different plant species. *Adv. Plant Nutr.* 2:123–218.

Sah, R. N., D. S. Mikkelsen, and A. A. Hafez. 1989. Phosphorus behavior in flooded-drained soils. I. Iron transformation and phosphorus sorption. *Soil Sci. Soc. Am. J.* 53:1723–1729.

Sahrawat, K. L. 2004. Iron toxicity in wetland rice and its role of other nutrients. *J. Plant Nutr.* 27:1471–1504.

Sahu, B. N. 1968. Bronzing disease of rice in Orissa as influenced by soil types and manuring and its control. *J. Indian Soc. Soil Sci.* 16:41–54.

Schmidt, W. 1999. Mechanisms and regulation of reduction-based iron uptake in plants. *New Phytol.* 141:1–26.

Schwab, A. P. and W. L. Lindsay. 1983. Effect of redox on the solubility and availability of iron. *Soil Sci. Soc. Am. J.* 47:201–205.

Seckback, J. 1982. Ferreting out the secrets of plant ferritin: A review. *J. Plant Nutr.* 5:369–394.

Sims, J. T. and G. V. Johnson. 1991. Micronutrient soil test. In: *Micronutrients in agriculture,* 2nd edition, J. J. Mortvedt, Ed., 417–476. Madison, WI: SSSA.

Stahl, R. S. and B. R. James. 1991. Zinc sorption by B horizon soils as a function of pH. *Soil Sci. Soc. Am. J.* 55:1592–1597.

Tadano, T. 1976. Studies on the methods to prevent iron toxicity in lowland rice. *Memoirs of the Faculty of Agriculture* 10:22–88.

Tadano, T. 1995. Akagare disease. In: *Science of rice plant: Physiology*, Vol. 2, T. Matsu, K. Kumazawa, R. Ishii, K. Ishihara, and H. Hirata, Eds., 939–953. Tokyo: Food and Agricultural Policy Research Center.

Tiffin, L. O. and R. L. Chaney. 1973. Translocation of iron from soybean cotyledons. *Plant Physiol.* 52:393–396.

Tisdale, S. L., W. L. Nelson, and J. D. Beaton. 1985. *Soil fertility and fertilizers,* 4th edition. New York: Macmillan Publishing Company.

Uren, N. C. 1984. Forms, reactions and availability of iron in soils. *J. Plant Nutr.* 7:165–176.

Wiersma, J. V. 2005. High rates of Fe-EDDHA and seed iron concentration suggest partial solutions to iron deficiency in soybean. *Agron. J.* 97:924–934.

Winslow, M. D., M. Yamauchi, K. Alluri, and T. M. Masajo. 1989. Reducing iron toxicity in rice with resistant genotype and ridge planting. *Agron. J.* 81:458–460.

Yoshida, S. 1981. *Fundamentals of rice science.* Los Banos, Philippines: IRRI.

Zuo, Y. M., F. S. Zhang, X. L. Li, and Y. P. Cao. 2000. Studies on the improvement in iron nutrition of peanut by intercropping with maize on a calcareous soil. *Plant Soil* 220:13–25.

11 Manganese

11.1 INTRODUCTION

J. S. McHague established the essentiality of manganese (Mn) for higher plants in 1922. However, most of the research on nutrition of Mn in crop plants was conducted in the second half of the 20th century. There are still many gaps in our understanding of the genetics of mineral nutrition of higher plants. Manganese deficiency as well as toxicity has been reported in different crops and soils around the world (Clark, 1982; Fageria, 2001; Fageria et al., 2002). Approximately one-fourth of the earth's soils are considered to produce some kind of mineral stress (Dudal, 1976; Baligar et al., 2001). Manganese deficiency has been reported for plants grown in coarse-textured and poorly drained coastal plains soils of the United States (Reuter et al., 1988) and in soils of Central America, Brazil, and Bolivia (Leon et al., 1985; Fageria et al., 2002). Manganese deficiency of rice occurs on high-pH organic soils used for rice production in the Florida Everglades Agriculture Area (Snyder et al., 1990). In Europe, Mn deficiency has been reported for plants grown in peaty (England and Denmark), coarse-textured (Sweden and Denmark), coarse/fine textured (Netherlands), and podzolic and brown forest (Scotland) soils (Welch et al., 1991). Manganese deficiency has also been reported on plants grown in semiarid regions of China, India, southeast and western Australia, Congo, Ivory Coast, Nigeria, and other western African countries (Fageria et al., 2002).

Most soils contain relatively high concentrations of total Mn, and pH, precipitation, and redox reactions control Mn availability. Adsorption reactions are of less importance for Mn availability (Mortvedt, 1994). When the soil pH drops below 5.5, Mn toxicity may be evident, whereas above ph 6.5, Mn deficiency may occur (Rengel, 2000; Ducic and Polle, 2005; Rosas et al., 2007). Manganese deficiencies occur in organic soils, in alkaline and calcareous soils, and in slightly acid soils with sandy textures (Fageria et al., 2002). Organic soils or mineral soils high in organic matter also may be Mn deficient (Mortvedt, 1994). In Brazilian Oxisols, Mn deficiency in upland rice, corn, soybean, dry bean, and wheat has been reported when soil pH is higher than 6.0 in water (Fageria, 2001). Figure 11.1 shows the response of soybean to manganese fertilization applied to a Brazilian Oxisol. Soybean growth increased when Mn levels increased from 0 to 10 mg kg^{-1} of soil. At 160 and 320 mg kg^{-1} Mn levels, soybean plants showed toxicity symptoms of leaf discoloration and reduced growth.

Table 11.1 shows uptake (concentration) of Mn by dry bean, wheat, corn, lowland rice, and soybean under different pH levels in a Brazilian Inceptisol. Concentration of Mn was significantly decreased when soil pH was raised from 4.9 to 7.0

FIGURE 11.1   Soybean response to Mn fertilization applied to Brazilian Oxisol. Left to right: 0, 10, 160, and 320 mg Mn kg^{-1} of soil. The pH of the soil used in the experiment was about 6.5.

TABLE 11.1
Manganese Uptake (mg kg^{-1}) in the Five Annual Crops as Influenced by Soil pH in a Brazilian Inceptisol

Soil pH (H$_2$O)	Dry Bean	Wheat	Corn	Lowland Rice	Soybean
4.9	4333	633	483	1433	1120
5.9	107	53	53	833	60
6.4	40	47	63	757	43
6.7	40	43	57	767	60
7.0	30	37	40	663	30
R^2	0.90**	0.97**	0.98**	0.89**	0.98**

** Significant at the 1% probability level.
Source: Adapted from Fageria and Baligar (1999).

in all five crop species. These results prove that soil pH is one of the main factors determining Mn availability to crop plants. Lindsay and Cox (1985) and Katyal and Sharma (1991) identified pH as the main factor influencing Mn availability in soils. The effect of pH >6.0 in lowering free metal ion activities in soils has been attributed to the increase in pH-dependent surface charge on oxides of Fe, Al, and Mn (Stahl and James, 1991), chelation by organic matter, or precipitation of metal hydroxides (Lindsay, 1979; Bolan et al., 2003). Adsorption reactions are of less consequence with Mn availability (Mortvedt, 1994). Marschner (1995) reported that the uptake of Mn is more closely associated to soil pH than the uptake of any other micronu-trients. Marked decreases in the bioavailability of Mn should be expected when soil pH decreases below 5.5 (Bolan et al., 2003). Manganese toxicity in acid soils can be easily corrected through adsorption and precipitation reactions by lime application at rates sufficient to raise the soil pH to about 6.5 (Jauregui and Reisenauer, 1982; Fageria and Baligar, 1999).

In some regions of the world, however, manganese toxicity is also reported in crop plants when soil pH is less than 5.0 (Foy, 1984; Welch et al., 1991). However, in modern agriculture, liming is a very common practice to raise soil pH of acid soils to desirable levels and correct Mn toxicity (Fageria et al., 2002). Smith et al. (1993) reported that liming acid soils to pH 5.5 decreases the solubility and uptake of Mn sufficiently to eliminate Mn toxicity in peanuts.

Deficiency of an element or nutrient is associated with or results from many soil, plant, and climatic factors. These factors are quantity of nutrient present in the soil or soil solution, form in which it is present, soil temperature (low and high or optimal), soil moisture, soil aeration, soil pH, organic matter content, root system development, soil type, balance with other nutrients, and crop species or genotypes within species (Fageria et al., 2002). All these factors are very dynamic in soil–plant systems; hence, nutrient availability and uptake by plants is very dynamic in nature (Fageria et al., 2006). Due to the wide importance of Mn in crop plants, we will attempt to summarize current knowledge on the Mn cycle in soil–plant systems, Mn functions and deficiency symptoms, and the Mn uptake process. In addition, management practices that can maximize uptake and utilization efficiency of this element by crop plants are also discussed.

11.2   CYCLE IN SOIL–PLANT SYSTEMS

As with iron, major Mn addition sources are organic manures, chemical fertilizers, release from parent materials, and microbial biomass. The principal Mn depletion sources are uptake by crop plants, loss through soil erosion, leaching in sandy soils, and adsorption on organic compounds or microbial biomass. Manganese uptake by plants is influenced by many factors that determine its solubility and transport behavior in soil–plant systems. The Mn content of the lithosphere is approximately 900 mg kg^{-1}, and soils generally contain from 20 to 3000 mg kg^{-1} with an average of 600 mg kg^{-1} (Lindsay, 1979). The chemistry of manganese in soils is complex because in natural systems, Mn commonly occurs in more than one oxidation state (Mn^{4+}, Mn^{3+}, and Mn^{2+}; Foth and Ellis, 1988). Manganese forms hydrated oxides with mixed valence states (Lindsay, 1979).

Manganese as Mn^{4+} and Mn^{3+} compound occurs in solid phases in oxidized environments, while Mn^{2+} is dominant in solution and in solid phases under reduced conditions. Manganese (Mn^{4+}) compounds are slightly soluble, and Mn^{3+} is unstable in solution. Consequently, Mn^{2+} represents the only important soluble form of Mn in soils. The redox status of the aqueous system greatly affects the solubility and availability of Mn (Schwab and Lindsay, 1983). Manganese-oxidizing microbes increase Mn oxidation rates in soils by up to five orders of magnitude (Brouwers et al., 2000). For this reason, soil microbes, especially bacteria and fungi, have a profound effect on the global cycling of Mn and its availability for plants (Ghiorse, 1988; Thompson and Huber, 2007).

TABLE 11.2

Extractable Manganese as Influenced by Manganese Rates and Extracting Solutions at Two Soil Depths in a Brazilian Oxisol

	Mehlich-1		DTPA	
Mn Rate (kg ha^{-1})	0–10 cm	10–20 cm	0–10 cm depth	10–20 cm Depth
0	13.89	13.07	3.58	3.03
10	18.19	13.85	5.01	3.54
20	21.26	13.33	6.59	3.67
40	25.78	14.63	8.60	4.10
80	35.59	17.19	11.68	4.76
160	46.85	18.93	19.70	5.74
$R^{2\dagger}$	0.9350**	0.5730**	0.8892**	0.8667**

Note: Values are averages of three-year field trial determined after harvest of dry bean crops; manganese was added as broadcast, and carrier was manganese sulfate.

** Significant at the 1% probability level.

Manganese oxides are the most common manganese minerals in soils, include pyrolusite (MnO_2), manganite (MnOOH), and hausmannite (Mn_3O_4) (Barber, 1995). Adding Mn^{2+} as manganese sulfate to the soil significantly increased exchangeable manganese in the 0–10 and 10–20 cm soil depths (Table 11.2). Data in Table 11.2 also show that extractable Mn was higher in the 0–10 cm depth compared to 10–20 cm depth. This may be due to increased levels of organic matter in the topsoil layer, and most of the manganese might be complexed with organic matter. It is reported that 80 to 90% of the Mn in the soil solution is complexed with organic matter (Tisdale et al., 1985; Foth and Ellis, 1988). Shuman (1985) also reported that Mn in the soil is mainly in the organic and Mn-oxide fractions. Data in Table 11.2 also show that Mehlich-1 extractable Mn was higher compared to DTPA extractable Mn in both soil depths.

In natural systems, Mn is present in discrete minerals in Mn-oxide nodules that are often admixed with Fe oxides and as coating on clays and other soil particles (McKenzie, 1989). The dynamics and availability of soil Mn are determined by many factors including pH, redox potential, the nature and concentration of cations and anions, clay mineralogical composition, organic matter content, and microorganisms. Manganese availability is higher in acid soils due to the higher solubility of Mn compounds under low-pH conditions. Lindsay (1972) reported that the soluble Mn^{2+} decreased 100-fold for each unit increase in pH. Fageria et al. (2006) reported that Mn uptake in dry bean shoot decreased from 4420 mg kg^{-1} at pH 4.9 to 174 mg kg^{-1} at pH 7.0 in a Brazilian Inceptisol. Adams and Wear (1957) stated that Mn solubility in aerated soils is extremely pH dependent, with appreciable amounts being brought into solution below pH 5.0. A pH value below 6.0 to 6.5 favors reduction of Mn and the formation of the more available divalent form (Mn^{2+}); higher pH values favor oxidation to the Mn^{4+} ion, from which insoluble oxides are formed (MnO_2, Mn_2O_3, and Mn_3O_4) (Stevenson, 1986).

Another physicochemical parameter that profoundly affects the solubility of Mn in soils is Eh (redox potential). Patrick and Reddy (1978) classified soils based on redox potential as aerated or well-drained soils (+700 to +500 mV), moderately reduced (+400 to +200 mV), reduced (+100 to –100 mV), and highly reduced (–100 to –300 mV). Transformation of Mn^{4+} to Mn^{2+} takes place when the redox potential is in the range of +200 to +400 mV (Tisdale et al., 1985). The reduction of Mn can be either chemical or microbiological, although microbiological reduction is likely to predominate in flooded rice soils that are at about pH 5.5–6.0 (Patrick and Reddy, 1978). The reduction of Mn is described by the following equation (Fageria et al., 2003b):

$$MnO_2 + 4H^+ + 2e^- \Leftrightarrow Mn^{2+} + 2H_2O$$

Patrick and Turner (1968) noted that an increase in water-soluble and exchangeable Mn was one of the first measurable effects of reducing conditions brought on by waterlogging. Moore and Patrick (1989b) measured the solubility of Mn in flooded acid sulfate soils cropped to rice. They concluded that Mn solid phases such as oxides, hydroxides, carbonates, phosphates, and silicates were probably not controlling Mn solubility in these soils. Plots of E–Mn (the ratio of Mn^{2+} to the sum of divalent cations in soil solution) vs. E–Mn (the ratio of exchangeable Mn to the sum of the exchangeable divalent cations) indicated that the divalent charge fraction attributable to Mn^{2+} in soil solution was highly correlated to that on the cation exchange capacity. They also concluded that Mn^{2+} concentration was correlated with pH (Mn^+/Fe^{2+}), indicating that a Fe/Mn interaction may be occurring. Leaf metal analysis also indicated that Mn toxicity in rice crop probably did not occur. Since rice is generally grown in soils made anaerobic by flooding, the crop is likely to be exposed to high levels of available Mn. Rice apparently is not adversely affected by excessive Mn even at leaf concentrations of 1000 to 4000 mg kg^{-1} (Tanaka and Navasero, 1966). Thermodynamic models predict that Mn is more easily reduced and solubilized than is Fe and therefore is more mobile in the soil system (McDaniel et al., 1992). Manganese and Fe ions have similar ionic radii, so Mn can readily substitute for Fe in the crystal structure (Klein and Hurlbut, 1985).

Stability of the various ions of Mn as a function of pH and redox potential has been reported by many researchers (Lindsay, 1979; Bohn et al., 1985). Therefore, redistribution of Mn in soil capes is closely tied to the oxidation-reduction dynamics of the system. The oxidation-reduction reactions involving Mn appear to occur largely as a result of soil microbial activity (Alexander, 1977). Oxidation and reduction of Mn are thermodynamically favored at relatively higher redox potentials than for Fe at a given pH. As a result, Collins and Buol (1970) were able to demonstrate a spatial relationship between Mn and Fe. Precipitation in horizontal sand columns occurred in response to increased redox potential. Iron precipitated at relatively lower redox potentials, while most Mn did not precipitate until reaching the more oxidized portions of the columns. Buol et al. (1980) have suggested that in well-drained soil profiles, secondary Mn is found deeper than Fe because Mn^{2+} remains in a reduced, soluble form longer than Fe^{2+} with increasing redox potential. Thus, Mn^{2+} would be expected to be leached deeper than Fe in a soil that becomes more oxidizing with depth or in which pH increases with depth.

Within soil fractions, exchangeable and organically bound forms of Mn appear to represent important dynamic pools in many soil systems (McDaniel and Buol, 1991). Several researchers have reported accumulation of Mn surface horizons and suggested that this Mn is associated with the organic fraction (McDaniel and Buol, 1991). Bloom (1981) suggested that fully hydrated Mn^{2+} ions form outer-sphere complexes with carboxyl groups of soil organic matter, a mechanism that explains the observation that Mn is more weakly associated with organic matter than are other metals such as Fe, Cu, and Zn (Bloom, 1981; Gambrell and Patrick, 1982).

11.3   FUNCTIONS AND DEFICIENCY SYMPTOMS

Manganese resembles iron in its function within plants. Its important functions in the plants are summarized as follows: (1) manganese is indirectly related to the formation of chlorophyll, (2) manganese is considered to be a constituent of some respiratory enzymes, (3) manganese activates several important metabolic reactions in the plants, (4) manganese accelerates germination and maturity, (5) manganese increases the availability of P and Ca, (6) it is involved in the evolution of O_2 in photosynthesis, (7) it is involved in the oxidation-reduction reactions and electron transport systems, (8) it is a structural component of certain metalloproteins, (9) it involved in iron metabolism and nitrate assimilation, (10) in Mn-deficient plants, soluble carbohydrates are reduced, particularly in the roots, (11) Mn-deficient plants are more susceptible to freezing injury, (12) in adequate amount, it provides disease resistance to plants, (13) in seeds, the oil content may decrease and the composition change with Mn deficiency, (14) manganese plays several critical roles in CO_2 fixation in C_4 plants, (15) it plays an important role in N_2 fixation in legumes, and (16) manganese contributes to the synthesis of important secondary metabolites that lead to the production of phenolics, cyanogenic glycosides, and lignin plant defense compounds (Burnell, 1988).

Manganese deficiency in crop plants in the beginning occurs in the newly developing leaves. Interveinal chlorotic streaks spread downward from the tip to the base of the leaves; the veins remain green. Necrotic brown spots develop as the deficiency becomes more severe. The newly emerging leaves become short, narrow, and light green. Stems are slender and weak. In the beginning, Mn deficiency symptoms are similar to those of iron deficiency. As deficiency advances, iron-deficient leaves become white and Mn-deficient leaves develop small, discrete dark reddish brown or black lesions within the chlorotic tissues. Various seed disorders in Mn-deficient legumes are common, such as dark discoloration on cotyledon of pea and some other legumes (marsh spots) and cracks of the testa in seeds of lupins (split seeds) (Romheld and Marschner, 1991). Although Mn deficiency is often associated with high soil pH, it may result from an imbalance with other nutrients such as Ca, Mg, and Fe. Soil moisture also affects Mn availability. Oats are very sensitive to Mn deficiency and are an excellent indicator of Mn deficiency in a given soil (Fageria and Gheyi, 1999).

Manganese toxicity can also occur and, depending on local soils and soil types, can cause more plant damage than deficiency commonly does. Symptoms of excess Mn are typically manifested as dark specks or spots on leaves. Other symptoms include crinkling and chlorosis of young leaves. High Mn concentration in the plant can result in symptoms associated with other nutrient deficiencies, through decreased uptake of other elements such as iron, copper, and calcium (Horst, 1988; Thompson and Huber, 2007).

11.4   CONCENTRATION AND UPTAKE

The nutrient concentration in plant tissues is an important indicator of its deficiency or sufficiency level. However, the concentration varied with the plant species or genotypes of the same species. Further, plant age and part analyzed significantly affect nutrient concentration in plant tissues. Genetic makeup and dynamic physiological and environmental factors interact to alter the concentrations of micronutrients in plants at which deficiencies or toxicities will occur (Welch, 1995). Figure 11.2 and Figure 11.3 show Mn concentration in the shoot of upland rice and dry bean, respectively. The Mn concentration significantly decreased with increasing plant age in both the crop species. In upland rice, variation in Mn concentration was about 68% with the advancement of plant age. Similarly, variability in Mn concentration in the shoot of dry bean was about 64% due to plant age variation. In upland rice shoot, Mn concentration was 250 mg kg^{-1} at 19 days after sowing and dropped to 150 mg kg^{-1} at 90 days after sowing.

Similarly, Mn concentration in the shoot of dry bean was about 90 mg kg^{-1} at 15 days after sowing and dropped to about 50 mg kg^{-1} at 96 days after sowing or at

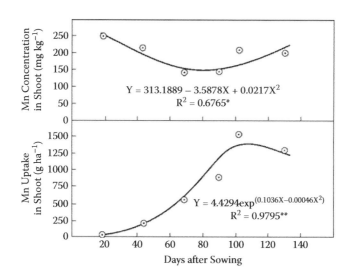

FIGURE 11.2   Manganese concentration and uptake in the shoot of upland rice during growth cycle.

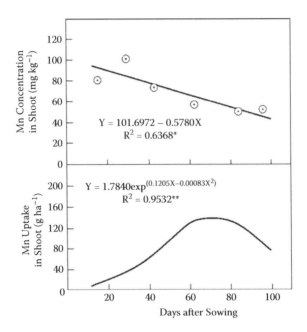

FIGURE 11.3 Manganese concentration and uptake in the shoot of dry bean during growth cycle.

harvest. Hence, plant age is an important factor in determining Mn concentration in the shoot of cereal and legume crops. Higher concentration of Mn in the shoot of upland rice indicates that cereal tolerance of Mn is higher than that of legumes. Adequate or sufficiency level of Mn in the plant species reported by Jones (1991) varied from 20 to 300 mg kg^{-1} dry plant tissues. Similarly, Stevenson (1986) reported a sufficiency level of Mn in the range of 20–50 mg kg^{-1} in plant tissues. Stevenson (1986) also reported that for many plants, Mn deficiency occurs when the plant tissue contains less than 20 mg kg^{-1}. Sufficiency levels of Mn in different crop species are presented in Table 11.3.

Critical deficient and critical toxic level also varied widely among crop species and environmental conditions (Marschner, 1995; Mengel et al., 2001). Even within a species, the critical toxic levels may vary by a factor of 3 to 5 between cultivars (Ohki et al., 1981; Edwards and Asher, 1982). Edwards and Asher (1982) reported that the critical toxic level of Mn in the corn of 200 mg kg^{-1} to 5300 mg kg^{-1} dry matter. For plant analysis to be effective in diagnosing Mn deficiency problems, critical levels need to be established at different growth stages under various conditions. Table 11.4 shows critical deficient and toxic levels of Mn in principal crop species grown on Brazilian Oxisols.

Soil manganese exists in three oxidation states: Mn^{2+}, Mn^{3+}, and Mn^{4+} (Barber, 1995). The Mn is, however, mainly absorbed as Mn^{2+} by crop plants, and the quantity absorbed depends on yield level, crop species, plant age, and nutrient concentration in the soil solution. It is reported in the literature that root exudates reduce Mn^{4+} to Mn^{2+}, complex it, and then make it more available to the plant (Godo and Reisenauer, 1980). These authors also reported that the effect was most evident at pH

TABLE 11.3
Adequate Level of Manganese in Plant Tissues of Crop Plants

Crop Species	Growth Stage and Plant Part Analyzed	Adequate Mn Level (mg kg^{-1})
Wheat	Mid tillering-stem elongation, whole tops	11–13
Barley	Heading, whole tops	25–100
Rice	Tillering, whole tops	30–600
Corn	30 to 45 days after emergence, whole tops	50–160
Sorghum	Seedling, whole tops	40–150
Sorghum	Bloom, 3rd blade below panicle	8–190
Soybean	Prior to pod set, upper fully developed trifoliate	21–100
Dry bean	Early flowering, uppermost blade	50–400
Peanut	Early pegging, upper stems and leaves	100–350

Source: Adapted from Fageria et al. (1997).

TABLE 11.4
Adequate and Toxic Levels of Manganese in the Plant Tissues of Five Crop Species Grown on Brazilian Oxisols

Crop Species	Adequate Level (mg kg^{-1})	Toxic Level (mg kg^{-1})
Upland rice	520	4560
Dry bean	400	1640
Corn	60	2480
Soybean	67	720
Wheat	173	720

Note: Adequate level was calculated at 90% of the maximum dry weight of shoot, and toxic level was calculated at the 10% reduction in shoot dry weight after achieving maximum weight. Plants were harvested at physiological maturity, and soil pH at harvest was about 6.0 in water in all five experiments.

Source: Adapted from Fageria et al. (2003a).

less than 5.5. Uptake of nutrients was generally measured at harvest to determine the quantity extracted from the soil by crop species. This information could prove useful in replenishing soil fertility for the succeeding crops or maintaining soil fertility at an adequate level for sustainable crop production. The Mn uptake in the shoot of upland rice and dry bean is shown in Figure 11.2 and Figure 11.3, respectively. In both crops, uptake followed a significant quadratic response in relation to plant age. In upland rice, maximum uptake occurs at 113 days after sowing, and variability in Mn uptake was about 98% due to plant age. In dry bean, Mn uptake was at a maximum at 73 days after sowing, and variability in Mn uptake was 95% in relation to plant age. Highly significant ($P < 0.01$) R^2 values (Mn uptake vs. plant age) in both crop species show that plant age is an important factor in determining Mn uptake in cereals as well as legume crops.

TABLE 11.5
Manganese Use Efficiency (MnUE) in Shoot and Grain of Upland Rice and Dry Bean during Crop Growth Cycle

Plant Age in Days	Upland Rice (kg g⁻¹)	Plant Age in Days	Dry Bean (kg g⁻¹)
19	4.6	15	13.6
43	5.2	29	9.8
68	10.4	43	14.6
90	7.2	62	20.3
102	5.3	84	21.3
130	5.4	96	21.0
130 (grain)	32.3	96 (grain)	83.4

Regression Analysis

Plant age (X) vs. MnUE in rice shoot (Y) = $1.7655 + 0.1580X - 0.00104X^2$, $R^2 = 0.2167^{NS}$

Plant age (X) vs. MnUE in dry bean shoot (Y) = $8.9094 + 0.1599X - 0.00023X^2$, $R^2 = 0.5102^{**}$

Note: Manganese use efficiency (MnUE kg g⁻¹) = (Shoot or grain dry weight in kg/Uptake of Mn in shoot or grain in g).

[**,NS] Significant at the 1% probability level and nonsignificant, respectively.

11.5 USE EFFICIENCY AND Mn HARVEST INDEX

Nutrient use efficiency (dry matter or grain yield/unit nutrient uptake) is an important index that reveals how the nutrient was utilized in dry matter or grain yield production by crop species under agroecological conditions. Similarly, nutrient harvest index is important in determining distribution of the nutrient in the shoot and grain. A nutrient that is mostly exported to grain will become highly depleted from the soil. However, this nutrient transfer may be beneficial in relation to human consumption. Table 11.5 shows Mn use efficiency of upland rice and dry bean for dry matter and grain yield. The Mn use efficiency displayed a quadratic association with plant age in both species. In dry bean, Mn use efficiency was significantly influenced by plant age. The Mn use efficiency was higher in grain of the two species compared to shoot dry weight. Similarly, Mn use efficiency for shoot dry matter production and grain yield of corn and soybean was also significantly influenced by plant age (Table 11.6). In both crops, Mn use efficiency followed a quadratic response in relation to plant age. The Mn use efficiency was higher in grain compared to shoot in corn as well as in soybean. The Mn uptake and Mn harvest index values for four crops are presented in Table 11.7. Exportation of Mn to grain was higher in legumes compared to cereals.

11.6 INTERACTION WITH OTHER NUTRIENTS

Fageria (2002) studied interactions of manganese with other nutrients in corn plants, and results are presented in Table 11.8. Manganese application significantly increased corn dry matter yield. In Oxisol of central Brazil, liming generally induced

TABLE 11.6
Manganese Use Efficiency (MnUE) in Shoot and Grain of Corn and Soybean during Crop Growth Cycle

Plant Age in Days	Corn (kg g⁻¹)	Plant Age in Days	Soybean (kg g⁻¹)
18	12.5	27	14.4
35	18.8	41	10.3
53	21.9	62	11.1
69	24.2	82	15.4
84	22.9	120	13.1
119	30.3	158	13.2
119 (grain)	100.0	158 (grain)	47.2

Regression Analysis

Plant age (X) vs. MnUE in corn shoot (Y) = $9.4955 + 0.2501X - 0.00069X^2$, $R^2 = 0.7700$**

Plant age (X) vs. MnUE in soybean shoot (Y) = $9.5317 + 0.2082X - 0.00055X^2$, $R^2 = 0.7005$**

Note: Manganese use efficiency (MnUE kg g⁻¹) = (Shoot or grain weight in kg/Uptake of Mn in shoot or grain in g).

*,** Significant at the 5 and 1% probability level, respectively.

TABLE 11.7
Manganese Uptake and Harvest Index in Principal Field Crops (Values Are Averages of Two-Year Field Trial)

Crop Species	Mn Uptake in Shoot (g ha⁻¹)	Mn Uptake in Grain (g ha⁻¹)	Total	Mn Harvest Index (%)
Upland rice	1318.51	284.03	1602.55	18
Corn	452.16	82.21	534.36	15
Dry bean	73.19	26.60	99.79	49
Dry bean	31.10	48.7	79.8	61
Soybean	117.40	120.10	237.50	51

Note: Mn harvest index = (Mn uptake in grain/Mn uptake in grain plus shoot).

micronutrient deficiencies and manganese application improved annual crops yields (Novais et al., 1989). Manganese application in corn significantly improved uptake of Mg and Zn. However, uptake of Ca and Fe was significantly decreased with the addition of Mn in the growth medium. The uptake of Mn has been reported to be inhibited by divalent cations such as Ca, Mg, and Zn (Maas et al., 1969; Robson and Loneragan, 1970). Foy (1984) reported that increased Ca in the growth medium may decrease Mn uptake and toxicity. Foy (1984) also reported that phosphorus additions could reduce the toxicity of Mn by rendering it inactive within the plant.

Chinnery and Harding (1980) have reported antagonistic effects of Mn on the uptake of Fe and vice versa. They reported that concentration of Mn in the soybean

TABLE 11.8

Influence of Manganese on Shoot Dry Weight and Uptake of Macronutrients and Micronutrients by Corn Plants Grown on Brazilian Oxisol

Mn Level (mg kg^{-1})	Shoot Dry Wt. (g 4 plants^{-1})	P (g kg^{-1})	K (g kg^{-1})	Ca (g kg^{-1})
0	17.93	1.93	23.66	4.36
10	20.83	1.63	23.00	4.26
20	17.70	1.90	24.33	4.20
40	20.00	1.70	22.00	3.93
80	20.40	1.73	22.33	4.30
160	19.80	1.50	23.33	3.43
320	20.20	1.50	22.66	3.53
640	13.77	1.63	25.00	2.93
R^2	0.8100*	0.6355NS	0.5256NS	0.8545**

Mn Level (mg kg^{-1})	Mg (g kg^{-1})	Zn (mg kg^{-1})	Cu (mg kg^{-1})	Fe (mg kg^{-1})
0	2.60	24	3.00	113
10	2.40	20	2.00	117
20	2.47	23	3.33	123
40	2.53	21	2.67	117
80	2.70	25	3.00	113
160	2.37	23	2.33	90
320	2.53	27	2.66	93
640	2.03	32	2.66	110
R^2	0.7099*	0.8326*	0.0366NS	0.7784*

*, **, NS Significant at the 5 and 1% probability level and nonsignificant, respectively.
Source: Adapted from Fageria (2002).

shoots decreased with increased iron concentration in the solution, probably because of oxidation of iron by Mn. Leach and Taper (1954) concluded that the optimum Fe/Mn ratios in plants ranged from 1.5 to 3.0 for kidney beans and from 0.5 to 5.0 for tomato. Iron deficiency developed at lower ratios and Mn toxicity at higher ratios. This means that the antagonistic interaction between Mn and Fe has some practical implications. For example, Fe toxicity is very common in flooded rice due to reduced conditions, and Mn toxicity is common in legumes in acid soils when pH is lower than 5.5 (Fageria et al., 2002). Therefore, Fe toxicity in flooded rice can be reduced by Mn application, and Mn toxicity in acid soil can be minimized by Fe application. There was no significant effect of Mn on uptake of P, K, and Cu in corn plants.

Zinc-deficient plants accumulate excess amounts of Mn, indicating interactions between these two elements (Brady and Weil, 2002). Schomberg and Weaver (1991) demonstrated that Mn decreased N_2 fixation in arrowleaf clover more than mineral uptake. Consequently, the potential impact of excess soluble Mn in the root zone was partially offset by the availability of mineral N for uptake and metabolism (Wilkinson et al., 2000). Increasing Mg level in Brazilian Oxisols significantly decreased

TABLE 11.9

Influence of Magnesium on Mn Concentration (mg kg^{-1}) in the Shoots of Upland Rice, Dry Bean, and Cowpea Grown on Brazilian Oxisols

Mg Level in Soil (cmol$_c$ kg^{-1})	Upland Rice	Dry Bean	Cowpea
0.30	693	238	403
1.05	520	160	330
1.15	588	193	240
1.33	468	195	238
3.52	415	110	138
6.22	333	95	130
Regression	Linear**	Linear**	Linear**

** Significant at the 1% probability level.

Source: Adapted from Fageria and Souza (1991).

Mn concentration in the shoot of upland rice, dry bean, and cowpea (Table 11.9). Foy (1984) reported that toxicity of a given level of soluble Mn in the growth medium, or even within the plant, depends on interactions between Mn and several other mineral elements, particularly Fe and Si.

11.7 MANAGEMENT PRACTICES TO MAXIMIZE Mn USE EFFICIENCY

Adopting appropriate management practices like use of adequate rate, appropriate source, and methods of application can improve Mn use efficiency by crop plants. In addition, use of soil test is an effective practice to make fertilizer recommendations. Use of some neutral chemical salts can also improve Mn use efficiency in crop plants. Planting Mn-efficient crop species or genotypes within species is a very economic and environmentally sound management practice. These practices are summarized in this section.

11.7.1 USE OF ADEQUATE RATE, APPROPRIATE SOURCE, AND METHODS

Use of adequate Mn rate is very important for a particular crop and soil to improve crop yields and maximize Mn use efficiency. Adequate rate of a nutrient is determined on the basis of crop response to a determined nutrient under different agroecological regions. Data in Table 11.10 show adequate and toxic rate of Mn fertilization for principal annual crops grown on Brazilian Oxisols. The most common inorganic fertilizer is MnSO$_4$ (26–28% Mn). It has proven effective at correcting Mn deficiency of most field crops grown on both acid and alkaline soils. Major Mn carriers are presented in Table 11.11.

Soil and foliar application of appropriate Mn fertilizers can correct manganese deficiency. Hatfield and Hickey (1981) suggested that measures to correct Mn deficiency in soybean are (1) broadcast application of 11 kg Mn ha^{-1}; (2) band application

TABLE 11.10
Adequate and Toxic Levels of Manganese for Annual Crops Applied to Brazilian Oxisols

Crop Species	Adequate Mn Rate (mg kg⁻¹)	Toxic Mn Rate (mg kg⁻¹)
Upland rice	2	560
Dry bean	12	112
Corn	12	400
Soybean	No response	72
Wheat	No response	10

Note: Adequate rate was calculated at 90% of the maximum dry weight of shoot, and toxic level was calculated at the 10% reduction in shoot dry weight after achieving maximum weight. Plants were harvested at physiological maturity, and soil pH at harvest was about 6.0 in water in all five experiments.

Source: Adapted from Fageria et al. (2003a).

TABLE 11.11
Principal Manganese Carriers

Common Name	Formula	Mn Content (%)
Manganese sulfate	$MnSO_4.3H_2O$	26–28
Manganese oxide	MnO	68–70
Manganese oxide	MnO2	63
Manganese EDTA chelate	MnEDTA	12
Manganese frit	Frit	10–25
Manganese carbonate	$MnCO_3$	31
Manganese chloride	$MnCl_2$	17
Manganese polyflavonoid	MnPF	8
Manganese methoxyphenyl-propane	MnPP	5

Source: Follett et al. (1981), Tisdale et al. (1985), Foth and Ellis (1988), Fageria (1989).

of one-half of the broadcast rate; or (3) timely foliar application of 1 to 2 kg Mn ha⁻¹. Mascagni and Cox (1985) conducted field studies to establish the most efficient rate for the broadcast and band methods of soil application and for foliar application of Mn for soybeans. Optimum applied Mn was 14 kg Mn ha⁻¹ with broadcast application and 3 kg Mn ha⁻¹ with band application, using manganese sulfate as a Mn source. Band application with acid-forming fertilizers results in increased efficiency of applied Mn because the rate of oxidation of the divalent Mn to the unavailable tetravalent form (MnO_2) is decreased (Mortvedt, 1994).

Mortvedt and Giordano (1975) reported higher forage yields and Mn uptake by oats with banded application than with mixed application of $MnSO_4$, MnO, or Mn EDTA applied with several granular or fluid fertilizers. Boswell et al. (1981) reported that a 239% yield increase of soybeans was obtained with 11.2 kg Mn ha^{-1} compared to no applied Mn. Hallock (1979) reported that highest yields of Virginia-type peanuts were obtained from 4.48 kg ha^{-1} Mn as MnSO$_4$ in three sprays (2.24 kg ha^{-1} in first spray), whereas the yield from a broadcast treatment (44.8 kg ha^{-1}) of Mn was nearly as high. Snyder et al. (1990) reported that Mn deficiency could occur in seedling rice grown in drained Histosols with pH values near or above 7. Affected seedlings contain <20 mg Mn kg^{-1} tissue, and the deficiency can be prevented by drilling MnSO$_4$ at seeding to provide approximately 15 kg Mn ha^{-1} as MnSO$_4$ and provide near maximum grain yield. There are low residual effects of applied Mn fertilizers because of ease of oxidation to the (MnO$_2$ or Mn^{4+}) unavailable forms; hence, annual applications are required.

11.7.2 Use of Acidic Fertilizers in the Band and Neutral Salts

Decreasing the soil pH in a localized zone by band application of acid-forming N or P fertilizers has resulted in increased Mn uptake and yields (Jackson and Carter, 1976). Increased levels of ammonium acetate–extractable Mn have also resulted from application of acid N or P fertilizers (Voth and Christenson, 1980). Petrie and Jackson (1984) showed that band application of acid-forming fertilizers was necessary to correct Mn deficiency and produce maximum barley and oat yields on alkaline, high-organic-matter mineral soils in the Klamath Lake area of South-Central Oregon. Hossner and Richards (1968) showed higher Mn uptake from MnSO$_4$ applied with monoammonium phosphate and acidic ammonium polyphosphate than with monocalcium phosphate, the predominant form in superphosphate. Band application of acid-forming phosphate fertilizers alone to soils often solubilized sufficient soil Mn to correct Mn deficiency in Mn-deficient soils (Randall et al., 1975).

Some neutral salts (KCl, NaCl, and CaCl$_2$) have been shown to increase bioavailability of soil Mn (Jackson et al., 1966; Krishnamurti and Huang, 1992), an effect that was hypothesized as due to an Mn^{4+}, Mn^{3+}, Cl$^-$ redox reaction (Westermann et al., 1971). Khattak et al. (1989) advocated ion exchange as the dominant mechanism to explain the salt-induced Mn release, from soils of California. Krishnamurti and Huang (1992) reported that increasing KCl concentration increased Mn release from selected soils of different taxonomic orders. They concluded that the ionic strength effect coupled with complexation was mainly responsible for the enhanced Mn release by KCl from the soils in the common soil pH range. Application of chloride (Cl) has also been shown to increase the availability and uptake of native soil Mn. Application of Cl as KCl or CaCl$_2$ increased the leaf Mn concentration in both common bean and sweet corn grown on acid soils; the increase uptake was such that Mn toxicity was evident on the common beans (Jackson et al., 1966). Application of KCl resulted in greater increases in Mg(NO$_3$)$_2$-extractable Mn than did K$_2$SO$_4$ (Westermann et al., 1971). The increase in extractable Mn was attributed to an oxidation-reduction couple involving Mn and Cl (Petrie and Jackson, 1984).

11.7.3 Use of Soil Test

Extractable Mn has been used successfully in many studies, particularly for soybean and small grains, to predict crop yield responses to Mn fertilization (Sims and

Johnson, 1991). Critical Mn level established by soil test is an important criterion for diagnosis of Mn disorder in crop plants. Cox (1987) reported that two values, Mehlich-1 extractable Mn and soil pH, were needed to explain the variation in soybean yield response. Mascagni and Cox (1985b) calibrated soil test Mn with soybean yield response using Mehlich-1 (0.05 M HCl + 0.0125 M H_2SO_4) and Mehlich-3 (0.2 M CH_3COOH + 0.25 M NH_4NO_3 + 0.015 M NH_4F + 0.013 M HNO_3 + 0.001 M EDTA) extractants. The critical level of soil Mn varied with soil pH. At pH 6 and 7, the critical levels were 4.7 and 9.7 mg Mn kg^{-1} for Mehlich-1 extracting and 3.9 and 8.0 mg Mn kg^{-1} for Mehlich-3 extracting, respectively. Gettier et al. (1985), in field research conducted on many Atlantic coastal plain soils, reported that Mn would not be recommended for soybeans when the Mehlich-1 extracting level in soils was 10 mg kg^{-1}. Sims and Johnson (1991) reported that critical Mn level for Mehlich-1, Mehlich-3, 0.0 M HCl, or 0.03 M H_3PO_4 extracting solutions varied from 4 to 20 mg kg^{-1}, relative to 1.4 mg kg^{-1} for DTPA (diethylenetriaminepentaacetic acid) (0.005 M DTPA + 0.01 M $CaCl_2$ + 0.1 M triethanolamine [TEA]), adjusted to pH 7.3 (Lindsay and Norvell, 1978). Lindsay and Cox (1985) reported that the critical Mn level in soil for the DTPA extracting solution is in the range of 1 to 5 mg kg^{-1} for the tropical soils.

Fageria et al. (2003) determined adequate and toxic levels of soil Mn extracted by Mehlich-1 and DTPA extracting solutions for five annual crops grown on Brazilian Oxisols (Table 11.12). The Mehlich-1 extractable adequate Mn level was 8 mg kg^{-1} for upland rice, dry bean, corn, soybean, and wheat. The toxic level of Mehlich-1 extractable Mn for these crops was 168 mg kg^{-1} for upland rice, 128 mg kg^{-1} for dry bean, 400 mg kg^{-1} for corn, 92 mg kg^{-1} for soybean, and 44 mg kg^{-1} for wheat. The adequate level of Mn by DTPA extracting varied from 3 to 6 mg kg^{-1} depending on crop species. It was maximum for dry bean and minimum for wheat (Table 11.12). The toxic level of Mn by DTPA extracting solution was 80 mg kg^{-1} for upland rice,

TABLE 11.12

Adequate and Toxic Level of Soil Manganese Extracted by Mehlich-1 and DTPA Extracting Solutions for Five Annual Crops Grown on Brazilian Oxisols

	Mehlich-1 Extracting Solution		DTPA Extracting Solution	
Crop Species	Adequate Mn Level (mg kg^{-1})	Toxic Mn Level (mg kg^{-1})	Adequate Mn Level (mg kg^{-1})	Toxic Mn Level (mg kg^{-1})
Upland rice	8	168	4	80
Dry bean	8	128	6	88
Corn	8	400	4	336
Soybean	8	92	4	56
Wheat	8	44	3	40

Note: Adequate level was calculated at 90% of the maximum dry weight of shoot, and toxic level was calculated at the 10% reduction in shoot dry weight after achieving maximum weight. Plants were harvested at physiological maturity, and soil pH at harvest was about 6.0 in water in all five experiments.

Source: Adapted from Fageria et al. (2003a).

88 mg kg^{-1} for dry bean, 336 mg kg^{-1} for corn, 56 mg kg^{-1} for soybean, and 40 mg kg^{-1} for wheat. These results indicate that corn was most tolerant to Mn toxicity and wheat was most susceptible to Mn toxicity among five crop species. In addition, adequate as well as toxic level of Mn was higher for Mehlich-1 extracting solution compared to DTPA extracting solution. Furthermore, adequate level of Mn was similar for five crop species. Overall, adequate level for Mehlich-1 extracting solution was 8 mg kg^{-1}, and for DTPA, it was 4 mg kg^{-1}.

11.7.4   USE OF EFFICIENT CROP SPECIES/GENOTYPES

Responses of crop species to applied Mn varied significantly. Similarly, Mn uptake and utilization varied from crop species to crop species and among genotypes within species. Oats is considered very sensitive to Mn deficiency, and rye is very efficient in uptake and utilization of Mn (Follett et al., 1981; Tisdale et al., 1985; Jones, 1991; Welch et al., 1991). Data in Table 11.13 show differences in dry matter production by 10 upland rice genotypes at two Mn levels. Shoot dry weight varied from 46.37 g 4 plants^{-1} to 91.23 g 4 plants^{-1} at lower Mn level depending on genotype, with an average value of 62.42 g 4 plants^{-1}. The variation was 97% between lowest-shoot-dry-weight-producing genotype Carajas and highest-shoot-dry-weight-producing genotype IR42. At the higher Mn level, shoot dry weight varied from 51.73 to 76.27 g 4 plants^{-1}, with an average value of 63.42 g 4 plants^{-1}.

TABLE 11.13
Shoot Dry Matter Yield of 10 Upland Rice Genotypes at Two Mn Levels Applied to a Brazilian Oxisol

	Shoot Dry Weight (g 4 plants^{-1})	
Genotype	0 mg Mn kg^{-1}	20 mg Mn kg^{-1}
Bonança	51.43cde	60.47abc
Caipó	68.03bc	76.27a
Canastra	65.20bcd	65.87abc
Carajas	46.37e	57.93bc
Carisma	61.00bcde	54.33c
CNA8540	55.70bcde	55.47c
CNA8557	64.57bcd	68.20abc
Guarani	48.83de	51.73c
Maravilha	71.80b	68.23abc
IR42	91.23a	75.70ab
Average	62.42	63.42

Note: Means in the same column followed by the same letter are statistically not different at the 5% probability level by the Tukey's test.

Source: Fageria and Barbosa Filho (2008).

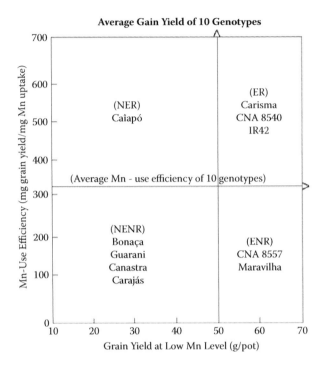

FIGURE 11.4 Classification of upland rice genotypes for Mn use efficiency (Fageria and Barbosa Filho, 2008).

Based on Mn use efficiency and grain yield at low Mn level, genotypes were classified into four groups (Figure 11.4). Fageria and Baligar (1993) suggested this type of classification for the nutrient use efficiency of crop genotypes. The first group comprised efficient and responsive (ER) genotypes. The genotypes that produced above-average yield at low Mn level and that had above-average Mn use efficiency were classified in this group. Genotypes Carisma, CNA8540, and IR42 fall into this group. The second classification included efficient and nonresponsive (ENR) genotypes. These genotypes produced above-average yield of 10 genotypes at low Mn level, but response to Mn application was lower than the average. The genotypes CNA8557 and Maravilha fell into this group. The third group comprised nonefficient and responsive (NER) genotypes. The genotypes that produced below-average grain yield of 10 genotypes at low Mn level but showed above-average Mn use efficiency are classified in this group. The only genotype that fell into this group was Caipó. The fourth group of genotypes included those that produced below-average yield at low Mn level and displayed a below-average response to applied Mn. These genotypes were classified as nonefficient and nonresponsive (NENR). The genotypes falling into this group were Bonança, Canastra, Carajas, and Guarani.

From a practical point of view, the genotypes that fall into the efficient and responsive group are the most desirable, because these genotypes can produce more at a low Mn level and also respond well to applied Mn. Thus, this type of genotype can be utilized under low as well as high technology with reasonably good yield

(Fageria and Baligar, 1993). The second most desirable group is efficient and non-responsive genotypes. Genotypes of this type can be planted under low Mn level and produce higher-than-average yield. The nonefficient and responsive genotypes sometimes can be used in breeding programs for their Mn-responsive characteristics. The most undesirable genotypes are the nonefficient and nonresponsive type. These results indicate that upland rice genotypes differ in Mn use efficiency. Both inter- and intraspecific variations in Mn nutrition have been recognized among cereal species and genotypes (Fageria, 1998; Fageria and Baligar, 1993; Fageria et al., 1997); this suggests that it may be possible to develop cultivars that are efficient at low-nutrient levels or are capable of using Mn more efficiently when applied as fertilizer.

Differences in Mn use efficiency among different plant species or genotypes within species are poorly understood (Graham, 1988; Huang et al., 1993, 1994, 1996; Rengel, 1999). However, Graham (1988) reported that better internal utilization, lower physiological requirements, better root geometry, and better absorption rate may be associated with higher Mn use efficiency. Rengel et al. (1996) also suggested that Mn-efficient genotype substances excreted by roots can solubilize unavailable Mn in the rhizosphere and improve Mn uptake. It is possible that the genetic control of Mn efficiency is expressed through the composition of root exudates encouraging a more favorable balance of Mn reducers to Mn oxidizers in the rhizosphere (Rengel, 1999). Mn-inefficient genotypes may contain more Mn-oxidizing microbes than Mn-efficient genotypes (Rengel 1999). The rhizosphere of wheat genotypes contained an increased proportion of Mn reducers under Mn-deficient compared to Mn-sufficient conditions (Rengel, 1999). Rengel (1997) concluded that the increase in the rate of Mn reducers and Mn oxidizers in the rhizosphere of wheat genotypes is one of the mechanisms underlying differential tolerance to Mn deficiency in some, but not all, Mn-efficient genotypes.

11.8 CONCLUSIONS

Manganese deficiency is widely reported in various parts of the world, especially in soils with higher pH (> 6.0 in water) and in calcareous, sandy, peat, or muck soils. Manganese availability to plants is mainly controlled by its concentration in the soil solution, soil pH, organic matter content, and oxidation-reduction reactions. Adsorption is of little importance to the availability of Mn at higher pH. However, precipitation as MnO_2 is mainly responsible for reducing Mn availability to plants in higher-pH soils. In some crops and soils, however, Mn toxicity can also occur if soil pH is not corrected (<5.0). This possibility is remote, because in modern agriculture, liming is an essential practice to improve crop yields in acid soils. Manganese plays many important roles in plants, such as photosynthesis, respiration, enzyme systems, nitrate assimilation, iron metabolism, and chlorophyll formation. Manganese is relatively immobile in the plant, and deficiency symptoms first occur in the younger leaves. Early signs of Mn deficiency in plants include interveinal chlorosis in the younger leaves, followed by long brown streaks that later turn dark reddish brown.

The Mn use efficiency is higher in grain production compared to grain production in cereals as well as legumes. A larger amount of Mn is exported to the grain of legumes compared to the grain of cereals. The Mn has antagonistic as well as synergistic interactions with macro- and micronutrients, depending on element, crop species, and its concentration in soil solution. Management practices that can maximize Mn use efficiency in crop plants are use of adequate rate, appropriate source, and methods of application. Soil test can be used successfully to correct Mn deficiency, especially in soybean and small grain crops. Overall, critical Mn level in tropical soils is about 4 mg kg^{-1} for DTPA extracting and 8 mg kg^{-1} for Mehlich-1 extracting for principal annual crops. However, soil pH should be defined for Mn soil tests for fertilizer recommendations. A number of Mn fertilizers can be used to correct Mn deficiency; however, manganese sulfate has been the most satisfactory material for most situations. Adequate rate varied with plant species, soil pH, and type of soils. However, 10–50 kg Mn ha^{-1} applied as broadcast soluble fertilizer is recommended for most annual crops to correct Mn deficiency. In band application, the rate may be reduced from one-half to one-third. Foliar spray of soluble $MnSO_4$ salts or chelated salts can correct Mn deficiency in crop plants. However, the higher cost of MnEDTA salts prohibits its use for foliar application. The manganese sulfate fertilizer rate for foliar spray varied from 1 to 5 kg Mn ha^{-1}, and the number of sprays depends on severity of deficiency and crop species. However, 1 to 3 sprays are generally sufficient to correct deficiency under all conditions.

There is no residual effect of Mn due to its conversion from available to unavailable form due to oxidation (Mn^{2+} to Mn^{4+}) and precipitation reaction in the soil. Hence, Mn fertilization should be applied each year or each crop. In addition, use of the soil test is an effective practice to make Mn fertilizer recommendations. Furthermore, planting Mn-efficient plant species or genotypes within species is an economical and environmentally sound management practice to improve Mn efficiency in crop plants. Root-induced changes in the rhizosphere are one of the mechanisms responsible for genotypical differences in Mn uptake and utilization. Recycling of crop residues and addition of animal manures can improve Mn availability to crops.

REFERENCES

Adams, F., and J. T. Wear. 1957. Manganese toxicity and soil acidity in relation to crinkle leaf of cotton. *Soil Sci. Soc. Am. Proc.* 21:305–308.

Alexander, M. 1977. *Introduction to soil microbiology*, 2nd edition. New York: John Wiley and Sons.

Baligar, V. C., N. K. Fageria, and Z. L. He. 2001. Nutrient use efficiency in plants. *Commun. Soil Sci. Plant Anal.* 32:921–950.

Barber, S. A. 1995. *Soil nutrient bioavailability: A mechanistic approach.* New York: John Wiley & Sons.

Bloom, P. R. 1981. Metal-organic matter interactions in soil. In:. *Chemistry in the soil environment,* eds., R. H. Dowdy et al. 129–149. Madison, WI: ASA.

Bohn, H. L., B. L. McNeal, and G. A. O'Connor. 1985. *Soil chemistry.* New York: John Wiley and Sons.

Bolan, N. S., D. C. Adriano, and D. Curtin. 2003. Soil acidification and liming interactions with nutrient and heavy metal transformation and bioavailability. *Adv. Agron.* 78:215–273.

Boswell, F. C., K. Ohki, M. B. Parker, L. M. Shuman, and D. O. Wilson. 1981. Methods and rates of applied manganese for soybeans. *Agron J.* 73:909–912.

Brady, N. C., and R. R. Weil. 2002. *The nature and properties of soils*, 13th edition. Upper Saddle River, NJ: Prentice-Hall.

Brouwers, G. J., E. Vijgenboom, P. L. A. M. Corstjens, J. P. M. Vrind, and E. W. Virnd-de Jong. 2000. Bacterial Mn^{2+} oxidizing systems and multicopper oxidases: An overview of mechanisms and functions. *Geomicrobiol. J.* 17:1024.

Buol, S.W., F. D. Hole, and R. J. McCracken. 1980. *Soil genesis and classification*. Ames, IA: Iowa State Univ. Press.

Burnell, J. N. 1988. The biochemistry of manganese in plants. *In: Manganese in soils and plants*, eds., R. D. Graham, R. J. Hannam, and N. C. Uren, 125–137. Dordrecht: The Kluwer Academic Publishers.

Chinnery, L. E., and C. P. Harding. 1980. The effect of ferrous iron on the uptake of manganese by *Juncus effusus. Ann. Bot.* 46:409–412.

Clark, R. B. 1982. Plant response to mineral element toxicity and deficiency. In: *Breeding plants for the less favorable environments*, eds., M. N. Christiansen and C. F. Lewis, 71–142. New York: John Wiley & Sons.

Collins, J. F., and S. W. Buol. 1970. Patterns of iron and manganese precipitation under specified Eh-pH conditions. *Soil Sci.* 110:157–162.

Cox, F. R. 1987. Micronutrient soil tests: Correlation and calibration. In: *Soil testing: Sampling, correlation, calibration, and interpretation,* ed., J. R. Brown, 97–117. Madison, WI: SSSA.

Ducic, T., and A. Dolle. 2005. Transport and detoxification of manganese and copper in plants. *Brazilian J. Plant Physiol.* 17:103–112.

Dudal, R. 1976. Inventory of the major soils of the world with special reference to mineral stress hazards. In: *Plant adaptation to mineral stress in problem soils,* ed., M. J. Wright, 3–13. Ithaca, NY: Cornell University Press.

Edwards, D. G. and C. J. Asher. 1982. Tolerance of crop and pasture species to manganese toxicity. In: *Proceedings of the ninth plant nutrition colloquium*, ed., A. Scaife, 145–150. Farnham Royal, Buckinghamshire, UK: Commonwealth Agriculture Bur.

Fageria, N. K. 1989. *Tropical soils and physiological aspects of crops*. Brasilia, Goiânia, Brazil: EMBRAPA-CNPAF.

Fageria, N. K. 1998. Optimizing nutrient use efficiency in crop production. *Rev. Bras. Eng. Agric. Ambien.* 2:6–16.

Fageria, N. K. 2001. Adequate and toxic levels of copper and manganese in upland rice, common bean, corn, soybean, and wheat grown on an Oxisol. *Commun. Soil Sci. Plant Anal.* 32:1659–1676.

Fageria, N. K. 2002. Influence of micronutrients on dry matte yield and interaction with other nutrients in annual crops. *Pesq. Agropec. Bras.* 37:1765–1772.

Fageria, N. K., and V. C. Baligar. 1993. Screening crop genotypes for mineral stresses. In: *Proceedings of the workshop on adaptation of plants to soil stress, August 1–4, 1993*, 142–159. INTSORMIL Publication No. 94-2. Lincoln, NE: University of Nebraska.

Fageria, N. K. and V. C. Baligar. 1999. Growth and nutrient concentrations of common bean, lowland rice, corn, soybean, and wheat at different soil pH on an Inceptisol. *J. Plant Nutr.* 22:1495–1507.

Fageria, N. K., V. C. Baligar, and R. B. Clark. 2002. Micronutrients in crop production. *Adv. Agron.* 77:185–268.

Fageria, N. K., V. C. Baligar, and R. B. Clark. 2006. *Physiology of crop production*. New York: The Haworth Press.

Fageria, N. K., V. C. Baligar, and C. A. Jones. 1997. *Growth and mineral nutrition of field crops*, 2nd edition. New York: Marcel Dekker.

Fageria, N. K., and M. P. Barbosa Filho. 2008. Screening upland rice genotypes for manganese use efficiency. *Commun. Soil Sci. Plant Anal.* 39 (in press)

Fageria, N. K., M. P. Barbosa Filho, and L. F. Stone. 2003a. Adequate and toxic levels of micronutrients in soil and plants for annual crops. Paper presented at the XXIX Brazilian Soil Science Congress, July 13–18, 2003, Ribeirão Preto, São Paulo, Brazil.

Fageria, N. K., and H. R. Gheyi. 1999. *Efficient crop production.* Campina Grande, Brazil: Federal University of Paraiba.

Fageria, N. K., N. A. Slaton, and V. C. Baligar. 2003b. Nutrient management for improving lowland rice productivity and sustainability. *Adv. Agron.* 80:63–152.

Fageria, N. K. and C. M. R. Souza. 1991. Upland rice, common bean, and cowpea response to magnesium application on an Oxisol. *Commun. Soil Sci. Plant Anal.* 22:1805–1816.

Follett, R. H., L. S. Murphy, and R. L. Donahue. 1981. *Fertilizers and soil amendments.* Englewood Cliffs, NJ: Prentice-Hall.

Foth, H. D. and B. G. Ellis. 1988. *Soil fertility.* New York: John Wily & Sons.

Foy, C. D. 1984. Physiological effects of hydrogen, aluminum and manganese toxicities in acid soils. In: *Soil acidity and liming*, 2nd edition, ed., F. Adams, 57–97. Madison, WI: ASA.

Gambrell, R. P., and W. H. Patrick, Jr. 1982. Manganese. In: *Methods of soil analysis*, Part 2, 2nd edition, ed., A. L. Page, 313–322. Madison, WI: ASA.

Gettier, S. W., D. C. Martens, and S. J. Donohue. 1985. Soybean yields response prediction from soil test and tissue manganese levels. *Agron. J.* 77:63–66.

Ghiorse, W. C. 1988. The biology of manganese transforming microorganisms in soil. In: *Manganese in soils and plants*, eds., R. D. Graham, R. J. Hannam, and N. C. Uren, 23–35. Dordrecht: The Kluwer Academic Publishers.

Godo, E. H., and H. M. Reisenauer. 1980. Plant effects on soil manganese availability. *Soil Sci. Soc. Am. J.* 44:993–995.

Graham, R. D. 1988. Genotype differences in tolerance to manganese deficiency. In: *Manganese in soils and plants,* eds., R. D. Graham, R. J. Hannam, and N. C. Uren, 261–276. Dordrecht: Kluwer Academic Publishers.

Hallock, D. L. 1979. Relative effectiveness of several Mn sources on Virginia-type peanuts. *Agron. J.* 71:685–688.

Hatfield, A. L., and J. M. Hickey. 1981. Crop fertilization based on North Carolina tests. Circular No.1 (revised) Agronomic Division, North Carolina Department of Agriculture, Raleigh, NC: University of North Carolina.

Horst, W. J. 1988. The physiology of manganese toxicity. In: *Manganese in soils and plants*, eds., R. D. Graham, R. J. Hannam, and N. C. Uren, 175–188. Dordrecht: The Kluwer Academic Publishers.

Hossner, L. R., and G. E. Richards. 1968. The effect of phosphorus source on the movement and uptake of band applied manganese. *Soil Sci. Soc. Am. Proc.* 32:83–85.

Huang, C., M. J. Webb, and R. D. Graham. 1993. Effect of pH on Mn absorption by barley genotypes in a chelate-buffered nutrient solution. *Plant Soil* 155/156:437–440.

Huang, C., M. J. Webb, and R. D. Graham. 1994. Manganese efficiency is expressed in barley growing in soil system but not in a solution culture. *J. Plant Nutr.* 17:83–95.

Huang, C., M. J. Webb, and R. D. Graham. 1996. Pot size affects expression of Mn efficiency in barley. *Plant Soil* 78:205–208.

Jackson, T. L. and G. E. Carter. 1976. Nutrient uptake by Russet Burbank potatoes as influenced by fertilization. *Agron. J.* 68:9–12.

Jackson, T. L., D. T. Westerman, and D. P. Moore. 1966. The effect of chloride and lime on the manganese uptake by bush beans and sweet corn. *Soil Sci. Soc. Am. Proc.* 30:70–73.

Jauregui, M. A., and H. M. Reisenauer. 1982. Calcium carbonate and manganese dioxide as regulators of available manganese and iron. *Soil Sci.* 134:105–110.

Jones, J. B., Jr. 1991. Plant tissue analysis in micronutrients. In: *Micronutrient in agriculture*, 2nd edition, eds., J. J. Mortvedt, F. R. Cox, L. M. Shuman, and R. M. Welch, 477–521. Madison, WI: SSSA.

Katyal, J. C., and B. D. Sharma. 1991. DTPA-extractable and total Zn, Cu, Mn and Fe in Indian soils and their association with some soils properties. *Geoderma* 49:165–179.

Khattak, R. A., W. M. Jarrell, and A. L. Page. 1989. Mechanism of native manganese release in salt treated soils. *Soil Sci. Soc. Am. J.* 53:701–705.

Klein, C., and C. S. Hurlbut, Jr. 1985. *Manual of mineralogy*, 20th edition. New York: John Wiley & Sons.

Krishnamurti, G. S. R. and P. M. Huang. 1992. Dynamics of potassium chloride induced manganese release in different soil orders. *Soil Sci. Soc. Am. J.* 56:1115–1123.

Leach, W., and C. D. Taper. 1954. Studies in plant mineral nutrition. II. The absorption of Fe and Mn by dwarf kidney beans, tomato, and onion from culture solutions. *Can. J. Bot.* 63:604–608.

Leon, L. A., A. S. Lopez, and P. L. G. Vlek. 1985. Micronutrient problem in tropical Latin America. *Fert. Res.* 7:95–129.

Lindsay, W. L. 1972. Zinc in soils and plant nutrition. *Adv. Agron.* 24:147–186.

Lindsay, W. L. 1979. *Chemical equilibria in soils*. New York: Wiley-Interscience.

Lindsay, W. L. and F. R. Cox. 1985. Micronutrient soil test for the tropics. *Fert. Res.* 7:169–200.

Lindsay, W. L. and W. A. Norvell. 1978. Development of a DTPA soil test for zinc, iron, manganese and copper. *Soil Sci. Soc. Am. J.* 42:421–428.

Marschner, H. 1995. *Mineral nutrition of higher plants*, 2nd edition. New York: Academic Press.

Maas, E. V., D. P. Moore, and B. J. Mason. 1968. Influence of calcium and manganese on manganese absorption. *Plant Physiol.* 44:796–800.

Mascagni, H. J. Jr., and F. R. Cox. 1985a. Effective rates of fertilization for correcting manganese deficiency in soybeans. *Agron. J.* 77:363–366.

Mascagni, H. J., Jr., and F. R. Cox. 1985b. Calibration of a manganese availability index for soybean soil test data. *Soil Sci. Soc. Am. J.* 49:382–386.

McDaniel, P. A., and S. W. Buol. 1991. Manganese distribution in acid soils of the North Carolina Piedmont. *Soil Sci. Soc. Am. J.* 55:152–158.

McDaniel, P. A., G. R. Bathke, S. W. Dual, D. K. Cassel, and A. L. Falen. 1992. Secondary manganese/iron ratios as pedochemical indicators of field-scale through flow water movement. *Soil Sci. Soc. Am. J.* 56:1211–1217.

McKenzie, R. M. 1989. Manganese oxides and hydroxides. In: *Minerals in soil environments*, 2nd edition, eds., J. B. Dixon and S. B. Weed, 439–465. Madison, WI: SSSA.

Mengel, K., E. A. Kirkby, H. Kosegarten, and T. Appel. 2001. *Principles of plant nutrition*, 5th edition. Dordrecht: Kluwer Academic Publishers.

Mortvedt, J. J. 1994. Needs for controlled availability of micronutrient fertilizers. *Fert. Res.* 38:213–221.

Mortvedt, J. J., and P. M. Giordano. 1975. Crop response to manganese sources applied with ortho and polyphosphate fertilizers. *Soil Sci. Soc. Am. Proc.* 39:782–785.

Moore, P. A., Jr., and W. H. Patrick, Jr. 1989. Manganese availability and uptake by rice in acid sulfate soils. *Soil Sci. Soc. Am. J.* 53:104–109.

Novais, R. F., J. C. L. Neves, N. F. Barros, and T. Sediyama. Manganese deficiency in soybean plants cultivated in cerrado soils. *Rev. Bras. Ci. Solo* 13:199–204.

Ohki, K., D. O. Wilson, and O. E. Anderson. 1981. Manganese deficiency and toxicity sensitivities of soybean cultivar. *Agron. J.* 72:713–716.

Patrick, W. H., Jr., and C. N. Reddy. 1978. Chemical changes in rice soils. In: *Soils and rice*, ed., International Rice Research Institute, 361–379. Los Bãnos, Philippines: IRRI.

Petrie, S. G. and T. L. Jackson. 1984. Effect of fertilization on soil pH and manganese concentration. *Soil Sci. Soc. Am. J.* 48:315–348.

Patrick, W. H., Jr., and F. T. Turner. 1968. Effect of redox potential on manganese transformations in waterlogged soil. *Nature* 220:476–478.

Randall, G. W., E. E. Schulte, and R. B. Corey. 1975. Soil Mn availability to soybeans as affected by mono and diammonium phosphate. *Agron. J.* 67:705–709.

Rengel, Z. 1997. Root exudation and microflora populations in rhizosphere of crop genotypes differing in tolerance to micronutrient deficiency. *Plant Soil* 196:255–260.

Rengel, Z. 1999. Physiological mechanisms underlying differential nutrient efficiency of crop genotypes. In: *Mineral nutrition of crops: Fundamental mechanisms and implications*, ed., Z. Rengel, 227–265. New York: The Haworth Press.

Rengel, Z. 2000. Uptake and transport of manganese in plants. In: *Metal ions in biological systems*, ed., A. Sigel and H. Sigel, 57–87. New York: Marcel Dekker.

Rengel, Z., R. Guterridge, P. Hirsch, and D. Hornby. 1996. Plant genotype, micronutrient fertilization and take-all infection influence bacterial populations in the rhizosphere of wheat. *Plant Soil* 183:269–277.

Reuter, D. J., A. M. Alston, and J. D. McFarland. 1988. Occurrence and correction of manganese deficiency in plants. In: *Manganese in soils and plants*, eds., R. D. Graham, R. J. Hannan, and N. C. Uren, 205–224. Dordrecht: Kluwer Academic Publishers.

Robson, A. D., and J. F. Loneragan. 1970. Sensitivity of annual species of manganese toxicity as affected by calcium and pH. *Aust. J. Agric. Res.* 21:223–232.

Romheld, V., and H. Marschner. 1991. Function of micronutrient in plants. In: *Micronutrient in agriculture*, 2nd edition, ed., J. J. Mortvedt, 297–328. Madison, WI: SSSA.

Rosas, A., Z. Rengel, and M. L. Mora. 2007. Manganese supply and pH influence growth, carboxylate exudation and peroxidase activity of ryegrass and white clover. *J. Plant Nutr.* 30:253–270.

Schomberg, H. H., and R. W. Weaver. 1991. Growth and N_2 fixation response of arrowleaf clover to manganese and pH in solution culture. *Develop. Plant Soil Sci.* 45:641–647.

Schwab, A. P., and W. L. Lindsay. 1983. Effect of redox on the solubility and availability of iron. *Soil Sci. Soc. Am. J.* 47:201–205.

Shuman, L. M. 1985. Fraction method for soil microelements. *Soil Sci.* 140:11–22.

Sims, J. T., and G. V. Johnson. 1991. Micronutrient soil test. In: *Micronutrients in agriculture*, 2nd edition, ed., J. J. Mortvedt, 417–476. Madison, WI: SSSA.

Smith, D. H., M. A. Wells, D. M. Porter, and F. R. Cox. 1993. Peanuts. In: *Nutrient deficiencies and toxicities in crop plants*, ed., W. F. Bennett, 105–110. St. Paul, MN: The ASP Press, The American Phytopathological Society.

Snyder, G. H., D. B. Jones, and F. J. Coale. 1990. Occurrence and correction of manganese deficiency in Histosol grown rice. *Soil Sci. Soc. Am. J.* 54:1634–1638.

Stahl, R. S., and B. R. James. 1991. Zinc sorption by B horizon soils as a function of pH. *Soil Sci. Soc. Am. J.* 55:1592–1597.

Stevenson, F. J. 1986. *Cycles of soil: Carbon, nitrogen, phosphorus, sulfur, micronutrients.* New York: John Wiley & Sons.

Tanaka, A., and S. A. Navasero. 1966. Interaction between iron and manganese in the rice plant. *Soil Sci. Plant Nutr.* 12:29–33.

Tisdale, S. L., W. L. Nelson, and J. D. Beaton. 1985. *Soil fertility and fertilizers*, 4th edition. New York: Macmillan Publishing Company.

Thompson, I. A., and D. M. Huber. 2007. Manganese and plant disease. In: *Mineral nutrition and plant disease*, eds., L. E. Datnoff, W. H. Elmer, and D. M. Huber, 139–153. St. Paul, MN: The American Phytopathological Society.

Voth, R. D., and D. R. Christenson. 1980. Effect of fertilizer reaction and placement on availability of manganese. *Agron. J.* 72:769–773.

Welch, R. M. 1995. Micronutrient nutrition of plants. *Critical Rev. Plant Sci.* 14:49–82.

Welch, R. M., W. A. Allaway, W. H. House, and J. Kubota. 1991. Geographical distribution of trace element problems. In: *Micronutrients in agriculture*, 2nd edition, eds., J. J. Mortvedt, F. R. Cox, L. M. Shuman, and R. M. Welch, 31–57. Madison, WI: SSSA.

Westermann, D. T., T. L., Jackson, and D. P. Moore. 1971. Effect of potassium salts on extractable soil manganese. *Soil Sci. Soc. Am. Proc.* 35:43–46.

Wilkinson, S. R., D. L. Grunes, and M. E. Sumner. 2000. Nutrient interactions in soil and plant nutrition. In: *Handbook of soil science*, ed., M. E. Sumner, 89–112. Boca Raton, FL: CRC Press.

12 Boron

12.1 INTRODUCTION

Boron (B) is an essential micronutrient for plants, and plant requirements for this nutrient are lower than the requirements for all other nutrients except molybdenum and copper. It is the only nonmetal among the micronutrients and also the only micronutrient present over a wide pH range as a neutral molecule rather than an ion (Epstein and Bloom, 2005). Although B uptake of crop plants is not higher than the uptake of other nutrients, its deficiency has been reported in many parts of the world. Sillanpaa's (1990) analysis of 190 soil samples from 15 countries revealed that 31% of these soils were low in B. Similarly, Asad et al. (2003) reported that B fertilizers increased sunflower yield in Australia. Xu et al. (2001) reported that more than 33 million ha cultivated land in China are B deficient. Mahler and Shafii (2007) reported that the Pacific Northwest is considered a B-deficient region in the United States. White and Zasoski (1999) reported that modern crop cultivars are highly sensitive to low micronutrient levels, including B. Shorrocks (1997) and Fageria et al. (2002) reported that B deficiency has been reported in at least 80 countries in 132 crop species. It is estimated that worldwide about 15 million ha of agricultural land are annually treated with B fertilizers (Shorrocks, 1997). Fageria (2000) reported responses of B application to dry bean, corn, soybean, and wheat grown on Brazilian Oxisols under greenhouse conditions. Similarly, Fageria et al. (2007) reported dry bean response to B fertilization in a Brazilian Oxisol in a field experiment. Figure 12.1 shows response of dry bean to B fertilization applied to a Brazilian Oxisol. Similarly, root growth of soybean and corn were also improved with B fertilization applied to a Brazilian Oxisol (Figure 12.2 and Figure 12.3). Mortvedt (1994) reported that B sorption on aluminum and iron oxide minerals plays a significant role in soils and helps explain the need for B fertilization in many tropical soils. Oplinger et al. (1993) reported that 0.28 kg B ha^{-1} applied foliarly increased soybean yields by 3% when averaged across 29 trials conducted on B-sufficient soils in the midwestern United States. Slaton et al. (2002) and Ross et al. (2006) reported that B deficiency in soybean in many counties of northeast Arkansas has been observed. Al-Molla (1985) reported a 15% yield increase from application of granular B at the R1 soybean stage (Fehr and Caviness, 1977) in Poinsett and Craighead counties of Arkansas, United States.

Two types of B deficiency are encountered in agricultural soils. One is a natural deficiency, due to a lack of boron in the soil-forming minerals, and the other is an induced deficiency, the result of overliming or other adverse environmental conditions. Data in Table 12.1 show that B concentration in dry bean and soybean significantly decreased with increasing pH in the range of 4.9 to 7.0. The variability in B

FIGURE 12.1 Dry bean growth with (left) and without (right) boron.

FIGURE 12.2 Influence of B on root growth of soybean.

uptake was 90% in dry bean and 84% in soybean due to change in pH. The reduction in B availability from increasing soil pH by liming is caused by B adsorption by iron and aluminum hydroxides. Keren and Bingham (1985), Moraghan and Mascagni (1991), and Goldberg et al. (2000), reported that availability of B to plants is affected by a variety of factors including soil solution pH, soil texture, soil moisture, temperature, oxide content, carbonate content, organic matter content, and clay mineralogy. Clay contents can also influence B adsorption. Barber (1995) reported that B adsorption by fine-textured soils is 2–3 times greater than by coarse-textured soils. Boron deficiency is frequently encountered on low-organic-matter, sandy soils. High rainfall and leaching losses reduce B availability. Dry weather can also trigger a deficiency (Fageria and Gheyi, 1999). This has been attributed partly to the reduced number of microorganisms that can release B from the parent materials (Gupta, 1979). There has been significant progress in the knowledge of B nutrition of crop plants in the last few decades. However, these data are scattered in several publications. Hence, the

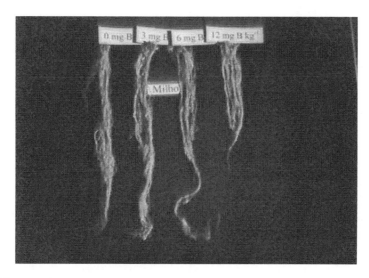

FIGURE 12.3 Corn root growth at (left to right) 0, 3, 6, and 12 mg B kg^{-1} of soil.

TABLE 12.1
Boron Concentration (mg kg^{-1}) in Shoot of Dry Bean and Soybean as Influenced by pH in Brazilian Inceptisol

Soil pH in H$_2$O	Dry Bean	Soybean
4.9	33	20
5.9	19	16
6.4	20	11
6.7	17	11
7.0	14	12
R^2	0.90**	0.84**

** Significant at the 1% probability level.
Source: Adapted from Fageria and Baligar (1999).

objective of this chapter is to compile the latest advances in the B nutrition of crop plants in the light of maximizing its uptake and use efficiency.

12.2 CYCLE IN SOIL–PLANT SYSTEMS

A basic understanding of the cycle of B in soil–plant systems is essential for the proper management of this nutrient in crop production. Major soil B addition sources are chemical fertilizers, organic matter, and parent materials. Boron is removed from soil–plant systems by crop uptake, adsorption on soil colloids, immobilization by microbiomass, and leaching in sandy soils. Three B fractions in soils are generally recognized: water soluble, acid soluble, and total B. The water-soluble fraction is considered related to plant response and is determined either in a soil saturation or in a boiling water extract. The latter is usually designated as *available B*. The acid-soluble fraction includes precipitated B and that incorporated in organic matter. This fraction was termed *maximum available B* to indicate its significance with respect to plant nutrition. A variety of soil properties have been identified as affecting behavior of B in soils. Soil pH, cation exchange capacity, sesquioxides, clay content, type of clay and specific surface, organic matter content, and soil salinity have been reported to influence the solubility and sorption of B in soils (Elrashidi and O'Connor, 1982; Keren and Bingham, 1985; Yermiyahu et al., 2001). Of the clays, illite is the most reactive with B and kaolinite the least reactive on a weight basis (Keren and Mezuman, 1981).

Marzadori et al. (1991) reported that soil organic matter appears to be responsible for occluding important adsorption sites and hinders possible hysteric behavior, i.e., plays a positive role in B release from soil surfaces by conferring reversibility characteristics on the adsorption processes. Boron adsorption is dependent on soil pH and boron solution concentration. Several investigators have reported that increasing pH enhances boron adsorption by clays and soils, showing a maximum in the alkaline pH range (Bingham et al., 1971; Sims and Bingham, 1967). Fox (1968) studied the effect of Ca and pH on B uptake from aqueous solutions by cotton and alfalfa. He found that at high pH, high-calcium solutions restricted B uptake by 50%.

Keren et al. (1981) explained the response of B adsorption to variation in pH as follows: Below pH 7, $B(OH)_3^0$ predominated, but because the affinity of the clay from this species is relatively low, the amount of adsorption is small. As the pH increased, the $B(OH)_4^-$ concentration increased rapidly. The amount of adsorbed B increased rapidly because of the relatively strong affinity of the clays. Further, increase in pH resulted in an enhanced OH^- concentration relative $B(OH)_4^-$, and B adsorption decreased rapidly due to the competition of OH^- for the adsorption sites. Peterson and Newman (1976) reported that B recovery by tall fescue varied from 30 to 50% of the added B by five cuttings of forage, indicating a soil fixation of B. A 2- to 5-fold drop in B uptake occurred at pH 7.4 as compared with the lower pH, indicating substantial fixation of B. With liming, retention of B by precipitant hydroxy Fe and Al compound is a possible mechanism (Peterson and Newman, 1976). Boron does not undergo oxidation-reduction reactions. Hence, B concentration is not appreciably affected when soil is flooded as in the case of lowland rice (Ponnamperuma, 1985). Boron is unique among the essential mineral nutrients because it is a nonmetal, and also it is the only element that is normally present in soil solution as a nonionized molecule over the pH range adequate for annual crops' growth and development.

Parent rock and derived sols are the primary sources of soil B for plants, and borate fertilizers are commonly used to correct B deficiency in field crops. The most common boron mineral is tourmaline, a complex borosilicate. The name *tourmaline* represents a group of minerals containing about 3% B (Goldberg, 1993). Phyllosilicate clay minerals (muscovite, biotite, illite, montmorillonite, kaolinite, and chlorite) also contain B, and these minerals possibly control solution phase activity of B. Tourmalines are highly resistant to weathering and virtually insoluble. Addition of finely ground tourmaline to soil did not correct B deficiency in crop plants (Goldberg, 1993).

12.3 FUNCTIONS AND DEFICIENCY SYMPTOMS

The important functions of boron in plants were compiled by Fageria and Gheyi (1999): (1) boron is essential for germination of pollen grains and growth of pollen tubes; (2) boron is essential for seed and cell wall formation; (3) boron is important in protein formation; (4) when B is deficient, the synthesis of cytokinins is depressed; (5) boron is of considerable importance in the synthesis of nucleic acids; (6) in plants poorly supplied with B, NO_3^- N accumulated in the roots, leaves, and stems, showing that NO_3^- reduction and amino acid synthesis were inhibited; (7) boron facilitates sugar translocation in plants; (8) boron appears to play a significant role in nutrient transport by plant membranes; (9) boron reduces pod abortion in legumes; (10) boron increases the number of pods per inflorescence in legumes; (11) boron influences cell development and elongation; (12) boron is involved in N and P metabolism; (13) boron improves seed germination and seed vigor; and (14) boron is mainly associated with cell wall pectin, and physical characteristics of the growing cell wall were altered under B deficiency (Loomis and Durst, 1992; Matoh et al., 1992; Hu and Brown, 1994; Brown and Hu, 1997).

Boron is known to play important roles in the structure of cell walls, membranes, and membrane-associated functions in plants (Power and Woods, 1997; Brown et al., 2002). Under B deficiency, increased levels of membrane permeability have been reported in sunflower (Cakmak et al., 1995) and soybean (Liu and Yang, 2000). Boron influences root development of cereals and legumes. B-deficient soybean plants produced shriveled and deformed seed, which is a characteristic of B-deficient soybean seed (Rerkasem et al., 1993). This element deficiency causes yield loss of wheat through grain sterility (Rerkasem et al., 1993; Abedin et al., 1994; Jahiruddin et al., 1995). Data in Table 12.2 show that dry bean and wheat root growth improved significantly with B addition in Brazilian Oxisols. Root dry weight of dry bean as well as wheat increased in a quadratic fashion with increasing B rate in the range of 0 to 12 mg kg^{-1} of soil. The variability in dry bean root was about 93% due to B addition. Similarly, variability in wheat root dry weight was about 88% when the B rate was increased from 0 to 12 mg kg^{-1} of soil. Hirsch and Torrey (1980) also reported that sunflower root growth was reduced in the absence of B from the nutrient solution. Gupta (1979) reviewed the role of B in crop plants and reported that the root growth of many crop plants was reduced when B was deficient in the growth medium.

Boron is not mobile in the plants; hence, boron deficiency symptoms first appear in the younger leaves or growing points (Fageria and Baligar, 2005). A deficiency of boron depresses growth and causes shortening of the internodes. Characteristic white or transparent lesions develop in interveinal tissues on the young leaves, often while they are still within the whorl in the case of sorghum and corn. Initially, the lesions develop in an intermittent pattern and appear firstly in the midsection of the leaf blade. Boron deficiency in alfalfa causes leaf yellowing and reddening of the upper leaves. The internodes of the top growth become progressively shorter, and

TABLE 12.2

Influence of Boron Fertilization on Root Growth of Dry Bean and Wheat Grown on Brazilian Oxisols

B Rate (mg kg⁻¹)	Dry Bean Root Dry Weight (g/4 plants)[a]	Wheat Root Dry Weight (g/4 plants)[a]
0	1.20	0.28
1	1.47	0.29
2	1.40	0.29
3	1.30	0.27
6	1.10	0.29
12	0.50	0.22

Regression Analysis

B rate vs. dry bean root dry wt. $(Y) = 1.3312 + 0.0155X - 0.0072X^2$, $R^2 = 0.9293*$

B rate vs. wheat root dry wt. $(Y) = 0.2797 + 0.t - 0.00083X^2$, $R^2 = 0.8796*$

[a] Bean plants were harvested 25 days after sowing, and wheat plants were harvested 21 days after sowing

* Significant at the 5% probability level.

Source: Adapted from Fageria (2000).

the short branches help to give the plant a rosette appearance (Dordas, 2006). At this stage, the growing point becomes dormant or dies, flowering is reduced, and the flower falls before setting seed (Shorrocks, 1997; Bell, 1997; Dell and Huang, 1997; Dordas, 2006). In corn, B-deficient leaves show small dead spots and are brittle in appearance (Voss, 1993). Other symptoms of B deficiency are crinkling of leaf blades and darkening and cracking of petioles in sugar beet (Ulrich et al., 1993).

Figure 12.4 shows deficiency symptoms of B in rice leaves. Deficiency of boron affects the development of the meristem or actively growing tissues; as a result, the

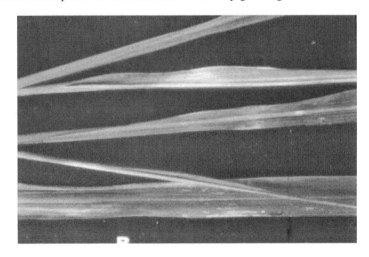

FIGURE 12.4   Boron deficiency in rice leaves.

FIGURE 12.5 Boron toxicity symptoms in dry bean grown on Brazilian Oxisol.

deficiency symptoms are the death of the growing points of the shoot and root, the failure of the flower bud to develop, and, ultimately, the blackening and death of these tissues (Fageria and Gheyi, 1999). Okuda et al. (1961) observed that the panicles of B-deficient rice plants failed to exert from the boot. Likewise, Fageria et al. (2003) reported that B deficiency produced similar symptoms as those associated with the physiological disorder straighthead that was induced by arsenic toxicity in greenhouse studies. Very little research has been conducted to verify or refute B deficiency as a possible cause of straighthead (Fageria et al., 2003).

The range of optimal and toxic level of B in crop plants is very low. Hence, care should be taken in correcting B deficiency by applying chemical fertilizers. Among a wide variety of plant species, the typical visible symptoms of B toxicity include leaf burn-chlorotic and/or necrotic patches, often at the margins and tips of older leaves (Eraslan et al., 2007). These symptoms reflect the distribution of B in most species, with B accumulating at the end of the transpiration stream. The chlorotic/necrotic patches have greatly elevated B concentrations compared with the surrounding leaf tissues, and some species (e.g., barley) show characteristic patterns for different genotypes (Nable et al., 1997). Figure 12.5 shows B toxicity symptoms in dry bean leaves. Bean leaves displayed necrotic edges and burned leaf tips from B toxicity. In addition, dead tissue spots also appeared on the leaves.

12.4 CONCENTRATION AND UPTAKE

Boron concentration in the plants is reported to reflect their B requirement (Jones, 1991). The B concentration in plants is significantly affected by plant age (Fageria et al., 1997; Fageria and Baligar, 2005; Fageria et al., 2006). Figure 12.6 shows B concentration in upland rice as a function of plant age. It was about 9 mg kg^{-1} at 19 days after sowing and dropped to 8 mg kg^{-1} 130 days after sowing or at harvest. In dry bean plants, the trend was a significant linear decrease with increasing plant age from 15 to 96 days' growth interval. Variation in B concentration was 97% in rice

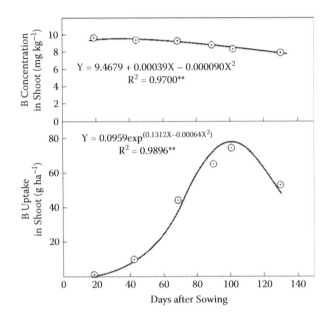

$$Y = 9.4679 + 0.00039X - 0.000090X^2$$
$$R^2 = 0.9700^{**}$$

$$Y = 0.0959\exp^{(0.1312X - 0.00064X^2)}$$
$$R^2 = 0.9896^{**}$$

FIGURE 12.6 Boron concentration and uptake in upland rice as influenced by plant age.

and 90% in dry bean plants due to plant age. Hence, plant age is an important factor affecting B concentration in cereals as well as legumes.

Plant concentration of B can be used as an index of its sufficiency or deficiency index. Jones (1991) reported that B deficiency occurs in many plants when the concentration in fully mature leaves is < 15 mg kg^{-1}; the B sufficiency range is between 20 and 100 mg kg^{-1}. Boron sufficiency level in crop plants is reported to be in the range of 20–100 mg kg^{-1} of dry weight (Fageria, 1992). Blamey et al. (1979) reported that critical concentration in the topmost mature leaf at flowering in two sunflower cultivars was 34 mg B kg^{-1} dry matter. However, Asad et al. (2003) reported that 95% of the maximum dry mass of the capitulum (sunflower) was associated with 20 mg B kg^{-1} in the youngest opened leaves, a value not significantly different from that of 25 mg kg^{-1} associated with 90% maximum yield (Asad et al., 2002). Ross et al. (2006) reported that a critical trifoliate leaf B concentration of soybean at the R2 growth stage of 20 mg B kg^{-1} seems to be reasonably accurate. Touchton et al. (1980) reported that although environmental conditions may play an important role in B fertilization, soybean leaf B concentration >25 mg B kg^{-1} may be sufficient. Melsted et al. (1969) reported that B critical concentration for corn was 10 mg kg^{-1}; for soybean, 25 mg kg^{-1}; for wheat, 15 mg kg^{-1}; and for alfalfa, 30 mg kg^{-1}. A summary of B deficiency and toxicity levels determined in the tissues of principal field crops is provided in Table 12.3. Grain set in wheat has been reported to be associated with B concentration in the ear and flag leaf (Rerkasem and Loneragan, 1994). Rerkasem et al. (1997) reported that B sufficiency for grain set in wheat was indicated by > 7 mg B kg^{-1} in the flag leaf at boot stage and > 6 mg B kg^{-1} in the ear just emerged.

Boron uptake by plants is controlled by the B level in soil solution rather than the total B content in soil (Yermiyahu et al., 2001). Plants absorb boron in the form H$_3$BO$_3$,

TABLE 12.3

Adequate and Toxic Levels of B in Tissues of Principal Field Crops

Crop Species	Adequate Level (mg kg⁻¹)	Toxic Level (mg kg⁻¹)
Upland rice	10	20
Dry bean	24	135
Corn	20	68
Soybean	75	155
Wheat	13	144

Note: Adequate level was calculated at 90% of the maximum dry weight of shoot, and toxic level was calculated at the 10% reduction in shoot dry weight after achieving maximum weight. Plants were harvested at 4 weeks after sowing, and soil pH at harvest was about 6.0 in water in all five experiments.

Source: Adapted from Fageria et al. (2003a).

and it moves to plant root mainly by mass flow and diffusion. Uptake of B in crop plants is mainly determined by yield level. Figure 12.6 and Figure 12.7 show that B uptake increased significantly in a quadratic fashion with increasing plant age in upland rice and dry bean, respectively. Variation in B uptake was about 99% in rice and 97% in dry bean with increasing plant age. This variation in B uptake may be associated with increasing dry matter of shoot in both crop species (Fageria et al., 2006). The decrease in B uptake at harvest was associated with translocation of this element to grain. Boron recovery under field conditions by annual crops is generally in the range of 5 to 15%

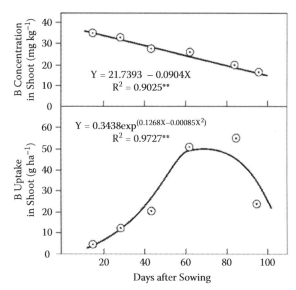

FIGURE 12.7   Boron concentration and uptake in dry bean as influenced by plant age.

in the year of application, and for most annual crops, uptake of 100 to 200 g B ha^{-1} of applied B could be expected to be sufficient (Shorrocks, 1997).

12.5 USE EFFICIENCY AND B HARVEST INDEX

Boron use efficiency (BUE) is an important indicator of its utilization by crop species under a given agroecological condition. Similarly, boron harvest index is an important indicator of the distribution of absorbed B to grain and straw. BUE in upland rice and dry bean grown on Brazilian Oxisols is presented in Table 12.4. In upland rice, BUE during growth cycle followed a quadratic decrease pattern; however, the effect of plant age was nonsignificant. BUE in dry bean was significantly affected by plant age and followed an exponential quadratic response. Variability in BUE was 97% in dry bean due to plant age, indicating the importance of plant age in determining BUE in this crop. In both crops, BUE was higher in grain compared to shoot at harvest. The higher BUE in grain was in both crops associated with lower B accumulation in grain compared to shoot. Similarly, BUE in corn and soybean shoot was significantly influenced by plant age (Table 12.5). In corn, variation in BUE efficiency due to plant age was about 91%; in soybean, variation was 84%. In corn, grain BUE was higher in grain compared to BUE in shoot. However, in soybean, BUE at harvest was higher in grain compared to shoot.

Boron harvest index (BHI) (proportion of B accumulation in grain/proportion of B accumulation in grain plus shoot) was less than 50% in four annual crops (Table 12.6). However, it was higher in legumes (dry bean and soybean) compared to cereals (upland rice and corn). The higher B exportation in legume grain indicates

TABLE 12.4
Boron Use Efficiency (BUE) in Shoot and Grain of Upland Rice and Dry Bean during Crop Growth Cycle (Values Are Averages of Two-Year Field Trial)

Plant Age in Days	Upland Rice (kg g^{-1})	Plant Age in Days	Dry Bean (kg g^{-1})
19	121.9	15	53.6
43	137.2	29	53.1
68	102.6	43	53.6
90	120.3	62	65.5
102	113.0	84	73.1
130	131.5	96	81.9
130 (grain)	177.1	96 (grain)	138.2

Regression Analysis

Plant age (X) vs. BUE in rice shoot (Y) = 140.6510 − 0.6914X + 0.0046X^2, R^2 = 0.2952^{NS}

Plant age (X) vs. BUE in dry bean shoot (Y) = 52.7482 Exp. −0.0011X + 0.000061X^2, R^2 = 0.9666**

**, NS Significant at the 1% probability level and nonsignificant, respectively.

Note: Boron use efficiency (BUE kg g^{-1}) = (Shoot or grain dry weight in kg/Uptake of B in shoot or grain in g).

TABLE 12.5
Boron Use Efficiency (BUE) in Shoot and Grain of Corn and Soybean during Crop Growth Cycle (Values Are Averages of Two-Year Field Trial)

Plant Age in Days	Corn (kg g^{-1})	Plant Age in Days	Soybean (kg g^{-1})
18	78.4	27	67.7
35	83.4	41	53.4
53	83.1	62	79.9
69	112.5	82	90.8
84	116.7	120	86.1
119	133.3	158	188.3
119 (grain)	199.5	158 (grain)	71.7

Regression Analysis

Plant age (X) vs. BUE in corn shoot (Y) = 63.5965 + 0.5974X, R^2 = 0.9082**

Plant age (X) vs. BUE in soybean shoot (Y) = 66.0501 Exp. $-0.00253X + 0.000054X^2$, R^2 = 0.8440*

*, ** Significant at the 5 and 1% probability level, respectively.

Note: Boron use efficiency (BUE kg g^{-1}) = (Shoot or grain weight in kg/uptake of B in shoot or grain in g).

TABLE 12.6
Boron Uptake and Harvest Index in Principal Field Crops (Values Are Averages of Two-Year Field Trial)

Crop Species	B Uptake in Shoot (g ha^{-1})	B Uptake in Grain (g ha^{-1})	Total	B Harvest Index (%)
Upland rice	53.05	29.52	82.57	36
Corn	103.12	42.62	145.75	29
Dry bean	19.74	13.94	33.68	41
Soybean	22.35	20.46	42.80	48

Note: Boron harvest index = (B uptake in grain/B uptake in grain plus shoot).

the superiority of these crops to cereals in human nutrition. Total B uptake in cereals (grain plus straw) was higher compared to legumes. This was associated with higher yields of cereals compared to legumes.

12.6 INTERACTION WITH OTHER NUTRIENTS

Positive relations have been noted between B and K and N fertilizers for improving crop yields (Hill and Morrill, 1975; Moraghan and Mascagni, 1991; Fageria et al., 2002). However, high B supplies resulted in low uptake of Zn, Fe, and Mn but increased uptake of Cu (Fageria et al., 2002). Miley et al. (1969) reported that

B fertilization in adequate amount enhances utilization of applied N in cotton by increasing the translocation of N compounds into the boll. Similarly, Smith and Heathcote (1976) found that when B deficiency occurred in cotton, the application of 250 kg N ha^{-1} depressed yield. However, this rate of N produced a yield increase when B was applied. Woodruff et al. (1987) reported that B fertilization might be necessary to prevent reduction in corn yield when heavy K fertilization and intensive production practices are used. Eraslan et al. (2007) reported that increasing levels of applied B resulted in significantly increased N concentrations of pepper (*Capsicum annuum* L.) and tomato (*Lycopersicon esculentum* L.) plants, except 5 mg B kg^{-1} level in tomato plant. Inal and Tarakcioglu (2001) and Lopez-Lefebre et al. (2002) reported positive effect of B on N uptake and metabolism.

High rate of Ca and Mg application may reduce B uptake. The effect of P and S on B uptake is not clear. These nutrients may reduce, enhance, or have no effect on B uptake (Gupta, 1993). Zinc deficiency enhances B accumulation (Graham et al., 1987), and Zn fertilization reduced B accumulation and toxicity on plants grown in soils containing adequate B (Graham et al., 1987; Moraghan and Mascagni, 1991; Swietlik, 1995). Boron deficiency reduced uptake of P by faba bean (Robertson and Loughman, 1974) and reduced uptake of Mn and Zn by cotton (Ohki, 1975). Boron became toxic to corn when grown under P deficiency conditions, and P applications alleviate B toxicity (Gunes and Alpaslan, 2000).

12.7   MANAGEMENT PRACTICES TO MAXIMIZE B USE EFFICIENCY

Adopting appropriate soil and crop management practices can improve B uptake and utilization efficiency by crop plants. These practices are use of appropriate source and methods of B fertilizers and adequate rate for B-deficient soils. In addition, use of soil test to define adequate B rate is another important management practice to improve B use efficiency in crop plants. Furthermore, use of B-efficient crop species or genotypes within species is also a very attractive strategy for maximizing B efficiency in crop plants. A synthesis of these management practices is discussed in this section.

12.7.1   Appropriate Source, Methods, and Adequate Rate

In B-deficient soils, use of appropriate source is fundamental to improve crop yields and B use efficiency. Principal B carriers are given in Table 12.7. Boric acid, borax, or sodium tetraborate are commonly used fertilizers for correcting B deficiency in crop plants. Boron fertilizers can be applied as broadcast, band, or foliar. However, soil application is more effective and desirable compared to foliar application in annual crops. Data in Table 12.8 show that B applied as broadcast through borax (11% B) significantly increased soil boron content in a Brazilian Oxisol. After three dry bean crops, there was sufficient B in the 0–10 and 10–20 cm soil depths when B was applied at the rate of 2–4 kg ha^{-1}. There were no leaching losses of B even from 0–10 to 10–20 cm soil depths, except slight leaching at 24 kg B ha^{-1}. The soil in the present study had a clay content of 47%. This means that in clay soils, B leaching is insignificant when B is applied at an adequate rate.

TABLE 12.7
Principal B Carriers

Common Name	Formula	B Content (%)
Boric acid	H_3BO_3	17
Borax	$Na_2B_4O_7.10H_2O$	11
Sodium borate (anhydrous)	$Na_2B_4O_7$	20
Sodium pentaborate	$Na_2B_{10}O_{16}.10H_2O$	18
Sodium tetraborate pentahydrate	$Na_2B_4O_7.5H_2O$	14
Boron frits	Fritted glass	10–17
Boron oxide	B_2O_3	31

TABLE 12.8
Hot Water Extracted B Content of Brazilian Oxisol at Two Soil Depths as Influenced by B Fertilization

B Rate (kg ha^{-1})	Soil B Content (mg kg^{-1})	
	0–10 cm	10–20 cm
0	0.81	0.76
2	1.03	0.93
4	1.24	1.11
8	1.75	1.43
16	2.40	2.02
24	3.07	3.15

Regression Analysis

B rate vs. B in 0–10 cm soil depth (Y) = 0.8711 + 0.0939X, R^2 = 0.9940**

B rate vs. B in 10–20 cm soil depth (Y) = 0.7072 + 0.0954X, R^2 = 0.9820**

Note: Values are averages of 3 years determined after harvest of dry bean crop; boron was applied as broadcast using borax (11% B) fertilizer.

** Significant at the 1% probability level.

The method of fertilizer application including B is important in determining its rate and efficiency. Band, broadcast, and foliar applications of B have been used with varying success on numerous crops. Martens and Westermann (1991) summarized results of various experiments in which different B rates were used. Legumes and certain root crops required 2 to 4 kg B ha^{-1}, while lower rates were necessary for maximum yields in other crops to avoid toxicity problems. Gupta (1979) reported that the rates of applied B to correct B deficiency vary from 0.5 to 3 kg ha^{-1}. The Na borates are generally used for soil application. Borox is a popular water-soluble B fertilizer, but it readily leaches in sandy soils. Boronated NPK fertilizers (B fertilizers incorporated at the factory) will ensure a more uniform application than most bulk blended fertilizers.

Foliar application of B to annual crops can also correct B deficiency; however, sometimes several applications are necessary to correct the deficiency and cost may be higher (Touchton and Boswell, 1975). Effective foliar application rates are 10 to 50% of that required by the broadcast method, although repeated applications may be necessary because of B immobility. Generally, a 0.25 to 0.5% solution of fertilizer salts is sufficient to correct B deficiency in crop plants, and solution concentration should not exceed 1%. Generally, 400 to 500 L of solution is required to cover 1 ha of foliage of annual crops. Phytotoxicity may be a problem at higher application rates (Martens and Westermann, 1991). Soluble inorganic salts generally are as effective as synthetic chelates in foliar sprays. Hence, the inorganic salts usually are chosen because of lower costs. Inclusion of sticker-spreader agents in the spray is suggested to improve adherence of the micronutrient source to the foliage. Application of B to seeds or even application in close proximity to seed is not advisable due to the toxic effect of this element (Shorrocks, 1997).

Fageria (2000) determined adequate and toxic rates of B for five annual crops grown on Brazilian Oxisol. Adequate B rates were 2 kg ha^{-1} for dry bean, 4.7 kg ha^{-1} for corn, and 3.4 kg ha^{-1} for soybean applied to a Brazilian Oxisol in a greenhouse experiment (Table 12.9). The toxic rates were 4.4 kg B ha^{-1} for upland rice and dry bean, 8.7 kg B ha^{-1} for corn, 6.8 kg B ha^{-1} for soybean, and 7.4 kg B ha^{-1} for wheat (Table 12.9). Results of this study show that corn requires maximum B and dry bean minimum B. Hewitt (1984) reported that corn is unusual in that it requires a larger supply of B than most other monocotyledonous plants. Mozafar (1989) reported that deficiency of B in corn subsequent to tasseling could cause poor ear development. Fageria et al. (2007) also studied response of dry bean to B fertilization under field conditions (Table 12.10). Grain yield was significantly affected by levels of soil applied B in the first year; however, applied B had no significant effects in the second

---

TABLE 12.9
Adequate and Toxic Levels of Boron for Annual Crops Applied to Brazilian Oxisols

Crop Species	Adequate B Rate (mg kg^{-1})	Toxic B Rate (mg kg^{-1})
Upland rice	No response	4.4
Dry bean	2	4.4
Corn	4.7	8.7
Soybean	3.4	6.8
Wheat	No response	7.4

Note: Adequate rate was calculated at 90% of the maximum dry weight of shoot, and toxic level was calculated at the 10% reduction in shoot dry weight after achieving maximum weight. Plants were harvested 4 weeks after sowing, and soil pH at harvest was about 6.0 in water in all five experiments.

Source: Adapted from Fageria et al. (2003a).

TABLE 12.10

Influence of B on Grain Yield of Dry Bean Grown on Brazilian Oxisol during Three Cropping Years

B Rate (kg ha^{-1})	1st Year	2nd Year	3rd Year	Average
0	3852	3295	2185	3111
2	4024	3432	2151	3202
4	4140	3556	2364	3353
8	3979	3475	2250	3232
16	3488	3448	2185	3041
24	2825	3472	2368	2888
F-test				
Year (Y)	**			
B rate (B)	**			
Y × B	**			
R^2	0.9688**	0.3294^{NS}	0.2318^{NS}	0.7816^{NS}

**,NS Significant at the 1% probability level and nonsignificant, respectively.

Source: Adapted from Fageria et al. (2007).

and third years (Table 12.10). Increasing B application up to 4 kg B ha^{-1} increased grain yield, but further increasing B reduced grain yields. Average increase in bean grain yield with the application of 4 kg B ha^{-1} was about 8% over control.

Reinbott and Blevins (1995) reported that 2.8 kg B ha^{-1} application to soil 8 weeks prior to planting increased soybean yield 11% during the first year and 13% the second year but had no effect on soybean yield by the third year after application. Fageria et al. (2007) also reported that grain yield of dry bean was significantly affected by levels of soil applied B in the first year; however, applied B had no significant effects in the second and third years. Increasing B application up to 4 kg B ha^{-1} increased grain yield of bean, but further increasing B reduced grain yields. Average increase in bean grain yield with the application of 4 kg B ha^{-1} was about 8% over control.

12.7.2 USE OF SOIL TEST

Soil test is an important index in assessing the level of available B and in determining possible residual effects (buildup). In soil test, use of extracting solution is very important in nutrient availability to crop plants. The hot water–soluble B soil test initially developed by Berger and Truog (1940), and subsequently modified by Wear (1965) and Gupta (1967), is still the prevalent method for assessing soil-available B. Li and Gupta (1991) compared various B extracting solutions, including hot water, 0.05 M HCl, 1.5 M CH$_3$COOH, and hot 0.01 M CaCl$_2$. The plants tested for B uptake were soybean, red clover, alfalfa, and rutabaga (*Brassica napobrassica*, Mill.). These authors concluded that the 0.05 M HCl solution was the best extracting (r = 0.82), followed by 1.5 M CH$_3$COOH, (r = 0.78), then hot water (r = 0.66), and lastly the hot

TABLE 12.11

Adequate and Toxic Level of Hot Water Extractable Soil Boron in Principal Crop Species Grown on Brazilian Oxisol

Crop Species	Adequate B Level (mg kg^{-1})	Toxic B Level (mg kg^{-1})
Upland rice	0.4	2.3
Dry bean	0.9	2.8
Corn	1.3	5.7
Soybean	2.6	5.2
Wheat	0.4	4.3

Note: Adequate rate was calculated at 90% of the maximum dry weight of shoot, and toxic level was calculated at the 10% reduction in shoot dry weight after achieving maximum weight. Plants were harvested 4 weeks after sowing, and soil pH at harvest was about 6.0 in water in all five experiments.

Source: Adapted from Fageria et al. (2003a).

0.01 M CaCl$_2$ ($r = 0.61$). Raza et al. (2002) compared four different extraction methods to determine available B using a four-step sequential B fraction as basis. The authors concluded that B pools of a series of Saskatchewan soils could be predicted by the soil cation exchange capacity and that hot water–extractable B was a good method to estimate available B.

Most of the soil B levels reported to be adequate or critical for the growth of annual crops are hot water–extractable B. Cox (1987) reported that the critical level of B, in soil extracted by hot water, ranged from 0.1 to 2 mg kg^{-1} and that the mean value is 0.7 mg B kg^{-1} of soil. According to Reisenauer et al. (1973), toxicity in crop plants is likely to occur when the level of hot water–soluble B in soil exceeds 5 mg B kg^{-1}, whereas soil levels of < 1 mg B kg^{-1} are generally not high enough for optimum plant growth. Fageria et al. (2003a) determined adequate and toxic levels of soil B for upland rice, dry bean, corn, soybean, and wheat (Table 12.11). Adequate and toxic levels varied from species to species. Upland rice required minimum hot water–extractable B (0.4 mg kg^{-1}) to produce 90% of the yield, and soybean required maximum (2.6 mg kg^{-1}) to produce 90% yield. The B toxic level also varied according to crop species. Corn was most tolerant (5.7 mg B kg^{-1}) to B toxicity, and upland rice (2.3 mg B kg^{-1}) was least tolerant to B toxicity. Mahler and Shafii (2007) reported that critical soil B value for optimum lentil (*Lens culinaris*) yield is 0.4 mg kg^{-1} in the Pacific Northwest region of the United States.

12.7.3 USE OF EFFICIENT CROP SPECIES/GENOTYPES

Large differences exist among crop species or genotypes of the same species in uptake and utilization of B (Fageria et al., 2002; Mengel et al., 2001; Jamjod et al., 2004). A wide range of genotype variation for response to low B has been reported in wheat (Rerkasem and Jamjod, 1997; Rerkasem et al., 1997; Jamjod et al., 2004; Kataki et al., 2001; Ahmed et al., 2007). Boron-efficient genotypes of wheat have been found to avoid male sterility in the field in low-B soils in Bangladesh, Nepal, and Thailand, where inefficient genotypes sustained serious yield losses (Rerkasem and Jamjod, 1989, 1997; Kataki et al., 2002; Jamjod et al., 2004). Jamjod et al. (2004) reported that B efficiency in wheat can be expressed as a partially dominant character, but the phenotypes of F_1 hybrids, relative to parents, indicated genetic control varying from recessive to additive to completely dominant with different cross combinations and B levels. Jamjod et al. (2004) also reported that B efficiency in wheat is controlled by major genes; the backcross method is the most efficient way for transferring B efficiency into locally adapted, inefficient cultivars.

12.8   CONCLUSIONS

Boron deficiency in crop plants reduces not only yields but also crop quality. The need for B has increased in recent years; this may be attributed to the use of high-purity fertilizers, less use of organic manures, increased plant populations, better management practices (which result in higher yields), and higher-yielding cultivars (which may require more B than the older cultivars). Deficiency of B is widely reported in various parts of the world, covering a large number of crop species. Boron requirement of crop plants is lower than that of all other essential micronutrients except molybdenum. In some crop species, Cu uptake is also lower than B uptake. Boron requirements varied between crop species and cultivars within species. Dicotyledons generally require 3–4 times more B than monocotyledons. It is commonly accepted that floral and fruiting organs are especially sensitive to B deficiency. The root system is also adversely affected by B deficiency.

Boron is absorbed as H_3BO_3 and exists in soil solution as an undissociated ion at pH less than 7.0. It disassociates to $B(OH)_4^-$ only at higher pH values (> 7.0). Soil pH, organic matter, clay content, oxide content, carbonate content, and parent material are important factors affecting the availability of B to plants. Boron becomes less available to plants with increasing soil pH, and this reduction in B availability is associated with B immobilization by liming acid soils. Coarse-textured soils often contain less available B compared to fine-textured soils. Similarly, organic matter is an important source of B in soils, and B in organic matter can be released for plant uptake by microbial action. Soil test (hot water extraction method) is an important index in defining B supplying capacity of a soil or B uptake by crop plants. On an average, 0.5 to 1 mg B kg^{-1} of soil extracted by hot water is considered an adequate B level for most crop species. Chemical fertilizers containing appropriate amounts of B are the most desirable sources of B to use to control B deficiency in crop plants. Application of 2 to 4 kg B ha^{-1} through chemical fertilizers is sufficient for most annual crop growth and development. The residual effect of B fertilizers depends on crop yield and clay content of the soil. In sandy soils, frequent B application is required due to leaching of this element. However, in clay soils (>30% clay content), residual effect of adequate rate of B (2 to 4 kg B ha^{-1}) may persist for three to four crops subsequently grown on the same area. Foliar spray of inorganic salts can be used to correct B deficiency in crop plants. However, sometimes foliar spray is not economical if more than one spray is required, unless the sprays can be combined with pesticide spray applications. Another disadvantage of foliar sprays is leaf burn, which may result if salt concentrations of the spray are high; in addition, nutrient demand often is high when the plants are small and leaf surface is insufficient for foliar absorption, and there is no residual effect from foliar sprays. Plant tissue test is an important criterion to determine B deficiency or sufficiency levels in crop plants. The B content of plant tissue of about 20 mg kg^{-1} can produce 90% of the maximum yield of most crop species. Use of B-efficient crop species or genotypes within species is the most attractive strategy for improving crop yields, reducing cost of production, and minimizing environmental pollution.

REFERENCES

Abedin, M. J., M. Jahiruddin, M. S. Hoque, M. R. Islam, and M. U. Ahmed. 1994. Application of boron for improving grain yield of wheat. *Progressive Agriculture* 5:75–79.

Ahmed, M., M. Jahiruddin, and M. M. Mian. 2007. Screening wheat genotypes for boron efficiency. *J. Plant Nutr.* 30:1127–1138.

Al-Molla, R. M. M. 1985. *Some physiological aspects of soybean development and yield as affected by boron fertilization.* PhD diss. Fayetteville, AR: University of Arkansas.

Asad, A. F., P. C. Blamey, and D. G. Edwards. 2002. Dry matter production and boron concentrations of vegetative and reproductive tissues of canola and sunflower plants grown in nutrient solution. *Plant Soil* 243:243–252.

Asad, A. F., P. C. Blamey, and D. G. Edwards. 2003. Effects of boron foliar applications on vegetative and reproductive growth of sunflower. *Ann. Bot.* 91:565–570.

Barber, S. A. 1995. *Soil nutrient bioavailability: A mechanistic approach*, 2nd edition. New York: Wiley.

Bell, R. W. 1997. Diagnosis and prediction of boron deficiency for plant production. *Plant Soil* 193:149–168.

Berger, K. C. and E. Truog. 1940. Boron deficiencies as revealed by plant and soil tests. *J. Am. Soc. Agron.* 3:297–301.

Bingham, F. T., A. L. Page, N. T. Coleman, and K. Flach. 1971. Boron adsorption characteristics of selected amorphous soils from Mexico and Hawaii. *Soil Sci. Soc. Am. Proc.* 35:546–550.

Blamey, F. P. C., D. Mould, and J. Chapman. 1979. Critical boron concentration in plant tissues of two sunflower cultivars. *Agron. J.* 71:243–247.

Brown, P. H., N. Bellaloiu, M. A. Wimmer, E. S. Bassil, J. Ruiz, H. Hu, H. Pfeffer, F. Dannel, and V. Romheld. 2002. Boron in plant biology. *Plant Biology* 4:205–223.

Brown, P. H. and H. Hu. 1997. Does boron play only a structural role in the growing tissues of higher plants? *Plant and Soil* 196:211–215.

Cakmak, I., H. Kurz, and H. Marschner. 1995. Short term effects of born, germanium, and high light intensity on membrane permeability in boron deficient leaves on sunflower. *Physiologia Plantarum* 95:11–18.

Cox, F. R. 1987. Micronutrient soil tests: Correlation and calibration. In: *Soil testing: Sampling, correlation, calibration, and interpretation,* J. R. Brown, Ed., 97–117. Madison, WI: SSSA.

Dell, B. and L. Huang. 1997. Physiological response of plants to low boron. *Plant Soil* 193:103–120.

Dordas, C. 2006. Foliar boron application improves seed set, seed yield and seed quality of alfalfa. *Agron. J.* 98:907–913.

Elrashidi, M. A. and G. A. O'Connor. 1982. Boron sorption and desorption in soils. *Soil Sci. Soc. Am. J.* 46:27–31.

Epstein, E. and A. J. Bloom. 2005. *Mineral nutrition of plants: Principles and perspectives*, 2nd edition. Sunderlands, MA: Sinauer Associates.

Eraslan, F., A. Inal, A. Gunes, and M. Alpaslan. 2007. Boron toxicity alters nitrate reductase activity, proline accumulation, membrane permeability, and mineral constituents of tomato and pepper plants. *J. Plant Nutr.* 30:981–994.

Fageria, N. K. 1992. *Maximizing crop yields.* New York: Marcel Dekker.

Fageria, N. K. 2000. Adequate and toxic levels of boron for rice, common bean, corn, soybean and wheat production in cerrado soil. *Rev. Bras. Eng. Agric. Ambien.* 4:57–62.

Fageria, N. K. and V. C. Baligar. 1999. Growth and nutrient concentrations of common bean, lowland rice, corn, soybean and wheat at different soil pH on an Inceptisol. *J. Plant Nutr.* 22:1495–1507.

Fageria, N. K. and V. C. Baligar. 2005. Nutrient availability. In: *Encyclopedia of soils in the environment,* D. Hillel, Ed., 63–71. San Diego, CA: Elsevier.

Fageria, N. K., V. C. Baligar, and R. B. Clark. 2002. Micronutrients in crop production. *Adv. Agron.* 77:185–268.

Fageria, N. K., V. C. Baligar, and R. B. Clark. 2006. *Physiology of crop production.* New York: The Haworth Press.

Fageria, N. K., V. C. Baligar, and C. A. Jones. 1997. *Growth and mineral nutrition of field crops,* 2nd edition. New York: Marcel Dekker.

Fageria, N. K., V. C. Baligar, and R. W. Zobel. 2007. Yield, nutrient uptake and soil chemical properties as influenced by liming and boron application in common bean in no-tillage system. *Commun. Soil Sci. Plant Anal.* 38:1637–1653.

Fageria, N. K., M. P. Barbosa Filho, and L. F. Stone. 2003a. Adequate and toxic levels of micronutrients in soil and plants for annual crops. Paper presented at the XXIX Brazilian Soil Science Congress, 13 to 18 July 2003, Ribeirão Preto, São Paulo, Brazil.

Fageria, N. K. and H. R. Gheyi. 1999. *Efficient crop production.* Campina Grande, Brazil: Federal University of Paraiba.

Fageria, N. K., N. A. Slaton, and V. C. Baligar. 2003b. Nutrient management for improving lowland rice productivity and sustainability. *Adv. Agron.* 80:63–152.

Fehr, W. R. and C. E. Caviness. 1977. Stages of soybean development. Spec. Rep. 80. Iowa Agric. Home Econ. Exp. Stan., Ames, Iowa: Iowa State University.

Fox, R. H. 1968. The effect of calcium and pH on boron uptake from high concentrations of boron by cotton and alfalfa. *Soil Sci.* 106:435–439.

Goldberg, S. 1993. Chemistry and mineralogy of boron in soils. In: *Boron and its role in crop production,* U. C. Gupta, Ed., 3–44. Boca Raton, FL: CRC Press.

Goldberg, S., S. M. Lesch, and D. L. Suarez. 2000. Predicting boron adsorption by soils using soil chemical parameters in the constant capacitance model. *Soil Sci. Soc. Am. J.* 64:1356–1363.

Graham, R. D., R. M. Welch, D. L. Grunes, E. E. Cary, and W. A. Norvell. 1987. Effect of zinc deficiency on the accumulation of boron and other mineral nutrients in barley. *Soil Sci. Soc. Am. J.* 51:652–657.

Gunes, A. and M. Alpaslan. 2000. Boron uptake and toxicity in maize genotypes in relation to boron and phosphorus supply. *J. Plant Nutr.* 21:541–550.

Gupta, U. C. 1967. A simplified method for determining hot-water soluble boron in podzol soil. *Soil Sci.* 103:424–428.

Gupta, U. C. 1979. Boron nutrition of crops. *Adv. Agron.* 31:274–307.

Gupta, U. C. 1993. Factors affecting boron uptake by plants. In: *Boron and its role in crop production,* U. C. Gupta, Ed., 87–123. Boca Raton, FL: CRC Press.

Hewitt, E. J. 1984. The essential and functional mineral elements. In: *Diagnosis of mineral disorder in plants,* Vol. 1, J. D. B. Robinson, Ed., 7–53. New York: Chemical Publ.

Hill, W. E. and L. G. Morrill. 1975. Boron, calcium and potassium interactions in Spanish peanuts. *Soil Sci. Soc. Am. Proc.* 39:80–83.

Hirsch, A. M. and J. G. Torrey. 1980. Ultrastructural changes in sunflower root cells in relation to boron deficiency and added auxin. *Can. J. Bot.* 58:856–866.

Hu, H. and P. H. Brown. 1994. Localization of boron in cell walls of squash and tobacco and its association with pectin: Evidence for a structural role of boron in the cell wall. *Plant Physiol.* 105:681–689.

Inal, A. and C. Tarakcioglu. 2001. Effects of nitrogen form on growth, nitrate accumulation, membrane permeability and nitrogen use efficiency of hydroponically grown bunch onion under boron deficiency and toxicity. *J. Plant Nutr.* 24:1521–1534.

Jahiruddin, M., M. S. Ali, M. A. Hossain, M. U. Ahmed, and M. M. Hoque. 1995. Effect of boron on grain set, yield and some other parameters of wheat cultivars. *Bangladesh J. Agric. Sci.* 22:179–184.

Jamjod, S., S. Niruntrayagul, and B. Rerkasem. 2004. Genetic control of boron efficiency in wheat (*Triticum aestivum* L.) *Plant Soil* 135:21–27.

Jones, J. B., Jr. 1991. Plant tissue analysis in micronutrients. In: *Micronutrient in agriculture*, 2nd edition, J. J. Mortvedt, F. R. Cox, L. M. Shuman, and R. M. Welch, Eds., 477–521. Madison, WI: SSSA.

Kataki, P. K., S. P. Srivastava, M. Saifuzzaman, and H. K. Upreti. 2002. Sterility of wheat and response of field crops to applied boron in the Indo-Gangetic Plains. *J. Crop Prod.* 4:133–165.

Kataki, P. K., H. K. Upreti, and M. R. Bhatta. 2001. Soil born deficiency induced wheat sterility in Nepal: Response to boron and nitrogen application. *J. New Seeds* 3:23–39.

Keren, R. and F. T. Bingham. 1985. Boron in water, soils, and plants. *Adv. Soil Sci.* 1:229–276.

Keren, R., R. G. Gast, and B. Bar-Yosef. 1981. pH-dependent boron adsorption by Na-montmorillonite. *Soil Sci. Soc. Am. J.* 45:45–48.

Keren, R. and U. Mezuman. 1981. Boron adsorption by clay minerals using a phenomenological equation. *Clay Miner.* 29:198–204.

Li, R. and U. C. Gupta. 1991. Extraction of soil B for predicting its availability to plants. *Commu. Soil Sci. Plant Anal.* 13:130–134.

Liu, P. and Y. A. Yang. 2000. Effects of molybdenum and boron on membrane lipid peroxidation and endogenous protective systems of soybean leaves. *Acta Botanica Sinica* 42:461–466.

Loomis, W. D. and R. W. Durst. 1992. Chemistry and biology of born. *Biofactors* 3:229–239.

Lopez-Lefebre, L. R., R. M. Rivero, P. C. Garcia, E. Sanchez, J. M. Ruiz, and L. Romero. 2002. Boron effect on mineral nutrients of tobacco. *J. Plant Nutr.* 25:509–522.

Mahler, R. L. and B. Shafii. 2007. Relationship between soil test boron and lentil yields in the inland Pacific Northwest. *Commun. Soil Sci. Plant Anal.* 38:1193–1202.

Martens, D. C. and D. T. Westermann. 1991. Fertilizer application for correcting micronutrient deficiencies In: *Micronutrients in agriculture,* 2nd edition, J. J. Mortvedt, Ed., 549–592. Madison, WI: SSSA.

Marzadori, C., L. Vittori Antisari, C. Ciavatta, and P. Sequi. 1991. Soil organic matter influences on adsorption and desorption of boron. *Soil Sci. Soc. Am. J.* 55:1582–1585.

Matoh, T., K. I. Ishigaki, M. Mizutani, W. Matsunaga, and K. Takabe. 1992. Boron nutrition of cultured tobacco BY-2 cells. I. Requirement for and intracellular localization of boron and selection of cells that tolerate low levels of boron. *Plant Cell Physiol.* 33:1135–1141.

Melsted, S. W., H. L. Motto, and T. R. Peck. 1969. Critical plant nutrient composition values useful in interpreting plant analysis data. *Agron. J.* 61:17–20.

Mengel K., E. A. Kirkby, H. Kosegarten, and T. Appel. 2001. *Principles of plant nutrition*, 5th edition. Dordrecht: Kluwer Academic Publishers.

Miley, W. N., G. W. Hardy, and M. B. Sturgis. 1969. Influence of boron, nitrogen and potassium on yield, nutrient uptake and abnormalities of cotton. *Agron. J.* 6:9–13.

Moraghan, J. T. and H. J. Mascagni, Jr. 1991. Environmental and soil factors affecting micronutrient deficiencies and toxicities. In: *Micronutrient in agriculture*, 2nd edition, J. J. Mortvedt, F. R. Cox, L. M. Shuman, and R. M. Welch, Eds., 371–425. Madison, WI: SSSA.

Mortvedt, J. J. 1994. Needs for controlled availability micronutrient fertilizers. *Fert. Res.* 38:213–221.

Mozafar, A. 1989. Boron effect on mineral nutrition of maize. *Agron. J.* 81:285–290.

Nable, R. O., G. S. Banuelos, and J. G. Paull. 1997. Boron toxicity. *Plant Soil* 193:181–198.

Ohki, R. 1975. Mn and B effect on micronutrients and P in cotton. *Agron. J.* 67:204–207.

Okuda, A., S. Hori, and S. Ida. 1961. Boron nutrition in higher plans. I. A method of growing boron deficient plants. *J. Sci. Soil Manure,* Japan 32:153–157.

Oplinger, E. S., R. G. Hoeft, J. W. Johnson, and P. W. Tracy. 1993. Boron fertilization of soybeans: A regional summary. In: *Proceedings of symposium: Foliar fertilization of soybeans and cotton*, PPI/FAR Tech. Bull. 1993-1, L. S. Murphy, Ed., 7–16. Norcross, GA: Potash & Phosphate Institute.

Peterson, L. A. and R. C. Newman. 1976. Influence of soil pH on the availability of added boron. *Soil Sci. Soc. Am. J.* 40:280–282.

Ponnamperuma, F. N. 1985. Chemical kinetics of wetland rice soils relative to soil fertility. In: *Wetland soils: Characterization, classification, and utilization,* International Rice Research Institute, Ed., 71–89. Los Banos, Philippines: IRRI.

Power, P. P. and W. G. Woods. 1997. The chemistry of boron and its speciation in plants. *Plant Soil* 193:1–13.

Raza, M., A. R. Mermut, J. J. Schoenau, and S. S. Mahli. 2002. Boron fractionation in some Saskatchewan soils. *Can. J. Soil Sci.* 82:173–179.

Reinbott, T. M. and D. G. Blevins. 1995. Response of soybean to foliar applied boron and magnesium and soil-applied boron. *J. Plant Nutr.* 18:179–200.

Reisenauer, H. M., L. M. Walsh, and R. G. Hoeft. 1973. Testing soils for sulphur, boron molybdenum and chlorine. In: *Soil testing and plant analysis,* L. M. Walsh and J. D. Beaton, Eds., 173–200. Madison, WI: SSSA.

Rerkasem, B., R. W. Bell, S. Loedkaew, and J. F. Loneragan. 1993. Boron deficiency in soybean (*Glycline max* L. Merr.), peanut (*Arachis hypogaea* L.), and black gram (*Vigna mungo* L. Hepper): Symptoms in seeds and differences among soybean cultivars in susceptibility to boron deficiency. *Plant Soil* 150:289–294.

Rerkasem, B. and S. Jamjod. 1997. Boron deficiency induces male sterility in wheat (*Triticum aestivum* L.) and implications for plant breeding. *Euphytica* 96:257–262.

Rerkasem, B. and S. Jamjod. 1989. Correcting boron deficiency induced ear sterility in wheat and barley. *Thai J. Soils and Fertility* 11:200–209.

Rerkasem, B. and J. F. Loneragan. 1994. Boron deficiency in two wheat genotypes in a warm, subtropical region. *Agron. J.* 86:887–890.

Rerkasem, B., S. Lordkaew, and B. Dell. 1997. Boron requirement for reproductive development in wheat. *Soil Sci. Plant Nutr.* 43:953–957.

Robertson, G. A. and B. C. Loughman. 1974. Reversible effects of boron on the absorption and incorporation of phosphate in *Vicia faba* L. *New Phytol.* 73:291–298.

Ross, J. R., N. A. Slaton, K. R. Brye, and R. E. DeLong. 2006. Boron fertilization influences on soybean yield and leaf and seed boron concentrations. *Agron. J.* 98:198–205.

Shorrocks, V. M. 1997. The occurrence and correction of boron deficiency. *Plant Soil* 193:121–148.

Sillanpaa, M. 1990. Micronutrient assessment at the country level: An international study. FAO Soils Bulletins 63. FAO/Finnish International Development Agency, Rome, Italy.

Sims, J. R. and F. T. Bingham. 1967. Retention of boron by layer silicates, sesquioxides, and soil minerals. I. Layer silicates. *Soil Sci. Soc. Am. Proc.* 31:728–732.

Slaton, N. A., L. Ashlock, J. McGee, E. Terhune, R. Wimberly, R. DeLong, and N. Wolf. 2002. Boron deficiency of soybeans in Arkansas. In: *Wayne E. Sabbe Arkansas soil fertility studies 2001.* Res. Ser. 490, N. A. Slaton, Ed., 37–41. Fayetteville, AR: Arkansas Agric. Exp. Stn.

Smith, J. B. and R. G. Heathcote. 1976. A new recommendation for the application of bor-onated superphosphate to cotton in northeastern Beune Plateau states. *Samarau Agric. Newsletter* 18:59–63.

Swietlik, D. 1995. Interaction between zinc deficiency and boron toxicity on growth and min-eral nutrition of sour orange seedlings. *J. Plant Nutr.* 18:1191–1207.

Touchton, J. T. and F. C. Boswell. 1975. Boron application for corn grown on selected south-ern soils. *Agron. J.* 67:197–200.

Touchton, J. T., F. C. Boswell, and W. H. Marchant. 1980. Boron for soybean grown in Geor-gia. *Commun. Soil Sci. Plant Anal.* 11:369–378.

Ulrich, A., J. T. Moraghan, and E. D. Whitney. 1993. Sugar beet. In: *Nutrient deficiencies and toxicities in crop plants*, W. F. Bennett, Ed., 91–98, St. Paul, MN: The APS Press, The American Phytopathological Society.

Voss, R. D. Corn. In: *Nutrient deficiencies and toxicities in crop plants,* W. F. Bennett, Ed., 11–14. St. Paul, MN: The APS Press, The American Phytopathological Society.

Wear, J. I. 1965. Boron. In: *Methods of soil analysis,* Part II., C. A. Black, Ed., 1059–1063. Madison, WI: ASA.

White, J. G. and R. J. Zasoski. 1999. Mapping soil micronutrients. *Field Crops Res.* 60:11–26.

Woodruff, J. R., F. W. Moore, and H. L. Musen. 1987. Potassium, boron, nitrogen and lime effects on corn yield and earleaf nutrient concentrations. *Agron. J.* 79:520–524.

Xu, F. S., Y. H. Wang, and J. L. Meng. 2001. Mapping boron efficiency genes in Brassica napus using RFLP and AFLP markers. *Plant Breeding* 120:319–324.

Yermiyahu, U., R. Keren, and Y. Chen. 2001. Effect of composted organic matter on boron uptake by plants. *Soil Sci. Soc. Am. J.* 65:1436–1441.

13 Molybdenum

13.1 INTRODUCTION

The importance of molybdenum (Mo) in plant nutrition is well established (Fageria, 1992; Marschner, 1995; Mengel et al., 2001). Molybdenum is required by higher plants in very small amounts (0.1 mg kg^{-1}) (Graham and Stangoulis, 2007). However, it has crucial roles in plants via molybdoenzymes (Yu et al., 1999). The essentiality of Mo for plants was first established in 1938 by D. I. Arnon and P. R. Stout, who used tomato as a test plant in nutrient solution (Marschner, 1995; Fageria et al., 1997). Molybdenum is the least abundant of all the micronutrients in the earth's crust, and the common range in soil is 0.2 to 5 mg kg^{-1}, with an average value of about 2 mg kg^{-1} (Lindsay, 1979). Molybdenum deficiency is common in acid soils because Mo is adsorbed on Fe hydrous oxides and hydroxides as MoO_4^{2-}. In several acid soils of Brazil, response of common bean to Mo application was observed only after the soil pH was raised above 5.5 (Franco and Day, 1980). This could be due to limitation of nodule or plant function or to inability of some bean cultivars to absorb Mo or move it to nodules under very acid soil conditions (Franco and Munns, 1981).

Research data on crop responses to Mo are not as extensive as those of other micronutrients like Zn, Cu, B, Fe, and Mn. However, Burmester et al. (1988) observed widespread Mo deficiency in northern Alabama in soybean. The Mo deficiency in Australia has been reported in crops grown on soils derived from sedimentary rocks, basalt, and granite (Anderson, 1970). Sillanpaa (1990) analyzed 190 soil samples from 15 countries and revealed that 15% of these soils were low in Mo. Molybdenum deficiency is most pronounced in high-rainfall areas, but the effect is generally related to adsorption of MoO_4^{2-} by soil constituents under acid conditions (Johansen et al., 1977). The objective of this chapter is to provide information on Mo nutrition of crop plants to maximize its uptake and use efficiency.

13.2 CYCLE IN SOIL–PLANT SYSTEMS

Molybdenum concentration in most soils is very low. Molybdenum normally occurs in soil solution as MoO_4^{2-}. Although Mo is not highly toxic, it has been the subject of considerable investigation from an environmental standpoint (Lindsay, 1979). Molybdenum reactions have been described as some of the most complex of any chemical element (Cotton and Wilkinson, 1972). Molybdenum is strongly adsorbed in acid soils on hydrous Fe oxides and hydroxides. Hence, its deficiency in acid soils may be corrected by liming. Molybdenum becomes more available as pH goes up, the opposite of other micronutrients. Maximum availability of Mo occurs at pH more than 6.5 (Jones, 1985). Molybdenum largely occurs in the soil as oxyanion (MoO_4^{2-}). However, other species may also present in small amounts. The solution species generally decrease in the order $MoO_4^{2-} > HMoO_4^- > H_2MoO_4^0 > MoO_2(OH)^+ > MoO_2^{2+}$ (Lindsay, 1979). In the pH range of 3 to 5, the first three species contribute significantly to total molybdenum in solution. The latter two ions can generally be ignored in soils (Lindsay, 1979).

This property clearly distinguishes Mo from the other heavy metals nutrients, and Mo more closely resembles phosphate or sulfate in its behavior in the soil. Of all the plant nutrient anions, molybdate ranks second after phosphate in its strength of adsorptive binding (Parfitt and Smart, 1978; Mengel et al., 2001). Reddy et al. (1997) reported that maximum adsorption of Mo occurs at pH 4. This effect of pH on adsorption may be explained by competition between OH^- and MoO_4^{2-} ions for adsorption sites and the increased positive charge on adsorbing surfaces at lower pH (Mengel et al., 2001).

Molybdenum may exist in several oxidation states, but molybdate predominates in the low concentrations found in agricultural soils (Karimian and Cox, 1978). Iron oxides found in acid soils carry positive charges and can react with molybdate. Jones (1956) reported that freshly prepared ferric oxides and a soil high in Fe oxides removed large amounts of Mo from aqueous solution. Aluminum oxides, halloysite, nontronite, and kaolinite are also capable of removing Mo from aqueous solutions, but their effectiveness is much less than that of Fe oxide under the same conditions (Jones, 1956). Accumulation of Mo in organic soils has been reported (Williams and Thornton, 1973). Organic matter is thus capable of removing Mo from solution and may affect plant availability of soil Mo.

13.3 FUNCTIONS AND DEFICIENCY SYMPTOMS

The function of molybdenum in plant metabolism is to reduce nitrate nitrogen to nitrite nitrogen. In legumes, Mo is a vital participant in nodule formation for symbiotic nitrogen fixation. Hence, Mo deficiency in a legume plant may be manifested as an N deficiency. The Mo ion is a component of several enzymes, including nitrate reductase and nitrogenase (Taiz and Zeiger, 1998). Nitrate reductase catalyzes the reduction of nitrate to nitrite during its assimilation by the plant cell. Nitrogenase converts N_2 to NH_3 in N-fixing microorganisms. It helps plants in protein synthesis and N metabolism. In soybean, Mo deficiency reduced plant growth, number of pods, number of seeds per pod, seed size, and nodulation, and the total N and protein contents of the seeds were reduced (Sinclair, 1993). Molybdenum reduces preharvest sprouting in wheat via the enhancement of abscisic acid concentration (Cairns and Kritzinger, 1992; Modi and Cairns, 1994; Modi, 2002). Molybdenum sufficiency increases cold tolerance of winter wheat seedlings under low temperature (Li et al., 2001). Molybdenum, like other heavy metals, deactivates viruses by denaturing their protein coat (Verma and Verma, 1967), and in this, Mo appears to be a particularly effective metal/metalloid (Graham and Stangoulis, 2007). Suppression of verticillium wilt of tomato by treatment of roots with Mo has been reported (Dutta and Bremmer, 1981). Jesus et al. (2004) found that a single application of Mo 25 days after sowing (sodium molybdate in solution, 20 g Mo ha^{-1}) dry bean suppressed angular leaf spot, caused by *Phaeoisariopsis griseola* (Sacc.) Ferraris, while increasing plant health, leaf photosynthesis, and yield (Graham and Stangoulis, 2007).

The role of Mo in control of diseases has been reported (Graham, 1983). Haque and Mukhopadhyaya (1983) reported that soil application of Mo decreased nematode population. The exact role of Mo in disease control is not known, but it may induce resistance to pathogens in the plant (Jesus et al., 2004).

Molybdenum deficiency symptoms appear first in older leaves (Gupta and Lipsett, 1981; Mengel et al., 2001) but may also be apparent in younger leaves (Clark, 1982). In cereals, Mo deficiency appears in younger leaves as dieback and curling of the emerging leaves, similar to symptoms of Ca deficiency (Clark, 1982). Molybdenum deficiency symptoms are similar to those of iron or manganese deficiency. However, in the latter case, symptoms are distributed over nearly all the leaves on a plant. Molybdenum-deficient leaf margins tend to curl or roll. Marginal chlorosis of older leaves may occur. In severe cases, necrosis follows and the entire plant is stunted. Young leaves become mottled, and their leaf margins are narrow (Foth and Ellis, 1988). The adverse effect of Mo deficiency on growth and yield is more pronounced during the reproductive growth stage compared to the vegetative growth stage (Mengel et al., 2001).

13.4 CONCENTRATION AND UPTAKE

Plants vary widely in their requirements for Mo and in their ability to extract the element from soils. Requirements for Mo are met at concentrations of 0.3 to 0.5 mg kg^{-1} in tissue of leguminous crops and at less than 0.1 mg kg^{-1} in most other plants (Johnson, 1966). Bennett (1993) reported that 0.1 to 0.5 mg kg^{-1} dry plant tissue Mo is sufficient for normal growth of crop plants. Cassman (1993) reported critical Mo concentration in 65-day-old cotton plants to be 0.5 mg kg^{-1}. Mengel et al. (2001) reported that Mo concentration of plant material is usually low and that plants are adequately supplied with less than 1 mg kg^{-1} dry matter. These authors also reported that deficiency of Mo is usually under 0.2 mg kg^{-1} dry matter. The low concentration of Mo in healthy plant tissues means that demand for this element is low. Molybdenum concentration in the seed is usually much higher than in the leaf (Barber, 1995).

Molybdenum toxicity is not generally observed under field conditions in crop plants (Jones, 1991). The Mo toxicity occurs at very high concentration in plant tissues. Romheld and Marschner (1991) reported that plants could absorb as much as 200–1000 mg Mo kg^{-1} dry matter without adverse effect on growth. Adriano (1986) reported that Mo toxicity in crop plants did not occur until Mo concentration exceeded 500 mg kg^{-1}. Adequate level of Mo in principal crop tissues is given in Table 13.1. Data in Table 13.1 show that adequate Mo concentration in plant tissue of legumes is much higher compared to cereals.

Molybdate (MoO_4^{2-}) ion is the most dominant ion above pH 4 in soil solution (Mengel et al., 2001). Plants in the form of molybdate (MoO_4^{2-}) ion absorb the Mo, and it moves mainly to plant roots by mass flow. Evidence for active uptake of Mo

TABLE 13.1

Molybdenum Adequate Level in Plant Tissue of Principal Crop Species

Crop Species	Plant Part Analyzed	Adequate Level (mg kg^{-1})
Alfalfa	Whole tops prior to bloom	1–5
Barley	Whole tops at heading	0.3–0.5
Barley	Whole tops at boot stage	0.1–0.2
Dry beans	Whole tops, 8 weeks old	0.4–0.6
Dry bean	Whole tops 56 days after sowing	0.4–0.8
Corn	Ear leaf prior to silking	0.1–2.0
Corn	Ear leaf at silk	0.6–1.0
Wheat	Whole tops at boot stage	1–2
Wheat	Leaf blade at stem extension	0.05–0.1
Rice	Uppermost mature leaves at tillering	0.5–2
Soybean	Upper fully developed trifoliate prior to pod set	1–5
Peanut	Upper stem and leaves at early pegging	1–5

Source: Adapted from Gupta and Lipsett (1981); Voss (1993); Fageria et al. (1997).

by roots has been limited (Moore, 1972; Clark, 1990). Mo typically moves through the xylem of plants as an anion, but Mo complexation with other organic molecules cannot be ruled out (Tiffin, 1972).

13.5 INTERACTION WITH OTHER NUTRIENTS

Uptake of Mo by plants is mainly in the form of MoO_4^{2-}; its interaction with other anions like SO_4^{2-} is expected. Reduction of Mo uptake by plants supplied with higher amounts of sulfate may be related to competition at root absorption sites due to the similar size and charge of these ions (Stout et al., 1951; MacLeod et al., 1997). Higher concentrations of NO_3^- in soil solution are reported to increase the Mo requirement by plants (Gupta and Lipsett, 1981). On the other hand, higher concentrations of NH_4^+ may decrease Mo uptake. The positive effect of NO_3^- ion on Mo uptake may be associated with release of OH^- ions and increase in pH. Similarly, higher uptake of NH_4^+ may release more H^+ ions in the rhizosphere and decrease pH, consequently decreasing Mo uptake. Plant uptake of Mo is usually enhanced by soluble P and decreased by available S (Mortvedt and Cunningham, 1971). Stout et al. (1951) reported that high P levels in solution culture increased uptake of Mo as much as 10-fold. Hence, soil application of Mo with P may be effective. Rebafka et al. (1993) reported that the choice of P fertilizer can be important in the Mo nutrition of groundnuts. These authors also reported that replacing the S-containing single superphosphate by the non-S-containing triple superphosphate enhanced Mo uptake, and this was the major reason for increased N fixation and hence increased crop yield. The Mo/S relationships have been reviewed by MacLeod et al. (1997).

Higher levels of Mo in the growth medium adversely affect Fe translocation from roots to tops (Hangar, 1965). Stunting and apical chlorosis of red clover due to high Mo levels were eliminated when chelated Fe was included in culture solutions (Hangar, 1965). Vellora et al. (2003) reported that increasing K concentration in the soil from 0.5 mM to 3 mm significantly decreased Mo content in the eggplants (*Solanum melongena* L.). Copper and Mn have been reported to have antagonistic effects on uptake of Mo. However, Mg has been reported to have a synergistic effect on uptake of Mo (Tisdale et al., 1985). Gupta (1979) summarized the interactive effects of P, S, N, and Mn with Mo as follows: (1) soluble P enhances plant uptake of Mo, (2) soluble S decreases Mo uptake, (3) N applications over time may decrease Mo uptake, and (4) there is an inverse relationship between leaf Mn and leaf Mo. Mutual antagonisms between Mo and Cu have been reported (MacKay et al., 1966). Excess Cu caused Mo deficiency, and excess Mo caused Cu deficiency (Clark, 1984). Extensive Mn-Mo interactions have been noted (Jackson, 1967). Some of the antagonistic effects may be used to reduce the toxic effects of Mo. For example, addition of sulfur calcareous soils containing toxic levels of Mo may reduce its availability and the toxicity (Brady and Weil, 2002).

13.6 MANAGEMENT PRACTICES TO MAXIMIZE Mo USE EFFICIENCY

Management practices that can improve uptake and use efficiency of Mo by crop plants include liming acid soils and using adequate source, method, and rate of application. Soil test is one of the most desirable criteria to determine adequate rate of Mo application. Use of Mo-efficient crop species or genotypes within species is another important management practice to improve uptake and use of Mo. These management practices are summarized in this section.

13.6.1 LIMING ACID SOILS

Availability of essential micronutrients decreased with increasing pH, except Mo. Since Mo becomes more available with increasing pH, liming will correct a deficiency in acid soils, if the soil contains enough of the nutrient. The adequate pH values, which can be defined as the pH to obtain maximum economic yield of a crop, depend on soil type, plant species, or genotypes within species. The soluble Mo fraction of submerged acid soils generally increases, presumably as a result of decreased adsorption of MoO_4 under the higher-pH conditions associated with waterlogging (Ponnamperuma, 1985). Molybdenum deficiency never occurs in neutral and calcareous soils (Moraghan and Mascagni, 1991).

13.6.2 USE OF APPROPRIATE SOURCE, METHOD, AND RATE OF APPLICATION

Molybdenum deficiency can also be corrected by applying Mo fertilizers. Ammonium and Na molybdates are soluble compounds used as sources of Mo fertilizers. Principal Mo carriers are given in Table 13.2. Soluble fertilizer applied in the soil as well as foliar application may correct Mo deficiency. Application rates of 0.01 to 0.5 kg ha^{-1} of Mo will generally correct Mo deficiencies (Martens and Westermann,

TABLE 13.2
Principal Molybdenum Carriers

Common Name	Formula	Mo Content (%)
Sodium molybdate	$Na_2MoO_4.2H_2O$	39
Ammonium molybdate	$(NH_4)_6Mo_7O_{24}.2H_2O$	54
Molybde trioxidenum	MoO_3	66
Molybdenum frits	Fritted glass	1–30
Molybdenum sulfide	MoS_2	60

Source: Follett et al. (1981); Tisdale et al. (1985); Foth and Ellis (1988); Fageria (1989).

1991). Soybean responded to Mo applications when the soil pH measured between 4.9 and 6.0. Soil application (226 g Mo ha^{-1}), foliar application (85 g Mo ha^{-1}), and seed treatments (7 g Mo ha^{-1}) were all effective (Franco and Day, 1980). Martens and Westermann (1991) reported that application rates to the soil of 100 to 500 g ha^{-1} are usual.

Vieira et al. (2005) studied the response of dry bean to foliar Mo application in a Brazilian Oxisol. These authors found a quadratic response of Mo application in the range of 0 to 200 g Mo ha^{-1} applied as ammonium molybdate. The equation was Y = 1699.8823 + 4.6713X – 0.0163X^2, R^2 = 0.8702**. Based on the equation, maximum grain yield of about 2000 kg ha^{-1} was obtained with the foliar application of 143 g Mo ha^{-1}.

Scott (1963) reported that application of 72 g Na_2MoO_4 ha^{-1} did not increase Mo concentration of pasture forage to levels that might cause toxicity in livestock. The disease molybdenosis in ruminants is usually associated with particular soils that are high in Mo. Ruminants in particular are susceptible to high Mo levels in the fodder, and a Mo concentration of 5 to 10 mg kg^{-1} in the dry matter is regarded as suspect for cattle and horses (Miller et al., 1991). Leaching does not easily cause Mo loss, and its recovery is less than 15%. The application of Mo at an adequate rate may have a long residual effect; hence, its application is not needed frequently. The residual effect of Mo fertilizer can last 2 to 3 years depending on soil type and crop intensity (Gupta and Lipsett, 1981). However, Scott (1963) reported that new application of Mo is required after 5 to 6 years on some New Zealand soils.

13.6.3 Soil Test

Soil test is a valid criterion for determining crop responses to applied Mo. For the comparison of soil test results, definition of an extracting solution is very important. The acid ammonium oxalate (AAO) procedure first proposed by Grigg (1953) is perhaps still the most commonly used soil test extracting for Mo (Sims and Johnson, 1991). Acid ammonium oxalate solution is composed of $(NH_4)_2C_2O_4 \cdot H_2O + H_2C_2O_4$, and detailed description of soil Mo determination by this solution is given by Reisenauer (1965). Critical Mo values ranging from 0.2 to 0.5 mg kg^{-1} have been reported for most crops (Fageria. 1992). Similarly, Sims and Johnson (1991) reported critical level of Mo in the range of 0.1 to 0.3 mg kg^{-1} for crop plants.

13.6.4   USE OF EFFICIENT CROP SPECIES/GENOTYPES

Use of Mo-efficient crop species or genotypes within species is an important strategy in maximizing Mo use efficiency in crop plants. Nautiyal and Chatterjee (2004) reported that legumes accumulate more Mo than cereals, as plant species differ in their ability to absorb Mo. Subterranean clover is one of the most sensitive legumes to Mo deficiency (Tisdale et al., 1985). Clark (1984) reported that the higher requirement of legumes for Mo is related to their involvement in N_2 fixation. Among legumes, Mo deficiency is reported to be in the order of alfalfa > soybean > clover (Clark, 1984). Yu et al. (1999) reported that wheat genotypes differ significantly in Mo uptake and use efficiency. Differences in Mo uptake and utilization have been reported in corn (Brown and Clark, 1974) and dry bean (Brodick and Giller, 1991). Clark (1984) reviewed literature on genotype differences in Mo uptake and use efficiency. He concluded that differences among genotypes for efficient use of Mo exist for corn, grasses, alfalfa, and pasture legumes.

13.7   CONCLUSIONS

The molybdenum requirement for plants is lower than the requirement for all other essential nutrients except nickel. Furthermore, cereals have low Mo requirements compared to legumes. Deficiency of Mo in crop plants generally occurs in acid soils, and liming is the most effective practice to improve Mo uptake in acid soils. Molybdenum is required for N_2 fixation by legumes and also by higher plants in reducing NO_3^-. Overall, molybdenum deficiency first appears in the older leaves. However, the deficiency symptoms in cereals first appear in the younger leaves. Solubility and uptake of Mo is highly pH dependent. Its solubility and availability increase with increasing pH up to a certain pH level. Soil solution Mo levels may increase 10-fold for each unit increase in soil pH. Unlike Fe and Mn availability, availability of Mo is lower under acidic soil conditions. At lower pH (<5.5), iron oxides can fix MoO_4^{2-} ion and reduce its availability to plants. In well-aerated soils, Mo is present in the soil solution as MoO_4^{2-} ion. This predominance of the anionic form makes Mo unique among metallic plant nutrients. Overall, critical deficiency level of Mo in soil is about 0.3 for most crop plants. The critical level of Mo in plants is less than 1 mg kg^{-1} tissue dry matter. The Mo toxicity in plants is rare; however, high concentration of Mo in forages may induce Cu deficiency.

Molybdenum deficiency in crop plants can be corrected with soil, foliar, or seed coating. Overall, Mo application rate in the soil may be 100 to 500 g ha^{-1} mixed with fertilizers or liming material. Ammonium molybdate and sodium molybdate are the most common carriers of Mo fertilizers. Foliar application rate may be about 200 g Mo ha^{-1} for annual crops like soybean and peanuts. Seed coating with Mo has been a common means of correcting Mo deficiency in legumes, and a rate of 50 to 100 g ha^{-1} may be used for crops like soybeans, peas, and alfalfa. In acid soils, the deficiency can be remedied by liming to raise the pH to around 6.5. However, liming could increase the amount of plant-available Mo only on soils that have a reserve of Mo. Incorporating crop residues into the soil leads to recycling of micronutrients including Mo, a factor of considerable importance in micronutrient-deficient soils. Use of Mo-efficient crop species or genotypes within species is an important strategy to maximize Mo use efficiency, reduce cost of crop production, and avoid environmental pollution.

REFERENCES

Adriano, D. 1986. *Trace elements in the terrestrial environment.* New York: Springer-Verlag.

Anderson, A. J. 1970. Trace elements for sheep pastures and fodder crops in Australia. *J. Aust. Inst. Agric. Sci.* 36:15–29.

Barber, S. A. 1995. *Soil nutrient bioavailability: a mechanistic approach,* 2nd edition. New York: John Wiley & Sons.

Bennett, W. F. 1993. Plant nutrient utilization and diagnostic plant symptoms. In: *Nutrient deficiencies and toxicities in crop plants,* W. F. Bennett, Ed., 1–7. St. Paul, MN: The ASP Press, The American Phytopathological Society.

Brady, N. C. and R. R. Weil. 2002. *The nature and properties of soils,* 13th edition. Upper Saddle River, NJ: Prentice-Hall.

Brodick, S. J. and K. F. Giller. 1991. Genotype difference in molybdenum accumulation affects N_2-fixation in tropical Phaseolus vulgaris L. *J. Exp. Bot.* 42:1339–1343.

Brown, J. C. and R. B. Clark. 1974. Differential response of two maize inbreeds to molybdenum stress. *Soil Sci. Soc. Am. Proc.* 38:331–333.

Burmester, C. H., J. F. Adams, and J. W. Odom. 1988. Response of soybean to lime and molybdenum on Ultisols in northern Alabama. *Soil Sci. Soc. Am. J.* 52:1391–1394.

Cairns, A. L. P. and J. H. Kritzinger. 1992. The effect of molybdenum on seed dormancy in wheat. *Plant Soil* 145:295–297.

Cassman, K. G. 1993. Cotton. In: *Nutrient deficiencies and toxicities in crop plants,* W. F. Bennett, Ed., 111–119. St. Paul, MN: The ASP Press, The American Phytopathological Society.

Clark, R. B. 1982. Plant response to mineral element toxicity and deficiency. In: *Breeding plants for less favorable environments,* M. N. Christiansen and C. F. Lewis, Eds., 71–142. New York: John Wiley & Sons.

Clark, R. B. 1984. Physiological aspects of calcium, magnesium, and molybdenum deficiencies in plants. In: *Soil acidity and liming,* 2nd edition, F. Adams, Ed., 99–170. Madison, WI: ASA, CSSA and SSSA.

Clark, R. B. 1990. Physiology of cereals for mineral nutrient uptake, use, and efficiency. In: *Crops as enhancers of nutrient use,* V. C. Baligar and R. R. Duncan, Eds., 131–209. San Diego, CA: Academic Press.

Cotton, F. A. and G. Wilkinson. 1972. *Advanced inorganic chemistry,* 3rd edition. New York: Interscience Publishers.

Dutta, B. K. and E. Bremmer. 1981. Trace elements as plant chemotherapeutants to control *Verticillium* wilt. *Z. Pflanzenkrankh. Pflanzenschutz* 88:405–412.

Fageria, N. K. 1989. *Tropical soils and physiological aspects of crops.* Brasilia: EMBRAPA.

Fageria, N. K. 1992. *Maximizing crop yields.* New York: Marcel Dekker.

Fageria, N. K., V. C. Baligar, and C. A. Jones. 1997. *Growth and mineral nutrition of field crops,* 2nd edition. New York: Marcel Dekker.

Foth, H. D. and B. G. Ellis. 1988. *Soil fertility.* New York: John Wiley & Sons.

Franco, A. A. and J. M. Day. 1980. Effects of lime and molybdenum on nodulation and nitrogen fixation of *Phaseolus Vulgaris.* L. in acid soils of Brazil. *Turrialba* 30:99–105.

Franco, A. A. and D. N. Munns. 1981. Response of *Phaseolus Vulgaris* L-to molybdenum under acid conditions. *Soil Sci. Soc. Am. J.* 45:1144–1148.

Follett, R. H., L. S. Murphy, and R. L. Donahue. 1981. *Fertilizers and soil amendments.* Englewood Cliffs, NJ: Prentice-Hall.

Graham, R. D. 1983. Effect of nutrient stress on susceptibility of plants to disease with particular reference to the trace elements. *Adv. Bot. Res.* 10:221–276.

Graham, R. D. and C. R. Stangoulis. 2007. Molybdenum and plant disease. In: *Mineral nutrition and plant disease*, L. E. Datnoff, W. H. Elmer, and D. M. Huber, Eds., 203–205. St. Paul, MN: The American Phytopathological Society.

Grigg, J. L. 1953. Determination of the available molybdenum of soils. *N.Z. J. Sci. Technol.* 34:405–414.

Gupta, U. C. 1979. Soil and plant factors affecting molybdenum uptake by plants. In: *Molybdenum in agriculture*, U. C. Gupta, Ed., 71–91. Cambridge: Cambridge University Press.

Gupta, U. C. and J. Lipsett. 1981. Molybdenum in soils, plants and animals. *Adv. Agron.* 34:73–115.

Haque, M. S. and M. C. Mukhopadhyaya. 1983. Influence of some micronutrients on *Rotylenchus reniformis. Indian J. Nematology* 13:115–116.

Hangar, B. C. 1965. The influence of iron upon the toxicity of manganese, molybdenum, copper and boron in red clover. *J. Aust. Inst. Agr. Sci.* 31:315–317.

Jackson, W. A. 1967. Physiological effects of soil acidity. In: *Soil acidity and liming*, R. W. Pearson and F. Adams, Eds., 43–124. Madison, WI: ASA.

Jesus, W. C., Jr., F. X. R. Vale, R. R. Coelho, B. Hau, L. Zambolim, and R. D. Berger. 2004. Management of angular leaf spot in common bean (*Phaseolus vulgaris* L.) with molybdenum and fungicide. *Agron. J.* 96:665–670.

Johansen, C., P. C. Kerridge, P. E. Luck, B. G. Cook, K. F. Lowe, and H. Ostrowski. 1977. The residual effect of molybdenum fertilizer on growth of tropical pasture legumes in a subtropical environment. *Aust. J. Exp. Agric. Anim. Husb.* 17:961–968.

Johnson, C. M. 1966. Molybdenum. In: *Diagnostic criteria for plants and soils*, H. D. Chapman, Ed., 286–301. Berkeley, CA: Univ. of California Div. of Agric. Sci.

Jones, L. H. D. 1956. Interaction of molybdenum and iron in soils. *Science* 123:116.

Jones, J. B. Jr. 1985. Soil testing and plant analysis: Guides to the fertilization of horticulture crops. *Horticulture Rev.* 7: 1–67.

Jones, J. B. Jr. 1991. Plant tissue analysis in micronutrients. In: *Micronutrients in agriculture*, 2nd edition, J. J. Mortvedt, Ed., 477–521. Madison, WI: SSSA.

Karimian, N. and F. R. Cox. 1978. Adsorption and extractability of molybdenum in relation to some chemical properties of soil. *Soil Sci. Soc. Am. J.* 42:757–761.

Li, W., Z. Wang, G. Mi, X. Han, and F. Zhang. 2001. Molybdenum in winter wheat seedlings as enhanced by freezing temperature. *J. Plant Nutr.* 24:1195–1203.

Lindsay, W. L. 1979. *Chemical equilibria in soils.* New York: Wiley-Interscience.

MacKay, D. C., E. W. Chipman, and U. C. Gupta. 1966. Copper and molybdenum nutrition of crops grown on acid sphagnum peat soil. *Soil Sci. Soc. Am. Proc.* 30:755–759.

MacLeod, J. E., U. C. Gupta, and B. Stanfield. 1997. Molybdenum and sulfur relationship in plants. In: *Molybdenum in agriculture*, U. C. Gupta, Ed., 229–249. Cambridge: Cambridge University Press.

Marschner, H. 1995. *Mineral nutrition of higher plants*, 2nd edition. New York: Academic Press.

Martens, D. C. and D. T. Westermann. 1991. Fertilizer application for correcting micronutrient deficiencies. In: *Micronutrients in agriculture*, 2nd edition, J. J. Mortvedt, Ed., 549–592. Madison, WI: SSSA.

Mengel, K., E. A. Kirkby, H. Kosegarten, and T. Appel. 2001. *Principles of plant nutrition*, 5th edition. Dordrecht: Kluwer Academic Publishers.

Miller, E. R., X. Lei, and D. E. Ullrey. 1991. Trace elements in animal nutrition. In. *Micronutrients in agriculture*, J. J. Mortvedt, P. M. Giordano, and W. L. Lindsay, Eds., 593–662. Madison, WI: SSSA.

Modi, A. T. 2002. Wheat seed quality in response to molybdenum and phosphorus. *J. Plant Nutr.* 25:2409–2419.

Modi, A. T. and A. L. P. Cairns. 1994. Molybdenum deficiency in wheat results in lower dormancy levels in reduced ABA. *Seed Sci. Res.* 4:329–333.

Moraghan, J. T. and H. J. Mascagni, Jr. 1991. Environmental and soil factors affecting micronutrient deficiencies and toxicities. In: *Micronutrient in agriculture*, 2nd edition, J. J. Mortvedt, P. M. Giordano, and W. L. Lindsay, Eds., 371–425. Madison, WI: SSSA.

Moore, D. P. 1972. Mechanisms of micronutrient uptake by plants. In: *Micronutrient in agriculture*, J. J. Mortvedt, P. M. Giordano, and W. L. Lindsay, Eds., 171–198. Madison, WI: SSSA.

Mortvedt, J. J. and H. G. Cunningham. 1971. Production, marketing, and use of other secondary and micronutrient fertilizer. In: *Fertilizer technology and use*, 2nd edition, R. A. Olson, Ed., 413–454. Madison, WI: SSSA.

Nautiyal, N. and C. Chatterjee. 2004. Molybdenum stress-induced changes in growth and yield of chickpea. *J. Plant Nutr.* 27:173–181.

Parfitt, R. L. and R. S. C. Smart. 1978. The mechanism of sulfate adsorption on iron oxides. *Soil Sci. Soc. Am. J.* 42:48–50.

Ponnamperuma, F. N. 1985. Chemical kinetics of wetland rice soils relative to soil fertility. In: *Wetland soils: Characterization, classification, and utilization*, International Rice Research Institute, Ed., 71–89. Los Banos, Philippines: IRRI.

Rebafka, F. P., B. J. Ndunguru, and H. Marschner 1993. Single superphosphate depresses molybdenum uptake and limits yield response to phosphorus in groundnut (*Arachis hypogaea* L.) grown on an acid soil in Niger West Africa. *Fert. Res.* 34:233–242.

Reddy, K. J., L. C. Munn, and L. Wang. 1997. Chemistry of mineralogy of molybdenum in soils. In: *Molybdenum in agriculture*, U. C. Gupta, Ed., 4–22. Cambridge: Cambridge University Press.

Reisenauer, H. M. 1965. Molybdenum. In: *Methods of soil analysis*, part 2, C. A. Black, Ed., 1050–1058. Madison, WI: ASA.

Romheld, V. and H. Marschner. 1991. Functions of micronutrients in plants. In: *Micronutrient in agriculture*, J. J. Mortvedt, P. M. Giordano, and W. L. Lindsay, Eds., 4297–328. Madison, WI: SSSA.

Scott, R. S. 1963. Long-term studies of molybdenum applied to pasture. III. Rates of molybdenum application in relation to pasture production. *Z. J. J. Agric. Res.* 6:567–577.

Sillanpaa, M. 1990. Micronutrients and the nutrient status of soils: A global study. FAO Soils Bulletin 63. FAO/Finnish International Development Agency, Rome, Italy.

Sims, J. T. and G. V. Johnson. 1991. Micronutrient soil test. In: *Micronutrient in agriculture*, 2nd edition, J. J. Mortvedt, P. M. Giordano, and W. L. Lindsay, Eds., 427–476. Madison, WI: SSSA.

Sinclair, J. B. 1993. Soybeans. In: *Nutrient deficiencies and toxicities in crop plants*, W. F. Bennett, Ed., 99–103. St. Paul, MN: The ASP Press, The American Phytopathological Society.

Stout, P. R., W. R. Meagher, G. A. Pearson, and C. M. Johnson. 1951. Molybdenum nutrition of crop plants. I. The influence of phosphate and sulfate on the absorption of molybdenum from soils and solution cultures. *Plant Soil* 3:51–87.

Taiz, L. and E. Zeiger. 1998. *Plant physiology*. Sunderland, MA: Sinauer Associates.

Tiffin, L. O. 1972. Translocation of micronutrients in plants. In: *Micronutrient in agriculture*, J. J. Mortvedt, P. M. Giordano, and W. L. Lindsay, Eds., 199–229. Madison, WI: SSSA.

Tisdale, S. L., W. L. Nelson, and J. D. Beaton. 1985. *Soil fertility and fertilizers*, 4th edition. New York: Macmillan Publishing Company.

Vellora, G., D. A. Moreno, and L. Romero. 2003. Potassium supply influences molybdenum, nitrate, and nitrate reductase activity in eggplant. *J. Plant Nutr.* 26:659–669.

Verma, H. N. and G. S. Verma. 1967. Inhibition of local lesion production by some chemical compounds. *Indian Phytopathol.* 20:176–178.

Vieira, N. M. B., J. A. Junior, B. P. Franzote, M. J. B. Andrade, and J. G. Carvalho. 2005. Foliar fertilization of molybdenum in dry bean. In: Annals of VIII National Congress on Bean, 922–925. October 18–20, 2005, Goiania, Brazil, National Rice and Bean Research Center of EMBRAPA, Santo Antônio de Goiás, Brazil.

Voss, R. D. 1993. Corn. In: *Nutrient deficiencies and toxicities in crop plants*, W. F. Bennett, Ed., 11–14. St. Paul, MN: The ASP Press, The American Phytopathological Society.

Williams, C. and I. Thornton. 1973. The use of soil extractants to estimate plant-available molybdenum and selenium in potentially toxic soils. *Plant Soil* 39:149–159.

Yu, M., C. Hu, and Y. Wang. 1999. Influences of seed molybdenum and molybdenum application on nitrate reductase activity, shoot dry matter, and grain yields of winter wheat cultivars. *J. Plant Nutr.* 22:1433–1441.

14 Chlorine

14.1 INTRODUCTION

Chlorine, or more correctly the chloride ion (Cl^-), is classified as a micronutrient; however, its uptake by crop plants is equal to that of macronutrients (Fageria et al., 2002). It is a nonmetal micronutrient like boron (B). Essentiality of Cl for higher plants was established by Broyer et al. (1954), who worked with tomato plants (Epstein and Bloom, 2005). Later on, the same group of scientists proved the essentiality of this element for corn, dry bean, alfalfa, barley, and sugar beets. All these studies were conducted in nutrient solutions. Other than nickel (Ni), chlorine is the most recently discovered essential micronutrient for higher plants. The potential role of Cl^- in crop production was not seriously considered until the 1970s, when research in the Philippines (Von Uexkull, 1972), Europe (Russell, 1978), and the northwestern United States (Powelson and Jackson, 1978) reported that Cl^- could play an important role in crop production. Although chlorine is not studied intensively, responses to this nutrient have been documented for corn (Heckman, 1995, 1998) and small grain crops in limited situations (Fixen et al., 1986ab; Engel et al., 1994). More importantly, reports of chloride deficiency in over 11 economic crops have been published (Gausman et al., 1958; Engel et al., 1997; Fixen, 1993). The first mention of chloride in crop management was a recommendation that NaCl be used as a topdressing on barley to prevent lodging, a condition that may have been precipitated by a root disease (Tottingham, 1919). Chloride deficiency was also demonstrated in sugar beet (Ulrich and Ohki, 1956) and in eight other plant species (Johnson et al., 1957).

Díaz-Zorita et al. (2004) reported that wheat yield increased significantly in the soils from the sandy pampas region of Argentina with chloride fertilization. These authors also reported that, averaged over the 10 locations, the grain yield response to Cl^- fertilization was 253 kg ha^{-1}, and it was mostly explained by a greater number of grains per square meter. Several studies conducted in the Great Plains and the Pacific Northwest regions of the United States suggest that Cl^- fertilization applied in the form of KCl improved the grain yield of wheat and other crops (Engel et al., 1997). Schumacher and Fixen (1989) also reported that many well-drained soils in the upper Midwest of the United States are low in this essential plant nutrient. For spring wheat, chloride fertilizer added to low chloride testing soil (<17 mg kg^{-1} soil or <34 kg ha^{-1} in the top 60 cm soil profile) results in a positive yield gain 70% of the time (Fixen et al., 1986a). A peculiar leaf spot complex that results in tissue necrosis and yield losses has been present in wheat fields of the Pacific Northwest and Great Plains for decades (Chester, 1944; Smiley et al., 1993a, 1993b; Engel et al., 2001). This nonpathogenic leaf spot complex results from inadequate Cl^- nutrition (Engel

TABLE 14.1

Influence of Cl⁻ Fertilization on Grain Yield and Physiological Leaf Spot Severity in Durum Wheat

Cl Rate (kg ha⁻¹)	Grain Yield (kg ha⁻¹)	Leaf Spot Severity (%)	
		Flag	Flag1
0	3308	86.8	95.8
45	4048	6.1	35.8
F-test	***	***	***

*** Significant at the 0.001 probability level.
Source: Adapted from Engel et al. (2001).

et al., 1994, 1997). Data in Table 14.1 show the influence of Cl fertilization on leaf spot severity and grain yield of durum wheat.

The improvement in yield of crops with Cl⁻ fertilization may be associated with several factors like improved nutrition, plant–water relations, and control of diseases (Fixen et al., 1986; Engel et al., 1994). The reduction of diseases in crop plant by Cl⁻ fertilization has been widely reported (Christensen et al., 1981; Engel and Grey, 1991; Xu et al., 2000). Chloride fertilization eliminated or prevented the occurrence of a leaf spot syndrome that is not disease related (Engel et al., 1997, 2001). Chloride may also increase wheat grain yields with an enhanced NH_4 supply attributed to lower leaf osmotic potentials and delayed nitrification in the soil (Christensen and Brett, 1985; Koening and Pan, 1996).

In addition to Cl⁻ deficiency under certain conditions (light-textured soils), toxicity of this nutrient may be a problem under certain circumstances. Cl⁻ is usually supplied to plants from various sources (soil reserves, rain, fertilizers, and air pollution) (Marschner, 1995). Under these situations, adverse effects of excess Cl accumulation should be ameliorated rather than correcting deficiency. The objective of this chapter is to discuss the role of chlorine in mineral nutrition of higher plants and how to maximize chlorine use efficiency or reduce its toxicity in annual crops.

14.2   CYCLE IN SOIL–PLANT SYSTEMS

In nature, the Cl⁻ ion is widely distributed and subject to rapid recycling (Mengel et al., 2001). The chloride cycle involves addition and depletion of this nutrient from the soil–plant system. The main Cl⁻ addition sources in soil–plant systems are release from the parent material, atmospheric deposition, addition by irrigation water, and fertilizers and organic manures. The Cl⁻ depletion may be related to soil erosion, plant uptake, and immobilization by microorganisms. Due to the high solubility of Cl⁻ salts, most soil Cl⁻ is found in the soil solution. Chloride in the soil is adsorbed by minerals at lower pH but negligible at pH higher than 7. At higher pH (>7), it is one of the most mobile ions, being easily lost by leaching under freely drained conditions (Mengel et al., 2001). Acid soils with clay mineralogy dominated by 1:1 clays and oxides may have significant anion exchange capacity. Significant Cl⁻ can

TABLE 14.2

**Chloride Content in the Top 0–20 cm Soil
Layer of 10 Locations in the Pampas Region
of Argentina—Method Used for Cl^- Analysis
Was Adriano and Doner (1982)**

Site	Location	Cl^- Content ($mg\ kg^{-1}$)
1	Drabble	14.3
2	América	5.6
3	América	15.0
4	Colonia Seré	7.7
5	Trenque Lauguen	8.1
6	América	17.2
7	América	8.1
8	Drabble	12.4
9	Drabble	11.1
10	América	11.1
Average		11.1

Source: Adapted from Díaz-Zorita et al. (2004).

be held in anion-exchange sites in such soils (Fixen, 1993). Díaz-Zorita et al. (2004) determined Cl^- content of soils of 10 locations in the sandy pampas of Argentina (Table 14.2). Data in Table 14.2 show that the average soil Cl^{-1} content of these soils was 11.1 mg kg^{-1} and that wheat responded to Cl^- fertilization on soils that had Cl^- content lower than 13 mg kg^{-1} (Díaz-Zorita et al., 2004).

14.3 FUNCTIONS AND DEFICIENCY SYMPTOMS

Chloride is essential for photosynthesis and as an activator of enzymes involved in splitting water. Although the amount of chlorine required for photosynthesis is provided by concentrations typical of a micronutrient (Broyer et al., 1954; Terry, 1977), higher rates of chlorides have marked effects on soil–root relations, such as inhibiting nitrification (Golden et al., 1981), enhancing Mn availability (Elmer 1995, 2003), and increasing beneficial microorganisms (Elmer, 2003, 2007). It also helps in osmoregulation of plants growing on saline soils (Fageria et al., 1997). Chloride crop nutrition improves the sanitary status of the crops under different environmental and production conditions (Fixen et al., 1986ab; Elmer, 2007). The presence of Cl^- was found to enhance both the evolution of O_2 and photophosphorylation (Terry, 1977; Mengel et al., 2001). Chloride activates the proton pumping ATPase at the tonoplast, and in some crop species or genotypes, it controls opening and closing of the stomata (Epstein and Bloom, 2005). Application of Cl-containing fertilizers to vegetables several days before harvest potentially lowers the NO_3-N concentration in the tissue by substitution of Cl for NO_3^-, thus increasing the nutritional quality of the vegetables (Blom-Zandstra and Lampe, 1983; Romheld and Marschner, 1991).

Engel et al. (1997, 2001) reported that a physiological leaf spot occurrence in wheat that results in tissue necrosis was linked to inadequate Cl⁻ nutrition and a possible osmotic imbalance in leaf mesophyll cells. The name *Cl-deficient leaf spot syndrome* was proposed to describe this physiological disorder in wheat (Wiese, 1977; Engel et al., 1997). Engel et al. (1994) reported that leaf spot disorder (portion of the leaf area affected by necrotic and chlorotic lesions) in wheat was reduced from 26.1 to 1.6% with 45 kg ha⁻¹ Cl⁻ added as KCl. Smiley et al. (1993a) also reported that a urea + $CaCl_2$ spray suppressed leaf spot and increased yield of winter wheat. Chlorine helps to control several plant diseases such as stalk rot in corn, take-all in wheat, stem rot and sheath blight in rice, and mildew in millet (Timm et al., 1986; Fixen, 1993; Brady and Weil, 2002). Chlorine also has an indirect effect on plant nutrition since it tends to suppress nitrification. This leads to a higher NH_4^+-to-NO_3^- ratio in soil solution, and as the NH_4^+ ion is taken up, rhizosphere pH can decrease. The decrease in acidity may improve availability of micronutrients. Higher uptake of Mn is reported to suppress take-all disease in wheat (Brady and Weil, 2002).

Christensen et al. (1981) reported that application of Cl⁻ reduced the osmotic potential in winter wheat leaves. They further suggested that take-all susceptibility in winter wheat could be reduced by lowering the chemical potential of water in the plant, achieved in part by Cl⁻ application.

Chlorine-deficient plants are chlorotic; growth is reduced, including that of the leaf area. Wilting of leaves at margins is a typical feature, and transpiration is affected (Mengel et al., 2001). Terry (1977) reported that Cl⁻ deficiency caused up to 60% decrease in dry weight of sugar beet. This author also reported that roots of sugar beet were also stunted and developed a "herringbone" appearance. Epstein and Bloom (2005) reported that on Cl-deficient plants, bronzing shows up on the leaves, followed by chlorosis and necrosis. Under severe deficiency conditions, plants are spindly and stunted. Whitehead (1985) reported that the first visible symptom in red clover was a curling of the leaflets on the younger leaves. Within a few days, some newly expanded leaflets shriveled completely, and older leaflets developed a brown necrosis that started at the tip and extended backward particularly at the margins. Subsequently, there was a general stunting of growth and the collapse of some petioles.

Compared with deficiency and sufficiency, toxicity of Cl⁻ is much more of a problem worldwide (Marschner, 1995). The Cl⁻ toxicity can be described as having an indirect effect on plant growth due to high osmotic potential in the soil solution, which inhibits water uptake and tissue desiccation. There is also a direct toxic effect of high Cl concentration in the plant tissue (Epstein and Bloom, 2005). Crops growing on salt-affected soils often show symptoms of Cl⁻ toxicity (Mengel et al., 2001). These include burning of leaf tips or margins, bronzing, premature yellowing, and abscission of leaves (Eaton, 1966).

14.4   CONCENTRATION AND UPTAKE

Concentration and uptake of Cl in crop plants depend on soil solution concentration of this ion, crop yield, and climatic conditions. Normally, plants accumulate higher concentration of Cl (2000 to 20000 mg kg⁻¹ tissue dry weight) (Fixen, 1993; Mengel et al., 2001), which is a typical macronutrient level. However, the demand of Cl^- for optimum growth for most crop species is in the range of 70 to 700 mg kg^{-1} tissue dry weight (Mengel et al., 2001), which is in the range of micronutrient levels. Whitehead (1985) reported that critical concentration of Cl^{-1} in the leaf plus petiole of red clover was at least 200 mg kg^{-1} dry weight. Romheld and Marschner (1991) reported that average Cl concentrations in plants range from 1000 to 2000 mg kg^{-1} and, thus, are in the range of macronutrients. These authors also reported that compared with the high tissue concentrations of Cl found in most plants, the requirement for optimal growth generally is much lower (150–300 mg kg^{-1}). Hence, there are significant differences in the literature regarding the adequate level of Cl in plant tissues. These differences may be associated with different crop species or genotypes within species, management practices, and variability in environmental conditions. The critical toxic concentration of Cl is reported to be in the range of 3000 to 5000 mg kg^{-1} tissue dry weight (Romheld and Marschner, 1991).

Critical Cl^- concentration for nonbiochemical functions like osmoregulation appears to be much higher than that for biochemical functions (Fixen, 1993; Evans and Riedell, 2006). Terry (1977) reported that there was no decrease in sugar beet top growth until the concentration of Cl^- in the leaf blade decreased to about 20 mg kg^{-1}. Below this concentration, plant dry weight decreased linearly with decrease in leaf Cl^-. The study was conducted in nutrient solution. Ulrich and Ohki (1956) reported that sugar beet growth was reduced when the concentration of Cl^- in the leaf blade decreased to about 25 to 50 mg kg^{-1} dry weight, and severe deficiency occurred at 3 to 5 mg kg^{-1} Cl^- concentration. Fixen (1993) reported that for biochemical functions like photosynthesis and enzyme activation, dry plant tissue Cl^- content should be around 100 mg kg^{-1}. Concentration of Cl^- in shoot is higher compared to grain in annual crops. De Datta and Mikkelsen (1985) reported that concentration of Cl in lowland rice straw was 0.65 g kg^{-1}; in grain, it was 0.42 g kg^{-1}. Engel et al. (2001) reported that a physiological leaf spot of wheat was closely related to shoot Cl concentration. These authors further reported that tissue necrosis was minor if CL concentration was >1.0 g kg^{-1} but increased exponentially below this level.

The Cl^- ion is highly mobile in the soil solution and easily taken up by plants. Uptake of Cl^- by plant roots is an active process that requires energy. The Cl^- ion retains its negative charge after absorption by roots, whereas SO_4^{2-} and NO_3^- are partly or completely reduced during metabolism in plant (Xu et al., 2000). Physiological mechanisms for the control of Cl accumulation in plant cells operate at the cell or organ level (Cram, 1988). Changes in root temperature and Cl concentration in the soil solution affect Cl absorption by plants (Cram, 1983, 1988). Uptake of Cl^- (concentration × dry weight) is a function of dry matter or grain yield of a crop. De Datta and Mikkelsen (1985) reported that in lowland rice, Cl accumulation in 8.3 Mg straw ha^{-1} was 54 kg ha^{-1}; in 9.8 Mg ha^{-1} grain, Cl accumulation was 41 kg ha^{-1}. Hence, lowland rice accumulated 95 kg Cl ha^{-1} in the straw and grain combined. Xu et al. (2000) reported that wheat crop can remove 18 and 61 kg ha^{-1} at low and high Cl content in the soil, respectively. The distribution of Cl in the grain of spring wheat was 2.15% of total Cl uptake by the crop; in soybean, 1.34%; and in rice, 1.62% (Xu et al., 2000).

14.5   INTERACTION WITH OTHER NUTRIENTS

Only limited information is available on interactions of Cl⁻ with other nutrients. High concentrations of Cl⁻ in soil solution may depress mineral nutrient activities and produce abnormal Na/Ca, Na/K, Ca/Mg, and Cl/NO_3-N ratios (Fageria et al., 2002). As a result, plants may become susceptible to osmotic injury as well as nutritional disorders that could reduce plant yield and quality (Grattan and Grieve, 1999). Increased levels of Cl reduced NO_3-N (Inal et al., 1995) as Cl competes with NO_3-N during uptake processes (Mengel et al., 2001). Evidence exists that if Cl rather than SO_4-S is dominant in saline soils, Ca deficiency can be alleviated, and Cl may increase Ca uptake independent of Ca addition (Curtin et al., 1993). Chloride enhancement of Ca may also be related to increases in cation activity from Cl in soil solution or from cotransport, resulting in neutralization of positive charges during cation uptake (Marschner, 1995).

Interaction between P and Cl⁻ is reported to be positive, negative, or not affected (Fixen, 1993). However, Gausman et al. (1958) reported that an optimum or critical level of Cl⁻ existed for maximum P uptake to occur, with uptake decreasing on either side of this level. In their studies, the optimum level was reported to be 100 mg Cl⁻ kg⁻¹ soil for potato. The reduction in P uptake at Cl⁻ levels exceeding the optimum was attributed to anion competition (Fixen, 1993). The Cl⁻ application in acid soils increased Mn concentration in the plants (Fixen, 1993). The increase in Mn uptake with the addition of Cl⁻ is reported to be associated with reduction of Mn oxides in acid soils and in the process increases extractable Mn (Westerman et al., 1971). Potassium chloride increases Mn release from 14 soils collected from six countries located in temperate and subtropical regions (Fixen, 1993).

14.6   MANAGEMENT PRACTICES TO MAXIMIZE Cl USE EFFICIENCY

Adopting several soil and plant management practices can maximize chloride use efficiency. These practices include using adequate source and rate, using soil test for chloride fertilization, and planting Cl-efficient crop species or genotypes within species.

14.6.1   USE OF APPROPRIATE SOURCE AND RATE

Use of appropriate source and rate is an important management practice to rational use of Cl⁻ by higher plants. The important sources of Cl are KCl and NH_4Cl fertilizers. These fertilizers can supply Cl⁻ as well as N and K. Principal Cl⁻ carriers are presented in Table 14.3. Díaz-Zorita et al. (2004) studied the influence of KCl and NH_4Cl fertilizers on wheat yield. These authors reported that there were no significant differences between these two fertilizer sources in wheat yield. They concluded that lack of significant differences between Cl⁻ sources (KCl vs. NH_4Cl) suggests that the wheat productivity responses to Cl fertilization could be mostly attributed to direct Cl effects of this ion on the behavior of the crops.

Christensen et al. (1981) reported that addition of ammonium chloride fertilizer (101 kg Cl ha⁻¹) increased wheat grain yield 40% over plots with no Cl⁻ applied.

TABLE 14.3
Principal Chlorine Carriers

Common Name	Formula	Cl (%)
Potassium chloride	KCl	47
Ammonium chloride	NH_4Cl	66
Calcium chloride	$CaCl_2$	65
Magnesium chloride	$MgCl_2$	74
Sodium chloride	$NaCl_2$	60
Zinc chloride	$ZnCl_2$	48
Manganese chloride	$MnCl_2$	56

TABLE 14.4
Influence of Chloride Fertilization on Wheat Yield (Data Averaged across Two Cropping Season and KCl and NH₄Cl Fertilizers)

Cl⁻ Fertilization Rate (kg ha⁻¹)	Grain Yield (kg ha⁻¹)[a]
0	3210a
23	3441b
46	3505b
69	3406b

[a] Values followed by the same letter in the column are not significantly different by LSD at 5% probability level.

Source: Adapted from Díaz-Zorita et al. (2004).

Engel et al. (1988) reported that application of 45 kg Cl ha⁻¹ significantly increased wheat yield in calcareous soil with pH 7.9. Data in Table 14.4 show the effects of mean Cl fertilization rate on mean wheat grain yield. The maximum grain yield was obtained with the application of 23 kg Cl ha⁻¹. The increase in grain yield with the application of 23 kg Cl⁻ was about 7% compared to control treatment. The increase in grain yield of wheat in this experiment was mainly due to the increase in the number of grains per square meter with the application of Cl⁻ (Díaz-Zorita et al., 2004). Engel et al. (1997) also reported that Cl applied to soil at the rate of 11 to 22 kg ha⁻¹ is sufficient to reduce leaf spot physiological disorder in wheat and improved grain yield.

14.6.2 Soil Test

Soil test is an important criterion to define deficient, adequate, or toxic levels of Cl⁻ in a soil. Water is a commonly used extractant for Cl⁻ analysis in the soil solution. However, Engel et al. (1997) used 0.01 M $Ca(NO_3)_2$ using a 2:1 extracting solution

TABLE 14.5
The Cl⁻ Soil Test Calibration for Wheat Cultivars in South Dakota (Values Are Averages of 36 Sites)

Soil Cl⁻ (mg kg⁻¹)	Interpretation	Yield Response (kg ha⁻¹)	Cl⁻ Required (kg ha⁻¹)
<17	Low	270	75
17–34	Medium	175	65
>34	High	20	0

Source: Adapted from Fixen et al. (1987).

to soil (25 g) ratio for Cl determination in soil samples. Soil Cl⁻¹ levels have been highly correlated with plant Cl⁻ levels in several studies (Ozanne, 1958; Fixen, 1993). Ozanne (1958) conducted greenhouse experiments with subterranean clover (*Trifolium subterranean* L.) in two sandy soils that had Cl⁻ levels equivalent to 9 mg kg⁻¹ of soil or 18 kg ha⁻¹ to 60 cm soil depth. Severe Cl⁻ deficiency was observed in clover plants after 3.5 months of growth, and yield was significantly reduced compared to pots in which Cl⁻¹ was added.

The critical Cl⁻ level in the soil varied with plant species and cultivars within species and type of soils. Díaz-Zorita et al. (2004) reported that a soil Cl⁻ level >13.2 mg kg⁻¹ was adequate for maximum grain yield of wheat. Fixen et al. (1986a) reported that a Cl⁻ level > 21.7 mg kg⁻¹ was adequate for maximum wheat yield in South Dakota. Soil test can also be used to determine residual effects of Cl⁻ fertilization. Schumacher and Fixen (1989) used soil test to evaluate residual effect of Cl⁻ fertilization in eastern South Dakota on a loam-textured soil using corn–wheat rotation. Both yield response to Cl⁻ fertilization and residual soil Cl⁻ persisted for 2 years following application. Data in Table 14.5 show soil test calibration for Cl⁻ using wheat as a test crop.

14.6.3 Planting Cl-Efficient/Tolerant Plant Species/Genotypes

Sensitivity to high Cl concentrations varies widely among plant species and cultivars (Eaton, 1966), but Cl toxicity is more extensive worldwide than Cl deficiency, particularly in arid and semiarid regions (Fageria et al., 2002). Plant tolerance to Cl has been reported. Pea is very sensitive to Cl toxicity, corn is moderately sensitive, wheat and ryegrass are slightly tolerant, and rape and barley are highly tolerant (Marschner, 1995; Fageria et al., 2002). Yield responses of hard red spring wheat to chloride fertilization are often cultivar specific (Evans and Riedell, 2006). In a low-chloride-testing soil, Fixen (1987) found up to a 20% yield increase in certain cultivars, while others were nonresponsive. Some, but not all, of these yield responses were explainable based on chloride interaction with disease susceptibility (Fixen, 1987, 1993). Xu et al. (2000) reported that a significant difference exists among barley cultivars in their tolerance to Cl⁻¹.

Chloride-tolerant plant species/cultivars restrict Cl translocation to shoot by a mechanism that resides in the roots (Grattan and Maas, 1985). Plant can avoid uptake of higher concentration of Cl or can tolerate high concentration in their tissues (Bar et al., 1997). Leaf concentrations of Cl in salt-tolerant soybean cultivars were low (Velagaleti et al., 1990).

14.7 CONCLUSIONS

Chlorine is grouped as a micronutrient in the classification of essential mineral nutrients for higher plants. However, uptake of Cl by crop plants is higher than that of macronutrients like Ca, Mg, S, and P in annual crops. The deficiency of Cl in crop plants had been reported in many regions of the world. The Cl^- deficiency symptoms in crop plants are characterized by wilting of leaves, curling of leaf margins, bronzing, chlorosis, and reduction in root growth. Under arid and semiarid conditions, Cl^- accumulates in the soils and causes toxicity to plants. Crop yield increases due to Cl fertilization have been reported in many crops. Improvement in crop yields due to Cl^- application may be associated with improved nutritional aspects, control of diseases, and improved plant–water relationships. The Cl^- ion functions as a nitrification inhibitor in acid soils (pH 5.0 to 5.5) and can improve N use efficiency by avoiding NO_3^- leaching or denitrification. The knowledge of functions of Cl in crop plants is still descriptive or speculative, and more research is needed.

Soil and plant tissue tests are important criteria to diagnose Cl^- deficiency in soils and plants. An adequate Cl^- soil test value for most annual crops is more than 13 mg kg^{-1} of soil. Similarly, adequate plant tissue values of Cl^- varied from 70 to 700 mg kg^{-1} depending on crop species, plant age, and plant part analyzed. Using adequate fertilizer rates containing Cl^- and planting Cl^--efficient crop species or genotypes among species are important management strategies to improve crop yields on Cl-deficient soils. Several soluble sources of Cl are available. However, the most common source of Cl is KCl, which can also supply both Cl^- and K. Application of about 20 kg Cl ha^{-1} is sufficient to correct Cl deficiency in crop plants. The Cl^- ion is not adsorbed in soils with optimum pH for crop growth and is easily leached to lower profile. Hence, the residual effect of Cl^- is practically negligible, and frequent application is necessary in Cl^--deficient soils.

REFERENCES

Adriano, C. and A. E. Doner. 1982. Bromine, chlorine and fluorine. In: *Methods of soil analysis, Part 2. Chemical and microbiological properties*. Ed. A. L. Page, R. H. Miller, and D. R. Kenney, 449–483. Madison, Wisconsin: ASA and SSSA.

Bar, Y., A. Apelbaum, U. Kafkafi, and R. Goern. 1997. Relationship between chloride and nitrate and its effect on growth and mineral composition of avocado and citrus plants. *J. Plant Nutr.* 20:715–731.

Blom-Zandstra, G. and J. E. M. Lampe. 1983. The effect of chloride and sulfate salts on the nitrate content in lettuce plants (*Lactuca sativa* L.) *J. Plant Nutr.* 6:611–628.

Brady, N. C. and R. R. Weil. 2002. *The nature and properties of soils*, 13th edition. Upper Saddle River, New Jersey: Prentice-Hall.

Broyer, T. C., A. B. Carlton, C. M. Johnson, and P. R. Stout. 1954. Chlorine: A micronutrient element for higher plants. *Plant Physiol.* 29:526–532.

Chester, K. S. 1944. A cause of physiological leaf spot of cereals. *Plant Disease Reporter* 60:497–499.

Christensen, N. W. and M. Brett. 1985. Chloride and liming effects on soil nitrogen form and take-all of wheat. *Agron. J.* 77:157–163.

Christensen, N. W., R. G. Taylor, T. L. Jackson, and B. L. Mitchell. 1981. Chloride effects on water potentials and yield of winter wheat infected with take-all root rot. *Agron. J.* 73:1053–1058.

Cram, W. J. 1983. Chloride accumulation as a homeostatic system: Set points perturbations. The physiological significance of influx isotherms, temperature effects and the influences of plant growth substances. *J. Exp. Bot.* 34:181–1502.

Cram, W. J. 1988. Transport of nutrient ions across cell membranes in vivo. *Adv. Plant Nutr.* 3:1–54.

Curtin, D., H. Stepphuhn, and F. Selles. 1993. Plant growth responses to sulfate and chloride salinity growth and ionic relations. *Soil Sci. Soc. Am. J.* 57:1304–1310.

De Datta, S. K. and D. S. Mikkelsen. 1985. Potassium nutrition of rice. In: *Potassium in agriculture*, R. D. Munson, Ed., 665–699. Madison, WI: ASA.

Díaz-Zorita, M., G. A. Duarte, and M. Barraco. 2004. Effects of chloride fertilization on wheat (*Triticum* aestivum L.) productivity in the sandy Pampas region, Argentina. *Agron. J.* 96:839–844.

Eaton, F. M. 1966. Chlorine. In: *Diagnostic criteria for plants and soils*, H. D. Chapman, Ed., 98–135. Riverside, CA: University of California.

Elmer, W. H. 1995. Association between Mn-reducing root bacteria and NaCl applications in suppression of Fusarium crown and root rot of asparagus. *Phytopathology* 85:1461–1467.

Elmer, W. H. 2003. Local and systemic effects of NaCl on root composition, rhizobacteria, and Fusarium crown and root rot of asparagus. *Phytopathology* 93:186–192.

Elmer, W. H. 2007. Chlorine and plant disease. In: *Mineral nutrition and plant disease*, L. E. Datnoff, W. H. Elmer, and D. M. Huber, Eds., 189–202. St. Paul, MN: The American Phytopathological Society.

Engel, R. E., L. Bruebaker, and T. J. Emborg. 2001. A chloride deficient leaf spot of durum wheat. *Soil Sci. Soc. Am. J.* 65:1448–1454.

Engel, R. E., P. L. Bruckner, D. L. Mathre, and S. K. Z. Brumfield. 1997. A chloride deficient leaf spot syndrome of wheat. *Soil Sci. Soc. Am. J.* 61:176–184.

Engel, R. E., J. Eckhoff, and R. Berg. 1994. Grain yield, kernel weight, and disease responses of winter wheat cultivars to chloride fertilization. *Agron. J.* 86:891–896.

Engel, R. E. and D. E. Matre. 1988. Effect of fertilizer nitrogen source and chloride on take-all of irrigated hard red spring wheat. *Plant Disease* 72:393–396.

Engel, R. E. and W. E. Grey. 1991. Chloride fertilizer effects on winter wheat inoculated with *Fusarium culmorum*. *Agron. J.* 83:204–208.

Epstein, E. and A. J. Bloom. 2005. *Mineral nutrition of plants: Principles and perspectives*, 2nd edition. Sunderland, MA: Sinauer Associates.

Evans, K. M. and W. E. Riedell. 2006. Response of spring wheat to nutrient solutions containing additional potassium chloride. *J. Plant Nutr.* 29:497–504.

Fageria, N. K., V. C. Baligar, and R. B. Clark. 2002. Micronutrients in crop production. *Adv. Agron.* 77:185–268.

Fageria, N. K., V. C. Baligar, and C. A. Jones. 1997. *Growth and mineral nutrition of field crops*, 2nd edition. New York: Marcel Dekker.

Fixen, P. E. 1987. Chloride fertilization? Recent research gives new answers. *Crops Soils* 39:14–16.

Fixen, P. E. 1993. Crop responses to chloride. *Adv. Agron.* 50:107–150.

Fixen, P. E., G. W. Buchenau, R. H. Gelderman, T. E. Schumacher, J. R. Gerwing, F. A. Cholick, and B. G. Farber. 1986a. Influence of soil and applied chloride on several wheat parameters. *Agron. J.* 78:736–740.

Fixen, P. E., R. H. Gelderman, J. R. Gerwing, and F. A. Cholick. 1986b. Response of spring wheat, barley and oats to chloride in potassium chloride fertilizers. *Agron. J.* 78:664–668.

Fixen, P. E., R. H. Gelderman, J. R. Gerwing, and B. G. Farber. 1987. Calibration and implementation of a soil Cl test. *J. Fertilizer Issues* 4:91–97.

Gausman, H. W., C. E. Cummingham, and R. A. Struchtemeyer. 1958. Effects of chloride and sulfate on ^{32}P uptake by potatoes. *Agron. J.* 50:90–91.

Golden, D. C., S. Sivasubramaniam, S. Sandanam, and M. A. Wijedasa. 1981. Inhibitory effects of commercial potassium chloride on the nitrification rates of added ammonium sulphate in an acid red yellow podzolic soil. *Plant Soil* 59:147–151.

Grattan, S. R. and C. M. Grieve. 1999. Salinity-mineral nutrient relations in horticulture crops. *Sci. Hort.* 78:127–157.

Grattan, S. R. and E. V. Maas. 1985. Root control of leaf phosphorus and chloride accumulation in soybean under salinity stress. *Agron. J.* 77:890–895.

Heckman, J. R. 1995. Corn responses to chloride in maximum yield research. *Agron. J.* 87:415–419.

Heckman, J. R. 1998. Corn stalk rot suppression and grain yield response to chloride. *J. Plant Nutr.* 21:149–155.

Inal, A., A. Gunes, and M. Aktas. 1995. Effect of chloride and partial substitution of reduced forms of nitrogen for nitrate in nutrient solution on the nitrate total nitrogen, and chloride contents of onion. *J. Plant Nutr.* 18:2219–2227.

Johnson, C. M., P. R. Stout, T. C. Broyer, and A. B. Carlton. 1957. Comparative chloride requirements of different plant species. *Plant Soil* 8:337–353.

Koening, R. T. and W. L. Pan. 1996. Chloride enhancement of wheat responses to ammonium nutrition. *Soil Sci. Soc. Am. J.* 60:468–505.

Marschner, H. 1995. *Mineral nutrition of higher plants*, 2nd edition. New York: Academic Press.

Mengel, K., A. Kirkby, H. Kosegarten, and T. Appel. 2001. *Principles of plant nutrition*, 5th edition. Dordrecht: Kluwer Academic Publishers.

Ozanne, P. G. 1958. Chlorine deficiency in soils. *Nature* 182:1172–1173.

Powelson, R. L. and T. L. Jackson. 1978. Suppression of take-all (*Gauemanomyces graminis*) root rot of wheat with fall applied chloride fertilizers. Proc. 29th Annual Northwest Fertilizer Conference, Beaverton, OR, July 11–13, 1978, 175–182.

Romheld, V. and H. Marschner. 1991. Function of micronutrient in plants. In: *Micronutrients in agriculture*, 2nd edition, J. J. Mortvedt, Ed., 297–327. Madison, WI: SSSA.

Russell, G. E. 1978. Some effects of applied sodium and potassium on yellow rust in winter wheat. *Ann. Applied Biol.* 90:163–168.

Schumacher, W. K. and P. E. Fixen. 1989. Residual effects of chloride application in a corn-wheat rotation. *Soil Sci. Soc. Am. J.* 53:1742–1747.

Smiley, R. W., L. M. Gillespie-Sasse, W. Uddin, H. P. Collins, and M. A. Stoltz. 1993a. Physiological leaf spot of winter wheat. *Plant Dis.* 77:521–527.

Smiley, R. W., W. Uddin, P. K. Zwer, D. J. Wysocki, D. A. Ball, and T. G. Chastain. 1993b. Influence of crop management practice on physiologic leaf spot of winter wheat. *Plant Dis.* 77:803–810.

Terry, N. 1977. Photosynthesis, growth, and the role of chloride. *Plant Physiol.* 60:69–75.

Timm, C. A., R. J. Goos, B. E. Johnson, F. J. Siobolik, and R. W. Stack. 1986. Effect of potassium fertilizers on malting barley infected with common root rot. *Agron. J.* 78:197–200.

Tottingham, W. E. 1919. A preliminary study of the influence of chloride on the growth of certain agricultural plants. *J. Am. Soc. Agron.* 11:1–32.

Ulrich, A. and K. Ohki. 1956. Chloride, bromine, and sodium as nutrients for sugarbeet plants. *Plant Physiol.* 31:171–181.

Velagaleti, R. R., S. Marsh, D. Kramer, D. Fleischman, and J. Corbin. 1990. Genotypic differences in growth and nitrogen fixation among soybean (*Glycine max* L. Merr.) cultivars grown under salt stress. *Trop. Agric.* 67:169–177.

Von Uexkull, H. R. 1972. Response of coconuts to potassium chloride in the Philippines. *Oleagineux* 27:13–19.

Westerman, D. T., T. L. Jackson, and D. P. Moore. 1971. Effect of potassium salts on extractable soil manganese. *Soil Sci. Soc. Am. J.* 35:43–46.

Whitehead, D. C. 1985. Chlorine deficiency in red clover grown in solution culture. *J. Plant Nutr.* 8:193–198.

Wiese, M. V. 1977. *Compendium of wheat diseases*, 2nd edition. St. Paul, MN: The American Phytopathology Society.

Xu, G., H. Magen, J. Tarchitzky, and U. Kafkafi. 2000. Advances in chloride nutrition of plants. *Adv. Agron.* 68:97–150.

15 Nickel

15.1 INTRODUCTION

Nickel (Ni) is the most recent addition to the list of essential micronutrients for higher plants. Nickel was suspected of being an essential plant nutrient in the early 20th century, when it was discovered to be a constituent of plant ash (Wood and Reilly, 2007). Foliar Ni sprays were noted to increase yields of wheat, potatoes, and broad beans as early as 1946 (Roach and Barclay, 1946), and responses in other crops were noted in subsequent years (Dixon et al., 1975; Welch, 1981). Its essentiality for higher plants was established in the 1980s (Welch, 1981; Eskew et al., 1983) using soybean as a test plant. Soybean plants were grown in highly purified nutrient solution (without Ni), and urea accumulated in the toxic level in the tips of leaflets, which become necrotic. When Ni was supplied to soybean plants in the concentration of 1 μg L^{-1}, no excess urea was accumulated in the leaf tips, and necrosis was also absent (Epstein and Bloom, 2005). Subsequent research by Brown et al. (1987ab) established the essentiality of Ni for other crop species such as barley. These authors reported that Ni is essential for plants supplied with urea. The Ni-deficient plants accumulate toxic levels of urea in leaf tips because of reduced urease activity (Daroub and Snyder, 2007). Daroub and Snyder (2007) also reported that the essentiality of Ni was established in 1987. In 1992, the U.S. Department of Agriculture's Agriculture Research Service added Ni to its list of essential plant nutrient elements (Wood and Reilly, 2007). Its essentiality was also recognized by the American Association of Plant Food Control Officials (AAPFCO), the umbrella organization that governs and influences regulation, labeling, ingredients, and amounts of elements in fertilizer products in the United States (Terry, 2004). Ni is now listed on fertilizer labels in the United States, and commercial Ni fertilizer products are now marketed (Wood and Reilly, 2007). Deficiency of Ni in crop plants is rarely observed under field conditions. Even in controlled conditions, creating deficiency symptoms is difficult. However, in nutrient solution experiments, Ni toxicity is frequently observed in crop plants (Table 15.1).

Ni is a heavy metal, and accumulation in higher concentrations in soils may be toxic to plants and create environmental pollution problems. Hence, care should be taken to avoid excess accumulation of Ni in the soils. Most soils naturally contain low amounts of Ni, rarely exceeding more than 100 mg kg^{-1}, well below the point at which toxicity occurs (Mengel et al., 2001). However, some soils derived from serpentine minerals may contain high amounts of Ni. Hence, human activity is the most common source of Ni contamination in soils. Use of sewage sludge for horticultural crops is the most common source of Ni in the soils. Ni toxicity is more common in acid soils. Use of organic residues or manures and liming can reduce Ni toxicity

TABLE 15.1

Influence of Nickel Concentration on Root and Shoot Length and Nodule Formation in Garden Pea Plants at 30 Days' Age in Nutrient Solution

Ni^{2+} Conc. (μmol L^{-1})	Root Length (cm)	Shoot Length (cm)	Nodule Number (3 Plants)
Control	11 ± 0.49	26 ± 1.17	27 ± 1.15
25	9 ± 0.40	25 ± 1.12	14 ± 1.17
100	8 ± 0.36	20 ± 0.40	4 ± 0.57
200	6 ± 0.30	10 ± 0.45	No nodule formation
300	6 ± 0.30	6.5 ± 0.29	No nodule formation
500	4 ± 0.11	No growth	No nodule formation
1000	4 ± 0.18	No growth	No nodule formation
R^2	0.67**	0.97**	0.87**

** Significant at the 1% probability level.
Source: Adapted from Singh et al. (2004).

in crop plants. The objective of this chapter is to discuss the role of Ni in the mineral nutrition of crop plants.

15.2 CYCLE IN SOIL–PLANT SYSTEMS

Addition of Ni in soil–plant systems is mainly through weathering of parent materials, liming of acid soils, use of sewage sludge, and municipality compost. Similarly, depletion from soil–plant systems is caused by soil erosion, sorption by soil colloids, immobilization by microbial mass, and uptake by plants. Nickel deficiency is rarely observed in crop plants. However, Ni contamination of soils is a serious problem as a result of industrial and mining activities. Since Ni is highly toxic to plants and animals, its fate and mobility in soils are of great concern. High nickel content of the soil accelerates the absorption of this element by the plants when other conditions are favorable (Mishra and Kar, 1971). Nickel sorption on soil minerals can result in both adsorbed (outer and inner sphere complexes) and precipitated phases (Scheidegger et al., 1997; Scheckel and Sparks, 2001). Use of a DTPA extracting solution (Lindsay and Norwell, 1978) composed of 0.005 M diethylene triamine pentaacetic acid (DTPA) with 0.01 M $CaCl_2$ and 0.1 M triethanolamine (TEA) at pH 7.3 is a good method for Ni determination in the soils (Echevarria et al., 1998).

The sorption of Ni onto soil surfaces controls the Ni distribution in soil and aquatic systems (Yamaguchi et al., 2001). The concentration of Ni in soil averages 5 to 500 mg kg^{-1}, with a range up to 53,000 mg kg^{-1} in contaminated soil near metal refineries and in dried sludges (USEPA, 1990). Agricultural soils contain approximately 3 to 1000 mg kg^{-1} Ni (World Health Organization, 1991). Similarly, Daroub and Snyder (2007) reported that most soils contain Ni in concentrations less than 100 mg kg^{-1}. Soil pH is one of the most important properties that determine Ni uptake by plants. Soil pH values below 5.6 seem to favor the absorption of Ni, while values above 5.6 do not (Mishra and Kar, 1971). This is largely due to the fact that the exchangeable Ni content of the soil increases with increasing soil acidity.

15.3 FUNCTIONS AND DEFICIENCY/TOXICITY SYMPTOMS

Essentiality of Ni was established in 1987 (Eskew et al., 1983; Brown et al., 1987a, 1987b), yet very little is known about the essential functions of Ni in plants (Brown et al., 1990). Brady and Weil (2002) and Daroub and Snyder (2007) reported that Ni is essential for urease, hydrogenases, and methyl reductase and for urea and ureide metabolism, to avoid toxic levels of these nitrogen fixation products in legumes. In addition, Ni deficiency results in a variety of physiological effects in plants (Brown et al., 1990; Mengel et al., 2001). Nickel is a constituent of plant enzyme urease, the enzyme that catalyzes the degradation of urea to carbon dioxide and ammonia (Dixon et al., 1975). The reaction can be written as follows:

$$CO(NH_2)_2 + H_2O \Leftrightarrow 2NH_3 + CO_2$$

Urease activity in garden pea was significantly increased with the addition of $100 \, \mu mol \, Ni^{2+} \, L^{-1}$ compared to control treatment (Singh et al., 2004). The increase in urease activity in peas with the addition of Ni indicates the analogy with the response of soybean regarding essentiality of Ni for expression of urease (Klucas et al., 1983). The Ni is not required for the synthesis of the enzyme protein but essential for the structure and functioning of the enzyme (Winkler et al., 1983; Dixon et al., 1980; Klucas et al., 1983; Marschner, 1995). Perhaps the most striking effects of Ni deficiency have been described for the cereals like wheat, barley, and oats (Brown et al., 1990). In these species, Ni deficiency results in growth reduction, premature senescence, decreased tissue Fe levels, inhibited grain development, and grain inviability (Brown et al., 1987b).

Nickel deficiency causes severe disruption in N metabolism and other metabolic process (Brown et al., 1990). Nickel stimulates proline biosynthesis in plants, which is responsible for osmotic balance in plant tissues (Salt et al., 1995; Singh et al., 2004). In studies with animals, Ni has also been found to replace Co (Underwood, 1971). Ni has been used as a systemic fungicide to control many plant diseases (Rowell, 1968). Salts of Ni control cereal rusts, rice blast and sheath spot disease, and cotton wilt (Mishra and Kar, 1971). Germination of seeds of several species was stimulated by treatment with Ni salts (Welch, 1981). Hydrogenase activity in soybean nodules diminished when the supply of nickel was inadequate (Klucas et al., 1983; Dalton et al., 1985). Nickel functions in the utilization of nitrogen translocated from roots to tops via guanidines or ureides that are subsequently used for anabolic reactions in growing tissues (Welch, 1981).

Nickel is mobile in plants; hence, deficiency symptoms first appear in the older leaves. Deficiency of Ni may be due to low levels of this element in the soil or be introduced by interaction with other elements. Nickel deficiency symptoms are characterized by marginal chlorosis of leaves, premature senescence, and diminished seed set (Epstein and Bloom, 2005). *Mouse ear* is the expression used to describe Ni deficiency symptoms in pecan (*Carya illinoinensis*) leaves (Malavolta and Moraes, 2007). Brown et al. (1987a) observed chlorosis of young leaves, reduced leaf area, and less upright growth of leaves as Ni deficiency symptoms in wheat, barley, and oats. Brown et al. (1987a, 1987b) also observed premature senescence in Ni-deficient oat plants. Ni deficiency also depresses growth, inhibits grain development, and causes grain inviability. It can also reduce iron levels in plant tissues (Wood and Reilly, 2007). The necrosis of leaf tips appears likely to be a key defining characteristic of Ni deficiency in all higher plant families and is a reliable diagnostic trait for the identification of the deficiency (Wood and Reilly, 2007). Nickel toxicity is expected if Ni has accumulated in higher concentrations in the soil. Excess Ni causes several physiological disturbances, of which the yellowing of leaves or chlorosis is usually a visual manifestation. When Ni is present in toxic levels in plants, anthocyanin accumulation in leaves has been observed (Someya et al., 2007). Chlorosis produced by excess Ni has been attributed to induced iron deficiency, since application of iron salts to the chlorotic plants has restored the green color (Mishra and Kar, 1971). Mishra and Kar (1971) also reported that chlorosis was severe at an Ni:Fe ratio value above 6 and usually negligible at values below 1. Reduction in root growth has been commonly observed in plants subjected to heavy metal toxicity (Woolhouse, 1983). It has been reported that alteration of plasma membrane integrity, through lipid peroxidation, was a primary effect of heavy metal toxicity (Pandolfini et al., 1992; Baccouch et al., 2001).

15.4   CONCENTRATION AND UPTAKE

Limited data are available for concentration and uptake of Ni in crop plants. Because most of the Ni uptake studies were conducted keeping in view the Ni toxicity problem rather than its deficiency, higher concentrations of Ni were used which do not represent the Ni transport system that functions to provide Ni at the very low levels needed to sustain growth (Kochain, 1991). Kochain (1991) suggested that there is a real need for detailed characterization of the physiology of Ni^{2+} uptake using intact plants and conducting experiments at realistic levels (< 1.0 mmol m^{-3}). However, Welch (1981) and Epstein and Bloom (2005) reported that the range of Ni concentration in crop plants varied from 0.05 to 5 mg kg^{-1} tissue dry weight. Mengel et al. (2001) reported that normally the Ni concentration of plant material is about 0.1 to 5 mg kg^{-1}. Marschner (1995) reported that Ni content of crop plants normally ranges from 0.1 to 1 mg kg^{-1} tissue dry weight. Brown et al. (1987ab) and Welch (1981) reported that Ni levels in Ni-adequate plant tissues are extremely low, ranging between 0.2 and 5 mg kg^{-1}. Welch (1995) reported that adequate levels of Ni for soybean varied from 0.05 to 0.1 mg kg^{-1}; for barley, wheat, and oat, they varied from 0.1 to 0.25 mg kg^{-1} tissue dry weight.

Nickel is readily transported into reproductive organs from phloem sap source organs and can accumulate in seeds and grains, organs that are dependent on phloem sap transport for their metal accumulation (Welch, 1986, 1995). There has been little systematic work on Ni uptake kinetics. However, Wild (1970) reported that for many plant species, the uptake pattern of Ni is similar to that of Cu and the level of Ni in the plant tops is low (about 100 mg kg^{-1}) irrespective of external concentration.

TABLE 15.2

Nickel Uptake in Root and Shoot of Garden Pea at Different Ni²⁺ Levels in Nutrient Solution

Ni²⁺ Level (µmol L⁻¹)	Uptake in Root (µg g⁻¹ Dry Weight)	Uptake in Shoot (µg g⁻¹ Dry Weight)
0	—	—
25	2.23 ± 0.10	1.58 ± 0.07
100	11.89 ± 0.59	6.22 ± 0.29
200	20.42 ± 0.98	14.20 ± 0.66
300	21.00 ± 1.00	14.45 ± 0.66
R^2	0.90**	0.92**

** Significant at the 1% probability level.
Source: Adapted from Singh et al. (2004).

However, this was not true for two species, *Vellozia equisetoides* and *Dicoma macrocephala*, where the uptake reached 1000 mg kg⁻¹ tissue dry weight. This means that mechanisms of Ni tolerance may differ among plant species.

Welch (1981) and Bollard (1983) reported that the critical toxic level of Ni for the sensitive species is more than 10 mg kg⁻¹; for moderately tolerant crop species, it is more than 50 mg kg⁻¹. In plants grown in serpentine soils, the Ni content might reach a level of 100 or 1000 times higher (Marschner, 1995). Such levels are highly toxic for most plant species. However, a number of adapted species (hyperaccumulators) in serpentine soils may contain more than 3% Ni in the shoot dry weight (Bollard, 1983; Marschner, 1995). A higher proportion of Ni is complexed to organic acids in these species, which may contribute to their tolerance. However, other mechanisms may also be involved (Woolhouse, 1983).

Higher amounts of Ni accumulated in roots compared to shoot in the garden pea (Table 15.2). In plants, most of the absorbed heavy metal content is retained in the roots. The higher amount of Ni retained in roots means that mechanisms exerted by plants have fewer toxic effects on tops.

Higher concentration has toxic effects on plant growth and development. Acute Ni toxicity induces chlorosis in plant leaves. In the cereals, pale yellow stripes along the length of the leaves and necrosis occur along the leaf margins (Mengel et al., 2001). In legumes, Ni toxicity appears as chlorosis between the leaf veins, the symptoms being similar to Mn deficiency (Hewitt, 1953). Toxic symptoms of Ni in oats, a Ni-sensible plant species, were observed when Ni concentration was higher than 100 mg kg⁻¹ tissue dry matter (Crooke, 1956). Rahman et al. (2005) reported that barley plants grown at an excess Ni level (100 µM) developed toxicity symptoms, such as interveinal chlorosis of younger leaves and necrosis of mature leaves. These authors also reported that plants grown in 100 µM Ni solution showed brown coloration of roots. Root browning, as a consequence of high metal availability, has been reported by Alam et al. (2000). Nickel is absorbed as divalent cation (Ni²⁺) by crop species. The uptake by less sensitive crops is lower. When present in higher amounts

TABLE 15.3

Influence of Ni on Uptake of Micronutrients in the Barley Plants Grown in Nutrient Solution

Ni Level (µM)	Zn (µg Plant⁻¹)	Cu (µg Plant⁻¹)	Fe (µg Plant⁻¹)	Mn (µg Plant⁻¹)
0	6.11	1.55	17.1	5.73
1.0	8.47	1.81	21.6	6.44
10	5.35	0.58	13.5	3.78
100	0.90	0.09	4.72	0.65

Source: Adapted from Rahman et al. (2005).

in the soil, a considerable amount of Ni is translocated to seeds and fruits (Cataldo et al., 1978; Mitchell et al., 1978).

15.5   INTERACTIONS WITH OTHER NUTRIENTS

Nickel is absorbed as Ni^{2+} and competes with other divalent cations like Ca^{2+}, Mg^{2+}, Fe^{2+}, Zn^{2+}, and Mn^{2+}. Korner et al. (1986) reported that Ca^{2+} and Mg^{2+} inhibited Ni^{2+} uptake noncompetitively by unknown mechanisms, whereas Cu^{2+}, Zn^{2+}, and Co^{2+} all competitively inhibited Ni^{2+} absorption. They suggested that all four divalent cations were transported via the same mechanisms, based on their similar ionic radii (Kochain, 1991). Hence, higher concentrations of Ni in the soils may contribute to deficiency of these divalent cations or nutrients. Data in Table 15.3 show that uptake of Zn, Cu, Fe, and Mn in barley shoot decreased with increasing Ni concentration from 1 to 100 µM in nutrient solution.

Maize grown in calcareous soil with Ni application decreased P concentration (Karimian, 1995), and high levels of Ni increased B and Mo in barley (Brune and Dietz, 1995). In wheat, application of phosphate fertilizer decreased the accumulation of Ni (Mishra and Kar, 1971). Simultaneous supplies of NO_3-N and NH_4-N reduced Ni toxicity in sunflower, and growth was enhanced by added Ni (Zornoza et al., 1999). Low-Ni plants became N deficient from lack of urease activity with a high accumulation of urea but low tissue N (Gerendas and Sattelmacher, 1997). Higher amounts of K have been reported to reduce uptake of Ni by crop plants and hence its toxicity (Mishra and Kar, 1971).

Data on the influence of Ni on concentrations of SO_4^{2-}, Cl^-, and NO_3^- in barley shoot are presented in Table 15.4. The SO_4^{2-} concentration in the barley shoot significantly increased with increasing Ni level from 0 to 1 µM, whereas concentration of Cl^- and NO_3^- significantly decreased with increasing Ni concentration. These results indicate that Ni deficiency may create anion imbalance in barley plants.

15.6   MANAGEMENT PRACTICES TO MAXIMIZE Ni USE EFFICIENCY AND REDUCE TOXICITY

Nickel is an essential element for plant growth; hence, if Ni is present in soil at levels lower than the critical level required by plants, its addition is necessary. In addition,

TABLE 15.4
Influence of Ni Treatment on Anion
Concentration in 30-Day-Old Barley
Shoots Grown in Nutrient Solution

Ni Level (µM)	SO_4^{2-}	Cl^-	NO_3^-
0	11.9a	9.8a	85.5a
0.6	17.8a	9.7a	65.9b
1.0	22.8b	5.7b	38.2c

Note: Values followed by the same letter in the
same column are not significantly different at 1% probability level by LSD test.
Source: Adapted from Brown et al. (1990).

other soils and plant management practices can be adopted to maximize its use efficiency by crops plants. Since Ni is highly toxic to plants if present in higher amounts
than necessary, soil and plant management practices can be adopted to minimize
its toxicity. Using appropriate source and rate, liming acid soils, improving organic
matter content of soils, planting Ni-efficient/tolerant plant species or genotypes
within species, and using adequate fertilizers may be important cultural practices to
improve its use efficiency and reduce toxicity in crop plants.

15.6.1 Appropriate Source and Rate

Nickel is required in trace amounts by higher plants to complete their life cycle. Most
soils contain Ni in sufficient amounts for normal plant growth (Mengel et al., 2001).
Hence, it is not necessary to apply Ni to soils for crop plants. However, Rahman
et al. (2005) reported that 1 to 10 µM Ni supplied as $NiSO_4$ is sufficient for normal
growth of barley plants in nutrient solution. Important Ni sources are nickel chloride
($NiCl_2 \cdot 6H_2O$, 25% Ni), nickel nitrate [$Ni(NO_3)_2 \cdot 6H_2O$, 20% Ni], and nickel oxide
(NiO, 79% Ni). The first two sources are soluble in water, whereas the last source
is insoluble in water (Fageria et al., 2002). Welch (1995) reported that the typical
Ni concentration in agricultural soils is about 10 µM. The Ni seed reserve may be
sufficient to supply some plants with adequate quantities of Ni for them to complete
their life cycle without absorption of Ni from the nutrient media (Eskew et al., 1983).
In general, there is greater concern about Ni toxicity in crop plants regarding, for
example, the application of sewage sludge, which is often high in Ni (Marschner,
1983; Juste and Mench, 1992).

15.6.2 Liming Acid Soils

Liming is an important and dominant practice to reduce toxicity of heavy metals,
including Ni, in acid soils (Bolan et al., 2003). Liming increased adsorption/precipi-
tation of heavy metals including Ni^{2+} at high pH (Pratt et al., 1964; Bisessar, 1989;
Brallier et al., 1996; Tyler and Olsson, 2001; Bolan et al., 2003). In general, with the
exception of Se and Mo, trace elements are more soluble in soils at low pH due to the

The Use of Nutrients in Crop Plants

TABLE 15.5
Influence of pH on Soil Extractable Heavy Metals (μg kg⁻¹) as Extracted by 1.0 M MH₄NO₃

Soil pH	Ni	Cd	Cr	Cu	Pb	Zn
<4.0	1000	80	50	300	3000	5000
4.0–4.5	1000	50	40	280	2000	4000
4.5–5.0	600	20	15	250	150	3000
5.0–5.5	300	15	12	250	30	1000
5.5–6.0	250	10	10	250	15	300
6.0–6.5	200	5	10	250	10	200
6.5–7.0	200	3	12	300	6	170
7.0–7.5	200	3	15	350	4	130
>7.5	200	3	15	400	3	100

Source: Adapted from Prueβ (1997) and McLaughlin et al. (2000).

dissolution of the carbonates, phosphates, and other solid phases. Low pH also lowers the CEC of organic matter and mineral surfaces, thereby weakening the sorption of metals to specific adsorption sites (Bolan et al., 2003). Reduction in Ni uptake by plants due to increasing soil pH of acid soils has been reported (Halstead et al., 1969; Mishra and Kar, 1971).

Agricultural soils treated with alkaline-stabilized biosolid help to overcome the phytotoxicity of Ni (Bolton, 1975). Alkaline-stabilized biosolids are becoming increasingly popular as a liming material because of the complementary mitigation effect from organic matter in immobilization metals in soils (Bolan et al., 2003). Data in Table 15.5 show the influence of pH on soil-extractable heavy metals including Ni. The Ni content in the soils decreased from 1000 μg kg⁻¹ at pH < 4.0 to 200 μg kg⁻¹ at pH > 7.0, a decrease of about 400%. Tyler and Olsson (2001) reported that uptake of Ni by common grass (*Agrostis capillaris* L.) was inversely related to soil solution pH in an acid Cambisol. Helyar and Anderson (1974) reported that liming increases the cation exchange capacity (CEC) of soils and that this would have a negative influence on the soil solution concentrations of cations other than the added Ca and Mg. Decrease in uptake of Ni by liming has been reported by Smith (1994) for Italian ryegrass.

15.6.3 IMPROVING ORGANIC MATTER CONTENT OF SOILS

Organic matter is an important component of alleviating mineral stresses (deficiency/toxicity) in crop plants. Improving organic matter content of soils is an important practice to reduce heavy metal toxicity of crop plants. The organic matter forms stable complexes with metal ions, thus making them unavailable to plants (Hodgson et al., 1966). Organic matter is extremely effective at alleviating the toxicity of metals in slagheaps, permitting healthy growth of various nontolerant species for several years (Hilton, 1967).

TABLE 15.6

Nickel-Tolerant Plant Species of Zimbabwe

Scientific Name of Crop Species	Family
Alyssum bertolonii	Cruciferae
Alyssum murale	Cruciferae
Albizia amara	Leguminosae
Dioma macrocephala	Compositae
Barleria aromatica	Acanthaceae
Combretum molle	Combretaceae
Dalbergia melanoxylon	Leguminosae
Eminia atennulifera	Leguminosae
Turraea nilotica	Meliaceae
Pterocarpus rotundifolias	Leguminosae

Source: Adapted from Antonovics et al. (1971).

15.6.4 PLANTING TOLERANT PLANT SPECIES

Because Ni toxicity is a major problem in contaminated soils, planting Ni-tolerant plant species is an attractive strategy. The plant species that are tolerant to heavy metal toxicity are called *indicator plants.* According to Antonovics et al. (1971), indicator plants are plants that grow well on soils contaminated with a particular heavy metal. Metal-tolerant plants are adaptive to specific agroecological regions. The Ni indicator plants adaptive to Zimbabwe are listed in Table 15.6. Most of the higher plants found on Ni-toxic soils are perennial herbs. The perennial habit probably makes colonization easier since it ensures persistence, and a low growing habit is an adaptation to exposure (Antonovics et al., 1971). In addition, the mechanisms whereby the plant species can grow on Ni-contaminated soils may prevent the heavy metals from reaching their sites of toxic action within the plant, or they may simply be external factors that prevent the absorption of metals by plants. Some plant species may also accumulate heavy metals in the tops without showing toxicity symptoms. Such plants are designated as hyperaccumulators of heavy metals.

In addition to higher plants, ferns also tolerate Ni toxicity. *Asplenium adulterium* is an indicator of Ni (Antonovics et al., 1971). Ferns are also reported to grow on serpentine soils that are high in Ni and Cr (Kruckeberg, 1964). Wild (1968) reported that the ferns *Pellaea calomelanos* and *Chelianthes hirta* are Ni and Cu tolerant.

15.6.5 USE OF ADEQUATE RATE OF FERTILIZERS

Use of adequate rate of N, K, Ca, Mg, and Mo to soils reduces toxic effects of Ni in crop plants (Mishra and Kar, 1971). Hunter and Vergnano (1952, 1953) reported that application of N and K fertilizers could correct nickel toxicity in crop plants. Similarly, Crooke and Inkson (1955) reported that in sand culture experiments with oats, the symptoms of nickel toxicity can be reduced by the application of N, K, Ca, or Mg to the culture solution. Mishra and Kar (1971) reported that application of Mo to the soil and foliage decreased the severity of chlorosis and other toxic effects of Ni. These authors reported that this corrective treatment for Ni toxicity was due to the antagonistic effect of Mo on nickel. Phosphorus fertilization tends to enhance the toxic effect of Ni (Halstead et al., 1969; Mishra and Kar, 1971).

15.6.6 USE OF RHIZOBACTERIUM

Many microorganisms colonize at the plant rhizosphere in nature, and it has been reported that specific rhizosphere microorganisms can reduce the toxicity of heavy metals to plants (Shilev et al., 2001; So et al., 2003; Christie et al., 2004; Someya et al., 2007). Burd et al. (1998, 2000) reported that a plant-growth-promoting bacterium (PGPB), *Kluyvera ascorbata* SUD 165, could alleviate Ni toxicity to canola plants by supplementing iron with the help of siderophores. It has been reported that siderophores play a role in the alleviation of heavy metal toxicity to the bacterial cells (Clarke et al., 1987; Rogers et al., 2001). Someya et al. (2007) also reported that Ni influx into plants was decreased by bacterial absorption in the rhizosphere.

15.7 CONCLUSIONS

There is an increasing interest in and awareness of the effect of Ni as an essential plant nutrient and heavy metal on higher plants and more specifically on crop plants. Major physiological functions of Ni in plants are urea and ureide metabolism, N fixation, reproductive growth, iron absorption, and seed viability. Deficiency of Ni in crop plants is rarely observed. However, Ni deficiency may occur in crop plants grown in soils with excessive heavy metals such as Ca, Mg, Mn, Fe, Cu, and Zn. The deficiency symptoms are dwarfed foliage with blunted apices and a necrotic margin at the leaf or leaflet tip. Nickel salts sprayed in appropriate concentrations can control cereal rusts and other plant diseases. However, toxicity of this heavy metal in greenhouse experiments is frequently observed in various studies for many crop species. Increasing applications of sewage sludge to agricultural soils and continuing release of industrial wastes cause a redistribution of heavy metals in the environment. Nickel toxicity to higher plants grown on serpentine soils is also common. Soil and plant tests for heavy metals, including Ni, have experienced significant advances in recent years due partly to improvements in analytical procedures. However, plant and soil calibration data are still not sufficient for heavy metal deficiency or toxicity diagnostic techniques. The adequate Ni concentration in plant tissue is in the range of 0.1 to 5 mg kg^{-1} dry weight depending on crop species.

Phytotoxicity of Ni can be ameliorated with liming. Liming enhances adsorption/precipitation of this heavy metal and reduces its uptake by plants. Addition of organic matter to heavy metal–contaminated soils is another strategy to ameliorate Ni toxicity. Generally, metal ions complexed to organic matter are unavailable to plants; hence, the effective metal concentration of the soil is reduced. Phytoremediation of Ni-contaminated soils is an attractive strategy. It is not only the most economic method but also environmentally sound. The species found on toxic soils vary widely and differ according to the local ecological conditions and geographical area. Hence, testing is necessary to evaluate adaptation of a particular plant species for an agroecological region. With the extensive amounts of genetic variability in plants for efficiency or tolerance to toxicity levels of elements, including Ni, plants should be developed to grow better on Ni-stressed soils. Soil scientists, physiologists, agronomists, and breeders should work together to achieve this noble goal.

REFERENCES

Antonovics, J., A. D. Bradshaw, and R. G. Turner. 1971. Heavy metal tolerance in plants. *Adv. Ecol. Res.* 7:1–85.

Alam, S., S. Kamei, and S. Kawai. 2000. Phytosiderophore release from manganese-induced iron deficiency in barley. *J. Plant Nutr.* 23:1193–1207.

Baccouch, S., A. Chaoui, and E. El Ferjani. 2001. Nickel toxicity induces oxidative damage in *Zea mays* roots. *J. Plant Nutr.* 24:1085–1097.

Bisessar, S. 1989. Effects of lime on nickel uptake and toxicity in celery grown on muck soil contaminated by a nickel refinery. *Sci. Total Environ.* 84:83–90.

Bolan, N. S., D. C. Adriano, and D. Curtin. 2003. Soil acidification and liming interactions with nutrient and heavy metal transformation and bioavailability. *Adv. Agron.* 78:215–272.

Bollard, E. G. 1983. Involvement of unusual elements in plant growth and nutrition. In: *Inorganic plant nutrition, Encyclopedia Plant Physiology*, New Series Vol. 15M, A. Lauchli and R. L. Bieleski, Eds., 695–744. New York: Springer-Verlag.

Boltan, J. 1975. Liming effects on toxicity to perennial ryegrass of a sewage sludge contaminated with zinc, nickel, copper and chromium. *Environ. Pollut.* 9:295–304.

Brady, N. C. and R. R. Weil. 2002. *The nature and properties of soils*, 13th edition. Upper Saddle River, NJ: Prentice Hall.

Brallier, S., R. B. Harrison, C. L. Henry, and X. Dongsen. 1996. Liming effects on availability of Cd, Cu, Ni, and Zn in a soil amended with sewage sludge 16 years previously. *Water Air Soil Pollut.* 86:195–206.

Brown, P. H., R. M. Welch, and E. E. Cary. 1987a. Nickel: A micronutrient essential for higher plants. *Plant Physiol.* 85:801–803.

Brown, P. H., R. M. Welch, E. E. Cary, and R. T. Checkai. 1987b. Beneficial effects of nickel on plant growth. *J. Plant Nutr.* 10:2125–2135.

Brown, P. H., R. M. Welch, and J. T. Madison. 1990. Effect of nickel deficiency on soluble anion, amino acid, and nitrogen levels. *Plant Soil* 125:19–27.

Brune, A. K. and K. J. Dietz. 1995. A comparative analysis of element composition of roots and leaves of barley seedlings grown in the presence of toxic cadmium, molybdenum, nickel and zinc concentrations. *J. Plant Nutr.* 18:853–868.

Burd, G. I., D. G. Dixon, and B. R. Glick. 1998. A plant growth-promoting bacterium that decreases nickel toxicity in seedlings. *Applied Environ. Microbiol.* 64:3663–3668.

Burd, G. I., D. G. Dixon, and B. R. Glick. 2000. Plant growth-promoting bacteria that decrease heavy metal toxicity in plants. *Canadian J. Microbiol.* 46:237–245.

Cataldo, D. A., T. R. Garland, R. E. Wildung, and H. Drucker. 1978. Nickel in plants: Distribution and chemical forms in soybean plants. *Plant Physiol.* 62:566–570.

Christie, P., X. Li, and B. Chen. 2004. Arbuscular mycorrhiza can depress translocation of zinc to shoots of host plants in soils moderately polluted with zinc. *Plant Soil* 261:209–217.

Clarke, S. E., J. Stuart, and J. Sanders-Loehr. 1987. Introduction of siderophore activity in *Anabaena spp.* and its moderation of copper toxicity. *Applied Environ. Microbiol.* 53:917–922.

Crooke, W. M. 1956. Effect of soil reaction on uptake of nickel from a serpentine soil. *Soil Sci.* 81:269–276.

Crooke, W. M. and R. H. E. Inkson. 1955. Relation between nickel toxicity and major nutrient supply. *Plant Soil* 6:1–15.

Dalton, D. A., H. J. Evans, and F. J. Hanus. 1985. Stimulation by nickel of soil microbial urease activity and urease and hydrogenase activities in soybeans grown in a low nickel soil. *Plant Soil* 88:245–285.

Daroub, S. M. and G. H. Snyder. 2007. The chemistry of plant nutrients in soil. In: *Mineral nutrition and plant disease*, L. E. Datnoff, W. H. Elmer, and D. M. Huber, Eds., 1–7. St. Paul, MN: The American Phytopathological Society.

Dixon, N. E., C. Gazzola, R. L. Blakeley, and B. Zerner. 1975. Jack bean urease (EC 3.5.1.5). A metalloenzyme. A simple biological role for nickel? *J. Am. Chem. Society* 97:4131–4133.

Dixon, N. E., R. L. Blakeley, and B. Zerner. 1980. Jack bean urease (EC 3.5.1.5). III. The involvement of active-site nickel ion in inhibition by β-mercaptoethanol, phosphoramidate, and fluoride. *Can. J. Biochem.* 58:481–488.

Echevarria, G., J. L. Morel, J. C. Fardeau, and E. Leclerc-Cessac. 1998. Assessment of phytoavailability of nickel in soils. *J. Environ. Quality* 27:1064–1070.

Epstein, E. and A. J. Bloom. 2005. *Mineral nutrition of plants: Principles and perspectives*, 2nd edition. Sunderland, MA: Sinauer Associates.

Eskew, D. L., R. M. Welch, and E. E. Cary. 1983. Nickel: An essential micronutrient for legumes and possibly all higher plants. *Science* 222:621–623.

Fageria, N. K., V. C. Baligar, and R. B. Clark. 2002. Micronutrient in crop production. *Adv. Agron.* 77:185–268.

Gerendas, J. and B. Sattelmacher. 1997. Significance of Ni supply for growth, urease activity and the concentration of urea, amino acids, and mineral nutrients of urea-grown plants. *Plant Soil* 190:153–162.

Halstead, R. L., B. J. Finn, and A. J. McLean. 1969. Extractability of nickel added to the soils and its concentration in plants. *Can J. Soil Sci.* 49:335–342.

Helyar, K. R. and A. J. Anderson. 1974. Effects of calcium carbonate on the availability of nutrients in an acid soil. *Soil Sci. Soc. Am. Proc.* 38:341–346.

Hewitt, E. J. 1953. Metal interrelationship in plant nutrition. *J. Exp. Bot.* 4:59–64.

Hilton, K. J. 1967. *The lower Swansea Valley Project.* Green, London: Longmans.

Hodgson, J. F., W. L. Lindsay, and J. F. Frierweiler. 1966. Micronutrient cation complexing in soil solution. II. Complexing of zinc and copper in displaced solution from calcareous soils. *Proc. Soil Sci. Soc. Am.* 3:723–726.

Hunter, J. G. and O. Vergnano. 1952. Nickel toxicity in plants. *Ann. Appl. Biol.* 39:279–281.

Hunter, J. G. and O. Vergnano. 1953. Trace element toxicities in oat plants. *Ann. Appl. Biol.* 40:761–777.

Juste, C. and M. Mench. 1992. Long-term application of sewage sludge and its effect on metal uptake by crops. In: *Biochemistry of trace metals*, D. C. Adriano and A. Arbor, Eds., 159–193. London: Lewis Publishers.

Karimian, N. 1995. Effect of nitrogen and phosphorus on zinc nutrition of corn in a calcareous soil. *J. Plant Nutr.* 18:2261–2271.

Klucas, R. V., F. J. Hanus, and S. A. Russell. 1983. Nickel. A micronutrient element for hydrogen-dependent growth of *Rhizobium japonicum* and for expression of urease activity in soybean leaves. *Proc. Nat. Acad. Sci.* USA 80:2253–2257.

Kochain, L. V. 1991. Mechanisms of micronutrient uptake and translocation in plants. In: *Micronutrient in agriculture*, 2nd edition, J. J. Mortvedt, Ed., 229–296. Madison, WI: SSSA.

Korner, L. E., I. M. Moller, and P. Jensen. 1986. Effects of Ca^{2+} and other divalent cations on uptake of Ni^{2+} by excised barley roots. *Physiol. Plant.* 71:49–54.

Kruckeberg, A. R. 1964. Ferns associated with ultramafic rocks in the Pacific Northwest. *Am. Fern J.* 54:113–126.

Lindsay, W. L. and W. A. Norwell. 1978. Development of a DTPA soil test for zinc, iron, manganese and copper. *Soil Sci. Soc. Am.* 42:421–428.

Malavolta, E. and M. F. Moraes. 2007. Nickel-from toxic to essential nutrient. *Better Crops* 91:26–27.

Marschner, H. 1983. General introduction to the mineral nutrition of plants. In: *Encyclopedia of plant physiology*, new series, A. Lauchli and R. L. Bieleski, Eds., 5–60. Berlin: Springer-Verlag.

Marschner, H. 1995. *Mineral nutrition of higher plants*, 2nd edition. New York: Academic Press.

McLaughlin, M. J., B. A. Zarcinas, D. P. Stevens, and N. Cook. 2000. Soil testing for heavy metals. *Commun. Soil Sci. Plant Anal.* 31:1661–1700.

Mengel, K., E. A. Kirkby, H. Kosegarten, and T. Appel. 2001. *Principles of plant nutrition*, 5th edition. Dordrecht: Kluwer Academic Publishers.

Mishra, D. and M. Kar. 1971. Nickel in plant growth and metabolism. *The Botanical Review* 40:395–452.

Mitchell, G. A., E. T. Bingham, and A. L. Page. 1978. Yield and metal composition of lettuce and wheat grown on soils amended by sewage sludge enriched with cadmium, copper, nickel, and zinc. *J. Environ. Qual.* 7:165–171.

Pandolfini, T., R. Gabbrielli, and C. Comparini. 1992. Nickel toxicity and peroxidase activity in seedlings of *Triticum* aestivum L. *Plant Cell Environ.* 15:719–725.

Pratt, P. F., F. L. Blair, and G. W. McLean. 1964. Reactions of phosphate with soluble and exchangeable nickel. *Soil Sci. Soc. Am. Proc.* 28:363–365.

Prueß, A. 1997. Action values for mobile (NH_4NO_3-extractable) trace elements in soils based on the German national standard DIN19730. In: *Contaminated soils*, 3rd International Conference on the Biogeochemistry of Trace Elements, R. Prost, Ed., 415–423. Paris: Institute National de la Recherche Agronomique.

Rahman, H., S. Sabreen, S. Alam, and S. Kawai. 2005. Effects of nickel on growth and composition of metal micronutrients in barley plant grown in nutrient solution. *J. Plant Nutr.* 28:393–404.

Roach, W. A. and C. Barclay. 1946. Nickel and multiple trace-element deficiencies in agricultural crops. *Nature* 157:696.

Rogers, N. J., K. C. Carson, A. R. Glenn, M. J. Dilworth, M. N. Hughes, and R. K. Poole. 2001. Alleviation of aluminum toxicity to *Rhizobium leguminosarum bv. viciae* by the hydroxamate siderophore vicibactin. *BioMetals* 14:59–66.

Rowell, J. B. 1968. Chemical control of the cereal rusts. *Annu. Rev. Phytopathology* 54:999–1008.

Salt, D. E., M. Blaylock, P. B. A. Kumar Nanda, V. Dushenkov, B. O. Ensley, L. Chet, and I. Raskin. 1995. I. Phytoremediation: A novel strategy for removal of toxic metals from the environment using plants. *Biotechnology* 13:468–478.

Scheckel, K. G. and D. L. Sparks. 2001. Dissolution kinetics of nickel surface precipitates on clay mineral and oxide surfaces. *Soil Sci. Soc. Am. J.* 65:685–694.

Scheidegger, A. M., G. M. Lamble, and D. L. Sparks. 1997. Spectroscopic evidence for the formation of mixed cation hydroxide phases upon metal sorption on clays and aluminum oxides. *J. Colloid Interface Sci.* 186:118–128.

Shilev, S. I., J. Ruso, A. Puig, M. Benlloch, J. Jorrin, and E. Sancho. 2001. Rhizosphere bacteria promote sunflower (*Helianthus annuus* L.) plant growth and tolerance to heavy metals. *Minerva Biotechnologia* 13:37–39.

Singh, S., A. M. Kayastha, R. K. Asthana, and S. P. Singh. 2004. Response of garden pea to nickel toxicity. *J. Plant Nutr.* 27:1543–1560.

Smith, S. R. 1994. Effect of soil pH on availability to crops of metals in sewage sludge-treated soils. I. Nickel, copper and zinc uptake and toxicity to ryegrass. *Environ. Pollut.* 85:321–327.

So, L. M., L. M. Chu, and P. K. Wong. 2003. Microbial enhancement of Cu^{2+} removal capacity of *Eichhornia crassipes* (Mart.) *Chemosphere* 52:1499–1503.

Someya, N., Y. Sato, I. Yamaguchi, H. Hamamoto, Y. Ichiman, K. Akutsu, H. Sawada, and K. Tsuchiya. 2007. Alleviation of nickel toxicity in plants by a rhizobacterium strain is not dependent on its siderophore production. *Commun. Soil Sci. Plant Anal.* 38:1155–1162.

Terry, D. 2004. AAPFCO Official Publication 57. Association of American Plant Food Control Officials, West Lafayette, Indiana.

Tyler, G. and T. Olsson. 2001. Plant uptake of major and minor mineral elements as influenced by soil acidity and liming. *Plant Soil* 230:307–321.

Underwood, E. J. 1971. *Trace elements in human and animal nutrition.* New York: Academic Press.

USEPA. 1990. Project summary health assessment document for nickel. EPA/600/S8-83/012. Office of Health and Environmental assessment, Washington, DC.

Welch, R. M. 1981. The biological significance of nickel. *J. Plant Nutr.* 3:345–356.

Welch, R. M. 1986. Effects of nutrient deficiencies on seed production and quality. *Adv. Plant Nutr.* 2:205–247.

Welch, R. M. 1995. Micronutrient nutrition of plants. *Critical Rev. Plant Sci.* 14:49–82.

Wild, H. 1968. Geobotanical anomalies in Rhodesia. I. The vegetation of copper-bearing soils. *Kirkia* 7:1–71.

Wild, H. 1970. Geobotanical anomalies in Rhodesia. III. The vegetation of nickel bearing soils. *Kirkia* 7:1–62.

Winkler, R. G., J. C. Polacco, D. L. Eskew, and R. M. Welch. 1983. Nickel is not required for apo-urease synthesis in soybean seeds. *Plant Physiol.* 72:262–263.

Wood, B. W. and C. C. Reilly. 2007. Nickel and plant disease. In: *Mineral nutritional and plant disease*, L. E. Datnoff, W. H. Elmer, and D. M. Huber, 215–231. St. Paul, MN: The American Phytopathological Society.

Woolhouse, H. W. 1983. Toxicity and tolerance in response of plants to metals. In: *Encyclopedia of plant physiology: Physiological plant ecology* III, 12C, O. L. Lange, P. S. Nobel, C. B. Osmond, and H. Ziegler, Eds., 245–300. New York: Springer Verlag.

World Health Organization. 1991. International program on chemical safety. Environmental health criteria 108: Nickel. WHO, Geneva, Switzerland.

Yamaguchi, N. U., A. C. Scheinost, and D. L. Sparks. 2001. Surface-induced nickel hydroxide precipitation in the presence of citrate and salicylate. *Soil Sci. Soc. Am. J.* 65:729–736.

Zornoza, P., S. Robles, and N. Martin. 1999. Alleviation of nickel toxicity by ammonium supply to sunflower plants. *Plant Soil* 208:221–226.

Index